Handbook of Research on Wireless Security

Yan Zhang
Simula Research Laboratory, Norway

Jun Zheng
City University of New York, USA

Miao Ma
Hong Kong University of Science and Technology, Hong Kong

Volume I

Information Science REFERENCE

INFORMATION SCIENCE REFERENCE

Hershey · New York

Acquisitions Editor:	Kristin Klinger
Development Editor:	Kristin Roth
Senior Managing Editor:	Jennifer Neidig
Managing Editor:	Sara Reed
Copy Editor:	Ashlee Kunkel, Holly J. Powell
Typesetter:	Jamie Snavely, Carole Coulson
Cover Design:	Lisa Tosheff
Printed at:	Yurchak Printing Inc.

Published in the United States of America by
Information Science Reference (an imprint of IGI Global)
701 E. Chocolate Avenue, Suite 200
Hershey PA 17033
Tel: 717-533-8845
Fax: 717-533-8661
E-mail: cust@igi-global.com
Web site: http://www.igi-global.com

and in the United Kingdom by
Information Science Reference (an imprint of IGI Global)
3 Henrietta Street
Covent Garden
London WC2E 8LU
Tel: 44 20 7240 0856
Fax: 44 20 7379 0609
Web site: http://www.eurospanonline.com

Library of Congress Cataloging-in-Publication Data

Handbook of research on wireless security / Yan Zhang, Jun Zheng, and Miao Ma, editors.

p. cm.

Summary: "This book combines research from esteemed experts on security issues in various wireless communications, recent advances in wireless security, the wireless security model, and future directions in wireless security. As an innovative reference source for students, educators, faculty members, researchers, engineers in the field of wireless security, it will make an invaluable addition to any library collection"--Provided by publisher.

Includes bibliographical references and index.

ISBN 978-1-59904-899-4 (hardcover) -- ISBN 978-1-59904-900-7 (ebook)

1. Wireless communication systems--Security measures. I. Zhang, Yan, 1962- II. Zheng, Jun, Ph.D. III. Ma, Miao. IV. Title.

TK5102.85.H35 2008

005.8--dc22

2007036301

British Cataloguing in Publication Data
A Cataloguing in Publication record for this book is available from the British Library.

Editorial Advisory Board

Table of Contents

Detailed Table of Contents

to enabling secure and private service discovery in these environments, including the dynamic population of participants, the lack of a universal trust mechanism, and the limited capabilities of the devices. To ensure secure service discovery while addressing privacy issues, trust-based models are inevitable. We survey secure service discovery in the broadband wireless environment. We include case studies of two protocols which include a trust mechanism, and we summarize future research directions.

The recent developments in the mobile technology (mobile phones, middleware, wireless networks, etc.) created a need for new methods of protecting the code transmitted through the network. The oldest and the simplest mechanisms concentrate more on the integrity of the code itself and on the detection of unauthorized manipulation. The newer solutions not only secure the compiled program, but also the data that can be gathered during its "journey," and even the execution state. Some other approaches are based on prevention rather than detection. In the chapter we present a new idea of securing mobile agents. The proposed method protects all components of an agent: the code, the data, and the execution state. The proposal is based on a zero-knowledge proof system and a secure secret sharing scheme, two powerful cryptographic primitives. Next, the chapter includes security analysis of the new method and its comparison to other currently most widespread solutions. Finally, we propose a new direction of securing mobile agents by straightening the methods of protecting integrity of the mobile code with risk analysis and a reputation system that helps avoiding a high-risk behavior.

The broad aim of identity management (IdM) is to manage the resources of an organization (such as files, records, data and communication infrastructure, and services) and to control and manage access to those resources in an efficient and accurate way. Consequently, identity management is both a technical and process orientated concept. The concept of IdM has begun to be applied in identities related applications in enterprises, governments, and Web services since 2002. As the integration of heterogeneous wireless networks becomes a key issue in towards the next generation (NG) networks, IdM will be crucial to the success of NG wireless networks. A number of issues, such as mobility management, multioperator, and securities require the corresponding solutions in terms of user authentication, access control, and so forth. IdM in NG wireless networks is about managing the digital identity of a user and ensuring that users have fast, reliable, and secure access to distributed resources and services of an NGN and the associated service providers, across multiple systems and business contexts.

Wardriving is the practice of searching wireless networks while moving. Originally, it was explicitly referred to people searching for wireless signals by driving on vans, but nowadays it generally identifies people searching for wireless accesses while moving. Despite the legal aspects, this "quest for connectivity" spawned a quite productive underground community, which developed powerful tools, relying on cheap and standard hardware. The knowledge of these tools and techniques has many useful aspects. First, when designing the security framework of a wireless LAN (WLAN), the knowledge of the vulnerabilities exploited at the basis of wardriving is a mandatory step, both to avoid penetration issues and to detect whether attacks are ongoing. Second, hardware and software developers can design better devices by avoiding common mistakes and using an effective suite for conducting security tests. Lastly, people who are interested in gaining a deeper understanding of wireless standards can conduct experiments by simply downloading software running on cost effective hardware. With such preamble, in this chapter we will analyze the theory, the techniques, and the tools commonly used for wardriving IEEE 802.11-based wireless networks.

The broadcast nature of wireless networks and the mobility features created new kinds of intrusions and anomalies taking profit of wireless vulnerabilities. Because of the radio links and the mobile equipment features of wireless networks, wireless intrusions are more complex because they add to the intrusions developed for wired networks, a large spectrum of complex attacks targeting wireless environment. These intrusions include rogue or unauthorized access point (AP), AP MAC spoofing, and wireless denial-of-service and require adding new techniques and mechanisms to those approaches detecting intrusions targeting wired networks. To face this challenge, some researchers focused on extending the deployed approaches for wired networks while others worked to develop techniques suitable for detecting wireless intrusions. The efforts have mainly addressed (a) the development of theories to allow reasoning about detection, wireless cooperation, and response to incidents, and (b) the development of wireless intrusion and anomaly detection systems that incorporate wireless detection, preventive mechanisms, and tolerance functions. This chapter aims at discussing the major theories, models, and mechanisms developed for the protection of wireless networks/systems against threats, intrusions, and anomalous behaviors. The objectives of this chapter are to (a) discuss security problems in wireless environment, (b) to present the current research activities, (c) study the important results already developed by researchers, and (d) to discuss

A lot of networks today are behind firewalls. In peer-to-peer networking, firewall-protected peers may have to communicate with peers outside the firewall. This chapter shows how to design peer-to-peer systems to work with different kinds of firewalls within the object-oriented action systems framework by combining formal and informal methods. We present our approach via a case study of extending a Gnutella-like peer-to-peer system (Yan et al, 2003) to provide connectivity through firewalls.

An ubiquitous access and pervasive computing concept is almost intrinsically tied to wireless communications. Emerging next-generation wireless networks enable innovative service access in every situation. Apart from many remote services, proximity services will also be widely available. People currently rely on numerous forms of identities to access these services. The inconvenience of possessing and using these identities creates significant security vulnerability, especially from network and device point of view in wireless service access. After explaining the current identity solutions scenarios, the chapter illustrates the on-going efforts by various organizations and the requirements and frameworks to develop an innovative, easy-to-use identity management mechanism to access the future diverse service worlds. The chapter also conveys various possibilities, challenges, and research questions evolving in these areas.

This chapter provides a survey of privacy enhancing techniques and discusses their effect using a scenario in which a charged location-based service is used. We introduce four protection levels and discuss an assessment of privacy enhancing techniques according to these protection levels.

This chapter shows that the security challenges posed by the 802.11 wireless networks are manifold and it is therefore important to explore the various vulnerabilities that are present with such networks.

Converging networks and mobility raise new challenges towards the existing authentication, authorisation, and accounting (AAA) systems. Focus of the research is towards integrated solutions for seamless service access of mobile users. Interworking issues between mobile and wireless networks are the basis for detailed research on handover delay, multidevice roaming, mobile networks, security, ease-of-use, and anonymity of the user. This chapter provides an overview over state-of-the-art in authentication for mobile systems, and suggests extending AAA-mechanisms to home and community networks, taking into account security and privacy of the users.

Wireless networks are gaining popularity that comes with the occurrence of several networking technologies raising from personal to wide area, from centralized to distributed, and from infrastructure-based to infrastructure-less. Wireless data link characteristics such as openness of transmission media make these networks vulnerable to a novel set of security attacks, despite those that they inherit from wired networks. In order to ensure the protection of mobile nodes that are interconnected using wireless protocols and standards, it is essential to provide an in-depth study of a set of mechanisms and security models. In this chapter, we present the research studies and proposed solutions related to the authentication, privacy, trust establishment, and management in wireless networks. Moreover, we introduce and discuss the major security models used in a wireless environment.

In the design and analysis of authentication and key agreement protocols, provable secure formal methods play a very important role, among which the Canetti-Krawczyk(CK) model and the universal composable(UC) security model are very popular at present. This chapter focuses on these two models and consists mainly of three parts. (1) There is an introduction to the CK model and the UC model. (2) There is also a study of these two models, which includes an analysis of the CK model and an extension of the UC security model. The analysis of the CK model presents its security analysis, advantages, and disadvantages, and a bridge between this formal method and the informal method (heuristic method) is established; an extension of the UC security model gives a universally composable anonymous hash certification model. (3) The applications of these two models are also presented. With these two models, the four-way handshake protocols in 802.11i and Chinese WLAN security standard WAPI are analyzed.

identified. In addition, the current research on the UMTS security and the proposed enhancements that aim at improving the UMTS security architecture are briefly presented and analyzed.

Motivated by the requirements for higher data rate, richer multimedia services, and broader radio range, wireless mobile networks are currently in the stage evolving from the second-generation (2G), for example, global system for mobile communications (GSM), into the era of third-generation (3G) or beyond 3G or fourth-generation (4G). Universal mobile telecommunications system (UMTS) is the natural successor of the current popular GSM. Code division multiple access 2000 (CDMA2000) is the next generation version for the CDMA-95, which is predominantly deployed in the North America and North Korea. Time division-sychrononous CDMA (TD-SCDMA) is in the framework of 3GPP2 and is expected to be one of the principle wireless technologies employed in China in the future. It is envisioned that each of three standards in the framework of international mobile telecommunications-2000 (IMT-2000) will play a significant role in the future due to the backward compatibility, investment, maintenance cost, and even politics. In all of the potential standards, access security is one of the primary demands as well as challenges to resolve the deficiency existing in the second generation wireless mobile networks such as GSM, in which only one-way authentication is performed for the core network part to verify the user equipment (UE). Such access security may lead to the "man-in-middle" problem, which is a type of attack that can take place when two clients that are communicating remotely exchange public keys in order to initialize secure communications. If both of the public keys are intercepted in the route by someone, that someone can act as a conduit and send in the messages with a fake public key. As a result, the secure communication is eavesdropped on by a third party.

The global system for mobile communications (GSM) is the most popular standard that implements second generation (2G) cellular systems. 2G systems combined with general packet radio services (GPRS) are often described as 2.5G, that is, a technology between the 2G and third (3G) generation of mobile systems. GPRS is a service that provides packet radio access for GSM users. This chapter presents the security architecture employed in 2.5G mobile systems, focusing on GPRS. More specifically, the security measures applied to protect the mobile users, the radio access network, the fixed part of the network, and the related data of GPRS, are presented and analyzed in details. This analysis reveals the security weaknesses of the applied measures that may lead to the realization of security attacks by adversaries. These attacks threaten network operation and data transfer through it, compromising end-users and network security. To defeat the identified risks, current research activities on the GPRS security propose a set of security improvements to the existing GPRS security architecture.

Security measures of mobile infrastructures have always been important from the early days of the creation of cellular networks. Nowadays, however, the traditional security schemes require a more fundamental approach to cover the entire path from the mobile user to the server. This fundamental approach is so-called end-to-end (E2E) security coverage. The main focus of this chapter is to discuss such architectures for IEEE 802.16e (Mobile-WiMAX) and major 3G cellular networks. The end-to-end implementations usually contain a complete set of algorithms and protocol enhancements (e.g., mutual identification, authentications, and authorization), including the VLSI implementations. This chapter discusses various proposals at the protocol level.

This chapter outlines how cellular authentication can be utilized for generic application security. It describes the basic concept of the generic bootstrapping architecture that was defined by the 3rd generation partnership project (3GPP) for current networks and outlines the latest developments for future networks.The chapter will provide an overview of the latest technology trends in the area of generic application security.

Network mobility (NEMO) enables seamless and ubiquitous Internet access while on board vehicles. Even though the Internet Engineering Task Force (IETF) has standardized the NEMO basic support protocol as a network layer mobility solution, few studies have been conducted in the area of the authentication, authorization, and accounting (AAA) framework that is a key technology for successful deployment. In this chapter, we first review the existing AAA protocols and analyze their suitability in NEMO environments. After that, we propose a localized AAA framework to retain the mobility transparency as the NEMO basic support protocol and to reduce the signaling cost incurred in the AAA procedures. The proposed AAA framework supports mutual authentication and prevents various threats such as replay

to known attacks. The distributed reputation system is incorporated for a trust management and malicious behavior detection in the network.

The increasing diffusion of wireless portable devices and the emergence of mobile ad hoc networks promote anytime and anywhere opportunistic resource sharing. However, the fear of exposure to risky interactions is currently limiting the widespread uptake of ad hoc collaborations. This chapter introduces to the challenge of identifying and validating novel security models/systems for securing ad hoc collaborations by taking into account the high unpredictability, heterogeneity, and dynamicity of envisioned wireless environments. We claim that the concept of trust management should become a primary engineering design principle, to associate with the subsequent trust refinement into effective authorization policies, thus calling for original and innovative access control models. The chapter overviews the state-of-the-art solutions for trust management and access control in wireless environments, by pointing out both the need for their tight integration and the related emerging design guidelines (e.g., exploitation of context awareness and adoption of semantic technologies).

Security has become a primary concern in mobile ad hoc networks (MANETs). The characteristics of MANETs pose both challenges and opportunities in achieving security goals, such as confidentiality, authentication, integrity, availability, access control, and nonrepudiation. Cryptographic techniques are widely used for secure communications in wired and wireless networks. Most cryptographic mechanisms, such as symmetric and asymmetric cryptography, often involve the use of cryptographic keys. However, all cryptographic techniques will be ineffective if the key management is weak. Key management is also a central component in MANET security. The purpose of key management is to provide secure procedures for handling cryptographic keying materials. The tasks of key management include key generation, key distribution, and key maintenance. Key maintenance includes the procedures for key storage, key update, key revocation, key archiving, and so forth. In MANETs, the computational load and complexity for key management are strongly subject to restriction by the node's available resources and the dynamic nature of network topology. A number of key management schemes have been proposed for MANETs. In this chapter, we present a survey of the research work on key management in MANETs according to recent literature.

Mobile-IP ad hoc networks (MANETs) have gained popularity in the past few years with the creation of a variety of ad hoc protocols that specifically offer quality of service (QoS) for various multimedia traffic between mobile stations (MSs) and base stations (BSs). The lack of proper end-to-end security coverage, on the other hand, is a challenging issue as the nature of such networks with no specific infrastructure is prone to relatively more attacks, in a variety of forms. The focus of this chapter is to discuss a number of attack scenarios and their remedies in MANETs including the introduction of two entities, ad hoc key distribution center (AKDC) and decentralize key generation and distribution (DKGD), which serve as key management schemes.

The integration of wireless communication and embedded video systems is a demanding and interesting topic which has attracted significant research efforts from the community of telecommunication. This chapter discusses the challenging issues in wireless video surveillance and presents the detailed design for a novel highly-secure video surveillance system over ad hoc wireless networks. To this end, we explore the state-of-the-art in the cross domains of wireless communication, video processing, embedded systems, and security. Moreover, a new media-dependent video encryption scheme, including a reliable data embedding technique and real-time video encryption algorithm, is proposed and implemented to enable the system to work properly and efficiently in an open and insecure wireless environment. Extensive experiments are conducted to demonstrate the advantages of the new systems, including high security guarantee and robustness. The chapter would serve as a good reference for solving the challenging issues in wireless multimedia and bring new insights on the interaction of different technologies within the cross application domain.

Intrusion detection in ad hoc networks is a challenge because of the inherent characteristics of these networks, such as, the absence of centralized nodes, the lack of infrastructure, and so forth. Furthermore, in addition to application-based attacks, ad hoc networks are prone to attacks targeting routing protocols,

Weiser envisioned ubiquitous computing with computing and communicating entities woven into the fabrics of every day life. This chapter deals with the survivability of ambient resource-constrained wireless computing nodes, from fixed sensor network nodes to small devices carried out by roaming entities, for example, as part of a personal area network of a moving person. First, we review the assets that need to be protected, especially the energy of these unplugged devices. There are also a number of specific attacks that are described; for example, direct physical attacks are facilitated by the disappearing security perimeter. Finally, we survey the protection mechanisms that have been proposed with an emphasis on cryptographic keying material and trust management.

Fault tolerance is one of the premier system design desiderata in wireless ad hoc and sensor networks. It is crucial to have a certain level of fault tolerance in most ad hoc and sensor applications, especially for those used in surveillance, security, and disaster relief. In addition, several network security schemes require that the underlying topology provide fault tolerance. In this chapter, we will review various fault tolerant techniques used in topology design for ad hoc and sensor networks, including those for power control, topology control, and sensor coverage.

Section IV
Security in Wireless PAN/LAN/MAN Networks

Security is always an important factor in wireless connections. As with all other existing radio technologies, the Bluetooth standard is often cited to suffer from various vulnerabilities and security inefficiencies, while attempting to optimize the trade-off between performance and complementary services including security. On the other hand, security protocols like IP secure (IPsec) and secure shell (SSH) provide strong, flexible, low cost, and easy to implement solutions for exchanging data over insecure communication links. However, the employment of such robust security mechanisms in wireless realms enjoins additional research efforts due to several limitations of the radio-based connections, for example link bandwidth and unreliability. This chapter will evaluate several Bluetooth personal area network (PAN) parameters, including absolute transfer times, link capacity, throughput, and goodput. Experiments shall employ both Bluetooth native security mechanisms, as well as the two aforementioned protocols. Through a plethora of scenarios, utilizing both laptops and palmtops, we offer a comprehensive in-depth comparative analysis of each of the aforementioned security mechanisms when deployed over Bluetooth communication links.

The electromagnetic energy source used by wireless communication devices in a vehicle can cause electromagnetic compatibility problems with the electrical and electronic equipment on board. This work is focused on the radiated susceptibility – EMS – issue and proposes a method for quantifying the electromagnetic influence of wireless RF transmitters on board vehicles. The key to the analysis is the evaluation of the relation between the electrical field emitted by a typical Bluetooth device operating close to the automobile's electrical and electronic systems and the field level specified by the EMC directive 2004/104/EC for radiated susceptibility tests. The chapter includes the model of a closed circuit structure emulating an automobile's electric wire system and the simulation of its behavior under electromagnetic fields' action. According to this a physical structure is designed and implemented, which is used for laboratory tests. Finally, simulated and experimental results are compared and the conclusions obtained are discussed.

The great promise of wireless LAN will never be realized unless there is an an appropriate security level. From this point of view, various security protocols have been proposed to handle WLAN security problems that are mostly due to the lack of physical protection in WLAN or because of the transmission on the radio link. The purpose of this chapter is (1) to provide the reader with a sample background in WLAN technologies and standards, (2) to give the reader a solid grounding in common security concepts and technologies, and (3) to identify the threats and vulnerabilities of WLAN communications.

Wireless local area networks (WLAN) are rapidly becoming a core part of network access. Supporting user mobility, more specifically, session continuation in changing network access points, is becoming an integral part of wireless network services. This is because of the popularity of emerging real-time streaming applications that can be commonly used when the user is mobile, such as voice-over-IP and Internet radio. However, mobility introduces a new set of problems in wireless environments because of handoffs between network access points (APs). The IEEE 802.11i security standard imposes an authentication delay long enough to hamper real-time applications. This chapter will provide a comprehensive study on fast authentication solutions found in the literature as well as the industry that address this problem. These proposals focus on solving the mentioned problem for intradomain handoff scenarios where the access points belong to the same administrative domain or provider. Interdomain roaming is also becoming common-place for wireless access. We need fast authentication solutions for these environments that are managed by independent administrative authorities. We detail such a solution that explores the use of local trust relationships to foster fast authentication.

Mass deployment of radio-frequency identification (RFID) technology is now becoming feasible for a wide variety of applications ranging from medical to supply chain and retail environments. Its main draw-back until recently was high production costs, which are now becoming lower and acceptable. But due to inherent constraints of RFID technology (in terms of limited power and computational resources) these devices are the subject of intensive research on how to support and improve increasing demands for security and privacy. This chapter therefore focuses on security and privacy issues by giving a general overview of the field, the principles, the current state of the art, and future trends. An improvement in the field of security and privacy solutions for this kind of wireless communications is described as well.

Wireless communications are becoming ubiquitous in homes, offices, and enterprises with the popular IEEE 802.11 wireless LAN technology and the up-and-coming IEEE 802.16 wireless MAN technology. The wireless nature of communications defined in these standards makes it possible for an attacker to snoop on confidential communications or modify them to gain access to home or enterprise networks much more easily than with wired networks. Wireless devices generally try to reduce computation overhead to conserve power and communication overhead to conserve spectrum and battery power. Due to these considerations, the original security designs in wireless LANs and MANs used smaller keys, weak message integrity protocols, weak or one-way authentication protocols, and so forth. As wireless networks became popular, the security threats were also highlighted to caution users. A security protocol redesign followed first in wireless LANs and then in wireless MANs. This chapter discusses the security threats and requirements in wireless LANs and wireless MANs, with a discussion on what the original designs missed and how they were corrected in the new protocols. It highlights the features of the current wireless LAN and MAN security protocols and explains the caveats and discusses open issues. Our aim is to provide the reader with a single source of information on security threats and requirements, authentication technologies, security encapsulation, and key management protocols relevant to wireless LANs and MANs.

An overview of the technical and business aspects is given for the corporate deployment of services over WiMAX. WiMAX is considered to be a strong candidate for the next generation of broadband wireless access; therefore its security is critical. This chapter provides an overview of the inherent and complementary benefits of broadband deployment over a long haul wireless pipe, such as WiMAX. In addition, we explore end-to-end (E2E) security structures necessary to launch secure business and consumer class services. The main focus of this chapter is to look for the best security practice to achieve E2E security in both vertical and horizontal markets. The E2E security practices will ensure complete coverage of the entire link from the client (user) to the server. This is also applicable to wireless virtual private network (VPN) applications where the tunneling mechanism between the client and the server ensures complete privacy and security for all users. The same idea for E2E security is applied to client-server-based multimedia applications, such as in IP multimedia subsystem (IMS) and voice over IP (VoIP), where secure client/server communication is required. In general, we believe that WiMAX provides the opportunity for a new class of high data rate symmetric services. Such services will require E2E security schemes to ensure risk-free high data-rate uploads and downloads of multimedia applications. WiMAX provides the capability for embedded security functions through the 802.16 security architecture standards. IEEE 802.16 is further subcategorized as 802.16d (fixed-WiMAX) and 802.16e (mobile-WiMAX). Due to the mobility and roaming capabilities in 802.16e and the fact that the medium of signal transmission is accessible to everyone, there are a few extra security considerations applied to 802.16e. These extra features include PKMv2, PKM-EAP authentication method, AES encryption wrapping, and so forth. The common security features of 802.16d and 802.16e are discussed in this chapter, as well as the highlights of the security comparisons between other broadband access, 3G technologies, and WiMAX.

During the last few years, mobile broadband access has been a popular concept in the context of fourth generation (4G) cellular systems. After the wide acceptance and deployment of the wired broadband connections, such as DSL, the research community in conjunction with the industry have tried to develop and deploy viable mobile architectures for broadband connectivity. The dominant architectures which have already been proposed are Wi-Fi, UMTS, WiMax, and flash-OFDM. In this chapter, we analyze these protocols with respect to their security mechanisms. First, a detailed description of the authentication, confidentiality, and integrity mechanisms is provided in order to highlight the major security gaps and threats. Subsequently, each threat is evaluated based on three factors: likelihood, impact, and risk.

Preface

Wireless networks have been seen unprecedented growth in the past few years. Wireless technologies provide users with a variety of benefits like portability, flexibility, increased productivity, and lower installation costs. Various wireless technologies, from wireless local area network (WLAN) and Bluetooth to WiMAX and third generation (3G) have been developed. Each of these technologies has its own unique applications and characteristics. For example, a WLAN can provide the wireless users with high bandwidth data communication in a restricted and dense area (hotpot). Ad hoc networks, like those enabled by Bluetooth, allow data synchronization with network systems and application sharing between devices. WiMAX can provide high-speed, high bandwidth efficiency, and high-capacity multimedia services for residential as well as enterprise applications.

However, any wireless technology is inherently risky. It has the same risks as the wired networks as well as new risks brought by the wireless connectivity. There have been many reports of security weaknesses and problems related to different wireless technologies, which make wireless security quite a hot research topic recently, both in the academia and industry.

Wireless security is a very broad area as there are so many different wireless technologies existing. Each wireless technology has its own architecture, algorithms, and protocols. Different wireless technologies have their own application areas and different security concerns, requirements, and solutions. To this end, we want to bring up the *Handbook of Research on Wireless Security* to serve as a single comprehensive reference in the field of wireless security.

In this book, the basic concepts, terms, protocols, systems, architectures, and case studies in the wireless security are provided. It identifies the fundamental problems, key challenges, and future directions in designing secure wireless systems. It covers a wide spectrum of topics in a variety of wireless networks, including attacks, secure routing, encryption, decryption, confidentiality, integrity, key management, identity management, and also security protocols in standards.

The chapters of this book are authoritatively contributed by a group of internationally renowned experts on wireless security. They are organized in four sections:

- Section I: Security Fundamentals
- Section II: Security in 3G/B3G/4G
- Section III: Security in Ad Hoc and Sensor Networks
- Section IV: Security in Wireless PAN/LAN/MAN

Section I introduces the basic concepts and fundamental mechanisms of wireless security. This section is able to provide the necessary background for readers and introduce all the fundamental issues on wireless security without previous knowledge on this area. Section II discusses all the security aspects in 3G/B3G/4G. It is well known that 3G mobile systems offer mobile users content rich services, wireless broadband access to Internet, and worldwide roaming. Future 4G mobile communication networks are expected to provide all IP-based services for heterogeneous wireless access technologies, assisted by mobile IP to provide seamless Internet access for mobile users. However the broadcast nature of the wireless communication and increased popularity of wireless devices introduce serious security vulnerabilities. A variety of security issues regarding 3G/B3g/4G will be introduced and addressed with effective solutions (e.g., identity management, confidentiality and integrity mechanisms, evaluation of the current 3G/B3G/4G security protocols, analysis of the impact of security deployment upon the network performance, etc.). Section III explores the security in ad hoc and sensor networks. In recent years, tremendous technological advances have been made in the areas of wireless ad hoc and sensor networks. Such networks have a significant impact on a variety of applications including scientific, military, medical, industrial, office, home, and personal domains. However, these networks introduce new security challenges due to their dynamic topology, severe resource constraints, and absence of a trusted infrastructure. Many aspects of security issues regarding the ad hoc and sensor networks will be covered, including key management, cryptographic protocols, authentication and access control, intrusion detection and tolerance, secure location services, privacy and anonymity, secure routing, resilience against different types of attacks, and so forth. Section IV exploits the security problems in wireless PAN/LAN/MAN. Nowadays we have continuously growing markets for the wireless PANs, wireless LANs, and wireless MANs, but there is a big black hole in the security of this kind of network. Diverse aspects of the security issues on these types of networks will be introduced. For instance, the threats and vulnerabilities in wireless LANs, access control in wireless LANs, evaluating security mechanisms in wireless PANs, the protocols and mechanisms to enhance the security of wireless LANs/MANs, security issues in WiMAX, and so forth are discussed. Practical examples will also be introduced to enhance the understanding.

This book can serve as an essential and useful reference for undergraduate and graduate students, educators, scientists, researchers, engineers, and research strategists in the field of wireless security.

We hope that by reading this book the reader can not only learn the basic concepts of wireless security but also get a good insight into some of the key research works in securing the wireless networks. Our goal is to provide an informed and detailed snapshot of this fast moving field. If you have any feedback or suggestion, please contact the editors.

Yan Zhang, Jun Zheng, and Miao Ma

Acknowledgment

The editors would like to acknowledge the help of all involved in the collation and review process of the handbook, without whose support the project could not have been successfully completed.

Deep appreciation and gratitude is first due to Editorial Advisory Board, whose suggestions and comments have greatly enhanced the quality of the book. Most of the authors of the chapters included in this handbook also served as referees for chapters written by other authors. We would like to thank them for their time, valuable comments, and hard work in reviewing the peers' work. Thanks also go to all the external reviewers who provided constructive and comprehensive reviews. Their critical suggestions and comments ensure the quality of the book.

Special thanks also go to the publishing team at IGI Global Inc., whose contributions throughout the whole process from inception of the initial idea to final publication have been invaluable. In particular to Kristin Roth, who continuously prodded via e-mail for keeping the project on schedule, to Jessica Thompson, whose support, patience, and professionalism during this project, and to Nicole Dean, for enhancing the book marketability. We are grateful for the staffs for the great efforts during the typesetting period. Last but not least, a special thank to the families and friends for their constant encouragement, patience, and understanding throughout this project.

In closing, we wish to thank all of the authors for their insights, excellent contributions, and professional cooperation to this handbook.

Co-Editors for *Handbook of Research on Wireless Security*

Yan Zhang, Ph.D.
Simula Research Laboratory, Norway

Jun Zheng, Ph.D.
CUNY, USA

Miao Ma, Ph.D.
HKUST

May 2007

Section I
Security Fundamentals

Chapter I
Malicious Software in Mobile Devices

Thomas M. Chen
Southern Methodist University, USA

Cyrus Peikari
Airscanner Mobile Security Corporation, USA

ABSTRACT

This chapter examines the scope of malicious software (malware) threats to mobile devices. The stakes for the wireless industry are high. While malware is rampant among 1 billion PCs, approximately twice as many mobile users currently enjoy a malware-free experience. However, since the appearance of the Cabir worm in 2004, malware for mobile devices has evolved relatively quickly, targeted mostly at the popular Symbian smartphone platform. Significant highlights in malware evolution are pointed out that suggest that mobile devices are attracting more sophisticated malware attacks. Fortunately, a range of host-based and network-based defenses have been developed from decades of experience with PC malware. Activities are underway to improve protection of mobile devices before the malware problem becomes catastrophic, but developers are limited by the capabilities of handheld devices.

INTRODUCTION

Most people are aware that malicious software (malware) is an ongoing widespread problem with Internet-connected PCs. Statistics about the prevalence of malware, as well as personal anecdotes from affected PC users, are easy to find. PC malware can be traced back to at least the Brain virus in 1986 and the Robert Morris Jr. worm in 1988. Many variants of malware have evolved over 20 years. The October 2006 WildList (www.wildlist.org) contained 780 viruses and worms found to be spreading "in the wild" (on real users' PCs), but this list is known to comprise a small subset of the total number of existing viruses. The prevalence of malware was evident in a 2006 CSI/FBI survey where 65% of the organizations reported being hit by malware, the single most common type of attack.

A taxonomy to introduce definitions of malware is shown in Figure 1, but classification is sometimes difficult because a piece of malware often combines multiple characteristics. Viruses and worms are characterized by the capability to self-replicate, but they differ in their methods (Nazario, 2004; Szor, 2005). A virus is a piece of software code (set of instructions but not a complete program) attached to a normal program or file. The virus depends on the execution of the host program. At some point in the execution, the virus code hijacks control of the program execution to make copies of itself and attach these copies to more programs or files. In contrast, a worm is a stand-alone automated program that seeks vulnerable computers through a network and copies itself to compromised victims.

Non-replicating malware typically hide their presence on a computer or at least hide their malicious function. Malware that hides a malicious function but not necessarily its presence is called a Trojan horse (Skoudis, 2004). Typically, Trojan horses pose as a legitimate program (such as a game or device driver) and generally rely on social engineering (deception) because they are not able to self-replicate. Trojan horses are used for various purposes, often theft of confidential data, destruction, backdoor for remote access, or installation of other malware. Besides Trojan horses, many types of non-replicating malware hide their presence in order to carry out a malicious function on a victim host without detection and removal by the user. Common examples include bots and spyware. Bots are covertly installed software that secretly listen for remote commands, usually sent through Internet relay chat (IRC) channels, and execute them on compromised computers. A group of compromised computers under remote control of a single "bot herder" constitute a bot net. Bot nets are often used for spam, data theft, and distributed denial of service attacks. Spyware collects personal user information from a victim computer and transmits the data across the network, often for advertising purposes but possibly for data theft. Spyware is often bundled with shareware or installed covertly through social engineering.

Figure 1. A taxonomy of malicious software

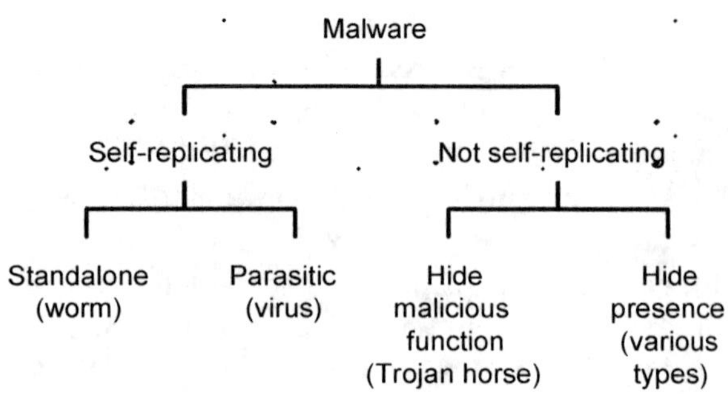

Since 2004, malware has been observed to spread among smartphones and other mobile devices through wireless networks. According to F-Secure, the number of malware known to target smartphones is approximately 100 (Hypponen, 2006). However, some believe that malware will inevitably grow into a serious problem (Dagon, Martin, & Starner, 2004). There have already been complex, blended malware threats on mobile devices. Within a few years, mobile viruses have grown in sophistication in a way reminiscent of 20 years of PC malware evolution. Unfortunately, mobile devices were not designed for security, and they have limited defenses against continually evolving attacks.

If the current trend continues, malware spreading through wireless networks could consume valuable radio resources and substantially degrade the experience of wireless subscribers. In the worst case, malware could become as commonplace in wireless networks as in the Internet with all its attendant risks of data loss, identity theft, and worse. The wireless market is growing quickly, but negative experiences with malware on mobile devices could discourage subscribers and inhibit market growth. The concern is serious because wireless services are currently bound to accounting and charging mechanisms; usage of wireless services, whether for legitimate purposes or malware, will result in subscriber charges. Thus, a victimized subscriber will not only suffer the experience of malware but may also get billed extra service charges. This usage-based charging arrangement contrasts with PCs which typically have flat charges for Internet communications.

This chapter examines historical examples of malware and the current environment for mobile devices. Potential infection vectors are explored. Finally, existing defenses are identified and described.

BACKGROUND

Mobile devices are attractive targets for several reasons (Hypponen, 2006). First, mobile devices have clearly progressed far in terms of hardware and communications. PDAs have grown from simple organizers to miniature computers with their own operating systems (such as Palm or Windows Pocket PC/Windows Mobile) that can download and install a variety of applications. Smartphones combine the communications capabilities of cell phones with PDA functions. According to Gartner, almost 1 billion cell phones will be sold in 2006. Currently, smartphones are a small fraction of the overall cell phone market. According to the *Computer Industry Almanac,* 69 million smartphones will be sold in 2006. However, their shipments are growing rapidly, and IDC predicts smartphones will become 15% of all mobile phones by 2009. Approximately 70% of all smartphones run the Symbian operating system, made by various manufacturers, according to Canalys. Symbian is jointly owned by Sony Ericsson, Nokia, Panasonic, Samsung, and Siemens AG. Symbian is prevalent in Europe and Southeast Asia but less common in North America, Japan, and South Korea. The Japanese and Korean markets have been dominated by Linux-based phones. The North American market has a diversity of cellular platforms.

Nearly all of the malware for smartphones has targeted the Symbian operating system. Descended from Psion Software's EPOC, it is structured similar to desktop operating systems. Traditional cell phones have proprietary embedded operating systems which generally accept only Java applications. In contrast, Symbian application programming interfaces (APIs) are publicly documented so that anyone can develop applications. Applications packaged in SIS file format can be installed at any time, which makes Symbian devices more attractive to both consumers and malware writers.

Mobile devices are attractive targets because they are well connected, often incorporating various means of wireless communications. They are typically capable of Internet access for Web browsing, e-mail, instant messaging, and applications similar to those on PCs. They may also communicate by cellular, IEEE 802.11 wireless LAN, short range Bluetooth, and short/multimedia messaging service (SMS/MMS).

Another reason for their appeal to malware writers is the size of the target population. There were more than 900 million PCs in use worldwide in 2005 and will climb past 1 billion PCs in 2007, according to the *Computer Industry Almanac.* In comparison, there were around 2 billion cellular subscribers in 2005. Such a large target population is attractive for malware writers who want to maximize their impact.

Malware is relatively unknown for mobile devices today. At this time, only a small number of families of malware have been seen for wireless devices, and malware is not a prominent threat in wireless networks. Because of the low threat risk, mobile devices have minimal security defenses. Another reason is the limited processing capacity of mobile devices. Whereas desktop PCs have fast processors and plug into virtually unlimited power, mobile devices have less computing power and limited battery power. Protection such as antivirus software and host-based intrusion detection would incur a relatively high cost in processing and energy consumption. In addition, mobile devices were never designed for security. For example, they lack an encrypting file system, Kerberos authentication, and so on. In short, they are missing all the components required to secure a modern, network-connected computing device.

There is a risk that mobile users may have a false sense of security. Physically, mobile devices feel more personal because they are carried everywhere. Users have complete physical control of them, and hence they feel less accessible to intruders. This sense of security may lead users to trust the devices with more personal data, increasing the risk of loss and appeal to attackers. Also, the sense of security may lead users to neglect security precautions such as changing default security configurations.

Although mobile devices might be appealing targets, there are certain drawbacks to malware for mobile devices. First, mobile devices usually have intermittent connectivity to the network or other devices, in order to save power. This fact limits the ability of malware to spread quickly. Second, if malware is intended to spread by Bluetooth, Bluetooth connections are short range. Moreover, Bluetooth devices can be turned off or put into hidden mode. Third, there is a diversity of mobile device platforms, in contrast to PCs that are dominated by Windows. Some have argued that the Windows monoculture in PCs has made PCs more vulnerable to malware. To reach a majority of mobile devices, malware writers must create separate pieces of malware code for different platforms (Leavitt, 2005).

EVOLUTION OF MALWARE

Malware has already appeared on mobile devices over the past few years (Peikari & Fogie, 2003). While the number is still small compared to the malware families known for PCs, an examination of prominent examples shows that malware is evolving steadily. The intention here is not to exhaustively list all examples of known malware but to highlight how malware has been developing.

Palm Pilots and Windows Pocket PCs were common before smartphones, and malware appeared first for the Palm operating system. Liberty Crack was a Trojan horse related to Liberty, a program emulating the Nintendo Game Boy on the Palm, reported in August 2000 (Foley & Dumigan, 2001). As a Trojan, it did not spread by self-replication but depended on being installed from a PC that had the "liberty_1_1_crack.prc" file. Once installed on a Palm, it appears on the display as an application, Crack. When executed, it deletes all applications from the Palm (www.f-secure.com/v-descs/lib_palm.shtml).

Discovered in September 2000, Phage was the first virus to target Palm PDAs (Peikari & Fogie, 2003). When executed, the virus infects all third-party applications by overwriting them (http://www.f-secure.com/v-descs/phage.shtml). When a program's icon is selected, the display turns gray and the selected program exits. The virus can spread directly to other Palms by infrared beaming or indirectly through PC synchronization.

Another Trojan horse discovered around the same time, Vapor is installed on a Palm as the application "vapor.prc" (www.f-secure.com/v-descs/vapor.shtml). When executed, it changes the file attributes of other applications, making them invisible (but not actually deleting them). It does not self-replicate.

In July 2004, Duts was a proof-of-concept virus, the first to target Windows Pocket PCs. It asks the user for permission to install. If installed, it attempts to infect all EXE files larger than 4096 bytes in the current directory.

Later in 2004, Brador was a backdoor for Pocket PCs (www.f-secure.com/v-descs/brador.shtml). It installs the file "svchost.exe" in the Startup directory so that it will automatically start during the device bootup. Then it will read the local host IP address and e-mail that to the author. After e-mailing its IP address, the backdoor opens a TCP port and starts listening for commands. The backdoor is capable of uploading and downloading files, executing arbitrary commands, and displaying messages to the PDA user.

The Cabir worm discovered in June 2004 was a milestone marking the trend away from PDAs and towards smartphones running the Symbian operating system. Cabir was a proof-of-concept worm, the first for Symbian, written by a member of a virus writing group 29A (www.f-secure.com/v-descs/cabir.shtml). The worm is carried in a file "caribe.sis" (Caribe is Spanish for the Caribbean). The SIS file contains autostart settings that will automatically execute the worm after the SIS file is installed. When the Cabir worm is activated, it will start looking for other (discoverable) Bluetooth devices within range. Upon finding another device, it will try to send the caribe.sis file. Reception and installation of the file requires user approval after a notification message is displayed. It does not cause any damage.

Cabir was not only one of the first malware for Symbian, but it was also one of the first to use Bluetooth (Gostev, 2006). Malware is more commonly spread by e-mail. The choice of Bluetooth meant that Cabir would spread slowly in the wild. An infected smartphone would have to discover another smartphone within Bluetooth range and the target's user would have to willingly accept the transmission of the worm file while the devices are within range of each other.

In August 2004, the first Trojan horse for smartphones was discovered. It appeared to be a cracked version of a Symbian game Mosquitos. The Trojan made infected phones send SMS text messages to phone numbers resulting in charges to the phones' owners.

In November 2004, the Trojan horse—Skuller—was found to infect Symbian Series 60 smartphones (www.f-secure.com/v-descs/skulls.shtml). The Trojan is a file named "Extended theme.SIS," a theme manager for Nokia 7610 smartphones. If executed, it disables all applications on the phone and replaces their icons with a skull and crossbones. The phone can be used to make calls and answer calls. However, all system applications such as SMS, MMS, Web browsing, and camera do not work.

In December 2004, Skuller and Cabir were merged to form Metal Gear, a Trojan horse that masquerades as the game of the same name. Metal Gear uses Skulls to deactivate a device's antivirus. This was the first malware to attack antivirus on Symbian smartphones. The malware also drops a file "SEXXXY.SIS," an installer that adds code to disable the handset menu button. It then uses Cabir to send itself to other devices.

Locknut was a Trojan horse discovered in February 2005 that pretended to be a patch for Symbian Series 60 phones. When installed, it drops a program that will crash a critical system service component, preventing any application from launching.

In March 2005, ComWar or CommWarrior was the first worm to spread by MMS among Symbian Series 60 smartphones. Like Cabir, it was also capable of spreading by Bluetooth. Infected phones will search for discoverable Bluetooth devices within range; if found, the infected phone will try to send the worm in a randomly named SIS file. But Bluetooth is limited to devices within 10 meters or so. MMS messages can be sent to anywhere in the world. The worm tries to spread by MMS messaging to other phone owners found in the victim's address book. MMS has the unfortunate side effect of incurring charges for the phone owner.

Drever was a Trojan horse that attacked antivirus software on Symbian smartphones. It drops non-functional copies of the bootloaders used by Simworks Antivirus and Kaspersky Symbian Antivirus, preventing these programs from loading automatically during the phone bootup.

In April 2005, the Mabir worm was similar to Cabir in its ability to spread by Bluetooth. It had the additional capability to spread by MMS messaging. It listens for any arriving MMS or SMS message and will respond with a copy of itself in a file named "info.sis."

Found in September 2005, the Cardtrap Trojan horse targeted Symbian 60 smartphones and was one of the first examples of smartphone malware capable of infecting a PC (www.f-secure.com/v-descs/cardtrap_a.shtml). When it is installed on the smartphone, it disables several applications by overwriting their main executable files. More interestingly, it also installs two Windows worms, Padobot.Z and Rays, to the phone's memory card. An autorun file is copied with the Padobot.Z worm, so that if the memory card is inserted into a PC, the autorun file will attempt to execute the Padobot worm. The Rays worm is a file named "system.exe" which has the same icon as the system folder in the memory card. The evident intention was to trick a user reading the contents of the card on a PC into executing the Rays worm.

Crossover was a proof-of-concept Trojan horse found in February 2006. It was reportedly the first malware capable of spreading from a PC to a Windows Mobile Pocket PC by means of ActiveSync. On the PC, the Trojan checks the version of the host operating system. If it is not Windows CE or Windows Mobile, the virus makes a copy of itself on the PC and adds a registry entry to execute the virus during PC rebooting. A new virus copy is made with a random file name at each reboot. When executed, the Trojan waits for an ActiveSync connection, when it copies itself to the handheld, documents on the handheld will be deleted.

In August 2006, the Mobler worm for Windows PCs was discovered (www.f-secure.com/v-descs/mobler.shtml). It is not a real threat but is suggestive of how future malware might evolve. When a PC is infected, the worm copies itself to different folders on local hard drives and writable media (such as a memory card). Among its various actions, the worm creates a SIS archiver program "makesis.exe" and a copy of itself named "system.exe" in the Windows system folder. It also creates a Symbian installation package named "Black_Symbian.SIS." It is believed to be capable of spreading from a PC to smartphone, another example of cross-platform malware.

At the current time, it is unknown whether Crossover and Mobler signal the start of a new trend towards cross-platform malware that spread equally well among PCs and mobile devices. The combined potential target population would be nearly 3 billion. The trend is not obvious yet but Crossover and Mobler suggest that cross-platform malware could become possible in the near future.

INFECTION VECTORS

Infection vectors for PC malware have changed over the years as PC technology evolved. Viruses initially spread by floppy disks. After floppy disks disappeared and Internet connectivity became ubiquitous, worms spread by mass e-mailing. Similarly, infection vectors used by malware for mobile devices have changed over the past few years.

Synchronization: Palm and Windows PDAs were popular before smartphones. PDAs install software by synchronization with PCs (Foley & Dumigan, 2001). For example, Palm applications are packaged as Palm resource (PRC) files installed from PCs. As seen earlier, Palm malware usually relied on social engineering to get installed. This is a slow infection vector for malware to spread between PDAs because it requires synchronization with a PC and then contact with another PC that synchronizes with another PDA. Much faster infection vectors became possible when PDAs and then smartphones started to feature communications directly between mobile devices without having to go through PCs.

E-mail and Web: Internet access from mobile devices allows users away from their desktops to use the most common Internet applications, e-mail and the World Wide Web. Most mobile devices can send and receive e-mail with attachments. In addition, many can access the Web through a microbrowser designed to render Web content on the small displays of mobile devices. Current microbrowsers are similar in features to regular Web browsers, capable of HTML, WML, CSS, Ajax, and plug-ins. Although e-mail and the Web are common vectors for PC malware, they have not been used as vectors to infect mobile devices thus far.

SMS/MMS messaging: Commonly called text messaging, SMS is available on most mobile phones and Pocket PCs. It is most popular in Europe, Asia (excluding Japan), Australia, and New Zealand, but has not been as popular in the U.S. as other types of messaging. Text messaging is often used to interact with automated systems, for example to order products or services or participate in contests. Short messages are limited to 140 bytes of data, but longer content can be segmented and sent in multiple messages. The receiving phone is responsible for reassembling the complete message. Short messages can also be used to send binary content such as ringtones or logos. While SMS is largely limited to text, MMS is a more advanced messaging service allowing transmission of multimedia objects—video, images, audio, and rich text. The ComWar worm was the first to spread by MMS (among Symbian Series 60 smartphones). MMS has the potential to spread quickly. ComWar increased its chances by targeting other phone owners found in the victim's address book. By appearing to come from an acquaintance, an incoming message is more likely to be accepted by a recipient. MMS will likely continue to be an infection vector in the future.

Bluetooth: Bluetooth is a short-range radio communication protocol that allows Bluetooth-enabled devices (which could be mobile or stationary) within 10-100 meters to discover and talk with each other. Up to eight devices can communicate with each other in a piconet, where one device works in the role of "master" and the others in the role of "slaves." The master takes turns to communicate with each slave by round robin. The roles of master and slaves can be changed at any time.

Each Bluetooth device has a unique and permanent 48-bit address as well as a user-chosen Bluetooth name. Any device can search for other nearby devices, and devices configured to respond will give their name, class, list of services, and technical details (e.g., manufacturer, device features). If a device inquires directly at a device's address, it will always respond with the requested information.

In May 2006, F-Secure and Secure Networks conducted a survey of discoverable Bluetooth devices in a variety of places in Italy. They found on average 29 to 154 Bluetooth devices per hour in discoverable mode in the different places. In discoverable mode, the devices are potentially open to attacks. About 24% were found to have visible OBEX push service. This service is normally used for transfer of electronic business cards or similar information, but is known to be vulnerable to a BlueSnarf attack. This attack allows connections to a cellular phone and access to the phone book and agenda without authorization. Another vulnerability is BlueBug, discovered in March 2004, allowing access to the ASCII Terminal (AT) commands of a cell phone. These set of commands are common for configuration and control of telecommunications devices, and give high-level control over call control and SMS messaging. In effect, these can allow an attacker to use the phone services without the victim's knowledge. This includes incoming and outgoing phone calls and SMS messages.

The Cabir worm was the first to use Bluetooth as a vector. Bluetooth is expected to be a slow infection vector. An infected smartphone would have to discover another smartphone within a 10-meter range, and the target's user would have to willingly accept the transmission of the worm file while the devices are within range of each other. Moreover, although phones are usually shipped with Bluetooth in discoverable mode, it is simple to change devices to invisible mode. This simple precaution would make it much more difficult for malware.

MALWARE DEFENSES

Practical security depends on multiple layers of protection instead of a single (hopefully perfect) defense (Skoudis, 2004). Fortunately, various defenses against malware have been developed from decades of experience with PC malware. A taxonomy of malware defenses is shown in Figure 2. Defenses can be first categorized as preventive or reactive (defensive). Preventive techniques help avoid malware infections through identification and remediation of vulnerabilities, strengthening security policies, patching operating systems and applications, updating antivirus signatures, and even educating users about best practices (in this case, for example, turning off Bluetooth except when needed, rejecting installation of unknown software, and blocking SMS/MMS messages from untrusted parties). At this time, simple preventive techniques are likely to be very effective because there are relatively few threats that really spread in the wild. In particular, education to raise user awareness would be effective against social engineering, one of the main infection vectors used by malware for mobile devices so far.

Host-Based Defenses

Even with the best practices to avoid infections, reactive defenses are still needed to protect mobile devices from actual malware threats. Reactive defenses can operate in hosts (mobile devices) or within the network. Host-based defenses make sense because protection will be close to the targets. However, host-based processes (e.g., antivirus programs) consume processing and power resources that are more critical on mobile devices than desktop PCs. Also, the approach is difficult to scale to large populations if software must be installed, managed, and maintained on every mobile device. Network-based defenses are more scalable in the sense that one router or firewall may protect a group of hosts. Another reason for network-based defenses is the possibility that the network might be able to block malware before it actually reaches a targeted device, which is not possible with host-based defenses. Host-based defenses take effect after contact with the host. In practice, host-based and network-based defenses are both used in combination to realize their complementary benefits.

The most obvious host-based defense is antivirus software (Szor, 2005). Antivirus does automatic analysis of files, communicated messages, and system activities. All commercial antivirus programs depend mainly on malware signatures which are sets of unique characteristics associated with each known piece of malware. The main advantage of signature-based detection is its accuracy in malware identification. If a signature is matched, then the malware is identified exactly and perhaps sufficiently for disinfection. Unfortunately, signature-based detection has two drawbacks. First, antivirus signatures must be regularly updated. Second, there will always be the possibility that new malware could escape detection if it does not have a matching signature. For that case, antivirus programs often include heuristic anomaly detection which detects unusual behavior or activities. Anomaly detection does not usually identify malware exactly, only the suspicion of the presence of malware and the need for further investigation. For that reason, signatures will continue to be the preferred antivirus method for the foreseeable future.

Several antivirus products are available for smartphones and PDAs. In October 2005, Nokia and Symantec arranged for Nokia to offer the option of preloading Symbian Series 60 smartphones with Symantec Mobile Security Antivirus. Other commercial antivirus packages can be installed on Symbian or Windows Mobile smartphones and PDAs.

In recognition that nearly all smartphone malware has targeted Symbian devices, a great amount of attention has focused on the vulnerabilities of that operating system. It might be argued that the system has a low level of application security. For example, Symbian allows any system application to be rewritten without requiring user consent. Also, after an application is installed, it has total control over all functions. In short, applications are totally trusted.

Although Windows CE has not been as popular a target, it has similar vulnerabilities. There are no restrictions on applications; once launched, an application has full access to any system function including sending/receiving files, phone functions, multimedia functions, and so forth. Moreover, Windows CE is an open platform and application development is relatively easy.

Symbian OS version 9 added the feature of code signing. Currently all software must be manually installed. The installation process warns the user if an application has not been signed. Digital signing makes software traceable to the developer and verifies that an application has not been changed since it left the developer. Developers can apply to have their software signed via the Symbian Signed program (www.symbiansigned.com). Developers also have the option of self-signing their programs. Any signed application will install on a Symbian OS phone without showing a security warning. An unsigned application can be installed with user consent, but the operating system will prevent it from doing potentially damaging things by denying access to key system functions and data storage of other applications.

Figure 2. A taxonomy of malware defenses

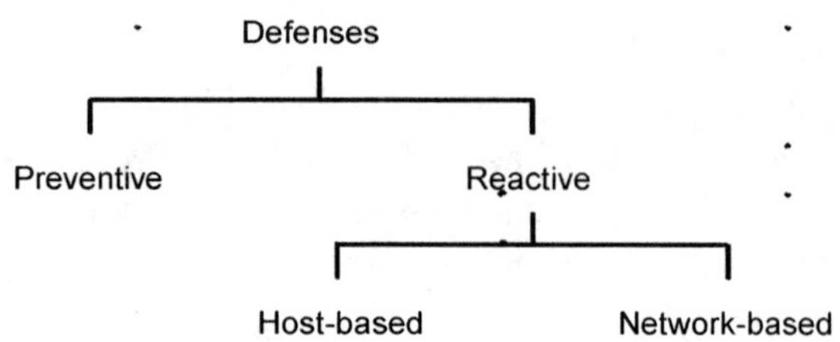

Network-Based Defenses

Network-based defenses depend on network operators monitoring, analyzing, and filtering the traffic going through their networks. Security equipment include firewalls, intrusion detection systems, routers with access control lists (ACLs), and antivirus running in e-mail servers and SMS/MMS messaging service centers. Traffic analysis is typically done by signature-based detection, similar in concept to signature-based antivirus, augmented with heuristic anomaly based detection.

Traffic filtering is done by configuring firewall and ACL policies.

An example is Sprint's Mobile Security service announced in September 2006. This is a set of managed security services for mobile devices from handhelds to laptops. The service includes protection against malware attacks. The service can scan mobile devices and remove detected malware automatically without requiring user action.

In the longer term, mobile device security may be driven by one or more vendor groups working to improve the security of wireless systems. For instance, the Trusted Computing Group (TCG) (www.trustedcomputinggroup.org) is an organization of more than 100 component manufacturers, software developers, networking companies, and service providers formed in 2003. One subgroup is working on a set of specifications for mobile phone security (TCG, 2006a). Their approach is to develop a Mobile Trusted Module (MTM) specification for hardware to support features similar to those of the Trusted Platform Module (TPM) chip used in computers but with additional functions specifically for mobile devices. The TPM is a tamper-proof chip embedded at the PC board level, serving as the "root of trust" for all system activities. The MTM specification will integrate security into smartphones' core operations instead of adding as applications.

Another subgroup is working on specifications for Trusted Network Connect (TCG, 2006b). All hosts including mobile devices run TNC client software, which collects information about that host's current state of security such as antivirus signature updates, software patching level, results of last security scan, firewall configuration, and any other active security processes. The security state information is sent to a TNC server to check against policies set by network administrators. The server makes a decision to grant or deny access to the network. This ensures that hosts are properly configured and protected before connecting to the network. It is important to verify that hosts are not vulnerable to threats from the network and do not pose a threat to other hosts. Otherwise, they will be effectively quarantined from the network until their security state is remedied. Remedies can include software patching, updating antivirus, or any other changes to bring the host into compliance with security policies.

FUTURE TRENDS

It is easy to see that mobile phones are increasingly attractive as malware targets. The number of smartphones and their percentage of overall mobile devices is growing quickly. Smartphones will continue to increase in functionalities and complexity. Symbian has been the primary target, a trend that will continue as long as it is the predominant smartphone platform. If another platform arises, that will attract the attention of malware writers who want to make the biggest impact.

The review of malware evolution suggests a worrisome trend. Since the first worm, Cabir, only three years ago, malware has advanced steadily to more infection vectors, first Bluetooth and then MMS. Recently malware has shown signs of becoming cross-platform, moving easily between mobile devices and PCs.

Fortunately, mobile security has already drawn the activities of the TCG and other industry organizations. Unlike the malware situation with PCs, the telecommunications industry has decades of experience to apply to wireless networks, and there is time to fortify defenses before malware multiplies into a global epidemic.

CONCLUSION

Malware is a low risk threat for mobile devices today, but the situation is unlikely to stay that way for long. It is evident from this review that mobile phones are starting to attract the attention of malware writers, a trend that will only get worse. At this point, most defenses are common sense practices. The wireless industry realizes that the stakes are high. Two billion mobile users currently enjoy a malware-free experience, but negative experiences with new malware could have a disastrous effect. Fortunately, a range of host-based and network-based defenses have been developed from experience with PC malware. Activities are underway in the industry to improve protection of mobile devices before the malware problem becomes catastrophic.

REFERENCES

Dagon, D., Martin, T., & Starner, T. (2004). Mobile phones as computing devices: The viruses are coming! *IEEE Pervasive Computing, 3*(4), 11-15.

Foley, S., & Dumigan, R. (2001). Are handheld viruses a significant threat? *Communications of the ACM, 44*(1), 105-107.

Gostev, A. (2006). *Mobile malware evolution: An overview.* Retrieved from http://www.viruslist.com/en/analysis?pubid=200119916

Hypponen, M. (2006). Malware goes mobile. *Scientific American, 295*(5), 70-77.

Leavitt, N. (2005). Mobile phones: The next frontier for hackers? *Computer, 38*(4), 20-23.

Nazario, J. (2004). *Defense and detection strategies against Internet worms.* Norwood, MA: Artech House.

Peikari, C., & Fogie, S. (2003). *Maximum wireless security.* Indianapolis, IN: Sams Publishing.

Skoudis, E. (2004). *Malware: Fighting malicious code.* Upper Saddle River, NJ: Prentice Hall.

Szor, P. (2005). *The art of computer virus research and defense.* Reading, MA: Addison-Wesley.

Trusted Computing Group (TCG). (2006a). *Mobile trusted module specification.* Retrieved from https://www.trustedcomputinggroup.org/specs/mobilephone/

Trusted Computing Group (TCG). (2006b). *TCG trusted network connect TNC architecture for interoperability.* Retrieved from https://www.trustedcomputinggroup.org/groups/network/

KEY TERMS

Antivirus Software: Antivirus software is designed to detect and remove computer viruses and worms and prevent their reoccurrence.

Exploit Software: Exploit software is written to attack and take advantage of a specific vulnerability.

Malware Software: Malware software is any type of software with malicious function, including for example, viruses, worms, Trojan horses, and spyware.

Smartphone: Smartphones are devices with the combined functions of cell phones and PDAs, typically running an operating system such as Symbian OS.

Social Engineering: Social engineering is an attack method taking advantage of human nature.

Trojan Horse: A Trojan horse is any software program containing a covert malicious function.

Virus: A virus is a piece of a software program that attaches to a normal program or file and depends on execution of the host program to self-replicate and infect more programs or files.

Vulnerability: Vulnerability is a security flaw in operating systems or applications that could be exploited to attack the host.

Worm: A worm is a stand-alone malicious program that is capable of automated self-replication.

Chapter II
Secure Service Discovery

Sheikh I. Ahamed
Marquette University, USA

Munirul M. Haque
Marquette University, USA

John F. Buford
Avaya Labs, USA

Nilothpal Talukder
Marquette University, USA

Moushumi Sharmin
Marquette University, USA

ABSTRACT

In broadband wireless networks, mobile devices will be equipped to directly share resources using service discovery mechanisms without relying upon centralized servers or infrastructure support. The network environment will frequently be ad hoc or will cross administrative boundaries. There are many challenges to enabling secure and private service discovery in these environments including the dynamic population of participants, the lack of a universal trust mechanism, and the limited capabilities of the devices. To ensure secure service discovery while addressing privacy issues, trust-based models are inevitable. We survey secure service discovery in the broadband wireless environment. We include case studies of two protocols that include a trust mechanism, and we summarize future research directions.

INTRODUCTION

Service orientation is widely used in client-server computing and is growing in importance for mobile wireless devices. In this way, a device's software and hardware components can be packaged as services for use by other devices. Many consumer electronics (CE) devices are specialized for specific uses. Due to form factor and cost considerations, devices vary in capability. With sufficiently high bandwidth network interfaces on these devices, such as 802.11, WiMax, and ultra-wideband (UWB), it is practical for sets of networked devices to share functionality. Service discovery and advertisement (SDA) is fundamental to service interoperability in pervasive computing applications.

Many service discovery protocols have been developed, including several for specific wireless networks. However, few of these protocols have been designed with security mechanisms and the majority use centralized enforcement and validation. Due to the emergence of mobile and large-scale peer-to-peer applications, there is growing

Figure 1. Different types of networks in a pervasive computing environment. (a) Ad hoc network in a pervasive environment with powerful device support. (b) Ad hoc network in a pervasive environment without powerful device support.

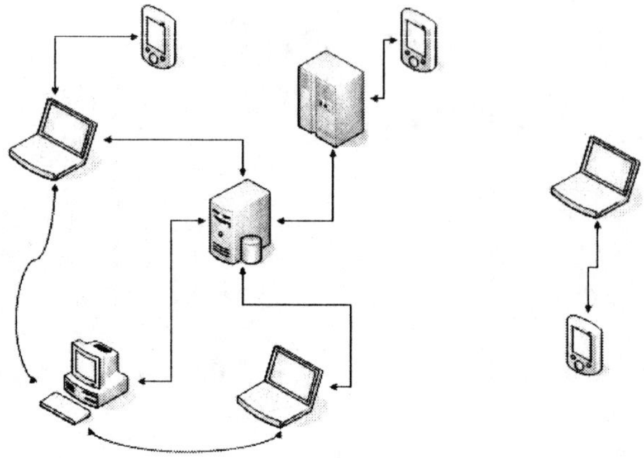

(a) *(b)*

interest in security mechanisms that do not require centralized enforcement and validation.

We present the current state of secure service discovery. Leading designs for secure service discovery are surveyed including industry standards and research systems. The types of security issues we are concerned with include: protecting the privacy of service advertisements and descriptions; authentication of service advertisements; secure distribution and updating of keys for service invocation; providing trust in service composition; and limiting vulnerability to attacks effecting the service discovery mechanism.

Pervasive computing environment focuses (Weiser, 1991, 1993) has evolved over the past few years with the availability of portable low-cost devices (such as PDAs, cell phones, smart phones, laptops, and sensors) and the emergence of short-range and low-power wireless communication networks. Pervasive computing environments focus on integrating computing and communications with the surrounding physical environment to make computing and communication transparent to the users in everyday contexts. In a broad sense, pervasive computing combines mobile computing, wireless networks, embedded computing, and context-aware sensor networks (Robinson, Vogt, & Wagealla, 2005).

The different kinds of networks in pervasive computing environments impact the design of secure service discovery mechanisms. On one end, there are smart spaces, or intelligent environments that provide devices with a variety of support for user awareness and context management, while at the other end there are networks that provide open network connectivity.

Figure 1 depicts two ad hoc networks in a pervasive computing environment. In Figure 1a, the devices communicate among themselves with the support of fixed, more powerful devices. These devices act as servers or proxies for the mobile devices. In Figure 1b, an ad hoc network is formed by mobile devices. There is no fixed infrastructure support. The devices communicate with each other directly or via another mobile device, and are responsible for performing computations by themselves.

In service discovery (Kindberg & Fox, 2002; Lee & Helal, 2002), a device searches for another device capable of offering a specific service or resource. An important trend is the adoption of a service-oriented architecture for resource discovery, not just for server systems accessed by mobile devices, but also for sharing of resources between devices. There are four elements found in the service-oriented approach: (1) service description, which provides an interchangeable way for devices to describe the service and its use; (2) service registration or advertisement on behalf of the service provider; (3) service discovery by devices seeking a service; and (4) service invocation, which is a protocol by which a service requester and service provider coordinate to deliver a service. Propagation of service advertisements can be using pull (query), push (announcement), or a combination of pull and push. In addition, the ability to dynamically discover and combine component services to form new services is referred to as service composition.

Broadband wireless technologies such as WiMax, UWB, and 802.11n are bringing broadband connectivity to mobile CE devices. These devices will be able to switch between different network access technologies. This has the following consequences for service discovery in pervasive computing:

- Due to broadband connectivity, devices will be able to participate in media-rich and sophisticated resource sharing.
- Wide-area service discovery and location-based discovery will grow in importance due to the combination of increased connectivity and wide-area roaming.
- The ability to act as multi-homed devices means that devices will have increased connectivity but also an increased rate of transitions due to roaming between different networks.
- Devices will be able to simultaneously participate in a personal area network (PAN), home networks, and wireless area networks (WANs) with different security and trust properties. In PANs and home networks, mediation of service discovery between networks is needed, in which devices such as gateways proxy or intermediate service discovery between network domains.
- Device-to-device interaction will grow in importance to users for applications such as content sharing, communication, and gaming.

Due to these trends, richer models of discovery are being considered such as federated discovery, meta discovery, and semantic discovery (Buford, Brown, & Kolberg, 2006; Buford, Celebi, & Frankl, 2006).

Consequently, it is important for wireless devices to securely participate in service discovery with other devices that are outside the immediate administrative security domain. Further, these devices interact with other devices in an ad hoc manner, and lack of fixed infrastructure support leads to the dependency on other devices for resources. The nature of devices, communication patterns, and dependency on other devices in turn causes security vulnerabilities. Due to the ad hoc connectivity and dynamic nature of the population of devices, access to specific devices may be intermittent and short-lived. Moreover, multiple devices may concurrently request one specific resource. These aspects demand a scalable, efficient, and responsive service discovery model.

Thus far, we have discussed the general view of and motivation for service discovery for mobile devices. The rest of the chapter is organized as follows: The next section summarizes the security goals for service discovery and presents a model for service discovery in pervasive computing. The third section surveys present unsecured service discovery models. The fourth section surveys existing secure service discovery models, organized into three different categories. Two case studies of service discovery protocols that incorporate trust-based mechanisms are described in the *Examples Using Trust Models* section. The final sections summarize important research issues and conclusions.

Security Goals in Service Discovery, Invocation, and Composition

The significance of security during service discovery is well established (Matsumiya et al., 2004; Stajano, 2002; Stajano & Anderson, 2002). Privacy, security, and trust issues in service discovery in the pervasive computing area are of utmost importance (Robinson et al., 2005). Thus, the service discovery process demands models that ensure the privacy and security of the user. In particular, this privacy and security should encompass:

- **Authentication:** Does the user and device actually have the indicated identity?
- **Authorization:** Does the user have access rights for issuing service advertisements, requesting services, and invoking services?
- **Trust:** Are the participating user and device trusted? Are the service and its components trusted?
- **Privacy:** Is only the approved information shared between the given users/devices during service discovery, advertisement and invocation (SDAI) operations? Is disclosure to unauthorized users prevented?
- **Vulnerability to attack and misuse:** Are the SDAI operations protected from attacks such as denial-of-service, spoofing, replay, and man-in-the-middle? Are the SDAI operations protected from misuse in enabling such attacks on other network components?

An important question is what security, privacy, and trust mechanisms are provided by the wireless network. IEEE 802.11i, also known as WiFi Protected Access 2 (WPA2), replaced Wired Equivalent Privacy (WEP) with stronger encryption and a new authentication mechanism incorporating an authentication server such as remote authentication dial in user service (RADIUS). This mechanism while suitable for enterprise deployment has had limited use in home networks because of complex administration and in public hot spots due to difficulty administering shared keys. Thus, in the best case, a set of devices are authenticated in a single administrative domain, and the authentication server can be used to support authorization policies including policies related to service discovery and use. Network packets between authenticated users are encrypted, providing communication privacy from non-authenticated parties. However, these security capabilities cover only a subset of the aforementioned security goals and are limited to single administrative domains. For interactions crossing administrative boundaries, or without infrastructure support, other mechanisms are needed.

Further, traditional security mechanisms do not work well in this environment because the devices are computationally limited and the notion of physical security is not applicable (Kagal, Finin, & Joshi, 2001). Then, considering the choices of totally sacrificing security versus imposing a full-fledged security structure similar to desktops and laptops, the question is whether there is any middle ground. Ensuring varying levels of security for various services is a research challenge. The insufficiency of user/device identity for trust is another concern in designing a discovery model, and techniques for peer trust and risk assessment (Chen, Jensen, Gray, Cahill, & Seigneur, 2003) are important tools to address this.

Desired characteristics of a secure and private service discovery model are summarized next.

- **Adaptive:** The trust value and security level should be adaptable depending on the service itself, the service provider, and the service requester.
- **Trust reliant:** The model should consider trust relationships among devices. Where no prior information is available, reputation, recommendation, or trust negotiation schemes can be used. If these are unsuitable, then risk assessment can be used.
- **Infrastructure independence:** No infrastructure support (e.g., powerful servers, proxies) should be required. Then the model should work independently without any external support, but be able to leverage infrastructure where it exists.
- **Lightweight:** The model should be lightweight in terms of executable file size.
- **Service oriented:** To control service security modularly, service discovery models should be service oriented.
- **Graceful performance degradation:** The model should not put much overhead on the performance of the device, and performance should degrade gracefully for more advanced security features.

- **Energy efficient:** Service discovery models should be energy conserving, for example, avoiding continuous broadcasting or polling.

A classification and detailed survey of service discovery models can be found in Zhu, Mutka, and Ni (2002). Service-oriented architectures (SOA) and their security are discussed in Cotroneo, Graziano, and Russo (2004). We classify existing service discovery models into two broad categories. First are service discovery models that do not address security issues (Balazinska, Balakrishnan, & Karger, 2002; Microsoft, 2000; Miller, Nixon, Tai, & Wood, 2001; Nidd, 2001; Winoto, Schwartz, Balakrishnan, & Lilley, 1999). Second, there are models that consider a full-fledged security mechanism with the help of infrastructure support (Czerwinski, Zhao, Hodes, Joseph, & Katz, 1999; Zhu, Mutka, & Ni, 2003, 2004). The next two sections discuss examples of these cases, and Table 1 compares the key features of the surveyed systems.

SERVICE DISCOVERY MODELS WITHOUT INHERENT SECURITY

We describe several designs that do not address security requirements. Nevertheless these models are important either because the systems are widely used, are representative approaches, or could be secured by additional mechanisms in a secure network. The designs we discuss are Bluetooth, DEAPSpace, and Intentional Naming System (INS).

Bluetooth (Bluetooth Special Interest Group [SIG], 2001a, 2001b) is a pull protocol. Device information, services, and the characteristics of the services are queried and connections between two or more Bluetooth devices are established. This facilitates user selection, scope-awareness, and both unicast and broadcast communication. A Bluetooth device returns all matched resource information.

Nidd (2001) developed the DEAPSpace service discovery method for ad hoc and mobile device applications. Each node broadcasts its advertisement of local services. After receiving a broadcast, each node updates its service list with information about the other nodes' services. This service information is included in that node's subsequent broadcast. Each node is a broadcaster and DEAPSpace uses contention timers at each node so that a node will randomly delay its broadcast after another broadcast is received. DEAPSpace can reduce service discovery time at the cost of increased bandwidth and power consumption.

INS (Winoto et al., 1999) supports both pull and push delivery of service advertisements. It also supports unicast, anycast, and broadcast methods. It offers the best-match resource information and also provides facilities for limited support of context information. In INS each device requests a central name resolver for the type of services it requires, and the resolver replies with the best matched device address.

Secure Service Discovery Models

Most contemporary service discovery models fall into this category. There are some models that include full-fledged security mechanisms, while others rely on simple algorithms for limited security. This category can be subdivided into infrastructure based, infrastructureless, hardware based, and smart-space-oriented security mechanisms. In the following subsections we discuss each of these categories.

Infrastructure-Based Security

UPnP is a specification for connecting multiple devices on a home network so that these devices can invoke services of each other. UPnP defines a set of protocols and a service description format. In addition, UPnP standardizes various service interfaces. UPnP relies on administratively scoped multicast IP address for service discovery, service advertisement, and event delivery. Each UPnP device broadcasts its advertisements when it first connects to the network. Thereafter, a UPnP device broadcasts advertisements in response to queries from other devices. These queries may be for all services on the network or a specific service on

Table 1. Comparison of secure service discovery models (SSDS): SSDS (Czerwinski et al., 1999), Ninja (Goldberg, Gribble, Wagner, & Brewer, 1999; Gribble et al., 2001), UPnP (Miller et al., 2001), SPDP (Almenarez & Campo, 2003), Progressive Exposure (Zhu et al., 2004; Zhu, Mutka, & Ni, 2006), Splendor (Kagal, Korolev, Chen, Joshi, & Finin, 2001), Jini (Sun Microsystems, 2001), CSAS (Minami & Kotz, 2005), CSM (Brezillon & Mostefaoui, 2004), AVCM (Shankar & Arbaugh, 2002), CSRA (Tripathi, Ahmed, Kulkarni, Kumar, & Kashiramka, 2004), TRAC (Basu & Callaghan, 2005), SME (Kopp, Lucke, & Tavangarian, 2005), HCA (Pearson, 2005), SSRD (Sharmin, Ahmed, & Ahamed, 2006a), SSRD+ (Sharmin, Ahmed, & Ahamed, 2006b), Centaurus2 (Undercoffer, Perich, Cedilnik, Kagal, & Joshi, 2003), SLP (Barbeau, 1999; Guttman, Perkins, Veizades, & Day, 1999), Sleeper (Buford, Celebi, et al., 2006)

Model	Adaptive	Infrastructure Support Needed	Lightweight	Service-Oriented	Trust Aware	Privacy Aware	Context Aware	Smart Space Needed
SSDS	No	Yes	No	No	N/A	N/A	N/A	No
Ninja	No	Yes	No	No	N/A	N/A	N/A	No
UPnP	No	N/A	No	No	No	Yes	No	Limited
SPDP	No	No	Yes	No	Yes	N/A	No	No
Progressive Exposure	No	Yes	No	No	No	Yes	Limited	No
Splendor	No	Yes	No	No	Yes	Yes	N/A	No
Jini	No	N/A	No	No	N/A	Yes	N/A	Limited
CSAS	No	No	Yes	No	N/A	N/A	Yes	No
CSM	Yes	No	Yes	No	N/A	N/A	Yes	No
AVCM	Limited	No	Yes	No	Yes	Yes	Yes	No
CSRA	No	Yes	No	No	N/A	N/A	Yes	Yes
TRAC	No	N/A	No	No	Yes	Yes	N/A	Yes
SME	Yes	N/A	N/A	Yes	N/A	Yes	No	N/A
HCA	No	N/A	Yes	No	No	Yes	No	N/A
SSRD	Yes	No	Yes	Yes	Yes	Yes	Limited	No
SSRD+	Yes	No	Yes	Yes	Yes	Limited	Yes	No
Centaurus2	Yes	Yes	No	No	No	N/A	Yes	No
SLP	No	Yes	Yes	Yes	No	No	No	No
Sleeper	Yes	No	Yes	Yes	Yes	Yes	No	No

the network. UPnP Device Security specification defines security mechanisms for simple object access protocol (SOAP)-based service invocation, but does not address simple service discovery protocol (SSDP) security.

As an extension of project Centaurus (Kagal, Korolev, Avancha, et al., 2001; Kagal, Korolev, Chen, et al., 2001), Centaurus2 (Undercoffer et al., 2003) provides a secure mechanism for service discovery and enables users to access services across heterogeneous network domains. The system uses a local certificate authority (CA) and each entity must be pre-registered in the system. The CA issues a certificate to each identified and verified entity. The design of Centaurus2 includes four components, and each component has a separate private key which is stored at the client using PKCS #11:

1. The local CA is responsible for issuing digital certificates and for validating these digital certificates.

2. The communication manager mediates communication between clients and networked services.
3. Group membership(s) is maintained and stored by the capability manager.
4. Each client is registered to a specific service manager that ensures security, access rights, and mediates between user client and service client. Service managers maintain a service registry.

Each domain has a root service manager. Static bridges are configured between service managers in different domains. Then clients in separate domains can access services across domains using the root service manager as the context.

In SSDS (Czerwinski et al., 1999), both service advertisement pull (query) and push (announcement) are supported. Service advertisements are stored in a hierarchy of servers. SSDS provides capability-based access control. All information passed between clients and servers is encrypted. A single copy of the resource information is stored and accessed, which makes the system vulnerable to single point failure. Subsequently, the Ninja project (Goldberg et al., 1999; Gribble et al., 2001) added the concept of secure identification of service through SSDS. In Ninja, the CA issues valid certificates and the capability manager authorizes user access to a particular resource. The service providers can also prescribe the conditions (capabilities) that are needed by a user in order to discover a particular service.

The context-sensitive authorization scheme (CSAS) (Minami & Kotz, 2005) provides authorization without a central server or CA. When a CSAS user wants to access a service from a resource, the associated server issues a logical authentication query and sends it to the host of the resource. Each host has a knowledge domain with which it attempts to prove the authorization query. If it fails, it distributes several portions of the proof to multiple hosts. Through this distribution CSAS reduces the computational overhead on any single node. After collecting the sub-proofs from the other hosts, the host of the resource can declare the result of the query to be true or false, thus indicating grant of access or denial respectively. This approach facilitates confidentiality, integrity, and scalability. To authorize access, CSAS uses previously stored information, which may be difficult to collect for users in an ad hoc network.

Splendor (Zhu et al., 2003) is a secure, private, and location-aware service discovery protocol. Splendor adapts depending on the network environment to use either a client-service model or client-service-directory model. Proxies are used to offload workload for mobile services. Mobile services authenticate with proxies and proxies handle registration. In these situations, proxies are considered to be trusted servers. However, if no trusted server is available in an environment, then there is no agent to handle the registration. Its security model is based on mutual authentication.

Progressive Exposure (Zhu et al., 2004, 2006) is a secure service discovery approach. It addresses privacy issues using a mutual matching technique. Progressive exposure addresses security and fairness by not exposing too much information. In each round of message exchange between communicating parties, it tries to find whether any mismatch occurs. In case of a mismatch, the communication stops. It uses one-time code words and a hash-based message authentication code. It considers the presence of one user and one service provider, but it does not address situations in which many users and many service providers are present. When a service provider leaves the network, the process of provider lookup and the authentication phase is restarted. It provides privacy for service information, requests, domain identity, and user credentials, and is based on the client-service-directory model.

Infrastructure-Less Security

SPDP (Almenarez & Campo, 2003) is a secure service discovery protocol based on the PTM (Almenarez, Marin, Campo, & Garcia, 2004; Almenarez, Marin, Dyaz, & Sanchez, 2006) model. The need for a centralized server is avoided by having each device act as its own CA. For a service request, this model uses broadcast messaging. The requesting device updates its cache after getting a reply from the devices (if any reply). It then stores the device identities that it believes trustworthy. The devices' user agents continually listen for messages, which in turn means continual energy consumption.

Narendar Sarkar et al. (Shankar & Arbaugh, 2002) propose an attribute vector calculus (AVCM) for modeling trust. Their model describes both identity-based trust and context-based trust and is one of the first models that discusses the importance of trust in a ubiquitous environment. Brezillion and Mostefaoui (2004) present a context-based security model (CSM) and they discuss the need for adaptive security based on the particular situation. Thomas and Sandhu (2004) present the challenges and research issues for secure pervasive computing. They express the need for a dynamic trust model as the pervasive computing environment poses new kinds of security challenges due to its diverse nature. They present a socio-technical view.

Smart Space Dependent Security

A smart space provides devices with complex computational support that supports context-awareness and collaboration. Components of the smart space can offload secure discovery tasks and relate them to other activities in the space. Examples include context-based secure resource access (CSRA) (Tripathi et al., 2004) and trust-based architecture (TRAC) (Basu & Callaghan, 2005).

CSRA (Tripathi et al., 2004) focuses on context-aware discovery of resources and how to access resources in a secure and unobtrusive manner. In a pervasive computing environment the rules and limitations imposed by the user, system, and the collaborative activity scenario have to be combined dynamically at runtime. CSRA uses a namespace related to each user and domain. These namespaces collect resources, services, and activities. The binding protocol defines the association of a user to a specific resource in the space. The binding changes based on the contextual information of the user including the location, activity, and role. A descriptor is associated with each namespace that combines functional attributes collected from resource descriptions in Web services description language (WSDL) and resource description framework (RDF) conditions for security, and policies for the binding protocol. The binding protocol specifies whether the binding of a resource is "shared" or "private," and whether the binding is "permanent" or "context-based."

Basu and Callaghan (2005) present a TRAC for increasing security and user confidence in pervasive computing systems. They use trust and role-based access control for ensuring security and privacy. However their model is aimed at an intelligent environment (IE) only. This policy-based model allows users to define policies for themselves and thus gives users control to define their own security level. This model works in an IE because every user is known beforehand. However, in a truly pervasive environment it is not possible to have prior information about every user and thus, this model is not applicable.

EXAMPLES USING TRUST MODELS

We next describe two service discovery protocols, Sleeper and SSRD, which incorporate trust models for infrastructure-less security.

Sleeper

Sleeper (Buford, Celebi, et al., 2006) is an energy-preserving service discovery protocol which features dynamic proxy selection for advertisement and discovery so that nodes can go to power standby while the proxy advertises on their behalf. The basic node states and transitions for Sleeper are shown in Figure 2. An off-line or disconnected node moves to an online state and broadcasts a *join* message that includes its advertisements and their popularity metrics. The current proxy caches these advertisements. Any proxy-candidate node may also cache these advertisements. An online node may broadcast a *leave* message prior to going off-line; if a leave message is not transmitted, advertisements may be purged from the proxy and other online nodes' cache by expiration. Transitions to/from standby state may also be indicated by broadcast messages.

An online node can be in one of four states (Figure 2). Every node initially goes online as a non-proxy node. A proxy-capable node becomes a proxy-candidate. There may be more than one proxy-candidate at any time. When no proxy is detected, for example by absence of a service advertisement broadcast or at the exit of a proxy, the first proxy-candidate to issue the proxy bootstrap becomes the proxy. A vacating proxy may transfer its cache to the new proxy, or the new proxy may collect advertisements from online nodes through the bootstrap. Nodes which are in standby state during the proxy change may be polled by the new proxy after the standby node transitions to online.

Sleeper uses property-based peer trust to secure service discovery operations. In property-based or credential-based trust (Hess et al., 2002; Seamons, Winslett, & Yu, 2001), each party has a set of certified attributes (e.g., credit card numbers, employee ID) that are exchanged to establish mutual trust. The typical components of a mechanism to provide property-based trust include:

- Trust negotiation protocol
- Trust negotiation policies
- Credentials

A method for trust negotiation has been defined for client-server context (Hess et al., 2002; Seamons et al., 2001). In this design, access control policies determine which credentials, services, and policies should be disclosed during a negotiation. Policies and credentials are secured locally at each node but are disclosed during negotiation to the remote party. Sleeper nodes establish mutual trust using the trust negotiation mechanism defined in Buford, Park, and Perkins (2006). Assuming that each peer caches public keys for certificate issuers that are relevant to its peer trust policies, then peer trust establishment can be performed without a centralized authority. A service discovery mechanism is *privacy preserving,* if a peer can discover the service description using the mechanism only if the peer satisfies the criteria C. Thus a mechanism that only distributes service descriptions to peers which are members of group G with criteria C is privacy preserving. Sleeper uses trust negotiation to create groups of peers that satisfy membership criteria C. Group management is provided by a group service (GS) that is available at every peer. The GS caches private service descriptions for each group and allows only group members to retrieve them. The GS publishes encrypted service descriptions that can only be decrypted by members of G. These encrypted service descriptions are broadcasted to all connected peers, but can only be decrypted by group members.

The secure agent technology (Buford, Park, et al., 2006) used in Sleeper for trust negotiation can also be used for enabling trust in service composition (Buford, Kumar, & Perkins, 2006).

Figure 2. Sleeper node states and state transitions; online nodes can be in one of four states (Buford et al., 2006)

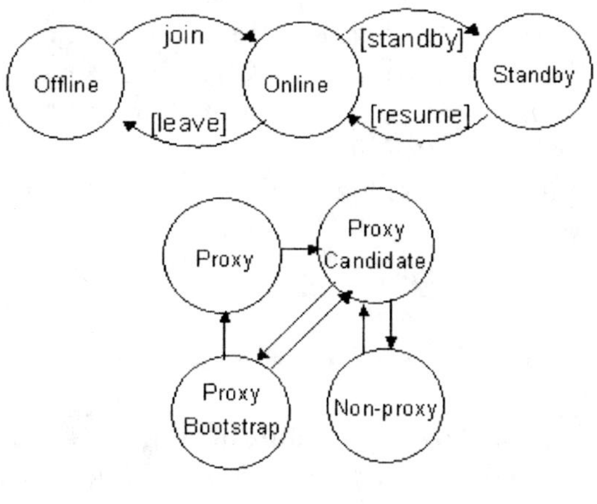

SSRD

With a view to ensure enhanced security through a lightweight solution for resource discovery in pervasive environment, simple and secure resource discovery (SSRD) has been proposed by the researchers in Sharmin et al. (2006a). The fundamental part of the solution is a trust–based, service-oriented adaptive security mechanism built on middleware adaptability for resource discovery, knowledge usability, and self-healing (MARKS), a middleware and framework developed for resource constrained pervasive devices for pervasive applications (Sharmin et al., 2006b). The SSRD unit of

Figure 3. Sleeper groups in broadcast of advertisements; symmetric keys are broadcast with public key encryption (Buford, Celebi, et al., 2006)

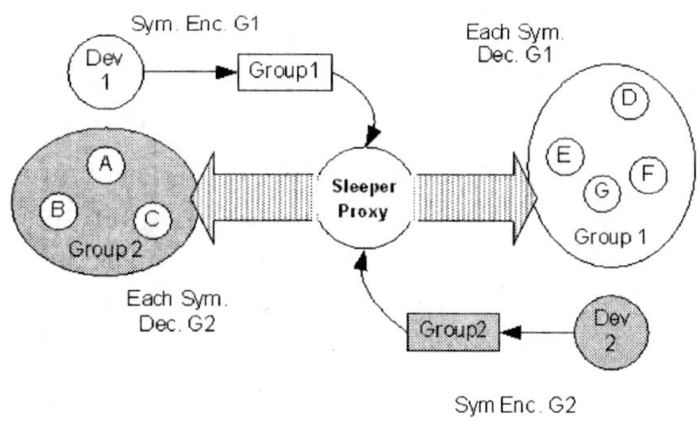

MARKS consists of *trust management* and *security management* sub units and it provides a resource discovery agent (Figure 4).

The *trust management unit* is responsible for maintaining trust relationships with other devices. It calculates trust values for the relationships between devices and also updates the trust values depending on the behavior of the service provider or requester. It maintains a list of service-specific average trust values and communicates to the security management unit whenever necessary. Trust values are quantified in the range of 0.0 to 1.0 to represent the degree of trustworthiness of a node. Complete trust and complete distrust are represented by 1.0 and 0.0 respectively. A new device with no prior interaction record is assigned a value of 0.5, which indicates a neutral condition. The dynamic property trust evolves over time and may possess the asymmetric transitive property depending on services. Each owner or manager of a device retains a table that indicates the security level (ranging from 1 to 10) required by each of the available services or applications. The resource manager consults "Service-trust" for all the neighboring nodes to decide whether the service could be provided. For example, for services with security level < 5, no trust calculation or secure communication is needed. For services with higher security levels, initially trust is calculated and then secure communication is established between the provider and the requester. This lessens both the computation cost and the communication overhead.

Figure 4. Resource discovery model (Sharmin et al., 2006a)

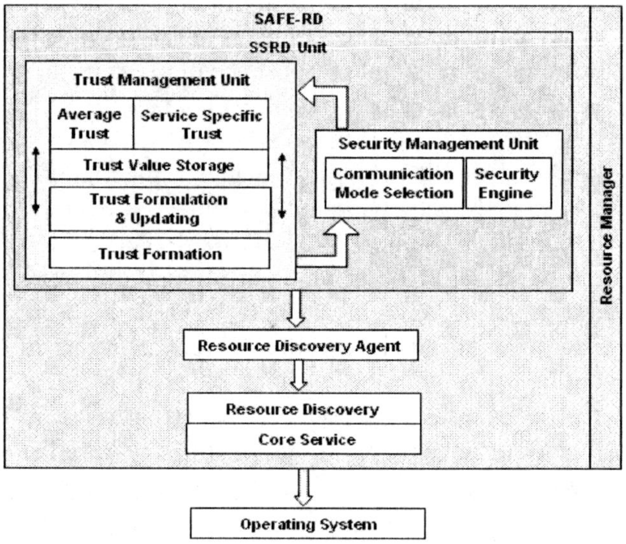

Trust models are designed to associate each device with a trust value based on past behavior with the requesting device. Also when we calculate a trust value for an unknown device, we consider the PGP (Zimmermann, 1995) based trust model. PGP is based on mutual certification of the validity of the keys. In case a new device joins the network or a device that never communicated with a service-providing device, the service providing device generates a multicast message to all devices that it has interacted with and asks for their recommendation about this device. From the recommendations the trust value is calculated for that service. The issue with dynamic update of trust values has been addressed more clearly with specific situations in the researchers' enhanced adaptation of this model named SSRD+ (Sharmin, Ahmed, & Ahamed, 2006c).

FUTURE RESEARCH

The open and dynamic nature of the pervasive computing environment requires a security mechanism that is unobtrusive to the user and makes it possible to securely provide and discover the services available for the user in a transparent manner. Some of the open issues regarding challenges in secure and private service discovery are highlighted in this section.

Privacy

Although contextual information plays a pivotal role in dynamic pervasive environments, it may also expose private information. When granting access to a service, a person's context information like location, time, and activity can be exposed. Further, policies and constraints are themselves subject to privacy protection. Private information management, such as the recursive constraint based security model in Hengartner and Steenkiste (2006), is one approach to prevent direct information leakage. However, such mechanisms are generally susceptible to attacks involving collusion and inference.

In a context- and location-sensitive medical application, researchers developed a system for practitioners to easily share context in their work tasks. Subsequently, questions of privacy led the designers to limit access to this information. As another example, the Gaia project has shown a privacy preserving hop by hop routing algorithm that carries information about the location of the user but does not reveal the exact location or identity of the user. Thus the privacy level and willingness of disclosure of personal information varies depending on information type, collection method, time, and other factors. In some scenarios users are reluctant to disclose identity information but do not care about location information. The situation might be reversed in other cases. Formulation of policies that are understood and can be managed by users is an important goal.

Trust

As discussed earlier, a key element for secure service discovery in ad hoc environments is the ability to establish a level of trust betweens peers. The trust life cycle can be narrated in short as trust formation, evolution, and exploitation. In general, trust is formed by experience through earlier interactions, verifiable properties of each party, recommendations from trusted entities, and reputation in a community. The challenges faced during trust establishment are due to the absence of a global trust framework, the large number of autonomous and anonymous entities, the large number of domains, and different trust requirements for large number of application contexts.

Recent context-aware trust models focus on dynamic trust values, which are updated over time and distance and incorporate behavioral models for evolution of trust. Risk analysis maps each action to possible outcomes associated with a cost/benefit. Decisions consider the likelihood of the risk and cost. Unresolved issues in trust establishment include detecting and prevent collusion, managing the trade-off between privacy and property disclosure, and efficient trust mechanisms in large communities.

Multi-Protocol Environments

The combination of multi-homed mobile devices and multiple service discovery protocols means that service access may cross not only administrative boundaries but also different service discovery domains with varying security properties. As an example, a mobile device may include protocol support for Bluetooth, SLP, and UPnP. Then the device can easily discover services in different domains that it roams to, if these domains use different service discovery protocols. As a multi-home device, it may simultaneously connect to domains with different service discovery protocols.

As a second example, a single user may have a set of personal mobile devices configured in a PAN. These devices can use the PAN security mechanism for security and privacy control, and identity-based authentication for mutual trust. The PAN may support a specific service discovery protocol. One or more of the devices in the PAN may also connect to outside networks with different service discovery protocols and security mechanisms.

These types of scenarios indicate that future mobile devices may need to operate in multiple

Figure5. Conceptual diagram of SSRD model (Sharmin et al., 2006b)

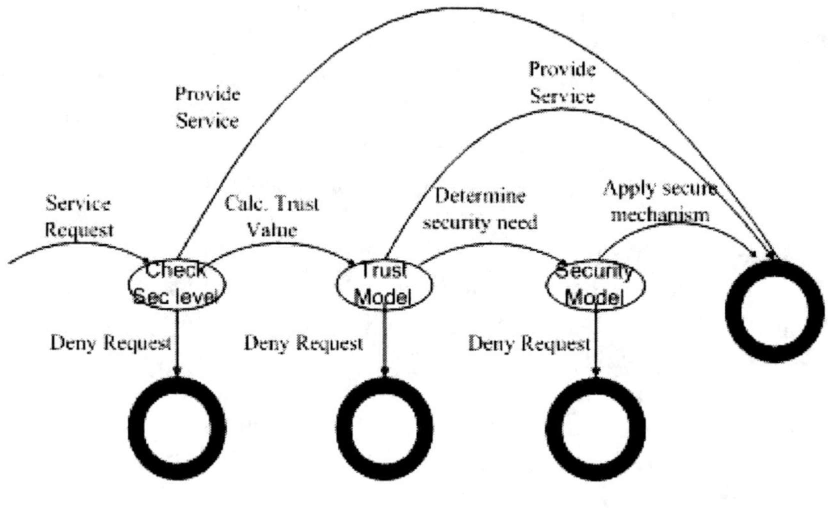

security contexts. In these cases there is the potential for conflicting access policies and unanticipated information flows between different regions. Further, there are challenges in managing groups across domains and mapping service semantics and identities between different domains.

Trust in Service Composition

A device in a pervasive computing environment may offer a service to other devices. The service may be aggregated from services offered by other devices. By aggregating service facilities across devices, a collection of limited-resource devices may be able to offer services that would otherwise not be available. However, devices which invoke or participate in these services may be concerned about the integrity and trustworthiness of the various components that are combined to provide these services. Existing service discovery mechanisms do not expose such nested or recursive relationships when a service is offered or invoked.

Conventional methods for assuring trustworthiness of software components are typically used to convey trustworthiness to the end user or developer. They provide no explicit representation of trust between distributed components. Further, these methods do not explicitly validate composite services that may be created from different service sources. Composition trust bindings (Buford, Kumar, et al., 2006) are one approach for providing trust in both control and data paths in peer-to-peer service composition.

CONCLUSION

The general availability of broadband-wireless-enabled devices is a key catalyst in enabling many powerful peer-to-peer usage patterns, which have been described as pervasive computing. However, these usage scenarios will frequently involve devices which are outside a single secure administrative boundary and may include ad hoc interactions where no prior trust relationship exists. Further, there is significant variation in basic authentication, authorization, and privacy mechanisms offered in wireless networks. Consequently many existing designs for service discovery have insufficient security, privacy, and trust support.

Assuming that most wireless networks will in the future provide encrypted transmission, user/device authentication, and authorization control in a given administrative domain, there remain important security related questions for service discovery in cross-domain cases, in ad hoc cases, and when the devices/users are not a priori mutually authenticated. Consequently, we do not expect that improvements in the security of wireless networks, while important, will be sufficient to address all the requirements identified here for secure service discovery.

Toward this end, after surveying a variety of approaches to secure service discovery today, we presented case studies of two recent service discovery protocols, which include trust establishment mechanisms to enable trust between a priori untrusted devices and peers. We also provided a summary of future research directions.

REFERENCES

Almenarez, F., & Campo, C. (2003). SPDP: A secure service discovery protocol for ad-hoc networks. In *Ninth Open European Summer School and IFIP Workshop on Next Generation Networks (EUNICE 2003)* (pp. 213-218).

Almenarez, F., Marin, A., Campo, C., & Garcia, C. (2004). PTM: A pervasive trust management model for dynamic open environments. In *Pervasive Security, Privacy, and Trust (PSPT 2004).*

Almenarez, F., Marin, A., Dyaz, D., & Sanchez, J. (2006). Developing a model for trust management in pervasive devices. In *Fourth Annual IEEE International Conference on Pervasive Computing and Communications Workshops (PERCOMW'06)* (pp. 267-271).

Balazinska, M., Balakrishnan, H., & Karger, D. (2002). INS/Twine: A scalable peer-to-peer architecture for intentional resource discovery. In *International Conference on Pervasive Computing* (pp. 195-210).

Barbeau, M. (1999). Service discovery in a mobile agent API using SLP. In *Global Telecommunications Conference (GLOBECOM '99)* (Vol. 1a, pp. 391-395).

Basu, J., & Callaghan, V. (2005). Towards a trust based approach to security and user confidence in pervasive computing systems. In *IEE International Workshop on Intelligent Environments 2005 (IE05)* (pp. 223-229).

Bluetooth Special Interest Group. (2001a). Specification of the Bluetooth system—Core [Version 1.1].

Bluetooth Special Interest Group. (2001b). Specification of the Bluetooth system—Core [Version 1.1.]. SDP specification (Vol. 1 part E).

Brezillon, P., & Mostefaoui, G. (2004). Context-based security policies: A new modeling approach. In *Second IEEE International Conference on Pervasive Computing and Communications-workshops* (pp. 154-158).

Buford, J., Brown, A., & Kolberg, M. (2006). Meta service discovery. In *Proceedings of the Fourth IEEE Conference on Pervasive Computing and Communications Workshops, Workshop on Mobile Peer-to-peer* (pp. 124-129).

Buford, J., Burg, B., Celebi, E., & Frankl, P. (2006). Sleeper: A power-conserving service discovery protocol. In *Third Annual International Conference on Mobile and Ubiquitous Systems, Networking, and Services* (*Mobiquitous 2006*) (pp. 1-10).

Buford, J., Celebi, E., & Frankl, P. (2006). Property-based peer trust in the sleeper service discovery protocol. In *30th Annual International Computer Software and Applications Conference (COMPSAC '06), Workshop on Security, Privacy, and Trust for Pervasive Applications (SPTPA 2006)* (Vol. 2, pp. 209-214).

Buford, J., Kumar, R., & Perkins, G. (2006). Composition trust bindings in pervasive computing service composition. In *Proceedings of the Fourth IEEE Conference on Pervasive Computing and Communications Workshops, Workshop on Pervasive Computing and Communication Security (PerSec)* (pp. 261-266).

Buford, J., Park, I., & Perkins, G. (2006). Social certificates and trust negotiation. In *Third IEEE Consumer Communications and Networking Conference (CCNC 2006)* (pp. 615-619).

Chen, Y., Jensen, C., Gray, E., Cahill, V., & Seigneur, J. (2003). *A general risk assessment of security in pervasive computing* (Tech. Rep. No. TCD-CS-2003-45). The University of Dublin, Trinity College, Department of Computer Science.

Cotroneo, D., Graziano, A., & Russo, S. (2004). Security requirements in service oriented architectures for ubiquitous computing. In *Proceedings of the Second Workshop on Middleware for Pervasive and Ad-hoc Computing* (pp. 172-177).

Czerwinski, S., Zhao, B., Hodes, T., Joseph, A., & Katz, R. (1999). An architecture for a secure service discovery service. In *Fifth Annual International Conference on Mobile Computing and Networks (MobiCom '99)* (pp. 24-35).

Ganu, S., Krishnakumar, A., & Krishnan, P. (2004). Infrastructure-based location estimation in WLAN networks. In *IEEE Wireless Communications and Networking Conference (WCNC)* (pp. 465-470).

Garlan, D., Siewiorek, D., Smailagic, A., & Steenkiste, P. (2002). Project Aura: Towards distraction-free pervasive computing. *IEEE Pervasive Computing, 1*(2), 22-31.

Goldberg, I., Gribble, S., Wagner, D., & Brewer, E. (1999). The Ninja jukebox. In *Proceedings of the Second USENIX Symposium on Internet Technologies and Systems (USITS-99)* (pp. 37-46).

Gribble, S., Welsh, M., Von Behren, R., Brewer, E., Culler, D., Borisov, N., et al. (1999). *Service location protocol version 2* (RFC 2608). Retrieved from http://www.faqs.org/rfcs/rfc2608.html

He, R., Niu, J., Yuan, M., & Hu, J. (2004). A novel cloud-based trust model for pervasive computing. In *The Fourth International Conference on Computer and Information Technology (CIT '04)* (pp. 693-700).

Hengartner, U., & Steenkiste, P. (2006). Avoiding privacy violations caused by context-sensitive services. In *Proceedings of the Fourth Annual IEEE International Conference on Pervasive Computer and Communications (PerCom 2006)* (pp. 222-233).

Hess, A., Jacobson, J., Mills, H., Wamsley, R., Seamons, K., & Smith, B. (2002). Advanced client/server authentication in TLS. In *Network and Distributed System Security Symposium.*

Joseph, A., Katz, R., Mao, Z., Ross, S., & Zhao, B. (2001). The Ninja architecture for robust Internet-scale systems and services. *Computer Networks, 35*(4), 473-497.

Kagal, L., Finin, T., & Joshi, A. (2001). Trust-based security in pervasive computing environments. *IEEE Computer, 34*(12), 154-157.

Kagal, L., Finin, T., Joshi, A., & Greenspan, S. (2006). Security and privacy challenges in open and dynamic environments. *IEEE Computer, 39*(6), 89-91.

Kagal, L., Korolev, V., Avancha, S., Joshi, A., Finin, T., & Yesha, Y. (2001). *Highly adaptable infrastructure for service discovery and management in ubiquitous computing* (Tech. Rep. No. TR CS-01-06). Baltimore: University of Maryland, Department of Computer Science and Electrical Engineering.

Kagal, L., Korolev, V., Chen, H., Joshi, A., & Finin, T. (2001). Project Centaurus: A framework for intelligent services in a mobile environment. In *International Workshop of Smart Appliances and Wearable Computing, International Conference of Distributed Computing Systems* (pp. 195-201).

Kindberg, T., & Fox, A. (2002). System software for ubiquitous computing. *IEEE Pervasive Computing, 1*(1), 70-81.

Kopp, H., Lucke, U., & Tavangarian, D. (2005). Security architecture for service-based mobile environment. In *Proceedings of the Third IEEE Conference on Pervasive Computing and Communications Workshops* (pp. 199-203).

Lee, C., & Helal, S. (2002). Protocols for service discovery in dynamic and mobile networks. *International Journal of Computer Research, 11*(1), 1-12.

Matsumiya, K., Tamaru, S., Suzuki, G., Nakazawa, J., Takashio, K., & Tokuda, H. (2004). Improving security for ubiquitous campus applications. In *Symposium on Applications and the Internet-Workshops (SAINT 2004)* (pp. 417-422).

Microsoft Corporation. (2000). Universal plug and play device architecture, Version 1.0.

Miller, B., Nixon, T., Tai, C., & Wood, M. (2001). Home networking with universal plug and play. *IEEE Communications Magazine, 39*(12), 104-109.

Minami, K., & Kotz, D. (2005). Secure context-sensitive authorization. In *Proceedings of the Third International Conference on Pervasive Computing and Communications Workshops (PerCom 2005)* (pp. 257-268).

Nidd, M. (2001). Service discovery in DEAPspace. *IEEE Personal Communications, 8*(4), 39-45.

Pearson, S. (2005). How trusted computers can enhance privacy preserving mobile applications. In *Proceedings of the Sixth International IEEE Symposium on a World of Wireless Mobile and Multimedia Networks (WoWMoM'05)* (pp. 609-613).

Robinson, P., Vogt, H., & Wagealla, W. (Eds.). (2005). *Privacy, security and trust within the context of pervasive computing.* Heidelberg, Germany: Springer-Verlag.

Saha, S., Chaudhuri, K., Sanghi, D., & Bhagwat, P. (2003). Location determination of a mobile device using IEEE 802.11b access point signals. In *IEEE Wireless Communications and Networking Conference (WCNC)* (pp. 1987-1992).

Satyanarayanan, M. (1996). Fundamental challenges in mobile computing. In *Fifteenth ACM Symposium on Principles of Distributed Computing* (pp. 1-7).

Seamons, K., Winslett, M., & Yu, T. (2001). Limiting the disclosure of access control policies during automated trust negotiation. In *Network and Distributed System Security Symposium.*

Shankar, N., & Arbaugh, W. (2002). On trust for ubiquitous computing. *Workshop on security in ubiquitous computing* (UBICOMP 2002).

Sharmin, M., Ahmed, S., & Ahamed, S. (2006a). MARKS (middleware adaptability for resource discovery, knowledge usability, and self healing) in pervasive computing environments. In *Third International Conference on Information Technology: New Generations* (pp. 306-313).

Sharmin, M., Ahmed, S., & Ahamed, S. (2006b). An adaptive lightweight trust reliant secure resource discovery for pervasive computing environments. In *Proceedings of the fourth annual IEEE international conference on pervasive computer and communications (PerCom 2006)* (pp. 258-263).

Sharmin, M., Ahmed, S., & Ahamed, S. (2006c). SSRD+: A privacy-aware trust and security model for resource discovery in pervasive computing environment. In *30th Annual International Computer Software and Applications Conference (COMPSAC 2006)* (pp. 67-70).

Smith, B., Seamons, K., & Jones, M. (2004). Responding to policies at runtime in TrustBuilder. In *Fifth International Workshop on Policies for Distributed Systems and Networks (POLICY 2004).*

Stajano, F. (2002). *Security for ubiquitous computing.* West Sussex, England: John Wiley and Sons.

Stajano, F., & Anderson, R. (2002). The resurrecting duckling: Security issues for ubiquitous computing. *IEEE Computer, 35*(4), 22-26.

Sun Microsystems. (2001). Jini™ technology core platform specification, version 1.2.

Thomas, R., & Sandhu, R. (2004). Models, protocols, and architectures for secure pervasive computing: challenges and research directions. In *Second IEEE International Conference on Pervasive Computing and Communications—Workshops (PerCom 2004)* (pp. 164-168).

Tripathi, A., Ahmed, T., Kulkarni, D., Kumar, R., & Kashiramka, K. (2004). Context-based secure resource access in pervasive computing environments. In *Second IEEE Annual Conference on Pervasive Computing and Communications—Workshops.* (p. 159).

Undercoffer, J., Perich, F., Cedilnik, A., Kagal, L., & Joshi, A. (2003). A secure infrastructure for service discovery and access in pervasive computing. *Mobile Networks and Applications, 8*(2), 113-125.

Want, R., & Pering, T. (2005). System challenges for ubiquitous and pervasive computing. In *Twenty-seventh International Conference on Software Engineering (ICSE 2005)* (pp. 9-14).

Weiser, M. (1991). The computer for the twenty-first century. *Scientific American, 265*(3), 94-104.

Weiser, M. (1993). Some computer science problems in ubiquitous computing. *Communications of the ACM, 36*(7), 75-84.

Winoto, W., Schwartz, E., Balakrishnan, H., & Lilley, J. (1999). The design and implementation of an intentional naming system. In *17th ACM Symposium on Operating Systems Principles (SOSP '99)* (pp. 186-201).

Winslett, M. (2003). An introduction to automated trust establishment. *First international conference on trust management.*

Wu, C., Fu, L., & Lian, F. (2004). WLAN location determination in e-home via support vector classification. In *IEEE Conference on Networking, Sensing & Control* (pp. 1026-1031).

Youssef, M., Agrawala, A., & Udaya, A. (2003). WLAN location determination via clustering and probability distributions. In *Proceedings of the First Annual IEEE International Conference on Pervasive Computer and Communications (PerCom 2003)* (pp. 143-150).

Yu, T., & Winslett, M. (2003). A unified scheme for resource protection in automated trust negotiation. In *IEEE Symposium on Security and Privacy* (pp. 110-122).

Zhu, F., Mutka, M., & Ni, L., (2002). *Classification of service discovery in pervasive computing environments* (Tech. Rep. No. MSU-CSE-02-24). East Lansing: Michigan State University.

Zhu, F., Mutka, M., & Ni, L. (2003). Splendor: A secure, private, and location-aware service discovery protocol supporting mobile services. In *Proceedings of the First IEEE Conference on Pervasive Computing and Communications (PerCom 2003)* (pp. 235-242).

Zhu, F., Mutka, M., & Ni, L. (2004). PrudentExposure: A private and user-centric service discovery protocol. In *Proceedings of the Second IEEE Conference on Pervasive Computing and Communications (PerCom 2004)* (pp. 329-340).

Zhu, F., Mutka, M., & Ni, L. (2005). Expose or not? A progressive exposure approach for service discovery in pervasive computing environments. In *Proceedings of the Third IEEE Conference on Pervasive Computing and Communications (PerCom 2005)* (pp. 225-234).

Zhu, F., Mutka, M., & Ni, L. (2006). A private, secure, and user-centric information exposure model for service discovery protocols. *IEEE Transactions on Mobile Computing, 5*(4), 418-429.

Zimmermann, P. (1995). *PGP source code and internals.* Cambridge, MA: MIT Press.

KEY TERMS

Context: Context is the location, time, and activity state of the user when performing a service-related operation such as discovery, advertisement, or invocation.

Federated Discovery: Federated discovery is a service discovery mechanism that incorporates two or more different service advertisement mechanisms.

Meta Discovery: Meta discovery is the discovery of a service discovery mechanism by using meta information about that mechanism (Buford, Brown et al., 2006).

Peer Trust: Peer trust is the degree to which a peer device is willing to disclose information or provide access to resources to another peer, and which may be determined by experience through earlier interactions, verifiable properties of each party, recommendations from trusted entities, and reputation in a community.

Pervasive Computing: Pervasive computing is the evolution of distributed computing in which networked computing devices are integrated throughout the personal and work environments in a connected way, also referred to as ubiquitous computing.

Secure Service Discovery: Secure service discovery is service discovery that enforces privacy and security policies of the devices participating in the service location process.

Service Composition: Service composition is the ability to dynamically discover and combine component services to form new services.

Service Discovery: Service discovery occurs when device resources and functions are packaged as services, in a networked environment, and a device finds another device capable of offering a specific service or resource.

Chapter III
Security of Mobile Code

Zbigniew Kotulski
Polish Academy of Sciences, Warsaw, Poland
Warsaw University of Technology, Poland

Aneta Zwierko
Warsaw University of Technology, Poland

ABSTRACT

The recent development in the mobile technology (mobile phones, middleware, wireless networks, etc.) created a need for new methods of protecting the code transmitted through the network. The oldest and the simplest mechanisms concentrate more on integrity of the code itself and on the detection of unauthorized manipulation. The newer solutions not only secure the compiled program, but also the data, that can be gathered during its "journey," and even the execution state. Some other approaches are based on prevention rather than detection. In this chapter we present a new idea of securing mobile agents. The proposed method protects all components of an agent: the code, the data, and the execution state. The proposal is based on a zero-knowledge proof system and a secure secret sharing scheme, two powerful cryptographic primitives. Next, the chapter includes security analysis of the new method and its comparison to other currently more widespread solutions. Finally, we propose a new direction of securing mobile agents by straightening the methods of protecting integrity of the mobile code with risk analysis and a reputation system that helps avoiding a high-risk behavior.

INTRODUCTION

A software agent is a program that can exercise an individual's or organization's authority, work autonomously toward a goal, and meet and interact with other agents (Jansen & Karygiannis, 1999). Agents can interact with each other to negotiate contracts and services, participate in auctions, or barter. Multi-agent systems have sophisticated applications, for example, as management systems for telecommunication networks or as artificial intelligence (AI)-based intrusion detection systems. Agents are commonly divided into two types:

- Stationary agents
- Mobile agents

The stationary agent resides at a single platform (host), the mobile one can move among different platforms (hosts) at different times.

The mobile agent systems offer new possibilities for the e-commerce applications: creating new types of electronic ventures from e-shops and e-auctions to virtual enterprises and e-marketplaces. Utilizing the agent system helps to automate many e-commerce tasks. Beyond simple information gathering tasks, mobile agents can take over all tasks of commercial transactions, namely, price negotiation, contract signing, and delivery of (electronic) goods and services. Such systems are developed for diverse business areas, for example, contract negotiations, service brokering, stock trading, and many others (Corradi, Cremonini, Montanari, & Stefanelli, 1999; Jansen & Karygiannis, 1999; Kulesza & Kotulski, 2003). Mobile agents can also be utilized in code-on-demand applications (Wang, Guan, & Chan, 2002). Mobile agent systems have advantages even over grid computing environments:

- Require less network bandwidth
- Increase asynchrony among clients and servers
- Dynamically update server interfaces
- Introduce concurrency

The benefits from utilizing the mobile agents in various business areas are great. However, this technology brings some serious security risks; one of the most important is the possibility of tampering with an agent. In mobile agent systems the agent's code and internal data autonomously migrate between hosts and can be easily changed during the transmission or at a malicious host site. The agent cannot itself prevent this, but different countermeasures can be utilized in order to detect any manipulation made by an unauthorized party. They can be integrated directly into the agent system, or only into the design of an agent to extend the capabilities of the underlying agent system.

Several degrees of agent's mobility exist, corresponding to possibilities of relocating code and state information, including the values of instance variables, the program counter, execution stack, and so forth. The mobile agent technologies can be divided in to two groups:

- **Weakly mobile:** Only the code is migrating; no execution state is sent along with an agent program
- **Strong mobile:** A running program is moving to another execution location (along with its particular state)

The protection of the integrity of the mobile agent is the most crucial requirement for the agent system. The agent's code and internal data autonomously migrate between hosts and can be easily changed during the transmission or at a malicious host site. A malicious platform may make subtle changes in the execution flow of the agent's code; thus, the changes in the computed results are difficult to detect. The agent cannot itself prevent this, but different countermeasures can be utilized in order to detect any manipulation made by an unauthorized party. They can be integrated directly into the agent system, or only into the design of an agent to extend the capabilities of the underlying agent system. However, the balance between the security level and solution implementation's cost, as well as performance impact, has to be preserved. Sometimes, some restrictions of agent's mobility may be necessary.

Accountability is also essential for the proper functioning of the agent system and establishing trust between the parties. Even an authenticated agent is still able to exhibit malicious behavior to the platform if such a behavior cannot later be detected and proved. Accountability is usually realized by maintaining an audit log of security-relevant events. Those logs must be protected from unauthorized access and modification. Also the non-repudiability of logs is a huge concern. An important factor of accountability is authentication. Agents must be able to authenticate to platforms and other agents and vice versa. An agent may require different degrees of authentication depending on the level of sensitivity of the data.

The accountability requirement needs also to be balanced with an agent's need for privacy. The platform may be able to keep the agent's identity secret from other agents and still maintain a form of revocable anonymity where it can determine the agent's identity if necessary and legal. The security policies of agent platforms and their auditing requirements must be carefully balanced with agent's privacy requirements.

Threats to security generally fall into three main classes: (1) disclosure of information, (2) denial of service, and (3) corruption of information (Jansen, 1999). Threats in agent system can be categorized with regard to agents and platform relations (e.g., agent attacking an agent, etc.). Another taxonomy of attacks in agent system was proposed in Man and Wei (2001). The article describes two main categories of attacks: purposeful and frivolous. The first kind is carefully planned and designed and can be further classified by the nature of attack (read or non-read) and number of attackers (solo or collaborative). During the second kind of attacks, the attacker may not know the effect of his/her actions or gain an advantage. These attacks can be random or total. Another category of attacks is connected with traffic analysis (Kulesza, Kotulski, & Kulesza, 2006) or called *blocking attacks* (when a malicious platform refuses to migrate the agent), as described by Shao and Zhou (2006). In this chapter we will focus on the threats from an agent's perspective.

Among the mentioned threats, the most important are connected with the agent platform since the most difficult to ensure is the agent's code/state integrity. There are two main concepts for protecting mobile agent's integrity:

- Providing trusted environment for agent's execution
- Detection or prevention of tampering

The first group of methods is more concentrated on the whole agent system than on an agent in particular. These seem to be easier to design and implement but, as presented in Oppliger (2000), mostly lead to some problems. The assumption that an agent works only with a group of trusted hosts makes the agent less mobile than it was previously assumed. Also an agent may need different levels of trust (some information should be revealed to host while in another situation it should be kept secret). Sometimes, it is not clear in advance that the current host can be considered as trusted. A method to provide such an environment is special tamper-resistant hardware, but the cost of such a solution is usually very high.

The second group of methods provides the agents' manager with tools to detect that the agent's data or code has been modified, or an agent with a mechanism that prevents a successful, unauthorized manipulation. In this chapter we concentrate on the "built-in" solutions because they enable an agent to stay mobile in the strong sense and, moreover, provide the agent with mechanisms to detect or prevent tampering. Detection means that the technique is aimed at discovering unauthorized modification of the code or the state information. Prevention means that the technique is aimed at preventing changes of the code and the state information in any way. To be effective, detection techniques are more likely than prevention techniques to depend on legal or other social framework. The distinction between detection and prevention can be sometimes arbitrary, since prevention often involves detection (Jansen, 2000).

BACKGROUND

Many authors proposed methods for protecting integrity of the mobile code. The most interesting of them are presented in this section.

Time Limited Black-Box Security and Obfuscated Code

These methods are based on a *black-box* approach. The main idea of the black-box is to generate executable code from a given agent's specification that cannot be attacked by read (disclosure) or modification attacks. An agent is considered to be black-box if at any time the agent code cannot be attacked in the previous sense, and if only its input and output can be observed by the attacker. Since it is not possible to implement it today, the relaxation of this notion was introduced Hohl (1998): it is not assumed that *the black-box protection* holds forever, but only for a certain known time. According to this definition, an agent has the time-limited black-box property if for a certain known time it cannot be attacked in the aforementioned sense. The *time limited black-box* fulfills two black-box properties for this limited time:

- Code and data of the agent specification cannot be read
- Code and data of the agent specification cannot be modified

This scheme will not protect any data that is added later, although the currently existing variables will be changeable. Thus, it cannot protect the state of an agent, which can change between different hosts or any data, which the agent gathered.

In order to achieve the black-box property, several conversion algorithms were proposed. They are also called obfuscating or mess-up algorithms. These algorithms generate a new agent out of an original agent, which differs in code but produces the same results.

The *code obfuscation* methods make it more complicated to obtain the meaning from the code. To change a program code into a less easy "readable" form, they have to work in an automatic and parametric manner. The additional parameters should make possible that the same original program is transformed into different obfuscated programs. The difficulty is to transform the program in a way that the original (or a similar, easily understandable) program cannot be re-engineered automatically. Another problem is that it is quite difficult to measure the quality of obfuscation, as this not only depends on the used algorithm, but on the ability of the re-engineering as well. Some practical methods of code obfuscation are described by Low (1998) and general taxonomy proposed by Coilberg, Thomborson, and Low (1997).

Since an agent can become invalid before completing its computation, the obfuscated code is suitable for applications that do not convey information intended for long-lived concealment. Also, it is still possible for an attacker to read and manipulate data and code but, as a role of these elements cannot be determined, the results of this attack are random and have no meaning for the attacker.

Encrypted Functions

The encrypted functions (EF) method is one step forward in implementing the perfect black-box security. It has been proposed initially by Sander and Tschudin (1998). Since then other similar solutions were introduced (Alves-Foss, Harrison, & Lee, 2004; Burmester, Chrissikopoulos, & Kotzanikolaou, 2000) and the method is believed to be one of the canonical solutions for preserving agent's integrity (Jansen, 2000; Oppliger, 2000).

The goal of the EF, according to Jansen (2000), is to determine a method, which will enable the mobile code to safely compute cryptographic primitives, such as digital signature, even though the code is executed in non-trusted computing environments and operates autonomously without interactions with the home platform. The approach is to enable the agent platform to execute a program assimilating an encrypted function without being able to extract the original form. This approach requires differentiation between a function and a program that implements the function.

The EF system is described as follows by Oppliger (2000):

A has an algorithm to compute function f. B has an input x and is willing to compute $f(x)$ for A, but A wants B to learn nothing substantial about f. Moreover, B should not need interacting with A during the computation of $f(x)$.

The function f can be, for example, a signature algorithm with an embedded key or an encryption algorithm containing the one. This would enable the agent to sign or encrypt data at the host without revealing its secret key.

Although the idea is straightforward, it is hard to find the appropriate encryption schemes that can transform arbitrary functions as shown. So far, the techniques to encrypt rationale functions and polynomials have been proposed. Also a solution based on the RSA cryptosystem was described (Burmester et al, 2000).

Cryptographic Traces

The articles by Vigna (1997, 1998) introduced cryptographic traces (also called execution traces) to provide a way to verify the correctness of the execution of an agent. The method is based on traces of the execution of an agent, which can be requested by the originator after the agent's termination and used as a basis for the execution verification. The technique requires each platform involved to create and retain a non-repudiation log or trace of the operations performed by the agent while resident there and to submit a cryptographic hash of the trace upon conclusion as a trace summary or fingerprint. The trace is composed of a sequence of statement identifiers and the platform signature information. The signature of the platform is needed only for those instructions that depend on interactions with the computational environment maintained by the platform. For instructions that rely only on the values of internal variables, the signature is not required and therefore is omitted.

This mechanism allows detecting attacks against code; state and control flow of mobile agents. This way, in the case of tampering, the agent's owner can prove that the claimed operations could never been performed by the agent. The technique also defines a secure protocol to convey agents and associated security-related information among the various parties involved, which may include a trusted third party to retain the sequence of trace summaries for the agent's entire itinerary. The approach has a number of drawbacks, the most obvious being the size and number of logs to be retained, and the fact that the detection process is triggered sporadically, based on suspicious results' observations or other factors.

Chained MAC Protocol

Different versions of chained message authentication code (MAC) protocol were described by Karjoth, Asokan, and Gulcu (1999) and Yee (1999). Some of them require existence of public key infrastructure, others are based on a single key. This protocol allows an agent to achieve strong forward integrity. To utilize this protocol, only the public key of the originator has to be known by all agent places. This can occur when the originator is a rather big company that is known by its smaller suppliers.

Assume that r_n is a random number that is generated by n^{th} host. This value will be used as a secret key in a MAC. The partial result o_n (single piece of data, generated on n host), r_n and the identity of the next host are encrypted with the public key of the originator K_{i_0}, forming the encapsulated message O_n:

$$O_n = \{r_n, o_n, id(i_{n+1})\}K_{i_0}$$

A *chaining relation* is defined as follows (here H denotes a hash-function and h denotes the digest):

$$h_0 = \{r_0, o_0, id(i_1)\}K_{i_0}$$

and

$$h_{n+1} = H\{h_n, r_n, o_n, id(i_{n+1}))$$

When an agent is migrating from host i_n to i_{n+1}:

$$i_n \rightarrow i_n+1 : \{O_0, ..., O_n, h_{n+1}\}$$

Similar schemes are also called *partial results encapsulation* methods (Jansen, 2000).

Watermarking

Watermarking is mainly used to protect the copyrights for digital contents. A distributor or an owner of the content embeds a mark into a digital object, so its ownership can be proven. This mark is usually secret. Most methods exploit information redundancy and some of them can also be used to protect the mobile agent's data and code.

A method of watermarking of the mobile code was proposed by Esparza, Fernandez, Soriano, Munoz, and Forne (2003). A mark is embedded into the mobile agent by using software watermarking techniques. This mark is transferred to the agent's results during the execution. For the executing hosts, the mark is a normal part of results and is "invisible." If the owner of the agent detects that the mark has been changed (it is different than expected), he or she has proof that the malicious host was manipulating the agent's data or code. Figure 1 illustrates how the mark is appended to data during the mobile agent's computations on various hosts.

The paper by Esparza et al. (2003) presents three ways of embedding the watermark into the agent:

- Marking the code
- Marking the input data
- Marking the obfuscated code

The mark or marks are validated after the agent returns to its originator.

Possible attacks against this method include:

- **Eavesdropping:** If the data is not protected in any way (e.g., not encrypted) it can be read by every host.
- **Manipulation:** The malicious host can try to manipulate either the agent's code or data to change the results and still keep the proper mark.
- **Collusion:** A group of malicious hosts can cooperate to discover the mark by comparing the obtained results.

Fingerprinting

Software fingerprinting uses watermarking techniques in order to embed a different mark for each user. Software fingerprinting shares weaknesses with software watermarking: marks must be resilient to manipulation and "invisible" to observers.

The method for fingerprinting was proposed by Esparza et al. (2003). Contrary to the watermarking methods presented previously here, the embedded mark is different for each host. When the agent returns to the owner, all results are validated and the malicious host is directly traced (see Figure 2).

The article presents two ways of embedding the mark into the agent:

- **Marking the code:** In this case, malicious hosts have the possibility of comparing their different codes in order to locate their marks.
- **Marking the input data:** The data are usually different for each host, so it is harder to identify the mark.

The procedure is similar to the mobile agent watermarking approach. However, the owner must know each mark for each host and their location. One of the possibilities of reconstructing the marks is to catch the information about the previously chosen places in the results.

Figure 1. Example of watermarking

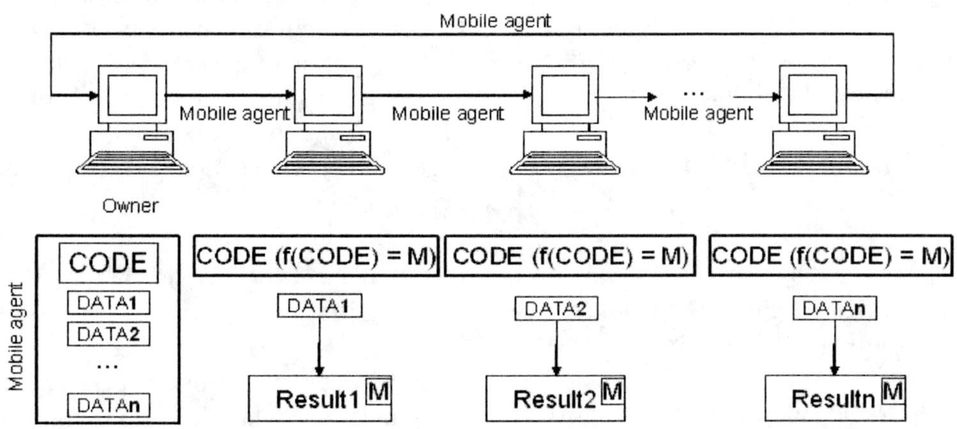

Figure 2. Example of fingerprinting

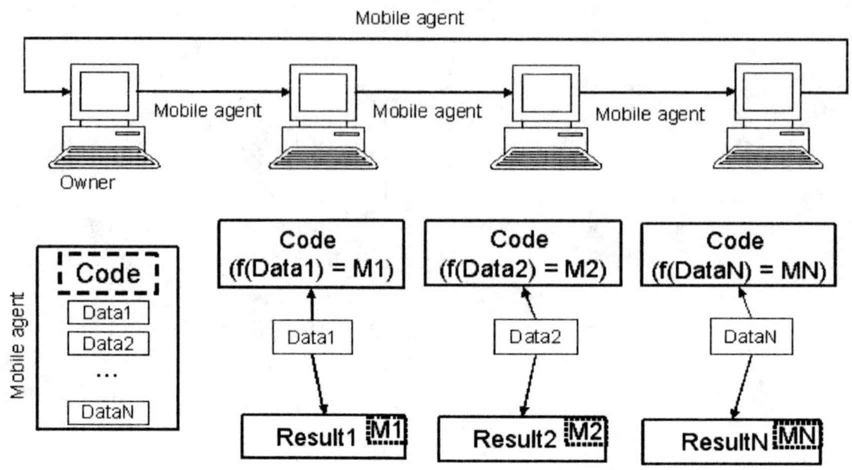

Possible attacks against this method include:

- **Eavesdropping:** If the data are not protected in any way (e.g., not encrypted) it can be read by every host.
- **Manipulation:** The malicious host can try to manipulate either the agent's code or data to change the results and still keep the proper mark.
- **Collusion:** Colluding hosts cannot extract any information about the mark comparing their data or results, because every host has a different input data and a different embedded mark.

The difference between mobile agent watermarking and fingerprinting is the fact that in the second case it is possible to detect collusion attacks performed by a group of dishonest hosts.

Publicly Verifiable Chained Digital Signatures

This protocol, proposed by Karjoth (1998) allows verification of the agent's chain of partial results not only by the originator, but also by every agent place. However, it is still vulnerable to interleaving attacks. This protocol makes it possible for every agent place, which receives an agent to verify that it has not been compromised. This saves computing power because if an agent has indeed been compromised, the agent place can reasonably refuse to execute the compromised agent.

Environmental Key Generation

This scheme allows an agent to take a predefined action when some environmental condition is true (Riordan & Schneier, 1998). The approach centers on constructing agents in such a way that upon encountering an environmental condition (e.g., via a matched search string), a key is generated, which is then used to cryptographically unlock some executable code. The environmental condition is hidden through either a one-way hash or public key encryption of the environmental trigger. This technique ensures that a platform or an observer of the agent cannot uncover the triggering message or response action by directly reading the agent's code.

Itinerary Recording with Replication and Voting

A faulty agent platform can behave similarly to a malicious one. Therefore, applying fault tolerant capabilities to this environment should help counter the effects of malicious platforms (Schneider, 1997). One such technique for ensuring that a mobile agent arrives safely at its destination is through the use of replication and voting. Rather than using a single copy of an agent to perform a computation, multiple copies are used. Although a malicious platform may corrupt a few copies of the agent, enough replicas avoid the encounter to successfully complete the computation. A slightly different method based on multiple copies of agent was proposed by Benachenhou and Pierre (2006). In this proposal, the copy of agent is executed on a trusted platform to validate results obtained on other platforms.

A METHOD BASED ON SECRETS AND PROOFS

In the proposed system we assume that there exist at least three parties:

- A manager
- An agent
- A host

The manager can be an originator of the agent. It plays a role of a verification instance in the scheme and creates initial countermeasures for the agent. The manager also plays a role of a trusted third party.

Outline of the Method

The zero-knowledge proof systems (Goldreich, 2002) enable the verifier to check validity of the assumption that the prover knows a secret. In our system the verifier would be the manager or owner of agents and, obviously, agents would be the provers. In the initial phase, the manager computes a set of secrets. The secrets are then composed into the agent, so that if the manager asks the agent to make some computations (denote them as a function *f*), the result of this would be a valid secret. This function should have the following property:

- If we have x_1 and $f(x_1)$ then it is computationally infeasible to find such x_2 that $f(x_1) = f(x_2)$

If the secret is kept within an agent, then also the host can use the zero-knowledge protocol to verify it. Every authorized change of agent's state results in such a change of the secret that the secret remains valid. On the other hand, every unauthorized change leads to loosing the secret, so at the moment of verification by host or manager, the agent is not able to prove possession of a valid secret. Since the host can monitor all agent's computations, the secret should not only change with agent's execution state, but should also be different for different hosts, so one host could only validate the secret prepared for operations that should be executed at this platform. In our system the host can tamper the agent and try to make such changes that so that he/she will be still able to obtain the proper secret, but the characteristics of function *f* will not allow doing this. Some possible candidates for the function *f* can be a hash function. Our approach is a detection rather than prevention (see Zwierko & Kotulski, 2007).

Specification of the Method

The Initial Phase

The initial phase has three steps:

1. The manager computes a set of so-called identities, denoted as ***ID***. It is public. For each identity, the manager computes appropriate secret, denoted as σ. The details for generating those values depend upon chosen zero knowledge system.
2. To compose σ into an agent, any secure secret sharing scheme (Pieprzyk, Hardjono, & Seberry, 2003) with threshold t can be used. The manager creates n shares, such that the reconstructed secret would be σ. The $t-1$ shares are composed into an agent and the rest are distributed among the hosts via secure channels (this is illustrated in Figure 3).
3. The manager now needs to glue the shares into an agent in such a way, that when the

agent is in a proper execution state, it is able to obtain from its code/state variables the correct shares. Since the agent is nothing more than a computer program, it can be described as a *finite state machine* (FSM). Assume, we have the agent of the form <Σ, S, S_I, S_F, δ>, where:

- Σ is the input alphabet
- $S = \{f_0, ..., f_n\}$ is a set of all possible states
- S_I is a subset of S with all initial states
- S_F is a subset of S with all finishing states, possibly empty
- δ: Σ × S → S is a state transition function.

Figure 4 shows an example of agent's FSM. It is obvious that only some execution states should be observed during the computation at the host platform (e.g., the ones connected with gathering and storing the data). If the state f_j is the first state of the agent's computations at the host platform, then it is natural that the shares should be generated only from this state. Additionally, some internal variables that differ for each host should be utilized to obtain different secrets for each host. Thus, to create agent's shares, f_j, $c_i \in \Sigma$, and the code should be used.

In other cases, where the pair f_j and c_i is not unique for each host, the previous states or other data should be used. It should be possible to obtain the proper shares for current host based on appropriate execution state and internal variables. If there is more than one unique combination of (f_j, c_i) for one host, then for each of them the host should obtain an ID and a share. The agent's code (in a certain form) should be a part of the data that are required to recreate the secret to enable detection of every unauthorized manipulation, which could be performed by previous host.

To create the shares from the mentioned data, the hash function or an encryption function with the manager's public key can be used.

The Validation Phase

1. The host, which wants to verify an agent's integrity, sends its share to the agent.

Figure 3. Distributing ID and shares to hosts

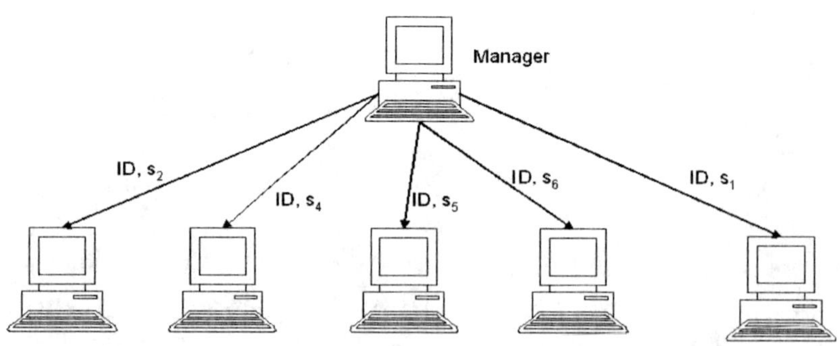

Figure 4. Mobile agent as an FSM

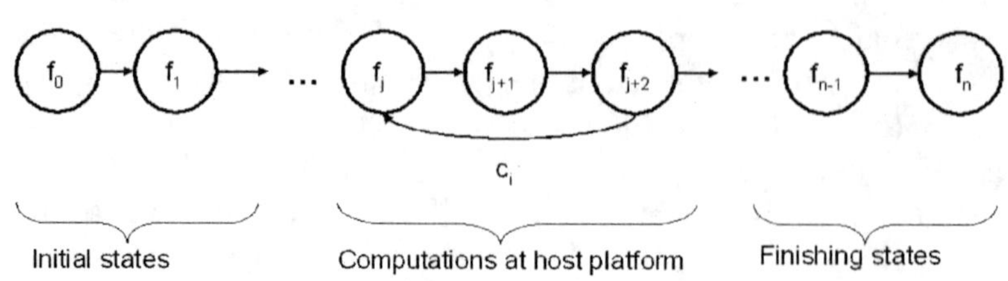

2. The agent creates the rest of the shares from its code and the execution state. It recreates the secret. The agent computes the secret σ and uses it for the rest of the scheme, which is a zero-knowledge identification protocol.
3. The agent and the host execute the selected zero-knowledge protocol, so that the host can confirm the correctness of σ.

The manager can compute many identities, which may be used with different execution states. In that situation the agent should first inform host which identity should be used, or the host can simply check the correctness of σ for all possible identities.

SECURITY AND SCALABILITY

Definitions and Notions

This section presents basic notions concerning agent's integrity that will be later used in description of the selected solutions. The integrity of an agent means that an unauthorized party cannot change its code or execution state, or such changes should be detectable (by an owner, a host or an agent platform, which want to interact with the agent). The authorized changes occur only when the agent has to migrate from one host to another. Next is a more formal definition:

Definition 1 (integrity of an agent). An agent's integrity is not compromised if no unauthorized modification can be made without the agent's owner noticing this modification.

The concept of forward integrity is also used for evaluation of many methods (Karjoth et al., 1999; Yee, 1999). This notion is used in a system where agent's data can be represented as a chain of partial results (a sequence of static pieces of data). Forward integrity can be divided into two types, which differ in their possibility to resist cooperating malicious hosts. The general goal is to protect the results within the chain of partial results from being modified. Given a sequence of partial results, the forward integrity is defined as follows:

Definition 2 (Karjoth et al., 1999; Yee, 1999). The agent posses the weak forward integrity feature if the integrity of each partial result m0, ..., mn-1 is provided when in is the first malicious agent place on the itinerary.

Weak forward integrity is conceptually not resistant to cooperating malicious hosts and agent places that are visited twice. To really protect the integrity of partial result, we need a definition without constraints.

Definition 3 [strong forward integrity (Karjoth et al., 1999)]. The agent system preserves strong forward integrity of the agent if none of the agent's encapsulated messages mk, with k < n, can be modified without notifying the manager.

In this chapter we refer to forward integrity as to strong forward integrity (when applicable). To make notion of forward integrity more useful, we define also publicly verifiable forward integrity, which enables any host to detect compromised agents:

Definition 4. The agent posses the publicly verifiable forward integrity if every host in can verify that the agent's chain of partial results mi0 ,..., min has not been compromised.

The other important notion concerning agent's integrity, a concept of black-box security (Hohl, 1998) was introduced in the Time Limited Black-Box Security and Obfuscated Code section.

Analysis

The proposed scheme should be used with more that one identity. This would make it very hard to manipulate the code and the data. The best approach is to use one secret for each host. We assume that the malicious host is able to read and manipulate an agent's data and code. He/she can try to obtain from an agent's execution state the proper shares. The host can also try to obtain a proper secret and manipulate the agent's state and variables in a way that the obtained secret would stay the same. But the host does not know other secrets that are composed into the agents; also he/she does not know more shares to recreate those secrets, so, any manipulation would be detected by the next host.

The protocol is not able to prevent any attacks that are aimed at destroying the agent's data or code, meaning that a malicious host can "invalidate" any agent's data. But this is always a risk, since the host can simply delete an agent.

- **Weak forward integrity:** The proposed method posses the *weak forward integrity* property: the malicious host cannot efficiently modify previously generated results.
- **Strong forward integrity:** The protocol provides the agent also with *strong forward integrity*, because the host cannot change previously stored results (without knowledge of secrets created for other hosts). He/she cannot also modify the agent in a way that could be undetectable by the next host on the itinerary or by the owner.
- **Publicly verifiable forward integrity:** Each host can only verify if the agent's code or the execution state has not been changed. They cannot check wherever the data obtained on other platforms has not been modified. The agent's owner, who created all secrets, can only do this.
- **Black-box security:** The proposed system is not resistant to read attacks. A malicious host can modify the code or data, but it is detectable by agent's owner, so it is resistant to manipulation attack. The system does not have full black-box property.

Comparison with Other Methods

It is a difficult task to compare systems based on such different approaches as presented here. We decided to split comparison into two categories:

- **Practical evaluation:** If the method is hard or easy to implement:
    - **Hard:** No practical implementation exists at the moment
    - **Medium:** The method has been implemented, with much effort
    - **Easy:** The method is widely used and has been implemented for different purposes

  and what elements of an agent it protects:
- **Theoretical evaluation:** If the method satisfies the security definitions from the *Definitions and Notions* section.

The theoretical evaluation is quite hard, because some methods that have the black-box property do not "fit" other definitions. If the code or data cannot be read or manipulated (the ideal case), then how we can discuss if it can be verifiable, or, if it fulfills the forward integrity.

As for evaluation of the black-box property, it is very hard to provide the code that cannot be read. In all cases, marked by *, (see Table 2) the adversary can modify the agent but not in a way that owner or other host would not notice. This means that no efficient manipulation attack can be made, so one part of the black-box property is satisfied.

In # case the *publicly verifiable forward integrity* is satisfied only partially, because the agent's code can be verified but the data cannot.

Scalability

The initialization phase. The first phase is similar to the bootstrap phase of the system. The hosts and the manager create a static network. It is typical for agents' systems that the manager or the owner of an agent knows all hosts, so distribution of all IDs and shares is efficient. We can compare this to sending a single routing update for entire network as in OSPF protocol (the flooding). Whenever a new agent is added to the system, the same amount of information to all hosts has to be sent. Since the messages are not long (a single share and few IDs) and are generated only during creating a new agent, that amount of information should not be a problem. The sizes of parameters (keys lengths, number of puzzles, and number of shares) are appropriately adjusted to the agents' network size.

The operating phase. During the validation phase no additional communication between the manager and the hosts is required.

Table 1. Practical comparison of the integrity protection methods

Method	Implementation	Protects code	Protects data	Protects execution state
Encryption functions	Hard	Yes	Yes	No
Obfuscated code	Medium	Yes	No	No
Cryptographic traces	Hard	Yes	No	Yes
Watermarking	Easy	Yes	Yes	No
Fingerprinting	Easy	Yes	Yes	No
Zero knowledge proof	Easy	Yes	Yes	Yes

Table 2. Theoretical comparison of integrity protection methods

Method	Weak forward integrity	Strong forward integrity	Publicly verifiable forward integrity	*Black-box* property
Encryption functions	No	No	No	Yes
Obfuscated code	Yes	Yes	No	Partially*
Cryptographic traces	Yes	Yes	Yes	No
Watermarking	Yes	No	No	Partially*
Fingerprinting	Yes	Yes	No	Partially*
Zero knowledge proof	Yes	Yes	No#	Partially*

Modifications

A similar scenario can be used to provide integrity to the data obtained by the agent from different hosts. A malicious host could try to manipulate the data delivered to the agent by the previously visited hosts. To ensure that this is not possible, the agent can use the zero-knowledge protocol to protect the data. For each stored piece of data, the agent can create a unique "proof," utilizing the zero-knowledge protocol. Any third party, who does not possess σ, is not able to modify the proof. So the manager knowing σ can be sure that the data was not manipulated.

An area for development of the proposed integrity solution is to find the most appropriate function for composing secrets into hosts: The proposed solution fulfills the requirements, but some additional evaluation should be done. The next possibility for the future work would be to integrate the proposed solution to some agents' security architecture, possibly the one that would also provide an agent with strong authentication methods and anonymity (Zwierko & Kotulski, 2005). Then, such a complex system should be evaluated and implemented as a whole. A good example of such a system would be an agent-based electronic elections system for mobile devices, where the code integrity together with the anonymous authentication is crucial for correctness of the system (Zwierko & Kotulski, in press).

FUTURE TRENDS

In this chapter we presented methods of protection of mobile agents against attacks on their integrity. The methods offer protection on a certain level, but the agents' security can be significantly increased by avoiding risky behavior, especially visiting suspicious hosts. This can be done by using mechanisms built into individual agents or by distributed solutions based on cooperation of agents and hosts. The most promising solutions for improvement of the mobile code security can be based on risk analysis or on reputation systems. The first one needs some built-in analysis tools while the second one requires trust management infrastructure.

Risk analysis is one of the most powerful tools used in economics, industry, and software engineering (Tixier, Dusserre, Salvi, & Gaston, 2002). Most of the business enterprises carry out such an analysis for all transactions. The multi-agent or mobile agent system can be easily compared with such an economic-like scenario: There are a lot of parties making transactions with other parties. The risk analysis could be utilized to estimate how high is the probability that selected agent platform is going to harm the agent. The biggest advantage of this solution is lack of any form of cooperation between different managers: Everyone can make its own analysis based on gathered knowledge. However, the cooperation between different managers can benefit in better analysis.

Reputation systems (Sabater & Sierra, 2005; Zacharia & Maes, 2000) are well known and utilized in different applications, especially in peer-to-peer environments. They enable the detection of malicious parties based on their previous behavior, registered, valuated, and published. We can imagine an agent system where managers and owners of agents would also rate agent platforms based on their previous actions towards the agents. Of course, such a system still requires some integrity protection mechanisms, which could be used to verify if results obtained by the agent are correct. However, the applied mechanism can be rather simple, not as complicated as some presented methods, for example, EFs.

CONCLUDING REMARKS

Among security services for stored data protection two are the most important: availability and integrity. The data unavailable is useless for a potential user. Also, the data illegally defected or falsified is a worthless source of information. No other protection has sense if the data's content is destroyed. In the case of executables we face analogous problems. Except others, the executables must be available and protected against falsification (that is unauthorized changes of the designed functioning, internal state and the carried data). The problem of availability has been successfully solved by a concept of mobile agents that simply go to the destination place and work in there. However, this solution made the problem of integrity of the mobile code or mobile agent even more important than in the case of the stored data. The falsified mobile agent is not only useless. It can be even harmful as an active party making some unplanned actions. Therefore, preserving agents' integrity is a fundamental condition of their proper functioning.

In this chapter we made an overview of the existing protocols and methods for preserving the agent's integrity. The basic definitions and notions were introduced. The most important mechanisms were presented and discussed. We also proposed a new concept for detection of the tempering of an agent, based on a zero-knowledge proof system. The proposed scheme secures both, an agent's execution state and the internal data along with its code. For the practical implementation the system requires some additional research and development work, but it looks to be a promising solution to the problem of providing an agent with effective and strong countermeasures against attacks on its integrity.

REFERENCES

Alves-Foss, J., Harrison, S., & Lee, H. (2004, January 5-8). The use of encrypted functions for mobile agent security. In *Proceedings of the 37th Hawaii International Conference on System Sciences*—Track 9 (pp. 90297b). US: IEEE Computer Society Press.

Benachenhou, L., & Pierre, S. (2006). Protection of a mobile agent with a reference clone. *Computer communications, 29*(2), 268-278.

Burmester, M., Chrissikopoulos, V., & Kotzanikolaou, P. (2000). Secure transactions with mobile agents in hostile environments. In E. Dawson, A. Clark, & C. Boyd (Eds.), Information security and privacy. *Proceedings of the 5th Australasian Conference ACISP* (LNCS 1841, pp. 289-297). Berlin, Germany: Springer.

Coilberg, Ch., Thomborson, C., & Low, D. (1997). *A taxonomy of obfuscating transformations* (Tech. Rep. No. 148). Australia: The University of Auckland.

Corradi, A., Cremonini, M., Montanari, R., & Stefanelli, C. (1999). Mobile agents integrity for electronic commerce applications. *Information Systems, 24*(6), 519-533.

Esparza, O., Fernandez, M., Soriano, M., Munoz, J. L., & Forne, J. (2003). Mobile agents watermarking and fingerprinting: Tracing malicious hosts. In V. Mařík, W. Retschitzegger, & O. Štěpánková (Eds.), *Proceedings of the Database and Expert Systems Applications (DEXA 2003)* (LNCS 2736, pp. 927-936). Berlin, Germany: Springer.

Goldreich, O. (2002). Zero-knowledge twenty years after its invention (E-print 186/2002). E-print, IACR.

Hohl, F. (1998). Time limited blackbox security: Protecting mobile agents from malicious hosts. In G. Vigna (Ed.), *Mobile agents and security* (LNCS 1419, pp. 92-113). Berlin, Germany: Springer.

Jansen, W. A. (2000). Countermeasures for mobile agent security. [Special issue]. *Computer Communications, 23*(17), 1667-1676.

Jansen, W. A., & Karygiannis, T. (1999). Mobile agents security (NIST Special Publication 800-19). Gaithersburg, MD: National Institute of Standards and Technology.

Karjoth, G., Asokan, N., & Gulcu, C. (1999). Protecting the computation results of free-roaming agents. In K. Rothermel & F. Hohl (Eds.), *Proceedings of the Second International Workshop on Mobile Agents (MA '98)* (LNCS 1477, pp. 195-207). Berlin, Germany: Springer.

Kulesza, K., & Kotulski, Z. (2003). Decision systems in distributed environments: Mobile agents and their role in modern e-commerce. In A. Lapinska (Ed.), *Proceedings of the Conference "Information in XXI Century Society"* (pp. 271-282). Olsztyn: Warmia-Mazury University Publishing.

Kulesza, K., Kotulski, Z., & Kulesza, K. (2006). On mobile agents resistant to traffic analysis. *Electronic Notes in Theoretical Computer Science, 142,* 181-193.

Low, D. (1998). Protecting Java code via code obfuscation. *Crossroads, 4*(3), 21-23.

Man, C., & Wei, V. (2001). A taxonomy for attacks on mobile agent. In *Proceedings of the International Conference on Trends in Communications, EUROCON'2001* (pp. 385-388). IEEE Computer Society Press.

Oppliger, R. (2000). *Security technologies for the World Wide Web.* Computer Security Series. Norwood, MA: Artech House Publishers.

Pieprzyk, J., Hardjono, T., & Seberry, J. (2003). *Fundamentals of computer security.* Berlin, Germany: Springer.

Riordan, J., & Schneier, B. (1998). Environmental key generation towards clueless agents. In G. Vinga (Ed.), *Mobile agents and security* (pp. 15-24). Berlin, Germany: Springer.

Sabater, J., & Sierra, C. (2005). Review on computational trust and reputation models. *Artificial Intelligence Review, 24* (1), 33-60.

Sander, T., & Tschudin, Ch. F. (1998, May 3-6). Towards mobile cryptography. In *Proceedings of the 1998 IEEE Symposium on Security and Privacy* (pp. 215-224). IEEE Computer Society Press.

Schneider, F. B. (1997). Towards fault-tolerant and secure agentry. In M. Mavronicolas (Ed.), *Proceedings 11th International Workshop on Distributed Algorithms* (pp. 1-14). Berlin, Germany: Springer.

Shao, M., & Zhou, J. (2006). Protecting mobile-agent data collection against blocking attacks. *Computer Standards & Interfaces, 28*(5), 600-611.

Tixier, J., Dusserre, G., Salvi, O., & Gaston, D. (2002). Review of 62 risk analysis methodologies of industrial plants. *Journal of Loss Prevention in the Process Industries, 15*(4), 291-303.

Vigna, G. (1997). Protecting mobile agents through tracing. In *Proceedings of the 3rd ECOOP Workshop on Mobile Object Systems*. Jyvälskylä, Finland.

Vigna, G. (1998). Cryptographic traces for mobile agents. In G. Vigna (Ed.), *Mobile agents and security* (LNCS 1419, pp. 137-153). Berlin, Germany: Springer.

Wang, T., Guan, S., & Chan, T. (2002). Integrity protection for code-on-demand mobile agents in e-commerce. *Journal of Systems and Software, 60*(3), 211-221.

Yee, B. S. (1999). A sanctuary for mobile agents. In J. Vitek & C. D. Jensen (Eds.), *Secure Internet programming: Security issues for mobile and distributed objects* (LNCS 1603, pp. 261-273). Berlin, Germany: Springer.

Zacharia, G., & Maes, P. (2000). Trust management through reputation mechanisms. *Applied Artificial Intelligence, 14*(9), 881-907.

Zwierko, A., & Kotulski, Z. (2005). Mobile agents: Preserving privacy and anonymity. In L. Bolc, Z. Michalewicz, & T. Nishida (Eds.), *Proceedings of IMTCI2004, International Workshop on Intelligent Media Technology for Communicative Intelligence* (LNAI 3490, pp. 246-258). Berlin, Germany: Springer.

Zwierko, A., & Kotulski, Z. (2007). Integrity of mobile agents: A new approach. *International Journal of Network Security, 2*(4), 201-211.

Zwierko, A., & Kotulski, Z. (2007). A lightweight e-voting system with distributed trust. *Electronic Notes in Theoretical Computer Science, 168*, 109-126.

KEY TERMS

Agent Platform (Host): Agent platform is a computer where an agent's code or program is executed. The software agent cannot perform its actions outside hosts. The host protects agents against external attacks.

Cryptographic Protocol: Cryptographic protocol is a sequence of steps performed by two or more parties to obtain a goal precisely according to assumed rules. To assure this purpose the parties use cryptographic services and techniques. They realize the protocol exchanging tokens.

Intelligent Software Agent: Intelligent software agent is an agent that uses artificial intelligence in the pursuit of its goals in contacts with hosts and other agents.

Mobile Agent: Mobile agent is an agent that can move among different platforms (hosts) at different times while the **stationary agent** resides permanently at a single platform (host).

Security Services: Security services guarantee protecting agents against attacks. During agent's transportation the code is protected as a usual file. At the host site, the agent is open for modifications and very specific methods must be applied for protection. For the agent's protection the following security services can be utilized:

- **Confidentiality:** Confidentiality is any private data stored on a platform or carried by an agent that must remain confidential. Mobile agents also need to keep their present location and the whole route confidential.
- **Integrity:** Integrity exists when the agent platform protects agents from unauthorized modification of their code, state, and data and ensure that only authorized agents or processes carry out any modification of the shared data.
- **Accountability:** Accountability exists when each agent on a given platform must be held accountable for its actions: must be uniquely identified, authenticated, and audited.
- **Availability:** Availability exists when every agent (local, remote) is able to access data and services on an agent platform, which responsible to provide them.

- **Anonymity:** Anonymity is when agents' actions and data are anonymous for hosts and other agents; still accountability should be enabled.

Software Agent: Software agent is a piece of code or computer program that can exercise an individual's or organization's authority, work autonomously at host toward a goal, and meet and interact with other agents.

Strong Mobility: Strong mobility of an agent means that a running program along with its particular (actual) state is moving from one host site to another.

Weak Mobility: Weak mobility of an agent means that only the agent's code is migrating and no execution state is sent along with an agent program.

Chapter IV
Identity Management

Kumbesan Sandrasegaran
University of Technology, Sydney, Australia

Mo Li
University of Technology, Sydney, Australia

ABSTRACT

The broad aim of identity management (IdM) is to manage the resources of an organization (such as files, records, data, and communication infrastructure and services) and to control and manage access to those resources in an efficient and accurate way. Consequently, identity management is both a technical and process-orientated concept. The concept of IdM has begun to be applied in identities-related applications in enterprises, governments, and Web services since 2002. As the integration of heterogeneous wireless networks becomes a key issue in towards the next generation (NG) networks, IdM will be crucial to the success of NG wireless networks. A number of issues, such as mobility management, multi-provider and securities require the corresponding solutions in terms of user authentication, access control, and so forth. IdM in NG wireless networks is about managing the digital identity of a user and ensuring that users have fast, reliable, and secure access to distributed resources and services of an next generation network (NGN) and the associated service providers, across multiple systems and business contexts.

INTRODUCTION

The broad aim of identity management (IdM) is to manage the resources of an organisation (such as files, records, data, and communication infrastructure and services) and to control and manage access to those resources in an efficient and accurate way (which in part usually involves a degree of automation). Consequently, IdM is both a technical and process-orientated concept.

The concept of IdM has begun to be applied in identities-related applications in enterprise, governments, and Web services since 2002. As the integration of heterogeneous wireless networks becomes a key issue in the fourth generation (4G) wireless networks, IdM will become crucial to the success of next generation (NG) wireless networks. A number of issues, such as mobility management, multi-provider, and securities require the corresponding solutions in terms of user authentication, access control, and so forth. Although IdM processes require the integration into existing business processes at several levels (Titterington, 2005), it remains an opportunity for NG wireless networks.

IdM in NG wireless networks is about managing the digital identity of a user and ensuring that users have fast, reliable, and secure access to distributed resources and services ofNG wireless networks and associated service providers across multiple systems and business contexts.

Definition

Given the open and currently non-standardised nature of IdM, there are varying views as to the exact definition of IdM. These include:

By HP (Clercq & Rouault, 2004)
Identity Management can be defined as the set of processes, tools and social contracts surrounding the creation, maintenance, utilization and termination of a digital identity for people or, more generally, for systems and services to enable secure access to an expanding set of systems and applications.

By Reed (2002)
The essence of Identity Management as a solution is to provide a combination of processes and technologies to manage and secure access to the information and resources of an organisation while also protecting users' profiles.

By Cisco Systems (2005)
Businesses need to effectively and securely manage who and what can access the network, as well as when, where, and how that access can occur...lets enterprises secure network access and admission at any point in the network, and it isolates and controls infected or unpatched (sic) devices that attempt to access the network.

Objectives

As IdM can be used in different areas such as enterprise, government, Web services, telecommunication networks and so forth, its objectives diversity in different contexts. Generally, the IdM system is expected to satisfy the following objectives (Reed, 2002):

- It should define the identity of an entity (a person, place, or thing).
- It should store relevant information about entities, such as names and credentials, in a secure, flexible, customisable store.
- It should make the information accessible through a set of standard interfaces.
- It should provide a resilient, distributed, and high performance infrastructure for identity management.
- It should help to manage relationships between the enterprise and the resources and other entities in a defined context.

Main Aspects

Authentication

Authentication is the process by which an entity provides its identity to another party, for example, by showing photo ID to a bank teller or entering a password on a computer system. This process is broken down into several methods which may involve something the user knows (e.g., password), something the user has (e.g., card), or something the user is (e.g., fingerprint, iris, etc.). Authentication can take many forms, and may even utilise combinations of these methods.

Authorisation

Authorisation is the process of granting access to a service or information based on a user's role in an organisation. Once a user is authenticated, the system then must ensure that a particular user has access to a particular resource.

Access Control

Access control is used to determine what a user can or cannot do in a particular context (e.g., a user may have access to a particular resource/file, but only during a certain time of day, e.g., work hours, or only from a certain device, e.g., desktop in the office).

Auditing and Reporting

Auditing and reporting involves creation and keeping of records, whether for business reasons (e.g., customer transactions), but also providing a "trail" in the event that the system is compromised or found faulty.

DIGITAL IDENTITY

What is Digital Identity?

In a business transaction, identity is used to establish a level of trust upon which business can be conducted. Trust in this context is the confidence that each party they are dealing with is who he/she claims to be. Traditionally, such trust was established with the use of an observable physical attribute of an entity. For example, business dealings were in person (appearance), on the phone (voice), or with the use of signatures on contracts (handwriting).

The identity of an individual is defined as the set of information known about that person (Pato & Rouault, 2003). For example, an identity in the real world can be a set of names, addresses, driver's licenses, birth certificate, and so forth.

With the development and widespread use of digital technologies, entities have been able to communicate with each other without being physically present. In some cases, the first meeting and possibly the entirety of the transaction between two parties is held over a digital medium. There is a growing need for trust to be established in transactions over the digital world.

Digital identity is the means that an entity can use to identify themselves in a digital world (i.e., data that can be transferred digitally, over a network, file, etc.). The aim of digital identity is to create the same level of confidence and trust that a face to face transaction would generate.

Composition of Digital Identity

A digital identity seeks to digitise an individual's identity to the extent that they cannot be mistaken for someone else and that it is difficult for another person to impersonate that individual. In a typical face to face situation, identity comprises of two parts: the actual identity of the entity (something that can be observed by human senses) and the credentials or what they use to prove their identity. In Reed (2002), the attributes of digital identity are given as follows:

Who You Are

"Who you are" is the attribute that in a real world context uniquely identify a single entity. These can include knowledge or data that is only known by that entity, unique physical characteristics of that entity, or items that the entity has.

Context

Context can refer to the type of transaction or organisation that the entity is identifying itself as well as the manner that the transaction is made. Different constraint on digital identity may be enforced depending on the context. For example, the sensitive transactions related to birth certificate information over phone or internet may be prohibited.

Profile

A profile consists of data needed to provide services to users once their identity has been verified. A user profile could include what an entity can do, what they have subscribed to, what groups they are a member of, their selected services, and so forth. The profile of a user will change during the course of interaction with a service provider.

Of particular consideration is the concept of "context." Depending on the context, we differ in the actions that we are able to do as individuals. In an Internet shopping context, we may only be able to browse or purchase items. In a corporate context, it may enable us to access files or otherwise do some other activity.

Context is also important from a digital identity context as it is likely to determine the amount and type of identity information that is needed in order for the determined level of "trust" to be available. For example, in an e-mail context, the amount of identifying information that is necessary is usually only two things: a username and password. However, with more security conscious applications, for example, bank transactions and governmental functions, more information is usually required (e.g., birth certificates, credit card numbers, and the like).

The digital identity of an individual user forms the main focus of security threats to any IdM system. As such, there are typical measures that must be taken to ensure that digital identities are kept securely.

Usage of Digital Identity

Digital identity can be used for authentication. It is where an entity must "prove" digitally that it is the one that it claims to be. It is at this stage that the credentials of digital identity are used. The simplest form of authentication is the use of a username and corresponding password. This is known as "single factor" authentication, since only a single attribute is used to determine the identity. Stronger authentication is usually obtained by not only increasing the number of attributes that are used, but also by including different types. To add to the previous example scheme, in addition to the password, an entity could also be called upon to have a particular piece of hardware plugged in, providing a "two factor" scheme (DIGITALIDWORLD, 2005).

Once an entity is authenticated, a digital identity is used to determine what that entity is authorised to do. This is where the profile of a digital identity is required. As an example, authorisation can be seen as the difference between an "administrator" and a "user" who share the same resource (for example, a computer). Both may be authenticated to use the computer, but the actions that each may do with that resource are determined by the authorisation. Authentication attempts to establish a level of confidence that a certain thing holds true, authorisation decides what the user is allowed to do.

Accounting provides an organisation with the ability of tracking unauthorised access when it occurs. Accounting involves the recording and logging of entities and their activities within the context of a particular organisation, Web site, and so forth.

PROS AND CONS OF IDENTITY MANAGEMENT

Benefits of Identity Management

Reduce Total Cost Ownership (TCO) for All Systems

Cost reduction by IdM usually is a result of more efficient use of personnel and resources, especially with regards to the following administrative bureaucracy. Examples include (Courion, 2005):

- Reducing the costs of auditing by providing real-time verification of user access rights and policy awareness enforcement
- Eliminating account administration such as account add/move/change and calls to information security staff for digital certificate registration
- Eliminating calls of password reset (the #1 support call) to internal or outsourced help desks
- Streamlining IT operations for more efficient management and reallocation to more strategic projects
- Reducing management overhead (Reed, 2002)

Competitive Advantage Through Streamlining and Automation of Business Processes

This competitive advantage is delivered by cutting down costs in areas with a high need for unnecessary support and being able to:

- Offer users a fast, secure way to access to revenue-generating systems, applications, and Web portals (Courion, 2005)
- Provide faster response to "password reset" and "insufficient access" user lockouts, thus increasing system and data availability (Courion, 2005)
- Provide 24x7x365, unassisted self-service for the most common of help desk calls (Courion, 2005)
- Improve customer and employee service; maintain confidentiality and control of customers, suppliers, and employees (Reed, 2002)

- Reduce time for new employees to gain access to required resources for work (Reed, 2002)

Increase Data Security

Data security includes the typical protection of data from unauthorised users as well as ensuring that the data being used is kept up to date across the organisation and is safe from inadvertent or intentional tampering by unauthorised users within the organisation.

- Minimise the "security gap" that exists between the time when employees leave a company and their accounts are disabled (Courion, 2005)
- Reduce the intrusion risk due to orphaned or dormant accounts (by ex-employees or those posing as ex employees) (Courion, 2005)
- Enforce the policies of consistent account provisioning to make sure that only those who need access get access (Courion, 2005)
- Enforce consistent password policies for stronger authentication (Courion, 2005)
- Reduce security threats (e.g., human error) through policy based automation (Courion, 2005)
- Ensure accurate audit trails for intrusion prevention and security reporting (Courion, 2005)
- Provide faster response to account access requests or password reset, thus reducing the need of proliferating "superuser" privileges (Courion, 2005)
- Increase the opportunity of adopting the Public Key Infrastructure by removing the biggest barrier (Courion, 2005)
- Reduce risk of incorrect information being used (Reed, 2002)

Support Legal Initiatives and Demonstrate Compliance (Courion, 2005; Reed, 2002)

In the case of legal initiatives, IdM can be used successfully to demonstrate a systematic and effective approach to safeguarding an organisation's assets and its business partners' (customers, suppliers, contractors, clients) assets. It also presents a method of ensuring that policies are enforced away from human effort and decision making (where often the process breaks down or is ignored). In summary, it can:

- Demonstrate policy enforcement
- Proactively verify the access right of a user
- Enable policy awareness testing
- Eliminate orphaned accounts systematically
- Increase protected data privacy

Additional benefits, mainly business centric, are described in more detail by Fujitsu (Locke & McCarthy, 2002):

- **Know who everyone is in the organisations:** Applied to the larger scale of the NG wireless networks, this prevents any user from "slipping through the cracks" whether they are employees or subscribers. Typically, telecommunications providers are adept at keeping customer records, but suffer the same problems with keeping track of staff. An IdM system will enable the organisation to keep stock of all their users.
- **Accurate and consistent people data in all systems:** This is particularly relevant to the existing telecommunications providers. Although services vary, the majority of providers have some lag between the time a record is changed, compared to when that change is made into the records that the company keeps. Typically, this results in undue delays when an existing or new subscriber wishes to get access to their new services. By speeding up the process by which data on users can be updated, this reduces the delay in service provisioning and offers a more significant level of quality of service.
- **Single source of data input/storage:** This feature has already been explored as one of the benefits of an IdM system. Although a distributed system must spread the location and access points for the data that it stores, by having one central system for organising

it, any additional processing that needs to be done, particularly when bridging between two different types of systems or departments, is avoided.

In general, IdM is used to provide an efficient system that covers all users within an organisation. It promotes a single system that does the entire task rather than several systems that conflict or compete with each other.

Drawbacks of Identity Management

IdM, while bringing several advantages to an organisation, may have several applicable drawbacks. These include:

- **Single point of vulnerability:** A feature that brings both advantages and disadvantages to IdM is the central system that is used. A central IdM system is used to avoid the vulnerabilities associated with competing or incompatible systems, as well as reducing the maintenance costs involved in running different types of systems. However, the flip side to this approach is that it represents a single point of vulnerability that, if compromised, can lead to the easy breach of all the data that the system is protecting. To counter this, IdM systems generally recommend that the additional resources that are saved by the organisation employing the IdM system are re-invested into providing more effective security measures. This will result in a system that is, overall, more secure than the existing mixture of systems that individually, are not as secure.
- **Migration from legacy systems and transition costs:** IdM systems are generally at odds with existing systems that manage and secure users and resources. The concept of IdM systems involves the replacing of existing systems with a single IdM system. For larger organisations with staff and hardware that are selected based upon a preference for an existing system or systems, this represents a significant along with all the associated costs of replacing or retraining staff, introducing new equipments and the like. It will also increase the reluctance and reduce the enthusiasm of the organisation to adopt the new IdM system.
- **Specific needs depending on the organisation:** IdM systems generally need to be customised for each particular organisation that intends to use one. This is particularly true for the areas where an IdM system must support the business processes that an organisation has set up. These are usually unique to the organisation. Other areas that would require customisation from system to system include hardware requirements, the nature of the organisations' distributed systems, and so on.
- **Extensive planning, designing, and implementation required:** An IdM system must be extremely well planned, designed, and executed if it is to avoid the disadvantages that it is trying to overcome over existing approaches to enterprise management. Due to the all-encompassing and authoritative control that an IdM system will have over an organisation, it is important that any such system caters or close to the exact specifications, outlined by the organisation. Otherwise, the system may be used incorrectly, resulting in the same inefficiencies from non-IdM systems.
- **Relatively new concept, lack of uniform standard:** IdM as a standardised concept and solution is yet to be finalised. This increases the likelihood of IdM systems to still be in various stages of development, and more importantly, different levels of effectiveness. This may lead to increased maintenance or upgrades in the near future, or lead to flawed development and implementation for the early adopters of IdM systems. Both these alternatives result in an inefficient outcome compared to IdM's claims.

STANDARDS AND SOLUTIONS

A number of IdM technologies and standards have emerged for enterprise networks, government, and Web services. The two main standard bodies to date are from the Liberty Alliance Project and the Web-Services (WS) Federation. However, the specifications produced by these organisations are mainly motivated by user profile management, single sign-on, and personalised services and do not address the requirements of NG wireless networks.

Relevant Standard Bodies

The standards organisations listed in Table 1 are involved in the development of standards for IdM.

IdM Standards

Directory Services
Directory services are considered a core part of any IdM system. The standards (with the standards body created by them) are:

- **X.500 (ISO):** Large global organisations/ governmental organisations
- **LDAP (IETF):** Core standard for systems relying on directory management
- **DSML (OASIS):** Web-orientated extending from LDAP

Web Services
Web services support IdM systems across private and public networks. They are aimed, as such, to connect heterogeneous systems. Several well known protocols, such as TCP/IP, belong here. The ones that have specific applications in IdM are:

- **SOAP (W3C, formerly Microsoft):** For transporting XML messages/remote procedure calls
- **WSDL (W3C):** Used to express the programming interface and location of a service
- **Universal Description, Discovery and Integration (UDDI):** Used to find and publish services

Security
Security protocols are used for protecting information:

- **SAML (OASIS):** XML-based security solution for Web services
- **Web Services Security (WSS) (Language):** Enhancements to SOAP protocol for security.

Federated Identity
Federated identity standards seek to standardise items that would make federated identities more feasible:

Table 1.

Standards Organisation	Area of Standards / Example Standards
Organisation for Advancement of Structured Information Standards (OASIS)	Private, worldwide organisation for XML standards. For example, Security Assertion Markup Language (SAML)
Web Services Interoperability (WS-I)	"Open, industry organisation to promote Web service interoperability across operating systems and programming languages" For example, Simple Object Access Protocol (SOAP)
World Wide Web Consortium (W3C)	Web Services Description Language (WSDL)
Internet Engineering Task Force (IETF)	Loose collection of organisations with internet standards as the main point of interest. For example, Light weight Directory Access Protocol (LDAP)
The Open Group	Sponsors sub groups, for example, Directory Interoperability Forum (DIF), Security Forum (SF)
International Organization for Standardization (ISO)	Well known international standards network. For example, International Telecommunication Union-Telecommunication Standardization Sector (ITU-T)

- **Liberty Alliance Project:** An organisation working mainly towards a solution/standard, they focus on the single sign on concept combined with federated identity.
- **Microsoft .NET Passport:** Primarily an organisational solution rather than standard. This provides a Microsoft managed authentication service for other web services/corporations.

Workflow
Workflow standards include:

- **Business Process Execution Language (BPEL):** Allows business processes (tasks) to be described by a combination of Web services and internal message exchanges.

Provisioning
Provisioning standards are hinted at from workflow standards (which ensure a process is followed by provisioning), but are otherwise not well covered, with one exception:

- **Service Provisioning Markup Language (SPML) (OASIS)**

IdM IN NG WIRELESS NETWORKS

Motivation

IdM issues were not critical in traditional telecommunication networks, because networks, applications, and billing for different services were not integrated. For example, if a service provider offers telephone, Internet access, and cable TV then all of these services are treated separately. Each service has its own subscriber database containing subscriber records and identity information.

IdM, in both concept and practice, has provided an effective alternative and complements to the existing security measures in enterprise networks. The NG wireless networks can be seen as a collective of organisations in addition to their customers. Considering its integrated nature, an IdM framework for NG wireless networks brings organisations closer than in the current telecommunications environment.

IdM in NG wireless networks will be more complex than enterprise and Web service solutions. It involves consolidation, management and exchange of identity information of users to ensure the users have fast, reliable, and secure access to distributed network resources across multiple service providers. Furthermore, NG wireless networks have to provide seamless and ubiquitous support to various services in a heterogeneous environment.

Carefully planned and deployed, IdM solutions in NG wireless networks can prevent fraud, improve user experience, assist in the rapid deployment of new services, and provide better privacy and national security. Conversely if it is not well planned and deployed, it can lead to identity theft, fraud, lack of privacy, and risk national security. In Australia, the cost of identity theft alone was estimated to be around $1.1 billion during 2001-2002 according to some 2003 SIRCA Research.

The digital identity information in NG wireless networks will be more complex because it has to cater to a number of mobility scenarios, access networks, and services. User identity could include a combination of names, unique user identifiers, terminal identifiers, addresses, user credentials, SLA parameters, personal profiles, and so forth. The digital identity information has to be exchanged between various entities in the networks for the purpose of authentication, authorisation, personalised online configuration, access control, accountability, and so forth. IdM in NG wireless networks is expected to provide a mechanism for controlling multiple robust identities in an electronic world, which is a crucial issue in developing the next generation of distributed services (Buell & Sandhu, 2003).

Let us have a look at a typical access scenario in traditional networks (shown in Figure 1). In these networks, one organisation is often isolated from another since each organisation is running and providing its services independently. Each customer has a number of identity credentials and each credential can only be used to access services from one subscribed organisation.

An expected access scenario in NG wireless networks is illustrated as Figure 2. The NG wire-

Figure 1. Typical access scenario in tradition networks

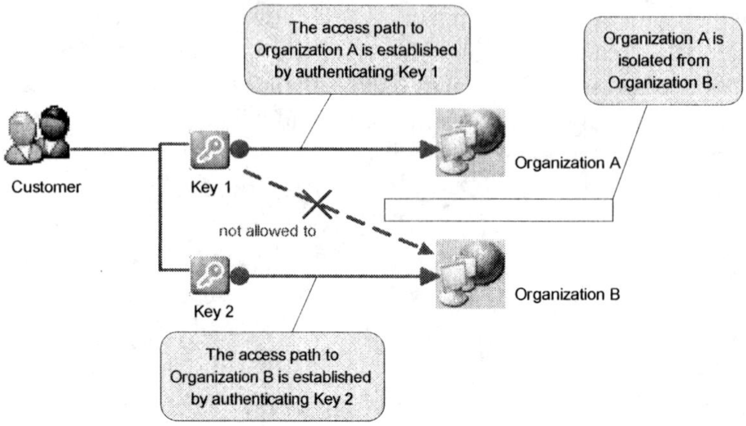

less network subscriber is expected to use the same credential to access multiple organisations. Without a well designed IdM solution, it will not be possible to cater to the following: (1) accessing the subscribed organisations frequently, (2) increased frequency of handoff between multiple organisations in NG wireless networks, and (3) mutual authentication between subscriber and service provider, or between various service providers. A security breach on any component of the NG wireless networks will result in more severe consequences for all the other business partners. Therefore, in order to maintain a similar level of trust, reliability and profitability for the NG wireless networks, integrated IdM measures in NG wireless networks must be taken.

Benefits in NG Wireless Networks

A carefully researched IdM framework for NG wireless networks has a number of benefits for NG wireless networks users, operators, and service providers.

1. User experience is often improved as users can ubiquitously access services and applications of their choice over a number of service providers without going through separate logins and avoiding the need to remember multiple usernames and passwords or use multiple tokens.
2. Service delivery can be improved, for example, the time required to get new subscriber access is reduced.
3. It supports flexible user requirements and personalisation.
4. As with enterprise networks, there are numerous benefits such as reduction in the cost of new service launch, operation and maintenance (O&M) and increased return on investment (ROI) for NG wireless network operators and service providers.
5. IdM is expected to support distributed network architectures where entities communicate through open but secure interfaces.
6. It is necessary for seamless user mobility across networks and terminals.
7. A carefully researched and implemented IdM solution improves the security of the NG wireless networks and the user confidence in the use of the services.
8. IdM will assist in the efficient implementation of current and new legal and compliance initiatives about user data, behaviour and privacy.
9. IdM is expected to support number and service portability of users in an NG wireless network environment.

However, introducing an IdM solution can bring new forms of security issues and threats. As you

Figure 2. Simple access scenarios in NG wireless networks

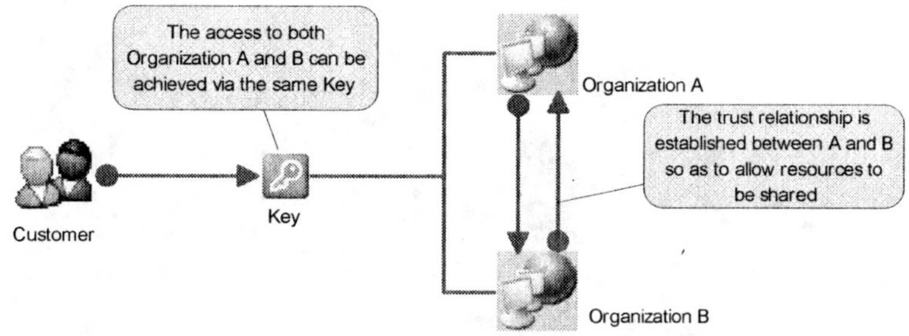

consolidate the identity-related information, you create a new target for security attacks. But the advantage of implementing IdM is that you do not have to worry about protecting disparate solutions. Now you are able to consolidate your defences to one point.

Requirements for IdM in NG Wireless Networks

In this section, an analysis of the requirements for IdM in NG wireless networks is presented. The analysis will be undertaken from three perspectives: user, network, and service. The requirement analysis is expected to cater to the needs of end users, network operators, and service providers in terms of some of NG wireless networks' key functional classifications such as operation, mobility, security, personalisation, and so forth.

Before we get started, a definition of various terms used in NG wireless networks is given:

- **User:** A user refers to a person or entity with authorised access (The Health Insurance Portability and Accountability Act (HIPAA), 2005). In describing NG wireless networks, the term *end user* is often used to refer to a person or entity that uses network resources or services.
- **User terminal:** The user terminal is the device that is used by an end user to access the services provided by the NG wireless networks. It can be a mobile station (MS) or a laptop.
- **Network operator:** Network operator is defined as a legal entity that operates, deploys, and maintains network infrastructure. In NG wireless networks, the networks provided by network operator become the intermediary broker between services and subscribers.
- **Service:** Besides the traditional legacy services, like telephony voice and data, the NG wireless networks can also offer new value-added services to accommodate increasing multimedia demands, for example, video conferencing.
- **Service provider:** The services in NG wireless networks can be provided by different service providers using a single network platform or separate network platforms offered by a network operator.

End User Requirements

Unique Identity for User and Terminal

A unique universal identity will have to be assigned to each individual user of the NG wireless networks and to each user terminal that a user may use to access services of the NG wireless networks. Examples of such identity in Global System for Mobile Communications (GSM)/Universal Mobile Telecommunications System (UMTS) networks include the International Mobile Subscriber Identity (IMSI) and International Mobile Equipment Identity (IMEI). Users should have a single identity regardless of the access technology or network being used.

Figure 3. An overview of IdM requirements and NG wireless networks

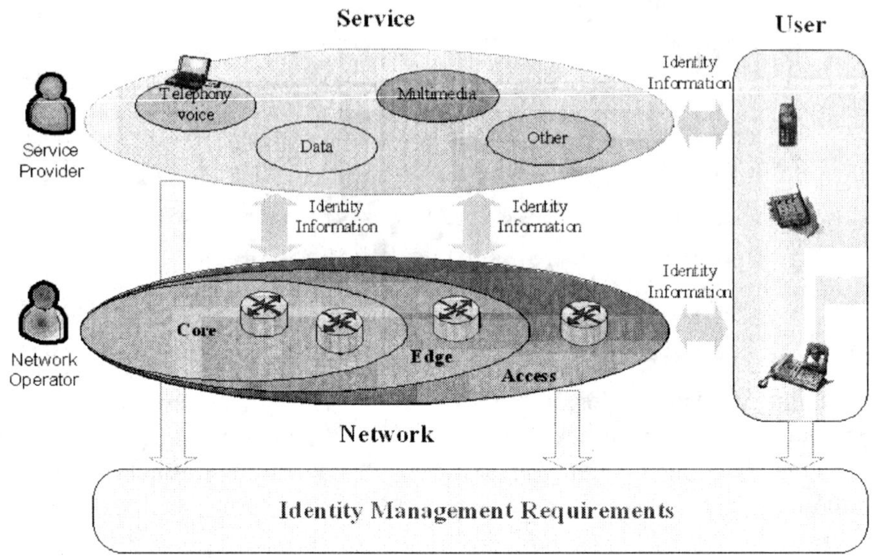

The user identity must possess sufficient features that enable it to be used in a variety of end user terminals (computer, mobile phone, landline phone). Additionally, the unique identity may be required to be compatible across several IdM systems.

Storage of User Information

User identity information may be stored in many locations: user card, home network, visited network, service providers, and so forth. Sometimes, the stored user information can be used as a credential for fast authentication, for example, HTTP cookies are adopted to facilitate quick access to protected Web sites. However, such kind of convenience can have a security risk as the security at user end is more likely to be compromised. NG wireless networks designers have to carefully decide how much information needs to be securely stored at user end. Any identity-related information stored at the user end has to be secure.

Exchange of User Identity

The unique identity allocated to a user should be treated confidentially. Sometimes, it is a risk to transmit the real identity of a user through radio or other public transmission mediums, like the Internet, or exchange it directly with unauthorised parties. Special measures must be taken to ensure that user identity is not disclosed during the exchanging process. One possibility to overcome this problem is to use a temporary user identity that is derived from the unique user identity and is valid for a fixed period of time. Once the validity of the temporary identifier is expired, a new temporary identity is generated. This way the real identity of a user is never compromised.

Self-Service

Self-service is the ability of a user to actively manage part of his or her records without requiring the intervention of help desk or support staff (Reed, 2002). This is an important requirement in all IdM systems. All NG wireless networks users should be able to securely manage some of their own identity information such as changing passwords, subscription status, choosing their mobility status, changing roaming authorisation, modifying user profiles, enabling location based services, and so forth. Users should also be able to modify content filtering options for upstream and downstream traffic.

Users should be able to view their up-to-date billing records and service usage patterns. To increase trust, users should be able to view their self-service activity journal, which displays all the self-service activities performed by a user.

An IdM system should be able to cater to situations where a user wants to delegate self-service privileges to another user such as maintaining accounts of family members.

Single Sign-On

An important user requirement of NG wireless networks is single sign-on. This means that once a user is authenticated, the user should have access to the entirety of their subscribed services without having to repeat the authentication process for each subscribed service.

Security and Privacy

To increase security, users should be able to choose end-to-end data encryption. Unauthorised users should not be allowed to access, view, or modify identity information.

With the growing awareness of privacy and the wish to protect it, users would be looking for more control over their privacy, in particular, what information is known about them and by whom. With an effective IdM system, a user should be able to exert some control as to how much identity data they want to release (which may consist of approval for sending some particular identity attributes) as well as being able to retrieve data concerning the location of their identity data and who is able to currently access it.

Users should also be able to stay anonymous while accessing some network services such as network time protocol (NTP).

Access Network Selection

NG wireless networks users should be able to choose between access networks based on a number of factors such as bandwidth, quality of services, cost, location, and so forth. The user should be able to move between the different access technologies with minimum configuration change and get access consistently to their services according to their user profiles.

Mobility

Mobility across heterogeneous environments requires service adaptation for terminal mobility as well as personal mobility (France Telecom, 2002). In the event of service difficulty during mobility, users should receive user friendly notification with choices of actions to restore the service without the need to contact support staff.

Another related implication is that a user, who is changing access networks during a session, should be able to continue to access the same service without repeated authentication. For example, a mobile user should be continuously attached to a network when there is a handover from a UMTS network to a wireless LAN (WLAN).

Network Operator Requirements

In the NG wireless networks, network operator will be responsible for maintaining and managing network infrastructure. In the ITU's general reference model for NG networks (ITU-T, 2004), network operator will be responsible for taking care of management plane, control plane, and user plane in the transport layer.

Interface to Other Network Operators

Because of the mobility of users, it is difficult for a single network operator to cover a vast geographical area. Thus national and global roaming among multiple network operators is needed in NG wireless networks. In order to support roaming between NG wireless networks, identities of users and networks need to be authenticated before access to resources is granted through a visited network. It may be cost effective for a roaming user to access services in the visited network than in the home network. A network operator should give choices to roaming users on the selection of services.

Interface to Trusted Third Party

It is possible that all of the IdM is performed by a third party that is different from the network operator or service provider. This third party will issue, authenticate, and control NG wireless networks user identities. A secure interface has to be provided between the NG wireless networks and the trusted third party.

Identity Requirements

The NG wireless networks operator should be able to maintain a unique identity for each user, terminal, network element, location area, and so forth, regardless of service and technologies used.

If the user is using faulty or dubious terminal equipment, it should be possible to bar services to the user.

The digital identity stored in a network should cater to various types of user identity information and data structures.

As in enterprise networks, proper implementation of account lifecycle management is required, that is, administrators should be able to manage the state of a user account for the complete span of that account. Even if an account is deleted or disabled, an audit history of the account should be maintained.

If necessary, the network operator should be able to remove self-service privilege of some users.

The IdM system should support open standards in order to interact with multi-vendor terminals and network elements. It should be compatible with existing legacy systems and be able to adapt to emerging technologies, methods, and procedures.

Scalability and Performance

The IdM system should be able to store, retrieve, and exchange billions of identity information in a highly seamless, scalable, quick, and efficient manner to facilitate multiple real-time service requests from users.

It should achieve a high level of availability by incorporating fault-tolerant redundant system implementation. Furthermore, it should implement

Figure 4. Network operator's position in the NG wireless networks

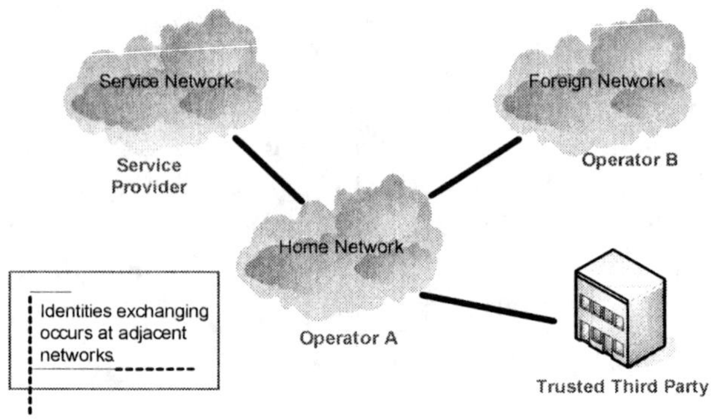

geographically distributed IdM servers in order to increase performance efficiency by load sharing and providing high availability. It should also maintain integrity and consistency of identity data across distributed identity information stores.

Mobility Management

NG wireless networks should be able to cater to the mobility requirements of users. This could include personal and/or terminal mobility, roaming, or nomadism. Mobility management may require a combination of identification, authentication, access control, location management, IP address allocation and management, user environment management, and user profile management functions. The network should cater for both foreign network IP address and home network IP address allocation scheme.

Security

Security requirements for NG wireless network operators should cover privacy, confidentiality, integrity, authenticity, non-repudiation, availability, intrusion detection, and maintenance of audit records as described later on.

Users and terminals should be reliably authenticated by the network operator using a nominated set of authentication credentials such as passwords, smart cards, biometrics, and other industry standard methods. All the identity data should be kept in a very secure and scalable manner. Unauthorised access to identity data should be prevented.

Intrusion detection is required to detect and prevent security breaches with the network operator. This can also be done to minimise the fraudulent use of resources in a network.

Network administrators should be granted different levels of access according to their authority within the organisation. For accountability and security reasons, consistent and reliable audit records of administrative activities must be kept.

In order to apply user and data security such as confidentiality, integrity, and authenticity, the IdM system should securely store and exchange relevant encryption keys.

Billing

Up-to-date, accurate, and detailed billing information should be maintained by the network operator. When there is more than one source sending billing data, the network operator has to consolidate this information from various sources.

Furthermore, when a subscriber is roaming in a foreign network, charging records from that foreign network has to be authenticated to prevent fraudulent usage of services.

The network operator should be able to support a number of charging mechanisms such as charging based on usage, access networks, time, geographical area, and so forth. All of these different charging mechanisms should be compatible with the IdM system.

Service Provider Requirements

A user may require services from a number of service providers. In this scenario, the home operator and the service provider(s) should support secure access and exchange of user identity and billing information.

The identity of each user should be uniquely and reliably identified by a service provider. The service providers may have to rely on third party IdM providers where the user has already established an account.

The IdM and related systems should support open standards with choices of number of technologies in order to interoperate with other entities.

Interface to Other Service Providers

Users may subscribe to the services offered by different service providers. Thus, the interoperability among service providers is important. User identity information may be exchanged between a group of service providers in order to improve "transparent user experience." This also requires trust to be established between these service providers.

Interface to Network Operator

A well-defined, open interface needs to be provided to the network operator at the service provider end. This would give service provider the necessary authentication, authorization and accounting (AAA) to access network resources offered by network operator.

Interface to Trusted Third Party

An interface to trusted third party would give service provider an opportunity to use external AAA services. By doing so, the complexity of implementation of services would be reduced. The authentication of users can be centralised.

Mobility Management

Some services require information about the current location and connectivity of subscribers. These are referred to as location-dependent or location-aware services. To provide such services to end users, a service provider must be able to access mobility-management-related information maintained by network operators. Subscribers have to consent to the release of this sensitive private information to service providers. Furthermore, when there are updates to location or mobility management data in the network operator, the update have to be passed to the subscriber.

Security

As one of the main holders of identity data about subscribers, service provider would have to exercise extra vigilance in ensuring that the data that they store is kept secure.

Additionally, in order to ensure a high degree of mobility and choice to the end user, this identity information must be able to be easily and securely transferred between different service providers depending on the end user's current choice.

Billing

A number of requirements pertaining to billing for network operators are equally applicable to service providers. Billing records of the user should be dynamically generated according to the usage.

Regulatory Requirements

It is expected that the NG wireless networks should support open standards and choices among a number of technologies to promote competition and flexibility. Thus, any IdM solution that favours a particular standard or technology can be deemed anti-competitive.

Privacy is an important issue that has to be addressed directly by IdM products and solutions. There are increasing concerns about the fact that enterprises, e-commerce sites, governments, and third parties can access and correlate people's identity information, sell this information, or misuse it. Current laws and legislation only partially address this problem. Despite the fact that many efforts have been made at the legislation level, there are still a lot of problems that have to be addressed. Furthermore, privacy laws can differ quite substantially depending on national and geographical aspects. All of the regulatory requirements pertaining to privacy and confidentiality of subscribers' personal information should be built into the IdM solution in NG wireless networks.

Identity subjects have little control over the management of their identity information. It is very hard (if not impossible) for the subjects of identity information to define their own privacy policies (or delegate this task to trusted third parties), check for their enforcement, track in real-time the dissemination and usage of their personal information be alerted when there are attempts to use or misuse it, and so forth. Because of emerging data protection laws, new legislation and the need of service providers to simplify the overall management, there is a tendency towards the delegation to users of the authoring of their identity profiles.

Figure 5. Service provider's position in the NG wireless networks

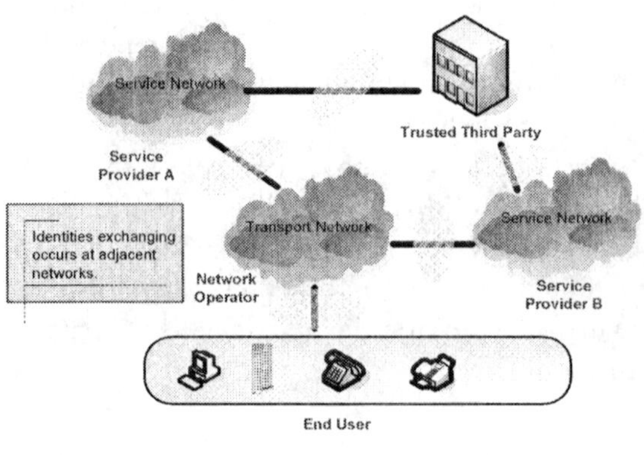

Legal Requirements

Privacy and confidentiality of subscribers' personal information and prevention of unauthorised access should be maintained at all times by network operators and service providers and any third party organisations involved in the NG wireless networks space.

If information has to be shared, subscribers should have a choice of the types of subscriber information that can be shared with various third parties. Reliable audit records of administrative and user activity should be kept, which could be retrieved and submitted to courts and other entities to meet legal requirements.

Legal interception of subscriber data should be possible. One of the new requirements for telecommunication network operators is to collect and pass on real-time transactions of target subscribers to law enforcement authorities (Council of Europe,

2001). Legal interception of subscriber data should be possible whichever network or service a subscriber is using.

REFERENCES

Buell, D. A., & Sandhu, R. (2003). Identity management. *IEEE Internet Computing, 7*(6), 26-28.

Cisco Systems. (2005). *Trust and identity management solutions.* Retrieved 2005, from http://www.cisco.com/en/US/netsol/ns463/networking_solutions_package.html

Clercq, J. D., & Rouault, J. (2004, June). *An introduction to identity management.* Retrieved 2005, from http://devresource.hp.com/drc/resources/idmgt_intro/index.jsp

Council of Europe. (2001). ETS No. 185—Convention on cybercrime. *Article 21, European Treaty Series (ETS).* Retrieved 2005, from http://conventions.coe.int/Treaty/en/Treaties/Html/185.htm

Courion. (2005). *Courion products overview: Enterprise provisioning.* Retrieved 2005, from http://www.courion.com/products/benefits.asp?Node=SuiteOverview_Benefits

DIGITALIDWORLD. (2005). *What is digital identity?* Retrieved July 2005, from http://www.digitalidworld.com/local.php?op=view&file=aboutdid_detail

France Telecom. (2002). Inter-network mobility requirements considerations in NGN environments. *Study Group 13—Delayed Contribution 322, Telecommunication Standardization Sector (WP 2/13)* Retrieved 2004.

The Health Insurance Portability and Accountability Act (HIPAA). (2005). *Glossary of HIPAA terms.* Retrieved 2005, from http://hipaa.wustl.edu/Glossary.htm

International Telecommunication Union-Telecommunication Standardization Sector (ITU-T). (2004). *NGN-related recommendations.* Study Group 13 NGN-WD-87.

Locke, M., & McCarthy, M. (2002). *Realising the business benefits of identity management*: FUJITSU SERVICES.

Pato, J., & Rouault, J. (2003, August). *Identity management: The drive to federation.* Retrieved 2006, from http://devresource.hp.com/drc/technical_white_papers/id_mgmt/index.jsp

Reed, A. (2002). *The definitive guide to identity management (e-Book).* Retrieved from http://www.rainbow.com/insights/ebooks.asp

Titterington, G. (2005, July). *Identity management: Time for action.* Ovum's Research Store.

KEY TERMS

Access Control: Access control is used to determine what a user can or cannot do in a particular context.

Auditing and Reporting: Auditing and reporting involves the creation and keeping of records, whether for business reasons (e.g., customer transactions), but also for providing a "trail" in the event that the system is compromised or found faulty.

Authentication: Authentication is the process by which an entity provides its identity to another party, for example, by showing photo ID to a bank teller or entering a password on a computer system.

Authorization: Authorisation is the process of granting access to a service or information based on a user's role in an organisation.

Context: Context can refer to the type of transaction or organisation that the entity is identifying itself as well as the manner that the transaction is made.

Digital Identity: Digital identity is the means that an entity can use to identify themselves in a digital world (i.e., data that can be transferred digitally, over a network, file, etc.).

Identity: The identity of an individual is the set of information known about that person.

Network Operator: Network operator is defined as a legal entity that operates, deploys, and maintains network infrastructure.

Profile: A profile consists of data needed to provide services to users once their identity has been verified.

User: A user refers to a person or entity with authorised access.

User Terminal: The user terminal is the device that is used by an end user to access the services provided by the NG wireless networks.

Chapter V
Wireless Wardriving

Luca Caviglione
Institute of Intelligent Systems for Automation (ISSIA)—Genoa Branch, Italian National Research Council, Italy

ABSTRACT

Wardriving is the practice of searching wireless networks while moving. Originally, it was explicitly referred to as people searching for wireless signals by driving in vans, but nowadays it generally identifies people searching for wireless accesses while moving. Despite the legal aspects, this "quest for connectivity" spawned a quite productive underground community, which developed powerful tools, relying on cheap and standard hardware. The knowledge of these tools and techniques has many useful aspects. Firstly, when designing the security framework of a wireless LAN (WLAN), the knowledge of the vulnerabilities exploited at the basis of wardriving is a mandatory step, both to avoid penetration issues and to detect whether attacks are ongoing. Secondly, hardware and software developers can design better devices by avoiding common mistakes and using an effective suite for conducting security tests. Lastly, people who are interested in gaining a deeper understanding of wireless standards can conduct experiments by simply downloading software running on cost effective hardware. With such preamble, in this chapter we will analyze the theory, the techniques, and the tools commonly used for wardriving IEEE 802.11-based wireless networks.

THE (ART OF) WARDRIVING

Owing to the absence of physicals barriers, the wireless medium, and consequently wireless (WLANs) are accessible in a seamless manner. Thus, checking for the presence of some kind of wireless connectivity is quite a natural instinct; it is sufficient to enable the wireless interface and wait. This action is a very basic form of wardriving, a term originally coined by Shipley (2000) to refer to the activity of "driving around, looking for wireless networks." This activity rapidly evolved, and nowadays it implies three basic steps: (1) finding a WLAN, (2) defining precisely its geographical coordinates by using GPS devices, and (3) publishing the location in specialized Web sites to enrich the wardriving community.

With the increasing diffusion of WLANs, especially those based on the cost effective IEEE 802.11 technologies, searching for wireless signals is a quite amusing and cheap activity. However, the IEEE 802.11 family originally relied (and still relies) on weak security mechanisms. In addition, many users unconsciously operate their wireless networks without activating any confidentiality, integrity, and availability (CIA) mechanisms: opportunity makes the thief. Then, wardriving becomes a less noble hobby, since many wardrivers try also to gain access to the discovered networks; many of them are only interested in cracking the network, while a portion will steal someone else's bandwidth. In this perspective, another basic step has been introduced: (4) trying to gain access to the WLAN.

It is also interesting that wardriving is becoming part of the urban culture. For instance, it spawned a strange fashion called *warchalking*, that is, *the drawing of symbols in public places to advertise wireless networks*, as defined by Matt Jones (as cited in Pollard, 2000).

Then, why is it important to know about wardriving?

Firstly, because you must become conscious that an active WLAN can trigger "recreational activities," even if it is solely employed to share a printer. Secondly, the coordinated effort of many people highlighted several security flaws in the IEEE 802.11 standards and produced effective tools to test (well, actually, to compromise) the security of access points (APs). Thirdly, while performing their "raids," *wardrivers* discovered flaws in the devices; consequently, this is a valuable knowledge that could be used to avoid further errors. Lastly, trying to be a wardriver is an instructive activity that will help to better understand WLANs technologies, develop your own auditing tools and procedures, and prevent, or at least, recognize attacks.

HARDWARE AND SOFTWARE REQUIREMENTS

In the basic form of searching for a WLAN, the act of wardriving could be simply performed by having a device equipped with an IEEE 802.11 air interface. Then, one can use a standard laptop, a wireless-capable console, or a handheld device. However, the typical gear consists of a laptop and a GPS device (even if not strictly necessary). Nevertheless, many wardrivers do prefer a Personal Computer Memory Card International Association (PCMCIA) wireless card that is capable to connect with an external antenna to sense a wider area. With this basic setup you should be able to enable the wireless interface and start scanning the air. But, in order to conduct more sophisticated actions, a deeper understanding of aspects related to hardware and software should be gained. A detailed breakdown follows.

Wireless Interfaces

Each model of wireless interface differs in some way. Regardless of different power consumption, better antennas, and so on, two major aspects must be taken into account: the chipset and the availability of ad hoc drivers. The chipset roughly represents the soul of a wireless interface and it is mostly responsible of its capability. For instance, some chipsets do not allow assembling ad hoc frames, preventing from exploiting particular attacks. The reasons are different: the chipset could lack the logic to deal with raw packets or its specification is not known, discouraging tool developers to exploit such functionalities. At the time of this writing, cards based on the Prism chipset are the most studied and documented, resulting in a variety of pre-made tools for preparing packets.[1] Lastly, being the interfaces engineered for providing connectivity and not such kind of tasks, manufacturers often change the internal chipset, even if maintaining the model or the brand name. This is why not all wireless cards are the same, and you should check their specifications carefully if you plan to use them for wardriving.

Device Drivers and Scanning

Device drivers provide the basic bridge between the user software and the hardware. Having a flexible device driver is mandatory to reach the soul of your interface. The best device drivers for wardriving are available for the aforementioned chipset, and for Unix systems. In addition, owing to its open source nature, Linux has the best available drivers.

The importance of drivers becomes evident when you scan the air for a network. About the totality of the bundled drivers does not allow to perform the so called *passive scan*. Passive scan implies that your interface operates in *passive mode*, often called *radio frequency monitoring* (rfmon) mode. While you operate in rfmon, you can scan APs and remain undetectable, since your card does not send any probe packets.

Conversely, when acting in *active mode*, which is the standard configuration, as soon as you start looking for an AP, you will be revealed. The ability of switching from active to passive mode and vice versa is provided by the drivers. Many drivers do not provide this functionality, while others have this functionality hidden and must be reverse engineered.

For the most popular chipsets, alternative drivers that allow the user to put the card in rfmon are available. If you plan to do undercover works, you should check the driver availability.

However, the active mode is faster than the passive mode. While operating in passive mode, the average time needed for scanning a channel is about 50 ms. Obviously, multiple channels scan requires $n \bullet 50$ ms. Conversely, when performing scanning operations in active mode, the needed time is lower. In fact, the operations required are: transmitting a probe request + waiting for a DCF IFS interval + transmitting a probe response. The overall time needed per channel is roughly equal to 0.45 ms. Again, scanning *n* channels increases the needed time accordingly (Ferro, 2005).

An Example of Driver Hacking

As said, the ability of enabling an air interface in rfmon could be available in the driver, but not documented. This is the case of the driver for the AirPort Extreme wireless adapters bundled with MacOS X. This example is introduced for didactical purposes, stressing how a simple "hack" can transform a partially closed platform in an excellent wardriving configuration.

In a nutshell, OSX drivers are implemented via *kernel extensions* (kexts) that are similar to Linux's modules. Every kext is bundled with a kind of configuration file called *Info.plist*. The *Info.plist* is a XML file containing a dictionary that describes the properties of the belonging kext. The "hack" consists in a simple operation (i.e., changing a string) but it took time to discover.

Firstly, the proper *Info.plist* must be located. In a console type:

```
Mud:Luca$ cd /System/Library/Extensions/AppleAirPort2.
kext/Contents/
```

Hence, you can see the content of the *kext* upon simply typing:

```
Mud:Luca$ ls
Info.plist          MacOS          version.plist
```

Then, it is possible to modify the *Info.plist*

```
Mud:Luca$ vim Info.plist
```

The key responsible of enabling the *rfmon* follows, in boldface:

```
<key>IOKitPersonalities</key>
    <dict>
        <key>Broadcom PCI</key>
        <dict>
            <key>APMonitorMode</key>
            <false/>
```

Switching the dictionary entry **<false/>** to **<true/>** enables the AirPort Extreme card in rfmon.

However, such a task could be performed programmatically.

This is the approach taken in KisMAC, which is popular among wardrivers. As an example, in the following, the Objective-C code snippet checking whether or not the wireless interface is rfmon is depicted in Snippet 1.

Roughly, the steps presented in Snippet 1 allow the user to: (1) obtain a handler to the proper *Info.plist* file; (2) prepare a dictionary for parsing the *Info.plist*; and (3) check if the **<APMonitorMode>** key is **<false/>** or **<true/>**.

The Operating System and Other Matters

Needles to say, the operating system (OS) plays a role. For instance, when processing data for

Snippet 1. How to programmatically retrieve if an AirPort card is configured in rfmon

```
fileData = [NSData dataWithContentsOfFile:
@"/System/Library/Extensions/AppleAirPort2.kext/Contents/Info.plist"];          1

dict = [NSPropertyListSerialization propertyListFromData:fileData
mutabilityOption:kCFPropertyListImmutable format:NULL errorDescription   :Nil]; 2

if ([[dict valueForKeyPath:@"IOKitPersonalities.BroadcomPCI.APMonitorMode"]     3
boolValue]) return YES;
```

bruteforcing an encrypted flow, a good symmetric multi process (SMP) support is a must (as well as a good multi-threaded implementation).

In addition, many APs can reject data from unrecognized MAC addresses: for this reason, having an OS that allows the user to change the MAC address of active interfaces is important. Lastly, many tools only run on *nix operating system. However, the traffic collection phase could be decoupled by the processing, hence allowing the user to collect data on a machine and process it on another. As a consequence, simple devices (e.g., with low computational power) could be employed to collect data and discover APs (e.g., PDAs and portable gaming devices), while a standard PC could be used for processing the collected traffic.

XOR Arithmetic and CRC32 in a Nutshell

In order to understand the security mechanisms, and possible attacks, a little remark about eXclusive OR (XOR) arithmetic and the properties of CRC_{32} functions, employed for data checking, are presented. Basically, the XOR operator respects the properties presented in Table 1.

Table 1. Basic XOR arithmetic (⊕ represents the XOR operator)

Operation	Result
$0 \oplus 0$	0
$1 \oplus 0$	1
$1 \oplus 1$	0
$(A \oplus B) \oplus A$	B
$(A \oplus B) \oplus B$	A

Concerning the CRC_{32}, it is employed to check data and to assure integrity. It has not the cryptographic strength of other hashing algorithms, such as the MD5 and the SHA1 (Schneier, 1996). The CRC_{32} employed in the wired equivalent privacy (WEP) algorithm has two major properties, as presented in Table 2.

ABOUT THE SECURITY OF IEEE 802.11

The IEEE 802.11 security framework has changed during the years: from the flawed WEP, to the wireless protected access (WPA) introduced by the Wi-Fi alliance in late 2002. However, since mid-2004, the IEEE 802.11i Working Group (WG) introduced a framework based on the 802.1X and the extensible authentication protocol (EAP), to bring the wireless security to the next level; such effort is known as WPA2.

Even if highly criticized, the security mechanisms proposed by different WGs have developed having in mind different operative contexts. For instance, the WEP (as the name suggests) has been developed to prevent simple connection attempts, while WPA has been developed to offer an adequate resistance to well-planned attacks. Currently, an average wardriver can: surely connect to an unprotected AP, spend 10 minutes to 1 hour to break the WEP, and crack a WPA-protected AP in some of its weak variants and well-suited circumstances. In order to understand the common technique employed by wardrivers, the commonly adopted security countermeasure will be briefly explained.

Table 2. Properties of the CRC_{32} function employed in the WEP

Property	Application
Linearity	$CRC_{32}(A \oplus B) = CRC_{32}(A) \oplus CRC_{32}(B)$
Independence of WEP Key	It is possible to flip bits without being recognized by the WEP

No Encryption

Many wireless networks operate without any encryption, and "security" is delegated to other mechanisms. It must be underlined that the lack of encryption allows everyone to listen to the channel and analyze the traffic (that flows in clear form if no security mechanisms at higher layers are adopted). Hence, for these users, "security" is solely a synonym of "preventing" the unauthorized usage. The most adopted methods are: MAC address filtering and hiding the service set identifier (SSID). They will be briefly explained, highlighting why they cannot be perceived as secure countermeasures.

MAC Address Filtering

MAC address filtering is a basic technique implemented in about the totality of the commercially available APs. Basically, before authorizing an association, the AP checks the allowed MAC addresses in a white list. The rationale under the approach relies on the uniqueness of the MAC address. As a matter of fact, this technique only discourages the occasional wardriver, but it is quite useless. In addition, it could be used jointly with WEP or WPA, in order to have another barrier if an attacker cracks the encryption mechanism. However, frame headers are never encrypted; hence, it is a simple task to retrieve some valid MAC addresses (e.g., by simply monitoring a channel). Then, there are a variety of tools for changing the MAC address of a wireless interface, performing the so-called MAC-spoofing.

Hiding the SSID

In order to advertise a network, it is possible to broadcast a special identifier called SSID. The standard allows the user to embed the SSID within beacons sent by APs or wireless routers. In order to "join" a WLAN, you must know its SSID. As a consequence, many users/administrators disable the SSID broadcasting, to prevent unauthorized accesses. However, this measure only prevents a minority of attempts. In fact, there are several tools and techniques that allow a user to uncloak a hidden SSID. A thorough discussion about such tools will be presented in the following sections, but we outline the basic procedures here. Specifically, it is possible to: (1) recover information about SSID contained in frames sent by other stations in the network; for instance, the SSID is contained in association request packets; and (2) if such frames are not available, it is possible to spoof the IEEE 802.11 de-authentication frames of target clients. This causes a client to start a new authentication and association round with the AP, providing the needed frames.

WEP Encryption

The scientific literature, as well as daily practice, commonly suggest that the WEP is a highly insecure encryption mechanism. No matter about the skill of the wardriver, or the quality of the implementation in the AP: a WEP-secured network can be cracked in a period varying from 5 minutes to 1 hour. Moreover, many tools implement automated procedures; thus, cracking the WEP is as simple as pressing a keyboard shortcut.

Understanding the Effective Strength of the WEP

Often, marketing collides with engineering: this is the case of the WEP. In order to understand the effective strength of the WEP, as well as its weak

Figure 1. The message in clear form to be encrypted with WEP

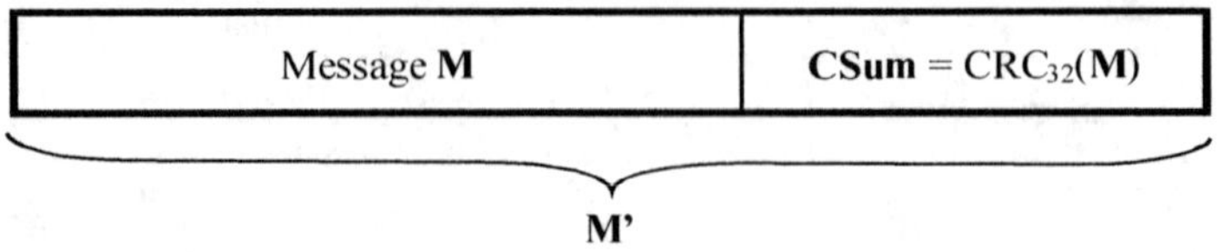

Figure 2. The seed used to encrypt packets with WEP

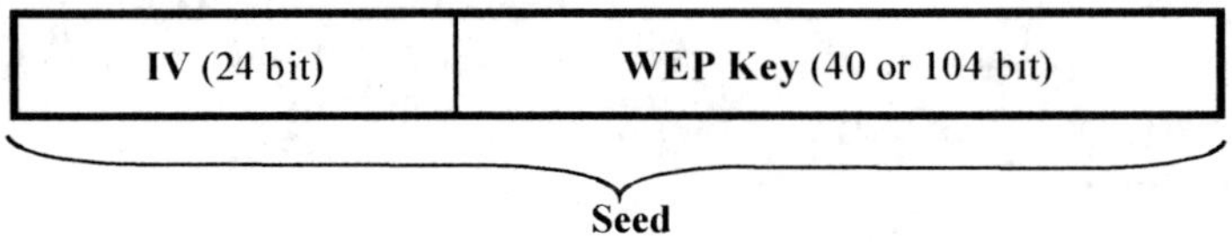

points, let us summarize its basic functionalities. The WEP performs the encryption per packet; let a given packet **M** represent a message in clear form to be sent. Hence, the following steps happen:

A 32-bit cyclic redundancy check (CRC) algorithm is applied to **M** in order to produce a checksum. Then: **CSum** = CRC_{32}(**M**). Basically, a CRC is introduced to assure message integrity. However, the use of CRC-like codes in this kind of environment has been proven to be very dangerous.

Let us define as **M'** the message actually processed by the WEP algorithm, hence to be really sent over the channel. **M'** is depicted in Figure 1.

Then **M'** is encrypted by using the RC4 algorithm, that relies on a stream cipher approach. Thus, the actual **Seed** used by the WEP is the combination of a 24-bit initialization vector (IV) and the WEP key, as depicted in Figure 2.

Referring to Figure 2, two different WEP keys are available: 40-bit long keys adopted in the *standard implementation*, or 104-bit long keys adopted in the *extended implementation*, which has been introduced to prevent brute force attacks. Here comes the marketing: a *"64 bit WEP secured network" actually relies only on 40-bit long keys,* since 24-bits represent the IVs. For the same reason, a *"128-bit WEP secured network" only relies on 104-bit keys.*

Splitting the **Seed** in two sub parts (the **IV** and the **WEP key**) is one of the major flaws of the procedure. However, the reason is rooted both in the nature of the RC4 and wireless channels. The RC4 has been used in the WEP since it is widely adopted and well studied. But its application over wireless channels poses some drawbacks. In fact, wireless channels frequently drop packets, thus maintaining a proper synchronization in the stream to allow the decryption operations is a challenging task. Consequently, to overcome the possibility of packet loss and stream de-syncing, each encrypted packet is sent along with the **IV** that generates its keystream. This represents another weakness in the algorithm, since it allows a wardriver (attacker) to seamlessly collect IVs.

Concluding, the ciphered text **C** is provided by:

$$\mathbf{C} = \mathbf{M'} \oplus \mathbf{RC4}_{\mathbf{Key}}$$

where, $\oplus$ represents the XOR operator, and $\mathbf{RC4}_{\mathbf{Key}}$ is the keystream generated by the RC4 algorithm by feeding it with **Seed**. Figure 3 depicts a WEP-encrypted packet that could be collected and exploited by a wardriver. Needles to say, IVs convey precious information, and in the following, we show how standard tools can exploit this.

Figure 3. A WEP encrypted frame. Notice that the IV is sent as cleartext.

WPA Encryption

As previously explained, WPA encryption schemes have been introduced to overcome the drawbacks in the WEP. The WPA exists in two different flavors: 802.1x jointly used with the temporal key integrity protocol (TKIP) that is intended for enterprises, and the less secure pre-shared key (PSK), possibly jointly used with the aforementioned TKIP. The 802.1x + TKIP is a quite secure protocol, and difficult to crack by *wardrivers*, but the PSK version still has some flaws. The WPA has been proved to be quite secure; thus, we will omit its details in this chapter.

Some Considerations about Layer 1 Security

All the aforementioned encryption mechanisms have been introduced to cope with the simplicity of sensing a WLAN, and consequently, to collect data. Then, it is hard to implement OSI-L1 security mechanisms, as it can be possible in wired networks. However, a basic countermeasure could be exploited: adjusting the wireless power. Conversely, wardrivers can adopt high gain antennas to intercept distant APs. Those concepts will be further discussed.

Wireless Power

Many APs allow changing the power employed for transmitting data. However, many users keep the default values or use more power than required. Despite the waste of energy, this raises also some security risks. For instance, if there is the need of covering a conference room, it is harmful to irradiate more than required power, resulting in the chance of detecting (and using) the WLAN also from the outdoor. This is at the basis of wardriving. In fact, wardrivers will seldom enter private areas; rather, they will station in streets and public places, capitalizing the unsolicited wireless coverage. Then, as a rule of thumb for protection, it could be useful to irradiate only the required power: no more, no less.

Antenna Gain

As said, wardrivers often utilize high gain antennas to reach distant networks. Thus, reducing the transmission power of the APs might not be enough.

Commonly, there are several techniques to replace the standard antenna available at the network interface, but they are out of the scope of this work. The simplest technique is to use an external PCMCIA wireless card equipped with a connector for an external antenna. One of the most interesting accessories is the *pigtail*. The pigtail is a converter allowing the user to connect high gain antennas with a wireless card, even if the terminal connectors are different (e.g., wireless cards often have MC-Card, MMCX or RP-MMCX connectors).

WEP ATTACKS

As discussed in the *Understanding the effective strength of the WEP* section, WEP offers different alternatives to be attacked and cracked. In this section, we will introduce the most popular attacks, and then we will present some practical examples. Besides, attacks could be roughly grouped in two categories: passive and active. A passive attack solely relies on the traffic collected, while an active attack consists also in injecting some additional traffic in the network. For instance, active attacks are employed to stimulate the traffic to collect if there are not any clients connected to an AP at a given time. The latter techniques will be presented when needed, then in the *Example* section.

Bruteforce Attacks

Every security algorithm is exposed to bruteforce attacks. The key point is if a bruteforce attack is feasible. As said, WEP exists in two variants. Concerning the 40 bit standard implementation, a bruteforce attack could be feasible. Probably, an occasional attacker will have a machine allowing to check 10,000 to 15,000 keys/second; hence, it is not sure that he/she will complete the attack (on an average laptop, 200 days are required). But an organization or a professional attacker can try to successfully bruteforce the WEP in the 40-bit variant. Nevertheless, nowadays there are several software libraries for parallelizing computations, as well as software tools for building clusters (e.g., Beowulf or Mosix for the Linux platform and XGrid for MacOS X). Owing to the availability of the source code of bruteforcing tools, porting them on such frameworks could be possible. Actually, bruteforce is never employed, since it is possible to successfully crack the WEP in simpler and quicker ways.

Conversely, the 104-bit long key available in the WEP extended implementation is immune against bruteforce attacks (with a standard gear, about 10^{19} years are needed).

The Tim Newsham's 21-Bit Attack

Tim Newsham is a well-known security expert and consultant. Among wardrivers he is very popular for inventing the 21-bit attack (Newsham, 2003), allowing to bruteforce some WEP implementations in minutes.

Basically, Newsham noticed that several vendors generate WEP keys from text, in order to make easy-to-use products and cover a wider market range. Usually, the user must insert a *pass phrase*, something like: "*Ken sent me*" and the wizard will automatically generate a WEP key. However, many generators appear to be flawed. He discovered that two steps in the generation process reduce the "strength" of the key; specifically:

1. The ASCII mapping reduces the entropy: usually ASCII strings are mapped to 32 bit value and the XOR operation guarantees four zero bits. In addition, the highest order bit of each character is equal to zero. Then, only seeds from 00:00:00:00 e 7f:7f:7f:7f can occur.
2. The use of Pseudo Random Number Generation (PNRG) reduces the entropy: for each 32bit output, only a portion of the available binary word is considered (e.g., bits 16 through 23). Besides, the generator has the properties of generating bits with different degrees of "randomness." For instance, a bit in position k has a cycle length of 2^k. Then, Newsham noticed that the produced bytes have a cycle length of 2^{24}, thus reflecting in seeds ranging from 00:00:00:00 and ff:ff:ff:ff.

In order to discover the key, it is sufficient to consider seeds ranging from 00:00:00:00 through 00:7f:7f:7f with zero highest order bits, hence reducing the space and only analyzing 2^{21} words. As a consequence, it is possible to bruteforce such flawed implementations in minutes. The most popular implementation of Newsham's 21-bit attack is available in the KisMAC tool. According to KisMAC documentation, Linksys and D-link devices appear, at the moment, the most vulnerable to this attack.

Weak IVs

This attack relies on how the RC4 is used to produce a WEP-encrypted stream. Basically, some IVs can reveal some information about the secret key embedded in the first byte of the keystream. Then it is enough to collect a sufficient number of weak IVs and, if the first byte of the keystream is known, it is possible to retrieve the key.

Regarding the collection of the first byte of the keystream, the IEEE 802 standard gives some useful hints. In fact, IEEE 802.11 frames always begin with the SNAP field, which most of the time is set to 0xAA. Then it is sufficient to collect weak IVs that come in the form of:

(Y+3, 256, X)

where Y is the portion of the key under attack, the second value is 256, since RC4 works on a modulo-256 arithmetic, and X can be any value. Fluhrer, Martin, and Shamir (FMS) have developed an efficient attack available in different tools. However, the core of the attack is out of the scope of this chapter.

As a concluding remark, new devices tend to avoid weak IVs' generation. In fact, hardware developers better engineer their devices, increasing attention to the IVs' generation mechanism.

Keystream Reuse

Suppose to be in the following scenario: two different cleartext messages, $\mathbf{M'_1}$ and $\mathbf{M'_2}$ must be transferred over the channel. Let us assume that both messages share the same keystream. Then:

$$\mathbf{C_1 = M'_1 \oplus RC4_{Key}(Seed)}$$
$$\mathbf{C_2 = M'_2 \oplus RC4_{Key}(Seed)}$$

$\mathbf{C_1}$ and $\mathbf{C_2}$ are the two WEP encrypted messages, and **Seed** is the one employed for the RC4, as depicted in Figure 2 of the *Understanding the effective strength of the WEP* section. Then, it is possible to perform the following operation:

$$\mathbf{C_1 \oplus C_2 = (M'_1 \oplus RC4_{Key}(Seed)) \oplus (M'_2 \oplus RC4_{Key}(Seed)) = M'_1 \oplus M'_2} \qquad (1)$$

As a consequence, knowing $\mathbf{M'_1}$ (or $\mathbf{M'_2}$), allows to recover $\mathbf{M'_2}$ (or $\mathbf{M'_1}$). One might argue that the knowledge of a message $\mathbf{M'_x}$ is a tight hypothesis. However, being messages packets generated by some well-known protocol, it is possible to craft packets and send them via the Internet to a target host on the WLAN. Hence, the AP will encrypt the data for the attacker.

This kind of attack relies on relation (1). However, the operations in (1) are possible since both messages have been encrypted with the same **Seed**. To overcome this, IVs have been introduced, being them the only portion of the **Seed** that varies. Alas, IVs are only 24-bit long, hence it is likely that the same **Seed** will be sent over the network again.

The Oracle

In order to recover a relevant amount of known plaintext, the AP could be used as an *Oracle*, a device that unconsciously encrypts well-crafted packets for the attacker.

Figure 4 depicts the basic operations performed to conduct the attack. This attack is nowadays unlikely, since as explained, there are several faster and simpler ways to crack the WEP. Basically, the attacker exploits an active connection targeting the victim. Then he/she sends (e.g., via General Packet Radio Service [GPRS], or Universal Mobile Telecommunications System [UMTS], or similar) known packets that will be encrypted by the AP before transmission over the WLAN.

It becomes clear that this attack exploits the fact that an AP could be used to connect a network to the Internet without any further protection mechanism (e.g., a firewall or a virtual private network [VPN] support). For completeness, in early days when GPRS was expensive, usually the attack was performed by cooperating with another wardriver, usually at home, with an active Internet connection remotely injecting packets to the AP.

Decryption Dictionary

This kind of technique is no longer employed, and there are not any proofs that it has ever been exploited in its basic form. However, it is interesting that this attack allows (at least theoretically) the user to decrypt all the traffic without knowing the WEP key. Basically, it is sufficient to build a table of the intercepted keystreams. Then, it is possible to compile a table of all the possible values (and also skip the RC4 phase). The drawback, preventing its proficient exploitation, is the space required for this kind of attack. In fact, the encrypted

Figure 4. Scenario when an oracle attack is performed

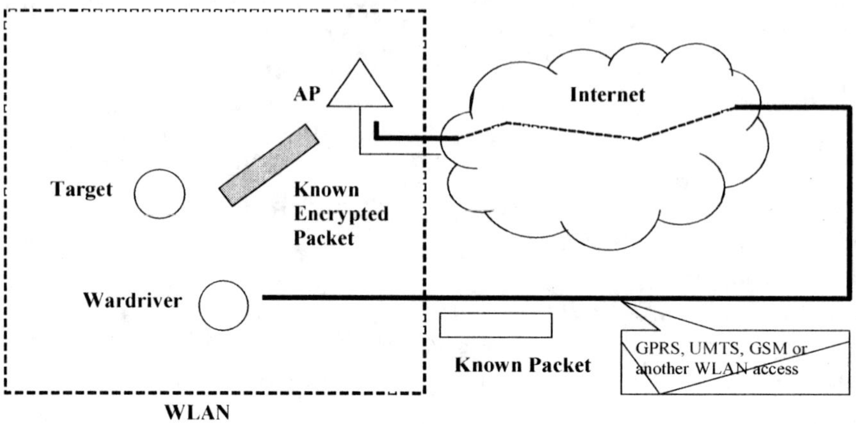

stream is 1,500 bytes long at maximum, owing to the maximum MTU available, and the adoption of a 24-bit IV produces 16,777,216 (2^{24}) possible streams. Hence, the required space is 16,777,216 • 1500 = 23.4 Gbytes.

With the advent of PCMCIA cards, and their poor implementation of the policies to generate IVs, the adoption of a dictionary-based attack became feasible. In fact, many PCMCIA wireless cards reset the IV to 0 each time they are re-initialized. Re-initialization happens each time they are activated (e.g., typically once a day in many circumstances). Then it is sufficient to build a dictionary only for the very first values of IVs, in order to decrypt most of the flowing traffic.

Examples

In this section, we will present briefly some possible attacks against a WEP-secured network. Firstly, we will show how to attack a network by using KisMAC, a tool running on MacOSX with a simple GUI. Then we will show how to use standard terminal-based tools commonly available for different Unix flavors. As a remark, we will not spend too much time on explaining bruteforce or dictionary attacks. In fact, WEP could be cracked in a more elegant way; conversely, owing to its better security, we will explain bruteforcing and dictionary attacks in the section devoted to WPA. Such concepts could be straightforwardly extended also to WEP.

WEP Attack via KisMAC

Let us show an attack performed to a WEP-secured network. Firstly, we show how to crack a network with KisMAC. This gives an idea of how simple it might be. After launching KisMAC, one can start the scanning. If supported, one can select whether or not to adopt *passive* or active *scanning*. Figure 5 depicts the result of a scan.

Then, if there is the need of cracking the WEP, different actions could be performed. Firstly, one can try the Newsham's 21-bit attack, or try to bruteforce the WEP, but owing to the "information" conveyed by the IVs, quicker solutions could be adopted.

Two things may happen: (1) the network is experiencing a huge amount of traffic, hence producing a huge amount of IVs. In this perspective, an attacker must only wait to collect a sufficient number of IVs to perform a suitable attack; or (2) the network is under a low load, hence the time needed to collect a sufficient amount of IVs is non-negligible. Then, it is possible to stimulate traffic by using the de-authentication attack or injecting well-crafted packets; Figure 6 depicts

Figure 5. Scan result provided by the KisMAC tool

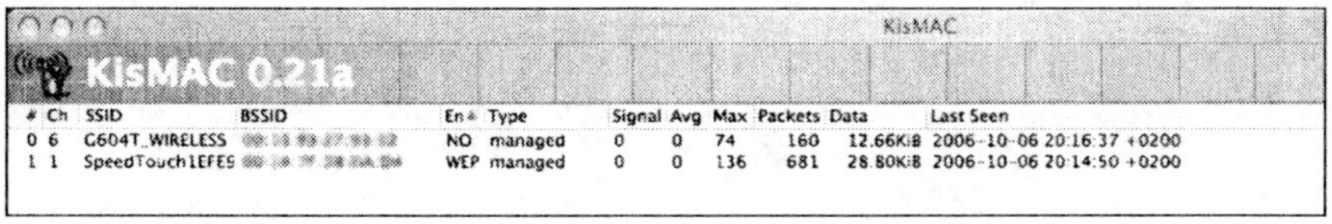

Figure 6. How to stimulate traffic in a WEP-secured network

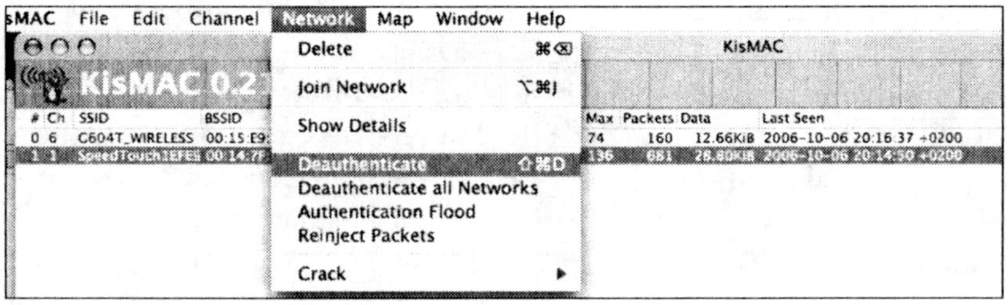

Figure 7. The network has been attacked with an authentication flood. Notice the random-generated MAC addresses.

Client	Vendor	Signal	sent Bytes	recv. Bytes
0C:80:15:93:38:8F	unknown	127	90B	0
02:ED:5E:0B:9A:76	unknown	131	90B	0
06:08:8D:F6:B5:D8	unknown	129	90B	0
06:4C:25:E3:D0:B1	unknown	135	90B	0
0E:6E:DB:5D:20:37	unknown	131	90B	0
0A:7F:45:75:5D:D6	unknown	127	90B	0
0C:0C:63:9E:E3:11	unknown	131	90B	0
06:3F:E2:81:E0:3E	unknown	135	90B	0
00:C3:4C:0E:8B:03	unknown	135	90B	0
0C:F0:C6:05:DE:19	unknown	129	90B	0
08:55:86:4B:11:4F	unknown	127	90B	0
02:6B:09:67:4F:25	unknown	129	90B	0
0E:65:B9:D5:0E:A8	unknown	127	90B	0
0C:A9:94:C5:F5:9B	unknown	127	90B	0
0E:75:3C:E8:14:88	unknown	132	90B	0
08:98:A7:C6:27:52	unknown	131	90B	0
0A:64:F0:52:79:94	unknown	135	90B	0
02:84:84:F6:D6:2C	unknown	129	90B	0
06:0D:6B:17:F4:97	unknown	129	90B	0
02:A5:6C:76:B3:18	unknown	127	90B	0
06:7F:D8:E8:03:BA	unknown	129	90B	0
02:BA:D8:C3:51:47	unknown	129	90B	0
00:8E:F0:B5:D0:80	unknown	129	30B	0
08:40:9B:03:30:2C	unknown	129	90B	0
0E:C5:EA:50:13:AF	unknown	127	90B	0
00:2C:A9:10:91:92	unknown	129	90B	0
0E:DB:BD:ED:67:C9	unknown	129	90B	0
04:C3:A5:77:11:AE	unknown	129	90B	0
04:04:01:92:1E:69	unknown	127	90B	0
02:2F:54:7C:C9:C2	unknown	129	90B	0
08:B2:EA:45:90:73	unknown	133	90B	0
FF:FF:FF:FF:FF:FF	Broadcast	0	0B	13.07Ki

possible attacks to stimulate traffic, while Figure 7 depicts the "fake" stations that populate the attacked wireless network.

WEP Attack via Terminal-Based Tools

Firstly, let us start searching a network. For doing this, let us use *airodump.* Airodump allows to collect traffic from a wireless interface. It could be possible that you have *airodump-ng* instead, since it represents the evolution of the *aircrack* wireless suite. We will refer to the classical tool, since it could be possible that you already have it, especially if your configuration is not up-to-date; however, the concepts, as well as its usage, are the same.

Supposing the tool properly installed, it is sufficient to type in a terminal:

```
Mud:Luca$ ./airodump cardName theTrafficFile 0 loggingMode
```

Here, ./airodump launches the tool, cardName is the name of the card used to monitor the air, theTrafficFile is the output file collecting data. The parameter 0 specifies that we want to hop channels, while loggingMode allows to switch between logging all traffic or only IVs.

If we have collected enough IVs, we can try to crack the WEP by using *aircrack*. Some couple of remarks: (1) the traffic collection and the cracking phases are decoupled. Then you can perform an attack off-line (not hidden in a parking lot); (2) it is possible to collect data with well-known sniffers, such as *Wireshark* (formerly known as *Ethereal*). For instance, under Linux it is possible to use *airmon-ng* to configure the wireless card, then using *Wireshark* to collect traffic. By using *ivstool* from the *aircrack-ng* suite you can convert IVs from *.pcap* format to *aircrack* one.

Then, you can crack a network by typing:

```
Mud:Luca$ ./aircrack -b MAC theTrafficFile
```

Here, -b MAC specifies the MAC address (or the BSSID) of the target network. In fact, your dump could have collected traffic from different networks. The needed number of IVs varies: if your traffic dump is blessed, collecting 100,000 IVs suffices. Usually, the needed number of IVs ranges from 250,000 to 500,000. However, some advanced APs have algorithms that avoid the generation of weak IVs, hence reflecting in a huge number of needed IVs (in the order of several millions).

If there is not enough traffic on the network, collecting IVs could be a tedious (or at least time consuming) task. Moreover, if a sophisticated AP is employed, collecting 5,000,000 IVs with a traffic of few packets per second could be impossible.

Then, it is possible to stimulate traffic on the WLAN, in order to increase the number of packets sent, hence speeding up the collection of IVs.

For instance, by using the *aircrack* suite, it is possible to exploit the so-called address resolution protocol (ARP) *replay.*[2] Roughly, ARP relies on broadcasting a request (an ARP Request) for an IP address, in order to discover the matching between L2 and L3 addressing. The device that recognizes its IP address sends back a query directly to the original requestor. Alas, WEP does not assure protection against replay attacks. So you can inject well-crafted ARP packets and generate answers containing valid IVs. Needles to say, the more aggressive your ARP generation strategy is, the more packets you will collect (thus, reducing the time needed to collect a certain x amount of valid IVs).

To perform an *ARP replay* attack you can use the tool as follows (notice, that you must have also a sniffer running in order to capture replies).

```
Mud: Luca$ ./aireplay-ng --arpreplay -b MACAP -h TMAC Interface
```

./aireplay-ng launches the tool, the flag --arpreplay specifies to perform the ARP replay attack, -b MACAP specifies the MAC address of the AP and -h TMAC specifies the MAC of the target (victim) host. Lastly, Interface tells the program which wireless interface must be used.

If everything is correct, the attack starts generating more traffic.

WPA-PSK ATTACKS

WPA exists in different flavors: for enterprises and for home security. It offers many improvements compared to the WEP. Firstly, IVs are still adopted, but IVs are 48-bit long, preventing from IVs reuse or IVs collision. Secondly, IVs are checked before using them to encrypt packets.

The solution that WPA proposes for Enterprises is barely adequate to discourage any wardriving activities. But the version for home security could be compromised. As a remark, the WPA suite does not offer the ultimate toolkit for security.

As said, a consumer version of the WPA exists, and it is called WPA-PSK. Roughly, WPA-PSK performs similar steps like WEP, but it is more robust. Needles to say, owing to its easy set-up and cost effective implementation, it is often adopted as the basis of corporate security infrastructure. The main characteristic of the WPA-PSK that could be exploited by wardrivers is the "PSK portion" of the procedure. In the PSK, as the acronym suggests, the secret key is pre-shared, hence known a priori and stored in the equipment. However, the WPA-PSK during normal operations has some logic to change the codes and making break into the system a harder work.

In order to stick with the topic of wardriving, we will only explain the unique attack proven to be effective for the WPA-PSK.

The Handshake Attack

The basic under this attack is rooted in how the PSK is engineered. The PSK relies on a user-defined password to initialize the TKIP. From the attacker point of view, the TKIP is quite strong, owing its "*per packet*" nature. Nevertheless, the wardriving community has not yet found out how to crack it. As a consequence, in order to gain access to a WPA-PSK network, a direct attack to the TKIP will not give any reasonable results.

However, there is a weak point in the chain: the authentication. In fact, during the authentication, the requestor sends the PSK, to spawn the TKIP procedure that will cover the rest of the transmission.

The core of the exploit is based on the handshake for the following reason. Prior to starting a secure communication, the key must be sent over an insecure channel. Needles to say, to avoid sending the password in cleartext, thus resulting in a huge security breach in the procedure, there are several mechanisms (outside the scope of this chapter) employed to transmit the *passphrase* over the channel.

However, if a complete handshake is collected, it is possible to bruteforce the handshake procedure, and to recover the password. This attack has two main drawbacks (or advantages, depending on the viewpoint):

1. It is based on a bruteforce technique. If the password is strong enough, it is quite impossible to retrieve;
2. A complete handshake is needed. Without such information, all the traffic collected (even if several Gbytes) is needless.

To overcome the previous drawbacks, some countermeasures are possible. Concerning 1), if in presence of a good dictionary that it is not limited to standard words, but also containing some well-known consumers' passwords, it is possible to bring into a feasible zone a bruteforce attack for some particular deployment (e.g., home network, where users tend to use weak passwords). Regarding 2), it is possible to force de-authentication of clients to collect the needed handshake traffic. However, at least one client must be present in the network to perform this attack. Besides, as explained previously, a wireless interface with packet injection capabilities is needed.

Example

In the *Weak IVs* section we showed some example by using the KisMAC software. KisMAC has a complete GUI, hence performing this attack solely implies to select it from the menu, as shown in Figure 8.

Notice that the Wordlist Attack against the WPA key is available only if a complete handshake has been collected.

Figure 8. WPA-PSK bruteforce attack when employing KisMAC

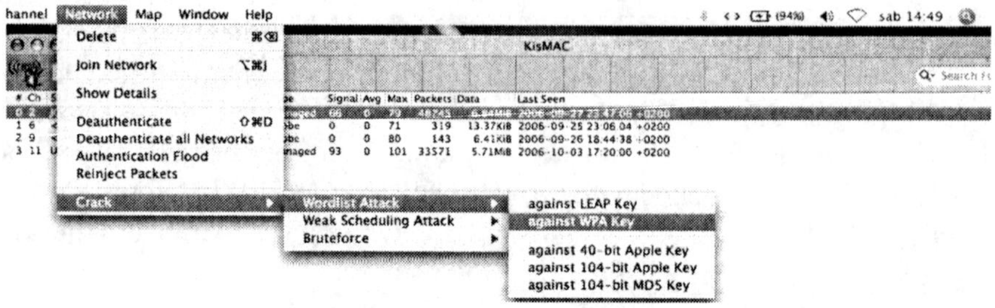

Instead of KisMAC, for pedagogical reasons, let us use *airodump*. Supposing the tool properly installed, it is sufficient to type in a terminal:

Mud:Luca$ ./airodump ath1 theTrafficFile 8

./airodump launches the program (./ to refer to a local path), ath1 specifies the interface where the traffic must be collected, TrafficFile specifies the file that will contain the traffic dump, and 8 is the channel to monitor. However, it is possible to force a de-authentication attack by specifying a flag to *airodump*.

Until there, we have only collected the traffic (and stimulated a complete handshake if needed). Now, it is possible to perform the off-line attack. Most of the tools can exchange data, so it is possible to collect data to *airodump* and perform the cracking procedure to KisMAC, cowpatty, …

Supposing we want to use *aircrack* we will use the tool (from the command line) in a form like:

Mud:Luca$ ./aircrack –a 2 –b MAC –w /Dictionary

./aircrack launches the tool, –a 2 specifies the attack, MAC is the MAC address of the AP to attack, –w specifies the path to a dictionary. For instance, many Unix systems have a minimal dictionary located in /usr/share/dict; you can preliminary start with this word collection. Notice that if the password is a standard dictionary, you should change it immediately, since it is very weak and predictable.

Lastly, another interesting tool (even if quite slow) is cowpatty. In order to crack a WPA key with cowpatty, you will use the tool like:

Mud:Luca$ ./cowpatty –f Dictionary –r theTrafficFile –s wpa

where ./cowpatty launches the software, –f specifies the dictionary (a file called Dictionary in this example), –r specifies where the traffic dump is located (theTrafficFile here) and –s wpa tells the program to crack against the WPA.

Concluding, if a proper handshake is collected, with the aforementioned tools it is possible to crack the WPA. As shown, the steps are not complex. Then, it is possible to understand the *importance of the password*, since it is the only barrier preventing your network to be cracked.

SOME THOUGHTS ABOUT THE WARDRIVERS COMMUNITY

In the following subsections, some ideas on why the wardriving community deserves attention are presented. Besides, always remember that many flaws of the WEP arose due to the fact that it has been developed without any "open" review.

Monitor the Internet Community

The Internet community does not only produce tools, but also important information regarding concepts of security and wardriving. Three major resources are suggested for periodic surveys:

1. Wardriving sites that publish the location of a network (that could be precisely located, as explained in Section I by using a GPS);

a smart step could be to investigate sites publishing WLANs, in order to discover if yours has been detected and cracked.

2. Check for (almost weekly) security bulletins (e.g., BugTraq). Gears are composed not only by hardware, but also software (e.g., the firmware) that could have vulnerabilities. For instance, one of the most famous was related to an AP that upon receiving a broadcast user datagram protocol (UDP) packet on port 27155 containing the string "getsearch" returned (in clear) the WEP keys, the MAC filtering database and the admin password (a big prize, indeed).
3. Periodically download and try the tools. It is useful, funny, and gives an idea of the activity of the underground community.

Avoid Default Configurations (Always)

It is widely known that default configurations are most of the time fine for normal users, but not particularly tweaked for security. For instance, in the *Wireless power* section we discussed some possible risks arising when too much transmission power is employed. Besides, another threat relies in default names for the SSID, which can be employed to uncloak a hidden network, even if without special tools. For instance, it is well known that many Cisco AP use "*tsunami*" as default SSID, and that Linksys uses "*linksys.*" Nevertheless, it is possible to retrieve them by performing a simple Web search (moreover it is possible to retrieve SSID naming schemas for hotels, retailers, and popular Internet cafès…). Lastly, a good suggestion is to change also the default password of your gear, since a malicious attacker (that normally is not a wardriver, but a vandal) can try to alter the AP configuration.

Browse the Source and Use the Tools

Owing to the availability of the tools, it is a better idea to try to be a wardriver sometimes, in order to test your own set-up, as well as the configuration made by your users (e.g., students, colleagues, or customer). Besides, studying the tools and collecting the traces is mandatory to discover possible attacks, for instance by recognizing unusual probes or excessive de-association requests.

Do Not Rely on Weak Passwords

As explained in previous section, bruteforcing a WLAN will be always possible. WEP makes bruteforcing to be useless (owing to its flaws), but WPA-PSK can be only exploited by using a dictionary attack. Hence, the strength of your WLAN depends on the password. Use a good policy to create and distribute passwords and change them often. Do not forget that hundreds of people collaborate to produce dictionaries with most popular passwords, also the most disparate ones (and also in leet variant – *l33t v4rlaNt*).

TOOLS

NetStumbler (www.netstumbler.com): NetStumbler is a program for the Windows™ operating system allowing to detect WLANs. It is a quite handy tool for locating WLANs but it has not all the features and the flexibility of the Aircrack-ng suite.

Kismet (www.kismetwireless.net): Kismet allows monitoring and sniffing traffic over a WLAN. In addition, it can also be adopted as an intrusion detection system. Kismet is able to identify networks both in active and passive mode. Besides, it also offers many other features, such as BSSID uncloaking. Kismet supports many wireless cards and many OSs, as well as many CPUs (e.g., x86, ARM, PPC, and X-Scale); however, some features are only available on the Linux-x86 version.

KisMAC (http://KisMAC.de/): KisMAC is the counterpart of Kismet, but it runs natively on MacOSX and it is easy to use, owing to its simple GUI.

Aircrack-ng (http://www.aircrack-ng.org): Aircrack-ng is a comprehensive suite of tools, ranging from analyzers, sniffers, and cracking tools. Sources and scripts are available, promoting aircrack-ng as one of the best tools and a starting

Table 3. Summary of wardriving threats and possible countermeasures

Attack - Detected Anomaly	Skills Needed	WLAN Affected	Security Risk	Countermeasures
SSID uncloack	None. Automatically done in several software	ALL	1	None at this level.
Active scan	None. Automatically done by interfaces' drivers	ALL	0	Forecasted in the standard. Check periodically MAC addresses of traffic flows.
Passive scan	None, but proper software and a proper interface is needed.	ALL	2	Reduce the transmission power.
WEP crack	Minimum	If WEP Protected	10	Avoid WEP. If WEP must be in place (for legacy support) change password often. Monitor traffic to detect peaks and activate MAC filtering (at least). Force users to adopt VPN and disable DHCPs.
MAC spoofing	Medium. Kernel patches could be needed.	ALL	8	MAC-based policies must be adopted jointly with encryption techniques.
Packet injection	Medium.	If WEP Protected	5	Tools for performing packet injection can also monitor the WLAN like IDS.
De-authentication flood	Medium	For WPA	7	The attacker could be "serious." Change the WPA password to avoid a dictionary attack.
Unsolicited traffic in indoor environments	Medium/High	ALL	9	When in presence of limited transmitting power, the attacker relies on high gain antennas, thus could be a prepared attacker.
Unrecognized	High	ALL	10	It could be a "false positive" or the attacker could be able to produce his/her own tools.

point for developing automated (e.g., *cron-drived*) or tweaked wardriving tools.

SUMMARY TABLE ABOUT WARDRIVING ATTACKS

In this section, we summarize many security threats deriving from wardrivers' activity, by offering a comprehensive table. In addition, we will also introduce some "security risks" in order to better calibrate the needed countermeasures. Security risks have been quantified on a range varying from 0 (none) to 10 (severe). However, the more security is employed in the WLAN, the better. But, being *wardriving* tightly mixed with people habits and urban culture, the exposures to risks may vary according where the WLAN is placed. Table 3 contains the summary.

CONCLUSION

In this chapter we introduced the concept of wardriving, and practices related to cracking wireless networks. As explained, cracking a WLAN is not a complex task: then, for your security you should rely on other techniques (e.g., RADIUS). In addition, by using examples, it is possible to produce your own penetration tests, as well as exercises to show some real world attack to students and engineers.

ACKNOWLEDGMENT

The author wishes to thank Prof. Franco Davoli for the technical suggestions and the thorough review, and Eng. Sergio Bellisario for the technical review.

REFERENCES

Ferro, E., & Potortì, F. (2005, February). Bluetooth and Wi-Fi wireless protocols: A survey and a comparison. *IEEE Wireless Communications,* 12-26.

Newsham, T. (2003). *Applying known techniques to WEP keys.* Retrieved December 12, 2006, from http://www.lava.net/~newsham/wlan/WEP_password_cracker.pdf

Pollard, D. (2002). *Write here, Right now.* Retrieved December 12, 2006, from http://news.bbc.co.uk/1/hi/in_depth/sci_tech/2000/dot_life/2070176.stm

Schneier, B. (1996). *Applied cryptography: Protocols, algorithms, and source code* (2nd ed.). John Wiley & Sons.

Shipley, P. M. (2000). *Peter M. Shipley personal homepage.* Retrieved December 12, 2006, from http://www.dis.org/shipley/

KEY TERMS

Active Mode: Active mode is an operative mode where scanning is done via probe packets. As a consequence, the scanner does not remain undetected.

MAC Address Filtering: MAC address filtering is a technique that allows/denies network accesses only for a predefined MAC address.

MAC Spoofing: MAC spoofing is changing the MAC of the L2 interface. Typically it is employed to by-pass MAC address filtering.

Packet Injection: Packet injection is the activity of inserting a packet in a network for some purpose. For instance, when attacking a WEP-protected network, to stimulate the traffic production to gain more data to be analyzed.

rfmon: rfmon is an operative mode of IEEE 802.11-based air interfaces, allowing to scan for access points while remaining undetectable, since the card does not send any probe packets.

Wardriving: Wardriving is the activity of "*driving around, looking for wireless networks.*"

Wired Equivalent Privacy (WEP): WEP is an encryption mechanism with many security flaws. Recognized as a real security issue, it has been replaced by wireless protected access (WPA).

ENDNOTES

1. However, if raw frames are supported by the internal chipset, you can always build your own tools and enabling drivers by investigating the data-sheets.
2. Many OSes or firmware clear the ARP cache upon disconnection. Then, it could be useful to use a more "aggressive" strategy, as suggested in aircrack documentation.

Chapter VI
Intrusion and Anomaly Detection in Wireless Networks

Amel Meddeb Makhlouf
University of the 7th of November at Carthage, Tunisia

Noureddine Boudriga
University of the 7th of November at Carthage, Tunisia

ABSTRACT

The broadcast nature of wireless networks and the mobility features created new kinds of intrusions and anomalies taking profit of wireless vulnerabilities. Because of the radio links and the mobile equipment features of wireless networks, wireless intrusions are more complex because they add to the intrusions developed for wired networks, a large spectrum of complex attacks targeting wireless environment. These intrusions include rogue or unauthorized access point (AP), AP MAC spoofing, and wireless denial of service and require adding new techniques and mechanisms to those approaches detecting intrusions targeting wired networks. To face this challenge, some researchers focused on extending the deployed approaches for wired networks while others worked to develop techniques suitable for detecting wireless intrusions. The efforts have mainly addressed: (1) the development of theories to allow reasoning about detection, wireless cooperation, and response to incidents; and (2) the development of wireless intrusion and anomaly detection systems that incorporate wireless detection, preventive mechanisms and tolerance functions. This chapter aims at discussing the major theories, models, and mechanisms developed for the protection of wireless networks/systems against threats, intrusions, and anomalous behaviors. The objectives of this chapter are to: (1) discuss security problems in a wireless environment; (2) present the current research activities; (3) study the important results already developed by researchers; and (4) discuss the validation methods proposed for the protection of wireless networks against attacks.

INTRODUCTION

Wireless has opened a new and exciting area for research. Its technology is advancing and changing every day. However, the biggest concern with wireless has been security. For some period of time, wireless has seen very limited security on the wide open medium. Along with improved encryption schemes, a new solution helping the problem resolution is the *wireless intrusion detection system* (WIDS). It is a network component aiming at protecting the network by detecting *wireless attacks*, which target *wireless networks* having specific features and characteristics. Wireless intrusions can belong to two categories of attacks. The first category targets the fixed part of the wireless network, such as MAC spoofing, IP spoofing, and denial of service (DoS); and the second category of these attacks targets the radio part of the wireless network, such as the access point (AP) rogue, noise flooding, and wireless network sniffing. The latter attacks are more complex because they are hard to detect and to trace-back.

To detect such complex attacks, the WIDS deploys approaches and techniques provided by intrusion detection systems (IDS) protecting wired networks. Among these approaches, one can find the signature-based and *anomaly* based approaches. The first approach consists in matching user's patterns with stored attack's patterns (or signatures). The second approach aims at detecting any deviation of the "normal" behavior of the network entities. The deployment of the aforementioned approaches in a wireless environment requires some modifications. The signature-based approach in wireless networks may require the use of a knowledge base containing the wireless attack signatures while an anomaly based approach requires the definition of profiles specific to wireless entities (mobile users and AP). Recently, efforts have focused on *wireless intrusion detection* to increase the efficiency of WIDS. Based on these efforts, models and architectures have been discussed in several research works.

The objective of this chapter is to discuss the major research developments in wireless intrusion detection techniques, models, and proposed architectures. Mainly, the chapter will: (1) discuss security problems in wireless environments; (2) present current research activities; (3) study important results already developed; and (4) discuss validation methods proposed for WIDS. The remaining part is organized as follows: The next section discusses *vulnerabilities*, threats, and attacks in wireless networks. The third section presents wireless intrusion and anomaly detection approaches. The fourth section introduces models proposed for detecting wireless intrusions. The fifth section presents WIDS architectures, proposed by researches papers. The sixth section presents the wireless distributed schemes for intrusion detection. The seventh section discusses mechanisms of *prevention* and *tolerance* provided to enhance the wireless intrusion detection. Finally, the last section concludes the chapter.

VULNERABILITIES, THREATS, AND ATTACKS IN WIRELESS NETWORKS

To present vulnerabilities, threats, and attacks targeting wireless networks, we have to discuss first the security requirements of wireless systems, including those concerning security policy. This section presents the concepts of wireless intrusion, anomaly, and attack scenario in wireless networks, in order to highlight intrusion and anomaly detection requirements. In particular, it discusses some attacks and attack classification that make security in wireless systems very special.

Security Requirements in Wireless Environments

Securing a communication channel should satisfy at least the following set of requirements: integrity, confidentiality, and availability. Moreover, wireless communications require authentication of the sender or/and the receiver and techniques that guarantee non-repudiation. In the following, we discuss technical security and security policy requirements which help reducing vulnerabilities and attack damages.

Because of their technical architecture, mobile communications are targets for a large set of threats and attacks that occur in wired networks, such as identity spoofing, authorization violations, data loss, modified and falsified data units, and repudiation of communication processes. Additionally, new security requirements and additional measures for wireless networks have to be added to the security requirements of wired networks (Schäfer, 2003). Vulnerabilities, threats, and attacks, existing in wireless networks represent a greater potential risk for wireless networks. One among technical requirements is the enforcement of security of the wireless links, because of the ease of gaining direct physical accesses. Moreover, new difficulties can arise in providing wireless security services. For example, the authentication of a mobile device has to be verified by (or for) all AP (or base station [BS]) under which the mobile changes its localization. Because of the handover, respective entities cannot be determined in advance, so the key management process is more complicated. Also, the difference with wired networks, in terms of confidentiality of mobile device location, reveals a number of threats against mobile communications. This appears because of the following conflict: In one hand, each mobile should be reachable for incoming communication requests while, on the other hand, any network entity should be able to get the current location of a mobile device in the network (Schäfer, 2003).

Wireless Vulnerabilities and Threats

A vulnerability is a weakness (or fault) in the communication medium or a protocol that allows compromising the security of the network component. Most of the existing vulnerabilities in the wireless medium are caused by the medium. Because transmissions are broadcast, they are easily available to anyone who has the appropriate equipment. Particular threats of the wireless communication are device theft, malicious hacker, malicious code, theft of service, and espionage (Boncella, 2006). There are numerous of wireless vulnerabilities and threats that are studied in the literature, for the purpose of detecting attacks exploiting them. In the following, we distinguish two categories of vulnerabilities and threats: those existing in a LAN-like wireless networks (WLAN) and those existing in cellular-like wireless networks (Hutchison, 2004).

WLAN Vulnerabilities and Threats

The following are typical vulnerabilities existing in the main component of WLAN, which is the AP.

- **Signal range of an authorized AP:** This vulnerability is about the possibility of the extension of AP signal strength beyond a given perimeter. Consequently, the AP's placement and signal strength have to be adapted to make sure that the transmitting coverage is just enough to cover the correct area.
- **Physical security of an authorized AP:** Because most APs are mounted by default, their placement is critical. An AP has to be correctly placed in order to avoid accidental damage, such as direct access to the physical network cable. To protect physically the access to the AP, many solutions were proposed; but all of them require a mandatory policy.
- **Rogue AP:** This vulnerability is a sort of man-in-the middle attack, where an attacker can place an unauthorized (or rogue) AP on the network and configure it to look legitimate to gain access to wireless user's sensitive data. This can be done because user's devices need to be connected to the strongest available AP signal.
- **The easy installation and use of an AP:** In order to use the advantages of internal networks, employees can introduce an unauthorized wireless network. The easy installation and configuration of the AP make this feasible for legitimate or illegitimate users.
- **The AP configuration:** If the AP is poorly configured or unauthorized, then it can provide an open door to hackers. This is caused by using a default configuration that annihilates the security controls and encryption mechanisms.
- **Protocol weaknesses and capacity limits on authorized AP's:** These limitations can cause DoS from hackers using unauthorized AP's when they can flood authorized AP with traffic forcing them to reboot or deny accesses.

Some of the attacks, exploiting the aforementioned vulnerabilities are discussed in the following section of this chapter.

Cellular System Vulnerabilities and Threats

This subsection presents cellular system vulnerabilities and threats that are categorized as follows (Nichols & Lekkas, 2002):

- **Service interruption:** The increased capacity provided by the high-speed technology has resulted in fewer cable routes necessary to meet capacity requirements. Consequently, this has decreased the number of switches. The lack of overall diversity in cabling and switching has increased the vulnerability of telecommunication infrastructures. This can cause DoS of an entire zone.
- **Natural threats:** Natural threats comprise the category of repeated threats caused by climatic, geological, or seismic events. Severe damage resulting from natural disaster can cause long-term damage to the telecommunication infrastructures.
- **Handset vulnerabilities:** Unlike computer systems, handsets are limited regarding the security features. Because wireless messages travel through the air by passing conventional wired network for transmission to the receiver, messages may need to be changed to another protocol (e.g., at the gateway, the *wireless transport layer security* message has to be changed to *Secure Socket Layer*). This operation presents vulnerability because anyone can access the network at this moment. Moreover, the use of encryption can add vulnerabilities, which can make confusion between mobile phones, since the node does not know its encrypted true location.

Wireless Attacks

Detecting a large set of attacks by a WIDS requires studying and developing the attacker's methods and strategies. We discuss in this subsection the typical attacks and malicious events that can be detected by a WIDS (Hiltunen, 2004; Vladimirov, Gavrilenko, & Mikhailovsky, 2004).

Illicit Use

Illicit use of a wireless network may involve an attacker connecting to the Internet or to the corporate network that lives behind the AP. Illicit use is a passive attack that does not cause damage to the physical network. It includes following attacks (Mateli, 2006):

- **Wireless network sniffing:** When wireless packets traverse the air, attackers equipped with appropriate devices and software can capture them. Sniffing attack methods include:
    - **Passive scanning:** This attack aims at listening to each channel. It can be done without sending information. For example, some radio frequency monitors can allow copying frames on a channel.
    - **Service set identifier (SSID) detection:** This consists in retrieving SSID by scanning frames of the following types: beacon, probe requests, probe responses, association requests, and re-association requests.
    - **MAC addresses collecting:** To construct spoofed frames, the attacker has to collect legitimate MAC addresses, which can be used for accessing AP filtering out frames with non registered MAC addresses.

To capture wireless packets, specific equipments should be used by the attackers, depending on the targeted wireless network interface card (Low, 2005).

- **Probing and network discovery:** This attack aims to identify various wireless targets. It uses two forms of probing: active and passive. Active probing involves the attacker actively sending probe requests with no identification using the SSID configured in order to solicit a probe response with SSID information and other information from any active AP. When an attacker uses passive probing, he is listening on all channels for all wireless packets, thus the detection capability is not limited by the transmission power (Low, 2005).

- **Inspection:** The attacker can inspect network information using tools like Kismet and Airodump (Low, 2005). He could identify MAC addresses, IP address ranges, and gateways.

Wireless Spoofing

Spoofing purpose is to modify identification parameters in data packets. New values of selected parameters can be collected by sniffing. Typical spoofing attacks include:

- **MAC address spoofing:** MAC spoofing aims at changing the attacker's MAC address by the legitimate MAC address. This attack is made easy to launch because some client-side software allows the user to view their MAC addresses.
- **IP spoofing:** IP spoofing attempts to change source or destination IP addresses by talking directly with the network device. IP spoofing is used by many attacks. For example, an attacker can spoof the IP address of host A by sending a spoofed packet to host B announcing the window size equal to 0; though, it originated from B (Mateli, 2006).
- **Frame spoofing:** The attacker injects frames having the 802.11 specification with spoofed containing. Due to the lack of authentication, spoofed frames cannot be detected.

Man in the Middle Attacks

This attack attempts to insert the attacker in the middle (man in the middle [MITM]) of a communication for purposes of intercepting client's data and modifying them before discarding them or sending them out to the real destination. To perform this attack, two steps have to be accomplished. First, the legitimate AP serving the client must be brought down to create a "difficult to connect" scenario. Second, the attacker must setup an alternate rogue AP with the same credentials as the original for purposes of allowing the client to connect to it. Two main forms of the MITM exist: the eavesdropping and manipulation. Eavesdropping can be done by receiving radio waves on the wireless network, which may require sensitive antenna. Manipulation requires not only having the ability to receive the victim's data but then be able to retransmit the data after changing it.

Denial of Service Attacks

DoS attacks can target different network layers as explained in the following:

- **Application layer:** DoS occurs when a large amount of legitimate requests are sent. It aims to prevent other users from accessing the service by forcing the server to respond to a large number of request's transactions.
- **Transport layer:** DoS is performed when many connection requests are sent. It targets the operating system of the victim's computer. The typical attack in this case is a SYN flooding.
- **Network layer:** DoS succeeds, if the network allows to associate clients. In this case, an attacker can flood the network with traffic to deny access to other devices. This attack could consist of the following tasks:
    - The malicious node participates in a route but simply drops several data packets. This causes the deterioration of the connection (Gupta, Krishnamurthy, & Faloutsos, 2002).
    - The malicious node transmits falsified route updates or replays stale updates. These might cause route failures thereby deteriorating performance.
    - The malicious node reduces the time-to-live (TTL) field in the IP header so that packets never reach destinations.
- **Data link layer:** DoS targeting the link layer can be performed as follows:

    - Since we assume that there is a single channel that is reused, keeping the channel busy in the node leads to a DoS attack at that node.
    - By using a particular node to continually relay spurious data, the battery life of that node may be drained. An end-to-end authentication may prevent these attacks from being launched.
- **Physical layer:** This kind of DoS can be executed by emitting a very strong RF interference on the operating channel. This will cause interference to all wireless networks that are operating at or near that channel.

WIRELESS INTRUSION AND ANOMALY DETECTION

This section discusses the major security solutions provided for wireless networks. In particular, the cases of WLAN and ad hoc networks will be addressed. The discussed methods include the radio frequency fingerprinting, cluster-based detection, mobile devices monitoring, and mobile profile construction.

Basic Techniques for Detection

Wireless intrusion detection protects wireless networks against attacks, by monitoring traffic and generating alerts. Two ways of detection are distinguished: signature based and anomaly based. The first category aims at detecting known attacks by looking for their signatures. The main disadvantage of such approaches is that they detect only known attacks. The anomaly based approaches are not often implemented, mostly because of the high amount of false alarms that have to be managed loosing a large amount of time. Anomaly based detection develops a baseline of the way of considering normal traffic. When an abnormal traffic is detected, an alert is generated. The advantage of such approach is that it can capture unknown attacks.

To take from the advantages of the previous two approaches, the hybrid approach consists on using in the same system the two approaches simultaneously. To be efficient, intrusion detection approaches has to be run online and in real time. Otherwise, the use of intrusion detection technique is useful for audit or postmortem digital investigation and it will not prevent an attack on time. Real-time intrusion detection has to be able to collect data from the network in order to store, analyze and correlate them, which can decrease network performance (Hutchison, 2004).

Wireless Detection Approaches

The main objective of wireless detection is to protect the wireless network by detecting any deviation with respect to the security policy. This can be done by monitoring the active components of the wireless network, such as the APs (Hutchison, 2004). Generally, the WIDS is designed to monitor and report on network activities between communicating devices. To do this, the WIDS has to capture and decode wireless network traffic. While some WIDSs can only capture and store wireless traffic, other WIDSs can analyze traffic and generate reports. Other WIDSs are able to analyze signal fingerprints, which can be useful in detecting and tracking rogue AP attack. As it is done for wired networks, the following classifications of IDSs can be distinguished according to several dimensions: the approach (signature based/anomaly based); the monitored system (network-based/host-based); and the way of response (active/passive).

Mobile Profiles Construction

The main objectives when using the anomaly based approach are to define user mobility profiles (UMPs) and design an appropriate system that permits the detection of any deviation with respect to UMP. The intrusion detection process begins with the data collection processing. Once the user location coordinates (LCs) are determined, a high-level mapping (HLM) is applied. The objective of the HLM is to decrease the granularity of the data in order to accommodate minor deviations or intra-user variability between successive location broadcasts. LCs features are extracted from each broadcast during feature extraction. A set of these chronologically ordered LCs are subsequently concatenated to define a mobility sequence (Hall, Barbeau, & Kranakis, 2005). This process continues until the creation of the mobility sequences. The training patterns from the first four of the six data set partitions are stored in the UMP, along with other user-related information. During the classification phase, a set of user mobility sequences are observed and compared to the training patterns in the user's profile to evaluate the similarity measure to profile (SMP) parameter. If the average of the SMP value exceeds predefined thresholds, then the mobility sequences are considered abnormal and an alert is generated (Hall et al., 2005).

The following parameters are defined for the mobility profiles: (1) the identifier representing the user identification; (2) the training patterns characterizing the user mobility behavior; (3) the window size representing the mobility sequence numbers (SN).

Monitoring Wireless Devices

Using a signature-based approach, the IDS bases its processing on the recognition of intrusion's patterns from the traffic outputs. This requires monitoring several parameters of the AP outputs and the wireless client. Monitoring APs is about monitoring their respective SSID, MAC address, and channel information. This requires listening wireless frames, such as beacons, probe response, and authentication/association frames at the AP outputs and comparing them to the predefined attack signatures. For example, in the case of MITM attack, the monitoring process would detect that there is a sudden introduction of an AP on another channel previously not present. Through the SSID, MAC address might be spoofed by the attacker in the process of setting up the rouge AP.

Because authorized clients cannot be listed, the information that may help detecting an attack cannot be totally available; nevertheless, the following aspects can be monitored (Low, 2005):

- The "blacklist" of wireless clients can be checked against all connecting clients. Any client within this list trying to access the network would be automatically denied and an alert can be sent off.
- All wireless clients with an "illegal" MAC address (MAC address ranges, which have not been allocated) are automatically denied access and an alert is sent off.
- A wireless client that just sends out probe requests or special distinguishable data packets after the initial probe request has not been authenticated can be flagged out as potential network discovery attack.
- Usually, when impersonation attacks are ongoing, the attacker will take on the MAC/IP address of the victim, but it will not be able to continue with the SN used previously by the victim. Thus, by monitoring the SN in these packets, potential impersonators could be identified.

Radio Frequency Fingerprinting (RFF)

The RFF is defined as the process identifying a cellular phone by the unique "fingerprint" that characterizes its signal transmission. It is used to prevent cloning fraud, because a cloned phone will not have the same fingerprint as the legal phone with the same electronic identification numbers. This approach aims to enhance the anomaly based wireless intrusion detection by associating a MAC address with the corresponding transceiver profile. The architecture of the corresponding IDS is shown by Figure 1, where the main objective is to classify an observed transceiver print as normal (belongs to the transceiver of a device with a given MAC address) or anomalous (belongs to another transceiver) (Barbeau, Hall, & Kranakis, 2006; Hall et al., 2005).

As illustrated in Figure 1, the information flow begins by converting the analog signal to a digital signal. This is done by the converter component. Second, the transient extractor extracts the transient portion from the digital signal. Then, the amplitude, phase, and frequency defining the transceiverprint are extracted by the feature extraction component. These features are compared to the transceiver profiles existing in the IDS. This operation is

Figure 1. The enhanced architecture of WIDS

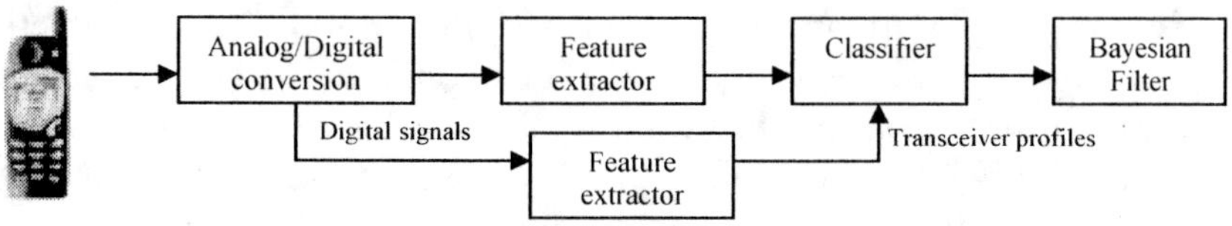

performed by the classifier component. To decide about the status of the transceiverprint, the Bayesian filter is applied. This process requires extracting predefined transceiver's profiles, which is detailed in the following sub-section.

- **Feature extractor:** In this step, amplitude and phase components are obtained using respectively, equations (1) and (2).

$$a(t) = \sqrt{i^2(t) + q^2(t)} \quad (1)$$

$$\theta(t) = \tan^{-1}[\frac{q(t)}{i(t)}] \quad (2)$$

Frequency extraction is done by applying the discrete wavelet transform (DWT), for example.

- **Classifier:** To classify a signal as anomalous, the probability of match has to be determined for each transceiver profile. Therefore, a statistical classifier using neural networks can be used, where the set of extracted features represent a vector and the outputs are a set of matching probabilities.
- **Bayesian filter:** To decide whether matching probabilities exceed threshold values, a Bayesian filter is applied because of the noise and interference, which are special characteristics of wireless environment. The Bayesian filter has to estimate the state of a system from noisy observations.
- **Feature selection/profile definition:** Before applying the detection process, the definition of transceiver's profiles has to be made. To do so, features that have low intra-transceiver variability and high inter-transceiver variability are selected. Examples of selected features include: deviations of normalized amplitude, phase and frequency, amplitude variance, and deviations of normalized in-phase data and normalized quadrate data.

Cluster-Based Detection in Ad Hoc Networks

Due to the distributed nature of wireless networks, especially ad hoc networks are vulnerable to attacks. In this case, intrusion detection provides audit and monitoring capabilities that offer local security to a node and helps to perceive specific trust levels of other nodes (Ahmed, Samad, & Mahmood, 2006; Samad, Ahmed, & Mahmood, 2005). Clustering protocols can be taken as an additional advantage in these processing constrained networks to collaboratively detect intrusions with less power usage and minimal overhead. Because of their relation with routes, existing clustering protocols are not suitable for intrusion detection. The route establishment and route renewal and route renewal affect clusters. Consequently, processing and traffic overhead increase, due to instability of clusters. Ad hoc networks present battery and power constraint. Therefore, the monitoring node should be available to detect and respond against intrusions in time. This can be achieved only if clusters are stable for a long time period. If clusters are regularly changed due to routes, the intrusion detection will not be efficient. Therefore, a generalized clustering algorithm, detailed in Ahmed et al. (2006) has been discussed. It is also useful to detect collaborative intrusions (Samad et al., 2005).

Cluster Formation

Clusters are formed to divide the network into manageable entities for efficient monitoring and low processing. Clustering schemes result in a special type of node, called the cluster head (CH) to monitor traffic within its cluster. It not only manages its own cluster, but also communicates with other clusters for cooperative detection and response. It maintains information of every member node (MN) and neighbor clusters. The cluster management responsibility is rotated among the cluster members for load balancing and fault tolerance and must be fair and secure. This can be achieved by conducting regular elections (Samad et al., 2005). Every node in the cluster must participate in the election process by casting their vote showing their willingness to become the CH. The node showing the highest willingness, by proving the set of criteria, becomes the CH until the next timeout period.

Intrusion Detection Architecture

Because ad hoc networks lack in centralized audit points, it is necessary to use the IDS in a distributed manner. This also helps reducing computation and memory overhead on nodes. The proposed clustering algorithm in Samad et al. (2005) can be related to the intrusion detection process as partial analysis of the incoming traffic is done at the CH and the rest of the analysis is done at the destination node. Traffic analysis at the CH and packet analysis at the MN is helpful in reducing processing at each node. If a malicious activity is found by the CH, it informs its members and the neighboring clusters to take a set of actions. It is the responsibility of CH to obtain help from and/or inform the MNs and neighboring clusters for a particular intrusion. Undecided node (UD) performs its own audit and analysis; however, it performs partial analysis immediately after becoming a CH or MN. Intrusion detection techniques can be anomaly based or signature based.

The host-based IDS (HIDS) observes traffic at individual hosts, while network-based IDS (NIDS) are often located at various points along the network. Since centralized audit points are not available in ad hoc networks, NIDSs cannot be used. Alternatively, if every host starts monitoring intrusions individually such as in HIDS, lot of memory and processing will be involved. Therefore, a distributed approach is used to perform monitoring, where both CH and MN collect audit data.

A flow model of intrusion detection architecture of cluster-based intrusion detection (CBID) is illustrated by Figure 2, which consists of four modules. Information collected during the training phase in the logging module is transferred to the intrusion information module to perceive a threshold value for the normal traffic. If it is the case, an alert is generated by the intrusion response module.

- **Logging:** The CH captures and logs all the traffic transferred through its radio range. It keeps the necessary fields and the data related to traffic such as number of packets sent, received, forwarded, or dropped in a database. The traffic can either be data traffic or control traffic. These logs can be helpful for the detection of many attacks, such as blackhole, wormhole, sleep deprivation, malicious flooding, packet dropping, and so forth.
- **Intrusion information:** If signature-based detection is used, every node must maintain

Figure 2. Intrusion detection process

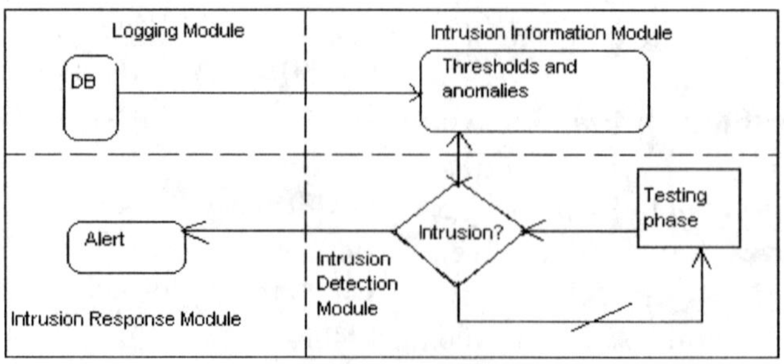

a database that contains all the intrusion signatures. For anomaly based detection, the anomalous behaviors must also be well defined.

- **Intrusion detection:** By this module, the node detects intrusions by analyzing and comparing the traffic patterns with the normal behavior. If anomaly is found, the CH generates an alarm and increases the monitoring level and analyzes the traffic in more detail to find out the attack type and identity of the attacker.
- **Intrusion response:** To inform about detected intrusions, nodes generate alerts. They also can provide responses to react against them.

DETECTION MODELS

To enhance IDS efficiency, theories and models have been developed to cope with intrusion correlation; action tracking and packet marking; digital investigation using evidences based on alerts; and attack reconstruction in wireless environments. The evidence is defined as a set of relevant information about the network state (Aime, Calandriello, & Lioy, 2006).

Intrusion and Anomaly Detection Model Exchange

This section discusses the anomaly model used in mobile ad hoc networks (MANET). It is based on the model distribution and model profiling and aggregation.

Model Distribution

Due to the lack of battery power or computation ability, MANET's model is required. Depending on the node location performing intrusion detection, the following distribution models can be adopted (Cretu, Parekh, Wang, & Stolfo, 2006):

- In the case of generating anomalies, training can be done by MANET nodes: (1) if the MANET nodes have WAN connectivity, the node can initiate download requests to obtain the latest model from the server; and (2) without WAN connectivity, MANET nodes can be initialized before deployment, where the default model is used.
- Another model consists in deploying a more powerful MANET node with sufficient processing and battery power to perform anomaly training. The node would listen promiscuously to all visible traffics on the MANET, generate anomalies, and distribute them to the peers.
- Use a pre-computed anomaly model. This scenario is worst case, but can be practical in situations where the MANET's behavior is well-defined and follows a standard protocol definition.

Model Aggregation/Profiling

The aggregation model was previously used in MANETs for alerts demonstrated that, by integrating security-related information at the protocol level from a wider area, the false positive rate and the detection rate can be improved (Cretu et al., 2006).

In addition, model aggregation enables peers to determine whether or not to communicate with a particular node n_l. If the peers' models are very similar to those used by n_l, it suggests that the node is performing similar tasks. A node with a dissimilar model is considered as suspicious and has a malicious content. For example, a node sending out worm packets will generate a substantially different content distribution. This can be done via comparison (Cretu et al., 2006).

Anomaly Based Detection Models

In this section, we discuss how to build anomaly detection models for wireless networks. Detection based on different kinds of activities may differ in the format and the amount of available audit data as well as the modeling algorithms. However, we admit that the principle behind the approaches will be the same. Therefore, we discuss in this section only one of these approaches, which is based on a routing protocol (Zhang, Lee, & Huang, 2003):

Building an Anomaly Detection Model

This method uses information-theoretic measures, namely, entropy and conditional entropy, to describe normal information flows and use classification algorithms to build anomaly detection models. When constructing a classifier, features with high information gain or reduced entropy are needed. Therefore, a classifier needs feature value tests to partition the original dataset into low entropy subsets. Using this framework, the following procedure for anomaly detection is applied (Zhang et al., 2003): (1) select audit data so that the normal dataset has low entropy; (2) perform appropriate data transformation according to the entropy measures; (3) compute classifier using training data; (4) apply the classifier to test data; and (5) post-process alarms to produce intrusion reports.

Detecting Abnormal Updates to Routing Tables

The main requirement of an anomaly detection model used by IDSs is a low false positive rate, calculated as the percentage of legitimate behavior variations detected as anomalies. Since the main concern for ad hoc routing protocols is that the false routing information generated by a compromised node will be disseminated to and used by the other nodes, the trace data can be designed for each node. A routing table contains, at the minimum, the next hop and the distance in hop number. A legitimate change in the routing table can be caused by the physical node movement or network membership changes. For a node, its own movement and the change in its own routing table are the only reliable and trustable information. Hence, used data exist on the node's physical movements and the corresponding change in its routing table as the basis of the trace data. The physical movement is measured mainly by distance and velocity. The routing table change is measured mainly by the percentage of changed routes (PCR), the percentage of changes in the sum of hops of all the routes (PCH), and the percentage of newly added routes (Zhang et al., 2003). These measurements are used because of the dynamic nature of mobile networks. The normal profile on the trace data specifies the correlation of physical movements of the node and the changes in the routing table.

Classification rules for PCR and PCH describe normal conditions of the routing table. These rules can be used as normal profiles. Checking an observed trace data record with the profile involves applying the classification rules to the record. Therefore, repeated trials may be needed before a good anomaly detection model is produced.

Detecting Abnormal Activities in Other Layers

Detecting anomalies for other entities of the wireless networks such as MAC protocols, or entities provided by the network (applications and services) follows a similar approach as in the physical layer. For example, the trace data for MAC protocols can contain the following features: for the past *s* seconds, the total number of channel requests, the total number of nodes making the requests, the largest, the mean, and the smallest of all the requests. The class can be the range of the current requests by a node. A classifier on this trace data describes the normal context of a request. An anomaly detection model can then be computed, as a classifier or clusters, from the deviation data. Similarly, at the mobile application layer, the trace data can use the service as the class (Zhang et al., 2003).

WIRELESS INTRUSION DETECTION SYSTEM ARCHITECTURES

This section discusses the proposed models, architectures, and methods to validate the used approaches.

Wireless Intrusion Tracking System

The wireless intrusion tracking system (WITS) deploys the Linksys WRT54G AP, Linux and other open source tools in order to track wireless intruders in a wireless cell. A WITS is designed to minimize the effect of the attacks against wireless networks. It combines technologies to produce a system that allows real-time tracking of intruders and extensive forensic data gathering (Valli, 2004).

- **Sacrificial access points (SAP):** WITS uses the concept of SAPs, which acts as a wireless honeypot and forensic logging device. The used SAP has conventional wired Ethernet capability. Its functionality is severely limited for deployment as a honeypot device. However, it permits the installation of customized firmware, which allows the reduction of installed facilities used as part of the routing and AP functionality for the WRT54G. The firmware can be upgraded to patch any new vulnerabilities or weaknesses. To be successful, the system must retain large, extensive and multiple log files that contain system statistics and sufficient network related data for forensic reconstruction of any traffic. The used data are data located in honeypot log files, snort data, and data provided by traffic analysis. The data in honeypot logfiles will indicate the level of probing and malicious activity. Traffic analysis provides an extensive analysis of the intruder activity.
- **Tracking the intruder:** Wireless intruders have the ability to be mobile and are not constrained to use predefined channels, which make them difficult to track. Furthermore, wireless attackers can manipulate layer 1 and layer 2 of the OSI model to mask activities and subsequent detection. WITS uses GPS techniques to locate and track intruders within the wireless cell. The resultant GPS data will be stored for later analysis or used by an immediate location process of the attacking device.

Agent-Based IDS for Ad Hoc Wireless Networks

This section introduces a multi-sensor IDS that employs a cooperative detection algorithm. A mobile agent implementation is chosen to support the wireless IDS features such as sensor mobility and intelligent routing of intrusion data throughout the network.

Modular IDS Architecture

The proposed IDS is built on a mobile agent framework. It employs several sensor types that perform specific functions, such as:

- **Network monitoring:** Only certain nodes will have sensor agents for network monitoring, in order to preserve the total computational power and the battery power of mobile hosts.
- **Host monitoring:** Every node on the ad hoc network will be monitored internally by a host-monitoring agent. This includes monitoring system-level and application-level operations.
- **Decision-making:** Every node will decide on the intrusion threat level on a host-level basis. Specific nodes will collect intrusion information and make collective decisions about intrusion level.
- **Reacting:** Every node can react in order to protect the host against detected attacks. Reactions can be predefined at that node.

To minimize power consumption and IDS-related processing time, the IDS must be distributed. A hierarchy of agents can be used to this end. A hierarchy of agents is composed of three agent classes, which are the monitoring agents, decision-making agents, and action agents. Some are present on all mobile hosts, while others are distributed to only selected nodes (Kachirski & Guha, 2003). Cluster heads, for example, are the typical nodes implementing the monitoring agents. The node selection is naturally dependent on the security requirements imposed to the mobile nodes.

Intrusion Response

The nature of an intrusion response for ad hoc networks depends on the intrusion type and the network protocols and applications types. Examples of responses can be:

- Re-initializing communication channels between nodes
- Identifying the compromised nodes and re-organizing the network to preclude the promised nodes
- Notifying the end user and take appropriate action
- Send a re-authentication request to all nodes in the network to prompt the end-users to authenticate themselves (Zhang et al., 2003)

DISTRIBUTED INTRUSION DETECTION

Any *distributed IDS* should enforce mechanisms that support the reliability of its nodes as well as the distributed analysis, integrity, and privacy of exchanged alerts. Several critical problems should be addressed to provide collaborative methods for wireless *distributed intrusion detection*. These problems include the reduction of the volume of alerts; the decrease the complexity of communication and bandwidth requirements; and the management of heterogeneity of formats and protocols.

IDS for PublicWiFi System

The IDS, used by the WIFI systems, bases its detection on network monitoring to produce evidences and share them among all nodes.

The monitor can be thought as an instance of the Ethereal network packet Sniffer. For each captured packet, Ethereal displays a complete view of the packet content and adds some general statistics as a timestamp, frame number, and length in bytes. By looking on the Ethernet level header and focusing on 802.11 frames, source, destination and BSSId addresses; SN; frame type and subtype; and the retry flag are distinguished. Other parameters are added such as counters for transmission retries and for frames received with wrong FCS, and packet transmission time. In this way, a list of events is built and matched, to detect in particular, jamming attacks and channel failures. Since all nodes participate in the detection process, multiple lists are matched to combine the two lists into a single list of events (Aime et al., 2006).

Multi-Layer Integrated Intrusion Detection and Response

Given that there are different kinds of vulnerabilities in mobile network layers, coordinating IDSs within layers is required. The following integration scheme can be investigated:

- If a node detects an intrusion that affects the entire network, it initiates the re-authentication process to exclude the compromised/malicious nodes from the network.
- If a node detects a local intrusion at a higher layer, lower layers are notified.

In this approach, the detection on one layer can be initiated from other layers. To do this, the lower layers need more than one anomaly detection model: one that relies on the data of the current layer and the one that considers information from the upper layer (Zhang et al., 2003).

WIRELESS TOLERANCE AND PREVENTION

Intrusion prevention is considered as an extension of intrusion detection technology, but it is actually another form of access control, like an application layer firewall. Intrusion prevention systems (IPSs) were developed to resolve ambiguities in passive network monitoring by placing detection systems online. Showing a considerable improvement upon firewall technologies, IPSs make access control decisions based on application content, rather than IP address or ports by denying potentially malicious activity. There are advantages and disadvantages to host-based IPS compared with network-based IPS.

Some IPSs can also prevent yet to be discovered attacks, such as those caused by a buffer overflow. Deployed to strengthen wireless security, wireless IPSs monitor radio frequencies in order to detect malicious traffic.

The development and support of intrusion aware survivable applications in wireless networks are key problems in the provision of wireless services.

Significant aspects of intrusion tolerance include: (1) the ability to adapt to changes in environmental and operational conditions for surviving intrusions; (2) the coordination and management of adaptation of changes in service provision; (3) the awareness of resource statuses to respond to attack symptoms effectively; and (4) the management of resource redundancy. The following are two approaches that deploy intrusion tolerance to prevent wireless attacks.

Intrusion Tolerance Based on Multiple Base Stations Redundancy

To provide fault tolerance, this research discusses a redundancy in the form of multiple base stations (BSs). Since an adversary can disallow delivery of sensor data that is routed over only one path to a given BS, a multi-path routing redundancy to improve intrusion tolerance of wireless nodes is introduced (Deng, Han, & Mishra, 2004).

The simplest way to set up multiple paths for each node to multiple BSs is to use a flooding message: each BS broadcasts a unique request message, called REQ. Upon the reception of REQ from a BS, it records the packet sender as its parent node for that BS, and re-broadcasts REQ to its neighbor and child nodes. The node then ignores all copies of the same REQ that it receives later. In this way, the REQ generated by a BS will be able to flood the entire network, even though the network nodes forward that message only once. If one BS broadcast its own REQ, every sensor node will have one path for it. However, this scheme cannot prevent a malicious compromised node from BS spoofing by sending forged REQ. Every node will think that the forged message is generated by the legitimate BS and will forward the forged REQ. To defend against such attack, each BS can authenticate the sent REQ (Deng et al., 2004).

INSENS: Intrusion-Tolerant Routing in Wireless Sensor Networks

INSENS (Deng, Han, & Mishra, 2003, 2005) can be used to prevent DoS attacks, where individual nodes are not allowed to broadcast routing data. Only the BS is allowed to broadcast (Deng et al., 2003). It proposes a BS authentication using a hash function. To prevent DoS/distributed denial of service (DDoS) broadcast attacks, unicast packets must first traverse through the BS. Second, the control routing information has to be authenticated and encrypted by using symmetric cryptography. To address the notion of compromised nodes, redundant multipath routing is built into INSENS to achieve secure routing.

INSENS proceeds through two phases, route discovery and data forwarding. The first phase discovers the sensor network topology, while the second deals with forwarding data from sensor nodes to the BS, and vice versa. Route discovery is performed in three rounds:

- During the first round, the BS floods a request message to all the reachable sensor nodes in the network. The BS broadcasts a request message that is received by all its neighbors. A sensor, receiving a request message for the first time, records the identity of the sender in its neighbor set and then broadcasts a request message. Two mechanisms are used to counter attacks. The first one identifies the request message initiated by the BS using hash. The second mechanism configures sensors with separate pre-shared keys by applying a keyed MAC algorithm to the complete path (Deng et al., 2005).
- During the second round, the sensor nodes send their local information using a feedback message to the BS. After a node has forwarded its request message, it waits a time period before generating a feedback message.
- In the third round, forwarding tables are computed by the BS for each sensor node based on the information received in the second round. Then, it sends them to the respective nodes using a routing update message and waits for a certain period to collect the connectivity information received via feedback messages in order to compute possible paths to each other node. The BS then updates the forwarding tables using entries of the form:

(destination, source, and immediate sender).

> Destination is the node ID of the destination node, source is the node ID of the node that created this data packet, and immediate sender is the ID of the node that just forwarded this packet. Once the data packet is received, a node searches for a matching entry in its forwarding table. If it finds a match, then it forwards the data packet (Deng et al., 2005).

CONCLUSION

We have shown in this chapter that WIDSs have an important role in securing the network by protecting its entities against intrusions and misuse. The protection is performed based on models capable of providing a framework for the description and correlation of attacks. Research works have focused on the development of techniques, approaches, and mechanisms, and WIDS architectures. Architectures include radio frequency fingerprinting, cluster-based detection, mobile devices monitoring, and mobile profile construction. Wireless intrusion prevention and tolerance are also discussed in this chapter; and systems such as INSENS are developed. In addition, we have shown that several challenges need to be addressed to enhance the efficiency of WIDSs.

REFERENCES

Ahmed, E., Samad, K., & Mahmood, W. (2006). Cluster-based intrusion detection (CBID) architecture for mobile ad hoc networks. In *Proceedings of AusCERT Asia Pacific Information Technology Security Conference (AusCERT),* Asia.

Aime, M. D., Calandriello, G., & Lioy, A. (2006, June 26-29). A wireless distributed intrusion detection system and a new attack model. In *Proceeding of the 11th Symposium in Computers and Communications* (pp. 35- 40). Italy.

Barbeau, M., Hall, J., & Kranakis, E. (2006, October 4-6). Detection of rogue devices in Bluetooth networks using radio frequency fingerprinting. In *Proceedings of the 3rd IASTED International Conference on Communications and Computer Networks.* Lima, Peru.

Boncella, R. J. (2006). Wireless threats and attacks. In H. Bidgoli (Ed.), *Handbook of information security* (pp. 165-175). John Wiley & Sons.

Cretu, G. F., Parekh, J. J., Wang, K., & Stolfo, S. J. (2006, January 10-12). Intrusion and anomaly detection model exchange for mobile ad-hoc networks. In *The third IEEE Consumer Communications & Networking Conference (CCNC).*

Deng, J., Han, R., & Mishra, S. (2003, May). INSENS: Intrusion-tolerant routing in wireless sensor networks. In *The 23rd IEEE International Conference on Distributed Computing Systems (ICDCS).* Providence.

Deng, J., Han, R., & Mishra, S. (2004, June 28-July 1). Intrusion tolerance and anti-traffic analysis strategies for wireless sensor networks. In *Proceedings of the 2004 International Conference on Dependable Systems and Networks (DSN'04)* (pp. 637- 646). Italy.

Deng, J., Han, R., & Mishra, S. (2005). INSENS: Intrusion-tolerant routing for wireless sensor networks. [Special issue]. *Computer Communications Journal, 29*(2), 216-230.

Farshchi, J. (2003). *Wireless policy development (part 1 & 2), Security focus.* Retrieved from http://www.securityfocus.com/print/infocus/1732 Retrieved from http://www.securityfocus.com/print/infocus/1735

Gupta, V., Krishnamurthy, S., & Faloutsos, M. (2002, October). *Denial of service attacks at the MAC layer in wireless ad hoc networks.* Anaheim, CA: MILCOM—Network Security.

Hall, J., Barbeau, M., & Kranakis, E. (2005, February 3-4). *Using mobility profiles for anomaly-based intrusion detection in mobile networks.* Paper presented at the 12th Annual Network and

Distributed System Security Symposium, San Diego, CA.

Hutchison, K. (2004). *Wireless intrusion detection systems.* Retrieved October 18, 2004 from http://www.sans.org/reading_room/whitepapers/wireless/

Kachirski, O., & Guha, R. (2003, January 6-9). Effective intrusion detection using multiple sensors in wireless ad hoc networks. In *Proceedings of the 36th Hawaii International Conference on System Sciences (HICSS'03).* Hawaii.

Low, C. (2005). *Understanding wireless attacks & detection.* Retrieved April 2005, from http://www.hackerscenter.com/public/Library/782_wireattacks.pdf

Mateli, P. (2006). Hacking techniques in wireless networks. In H. Bidgoli (Ed.), *Handbook of information security* (pp. 83-93). John Wiley& Sons.

Nichols, R. K., & Lekkas, P. C. (2002). *Telephone system vulnerabilities.* McGraw-Hill.

Phifer, L. (2006). *Wireless attacks, A to Z.* Retrieved April 10, 2006, from http://searchsecurity.techtarget.com/generic/0,295582,sid14_gci1167611,00.html

Samad, K., Ahmed, E., & Mahmood, W. (2005, September 15-17). Simplified clustering approach for intrusion detection in mobile ad hoc networks. In *13th International Conference on Software, Telecommunications and Computer Networks (SoftCOM 2005).* Split, Croatia.

Schäfer, G. (2003). *Security in fixed and wireless networks, An introduction to securing data communications.* John Wiley and Sons.

Valli, C. (2004, June 28-29). WITS—Wireless intrusion tracking system. 3rd European Conference on Information Warfare and Security. UK.

Vladimirov, A. A., Gavrilenko, K. V., & Mikhailovsky, A. A. (2004). Counterintelligence: Wireless IDS systems. In *WI-Foo: The secrets of wireless hacking* (pp. 435-456). Pearson/Addison-Wesley.

Zhang, Y., Lee, W., & Huang, Y. (2003). Intrusion detection techniques for mobile wireless networks. *Wireless Networks Journal, 9*(5), 545-556.

KEY TERMS

Access Point (AP): Access point in the base station in a wireless LAN. APs are typically standalone devices that plug into an Ethernet hub or switch. Like a cellular phone system, users can roam around with their mobile devices and be handed off from one AP to the other.

Ad Hoc Networks: Ad hoc networks are local area networks or other small networks, especially ones with wireless or temporary plug-in connections, in which some of the network devices are part of the network only for the duration of a communications session or, in the case of mobile or portable devices, while in some close proximity to the rest of the network.

Intrusion Prevention System (IPS): IPS is the software that prevents an attack on a network or computer system. An IPS is a significant step beyond an intrusion detection system (IDS), because it stops the attack from damaging or retrieving data. Whereas, an IDS passively monitors traffic by sniffing packets off a switch port, an IPS resides inline like a firewall, intercepting and forwarding packets. It can thus block attacks in real time.

Intrusion Tolerance: Intrusion tolerance is the ability to continue delivering a service when an intrusion occurs.

Wireless Attack: A wireless attack is a malicious action against wireless system information or wireless networks; examples can be denial of service attacks, penetration, and sabotage.

Wireless Intrusion Detection System (WIDS): The WIDS is the software that detects an attack on a wireless network or wireless system. A network IDS (NIDS) is designed to support multiple hosts, whereas a host IDS (HIDS) is set up to detect illegal actions within the host. Most

IDS programs typically use signatures of known cracker attempts to signal an alert. Others look for deviations of the normal routine as indications of an attack. Intrusion detection is very tricky.

Wireless Sensors Networks (WSN): WSN is a network of RF transceivers, sensors, machine controllers, microcontrollers, and user interface devices with at least two nodes communicating by means of wireless transmissions.

Wireless Traffic Anomaly: Wireless traffic anomaly is a deviation from the normal wireless traffic pattern. An intrusion detection system (IDS) may look for unusual traffic activities. Wireless traffic anomalies can be used to identify unknown attacks and DoS floods.

Wireless Vulnerability: Wireless vulnerability is a security exposure in wireless components. Before the Internet became mainstream and exposed every organization in the world to every attacker on the planet, vulnerabilities surely existed, but were not as often exploited.

Chapter VII
Peer-to-Peer (P2P) Network Security: Firewall Issues

Lu Yan
University College London, UK

INTRODUCTION

A lot of networks today are behind firewalls. In peer-to-peer (P2P) networking, firewall-protected peers may have to communicate with peers outside the firewall. This chapter shows how to design P2P systems to work with different kinds of firewalls within the object-oriented action systems framework by combining formal and informal methods. We present our approach via a case study of extending a Gnutella-like P2P system (Yan & Sere, 2003) to provide connectivity through firewalls.

PROBLEM DEFINITION

As firewalls have various topologies (single, double, nested, etc.) and various security policies (packet filtering, one-way-only, port limiting, etc.), our problem has multiple faces and applications have multitude requirements. A general solution that fits all situations seems to be infeasible in this case. Thus we define the problem as shown in Figure 1: How to provide connectivity between private peers and public peers through a single firewall?

We select the object-oriented action systems framework with Unified Modeling Language (UML) diagrams as the foundation to work on. In this way, we can address our problem in a unified framework with benefits from both formal and informal methods.

Action systems is a state based formalism. It is derived from the guarded command language of Dijkstra (1976) and defined using *weakest precondition* predicate transformers. An action, or guarded command, is the basic building block

Figure 1. Problem definition

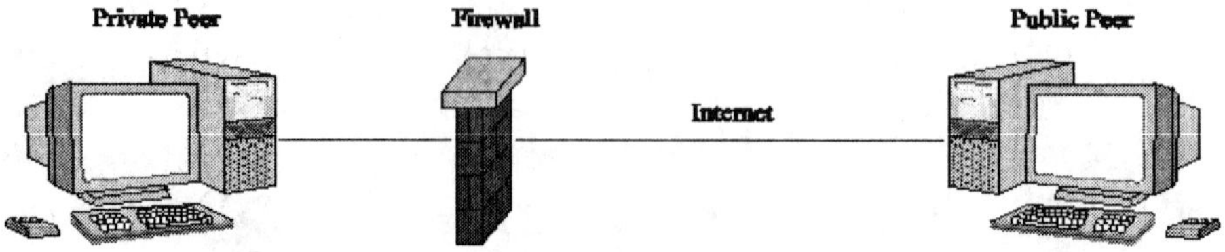

in the formalism. An action system is an iterative composition of actions. The action systems framework is used as a specification language and for the correct development of distributed systems.

Object-oriented (OO)-action system is an extension to the action system framework with OO support. An OO-action system consists of a finite set of classes, each class specifying the behavior of objects that are dynamically created and executed in parallel. The formal nature of OO-action systems makes it a good tool to build reliable and robust systems. Meanwhile, the OO aspect of OO-action systems helps to build systems in an extendable way, which will generally ease and accelerate the design and implementation of new services or functionalities. Furthermore, the final set of classes in the OO-action system specification is easy to be implemented in popular OO languages like Java, C++ or C#.

In this chapter, however, we skip the details of semantics of action systems (Back & Sere, 1996) and its OO extension (Bonsangue, Kok, & Sere, 1998).

GNUTELLA NETWORK

Gnutella (Ivkovic, 2001) is a decentralized P2P file-sharing model that enables file sharing without using servers. To share files using the Gnutella model, a user starts with a networked computer A with a Gnutella *servent*, which works both as a server and a client. Computer A will connect to another Gnutella-networked computer B and then announce that it is *alive* to computer B. B will in turn announce to all its neighbors C, D, E, and F that A is alive. Those computers will recursively continue this pattern and announce to their neighbors that computer A is alive. Once computer A has announced that it is alive to the rest of the members of the P2P network, it can then search the contents of the shared directories of the P2P network.

Search requests are transmitted over the Gnutella network in a decentralized manner. One computer sends a search request to its neighbors, which in turn pass that request along to their neighbors, and so on. Figure 2 illustrates this model. The search request from computer A will be transmitted to all members of the P2P network, starting with computer B, then to C, D, E, F, which will in turn send the request to their neighbors, and so forth. If one of the computers in the P2P network, for example, computer F, has a match, it transmits the file information (name, location, etc.) back through all the computers in the pathway towards A (via computer B in this case). Computer A will then be able to open a direct connection with computer F and will be able to download that file directly from computer F.

UNIDIRECTIONAL FIREWALLS

Most corporate networks today are configured to allow outbound connections (from the firewall protected network to Internet), but deny inbound connections (from Internet to the firewall protected network) as illustrated in Figure 3.

Figure 2. Gnutella peer-to-peer model

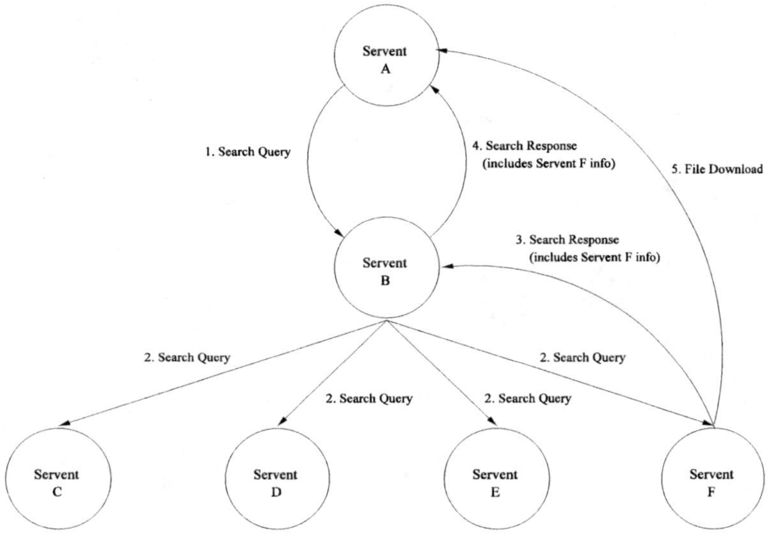

These corporate firewalls examine the packets of information sent at the transport level to determine whether a particular packet should be blocked. Each packet is either forwarded or blocked based on a set of rules defined by the firewall administrator. With packet-filtering rules, firewalls can easily track the direction in which a TCP connection is initiated. The first packets of the TCP three-way handshake are uniquely identified by the flags they contain, and firewall rules can use this information to ensure that certain connections are initiated in only one direction. A common configuration for these firewalls is to allow all connections initiated by computers inside the firewall, and restrict all connections from computers outside the firewall. For example, firewall rules might specify that users can browse from their computers to a web server on Internet, but an outside user on Internet cannot browse to the protected user's computer.

In order to traverse this kind of firewall, we introduce a *Push* descriptor and routing rules for servents: Once a servent receives a QueryHit descriptor, it may initiate a direct download, but it is impossible to establish the direct connection if the servent is behind a firewall that does not permit incoming connections to its Gnutella port. If this direct connection cannot be established, the servent attempting the file download may request that the servent sharing the file *push* the file instead. That is, A servent may send a Push descriptor if it receives a QueryHit descriptor from a servent that does not support incoming connections.

Intuitively, Push descriptors may only be sent along the same path that carried the incoming QueryHit descriptors as illustrated in Figure 4. A servent that receives a Push descriptor with *ServentID* = *n*, but has not seen a QueryHit descriptor with *ServentID* = *n* should remove the Push descriptor from the network. This ensures that only those servents that routed the QueryHit descriptors will see the Push descriptor.

We extend our original system specification (Yan & Sere, 2003) to adopt unidirectional firewalls by adding a Push router *Rf*, which is a new action system modeling Push routing rules as shown in Table 1. We compose it with the previous two action systems (Yan & Sere, 2003) *Rc* modeling Ping-Pong routing rules and *Rl* modeling Query-QueryHit routing rules together, to derive a new specification of router

$R = |[\ Rc\ //\ Rl\ //\ Rf\ ]|$

Figure 3. Unidirectional firewall

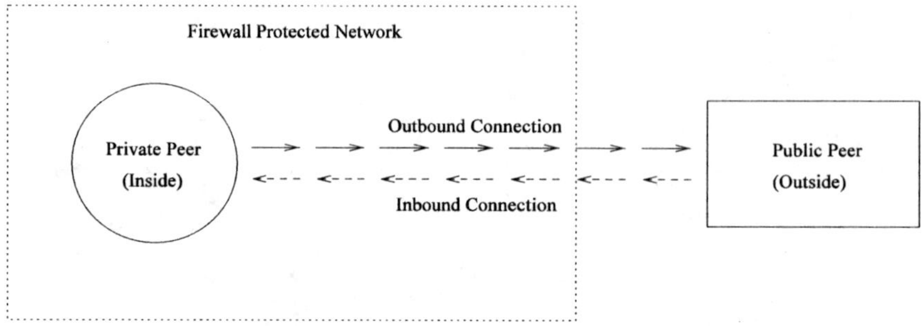

Figure 4. Push routing

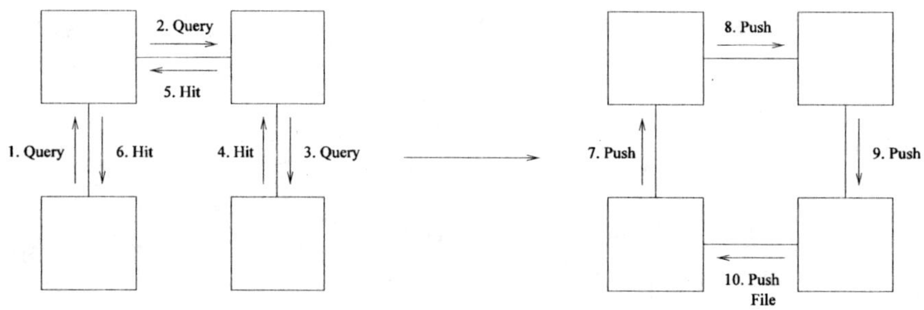

where on the higher level, we have components of a new router

{<*Router, R*>, <*PingPongRouter, Rc*>, <*QueryRouter, Rl*>, <*PushRouter, Rf*>}.

A servent can request a file push by routing a Push request back to the servent that sent the QueryHit descriptor describing the target file. The servent that is the target of the Push request should, upon receipt of the Push descriptor, attempt to establish a new TCP/IP connection to the requesting servent. As specified in the refined file repository in Table 2, when the direct connection is established, the firewalled servent should immediately send a HTTP GIV request with *requestIP, filename* and *destinationIP* information, where requestIP and destinationIP are IP address information of the firewalled servent and the target servent for the Push request, and filename is the requested file information. In this way, the initial TCP/IP connection becomes an outbound one, which is allowed by unidirectional firewalls. Receiving the HTTP GIV request, the target servent should extract the requestIP and filename information, and then construct an HTTP GET request with the above information. After that, the file download process is identical to the normal file download process without firewalls. We summarize the sequence of a Push session in Figure 5.

PORT-BLOCKING FIREWALLS

In corporate networks, other kinds of common firewalls are port-blocking firewalls, which usually do not grant long-time and trusted privileges to ports and protocols other than port 80 and

Table 1. Specification of push router

```
Rf = |[ attr serventDB := null; cKeyword := null;
        filename := null; target := null;
        pushTarget := null
    obj receivedMsg : Msg; newMsg : Msg;
        f: FileRepository
    meth SendPush( ) =
          (newMsg := new(Msg(Push));
           newMsg.info.requestIP := ThisIP;
           newMsg.info.filename :=
             receivedMsg.info.filename;
           newMsg.info.destinationIP :=
             receivedMsg.info.IP;
           OutgoingMessage := newMsg);
         ReceiveMsg( ) = receivedMsg :=
           IncomingMessage;
         ForwardMsg(m) = (m.TTL > 0 →
           m.Transmit( );
           OutgoingMessage := m)
    do
      true→
        ReceiveMsg( );
        if receivedMsg.type = QueryHit→
          serventDB := serventDB U
          receivedMsg.serventID;
         if receivedMsg.info.keyword =
           cKeyword→
             target := receivedMsg.info.filename
             @receivedMsg.info.IP;
             if f.firewall→
                SendPush( )
             fi
             cKeyword := null
         [] receivedMsg.info.keyword≠
           cKeyword ^
           receivedMsg.descriptorID ∈
           descriptorDB→
                              ForwardMsg(receivedMsg)
```

Table 1. continued

```
         fi
             [] receivedMsg.type = Push→
               if receivedMsg.info.destinationIP =
                 ThisIP→
                   PushTarget :=
                     receivedMsg.info.requestIP®
                     receivedMsg.info.filename@
                     receivedMsg.info.destinationIP
               [] receivedMsg.info.destinationIP≠
                 ThisIP ^
                 receivedMsg.serventID ∈
                 serventDB→
                   ForwardMsg(receivedMsg)
               fi
             fi
         od
     ]|
```

HTTP/HTTPS. For example, port 21 (standard FTP access) and port 23 (standard Telnet access) are usually blocked and applications are denied network traffic through these ports. In this case, HTTP (port 80) has become the only entry mechanism to the corporate network. Using HTTP protocol, for a servent to communicate with another servent through port-blocking firewalls, the servent has to *pretend* that it is an HTTP server, serving WWW documents. In other words, it is going to mimick an *httpd* program.

When it is impossible to establish an IP connection through a firewall, two servents that need to talk directly to each other solve this problem by having SOCKS support built into them, and having SOCKS proxy running on both sides. As illustrated in Figure 6, it builds an HTTP tunnel between the two servents.

After initializtion, the SOCKS proxy creates a *ProxySocket* and starts accepting connections on the Gnutella port. All the information to be sent by the attempting servent is formatted as a URL message (using the GET method of HTTP) and an *URLConnection* via HTTP protocol (port 80) is

Figure 5. Sequence diagram of a push session

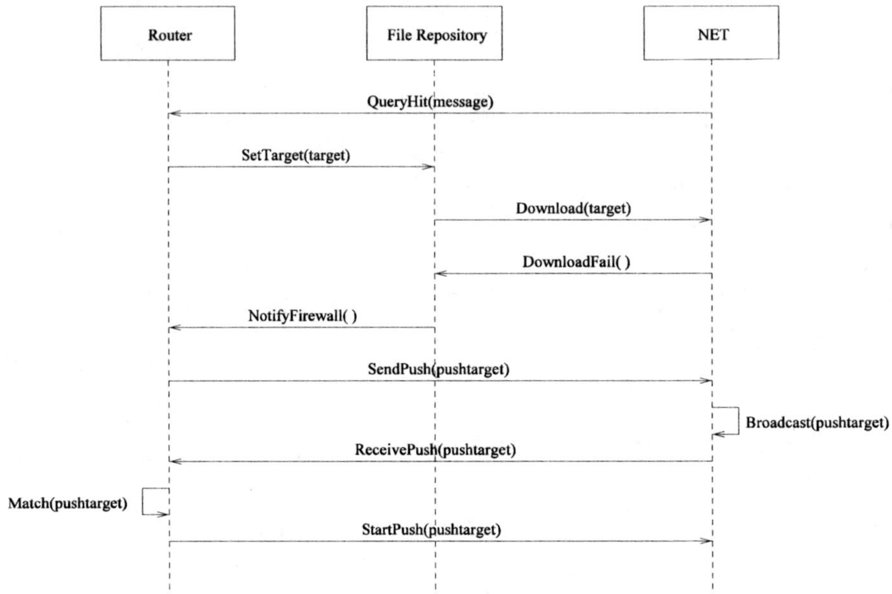

made. On the other side, the target servent accepts the request and a connection is established with the attempting servent (actually with the SOCKS proxy in the target servent). The SOCKS proxy in the target servent can read the information sent by the attempting servent and write back to it. In this way, transactions between two servents are enabled.

We extend our original system specification (Yan & Sere, 2003) to adopt port-blocking firewalls by adding a new layer to the architecture of servent in Figure 7. This layer will act as a tunnel between servent and Internet.

As specified in Table 3, after receiving messages from the attempting servent and encoding them into HTTP format, the SOCKS proxy sends the messages to the Internet via port 80. In the reverse way, the SOCKS proxy keeps receiving messages from HTTP port and decoding them into original format. With this additional layer, our system can traverse port-blocking firewalls without any changes in its core parts. We summarize the sequence of a SOCKS proxy session in Figure 8.

CONCLUSION

The corporate firewall is a double-edged sword. It helps prevent unauthorized access to the corporate Web, but may disable access for legitimate P2P applications. There have been protocols such as Point-to-Point Tunneling Protocol (PPTP) (Hamzeh et al., 1999), Universal Plug and Play (UPNP) (Microsoft, 2000), Realm Specific IP (RSIP) (Borella & Montenegro, 2000) and Middlebox protocol (Reynolds & Ghosal, 2002) to address the firewall problems in P2P networking. A recent protocol, JXTA (Gong, 2001) from Sun has provided an alternative solution to the firewall problem by adding a publicly addressable node, called *rendezvous server*, which a firewalled peer can already talk to. The scheme is that peers interact mostly with their neighbors who are on the same side of the firewall as they are and one or a small number of designated peers can bridge between peers on the different sides of the firewall. But the problem posed by firewalls still remains when configuring the firewalls to allow traffic through these bridge peers.

We have specified a Gnutella-like P2P system within the OO-action systems framework by combining UML diagrams. In this chapter, we have

Table 2. Specification of file repository

```
F = |[ attr firewall* := false; fileDB := FileDB;
          cFileDB; filename := null; target := null;
          pushTarget := null
     meth SetTarget(t) = (target := t);
            PushTarget(t) = (pushTarget := t);
            Has(key) = ({key} є dom(fileDB));
            Find(key) = (filename := file ^
               {file} є ran({key} ◀ fileDB))
     do
          target≠ null→
             cFileDB := fileDB;
             HTTP_GET(target);
             target := null;
             Refresh(fileDB);
             if fileDB = cFileDB→
                firewall := true
             [] fileDB≠ cFileDB→
                firewall := false
             fi
        [] pushTarget≠ null→
             HTTP_GIV(pushTarget);
             pushTarget := null;
             Refresh(fileDB)
     od
  ]|
```

Figure 7. Refined architecture of servent

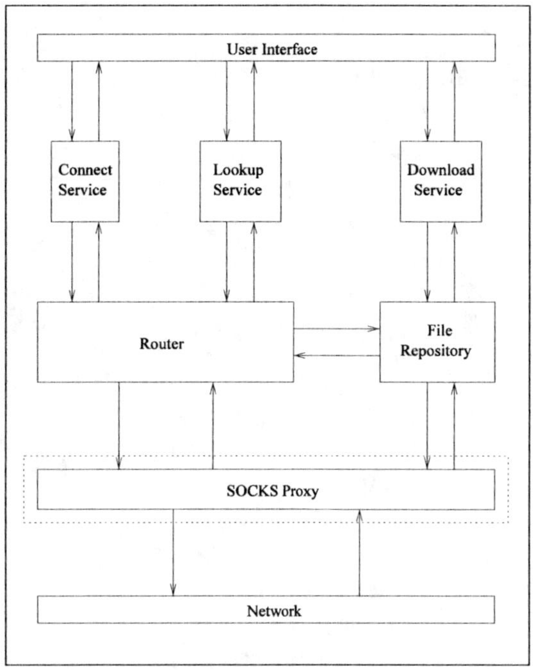

Figure 6. Firewall architecture and extendable socket

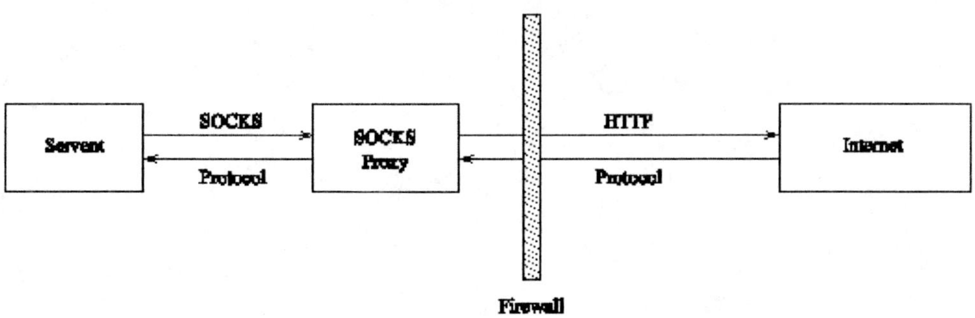

Table 3. Specification of SOCKS proxy

```
S = |[ attr listenPort := GnutellaPort;
        DestinationPort := 80
     obj ProxySocket : Socket;
        HTTPSocket : Socket;
        imsg : Msg; omsg : Msg
     init ProxySocket = new(Socket(listenPort));
        HTTPSocket =
          new(Socket(destinationPort))
     do
        IncomingRequest ≠ null→
          imsg := EncodeSOCK(DecodeHTTP
            (HTTPSocket.Read( )));
          IncomingMessage :=
            ProxySocket.Write(imsg)
       [] OutgoingRequest≠ null→
          omsg := EncodeHTTP(DecodeSOCK
            (ProxySocket.Read( )));
          OutgoingMessage :=
            HTTPSocket.Write(omsg)
     od
  ]|
```

presented our solution to traverse firewalls for P2P systems. We have extended a Gnutella-style P2P system to adopt unidirectional firewalls and port-blocking firewalls using OO-action systems. During the extending work, our experiences show that the OO aspect of OO-action systems helps to build systems with a reusable, composable, and extendable architecture. The modular architecture of our system makes it easy to incorporate new services and functionalities without great changes to its original design.

REFERENCES

Back, R. J. R., & Sere, K. (1996). From action systems to modular systems. *Software Concepts and Tools*.

Bonsangue, M., Kok, J. N., & Sere, K. (1998). An approach to object-orientation in action systems. In *Proceedings of Mathematics of Program Construction (MPC'98)* (LNCS 1422). Springer-Verlag.

Borella, M., & Montenegro, G. (2000). RSIP: Address sharing with end-to-end security. In *Proceedings of the Special Workshop on Intelligence at the Network Edge,* CA.

Figure 8. Sequence diagram of a SOCKS proxy session

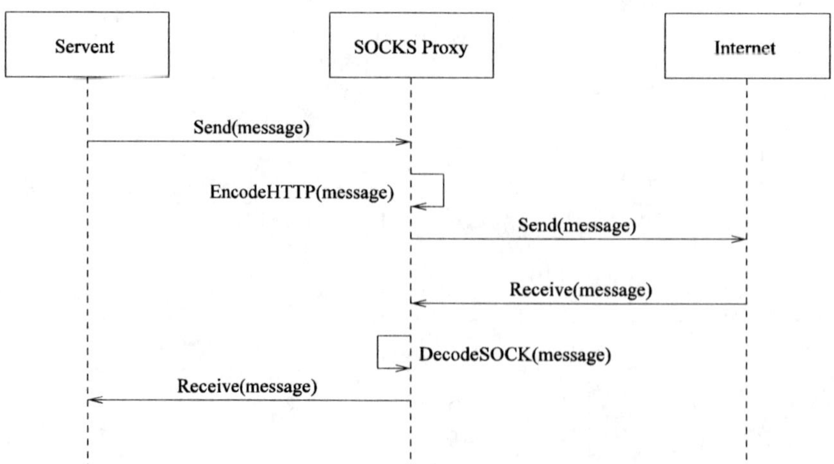

Dijkstra, E. W. (1976). *A discipline of programming.* Prentice-Hall International.

Gong, L. (2001). JXTA: A network programming environment. *IEEE Internet Computing.*

Hamzeh, K., Pall, G., Verthein, W., Taarud, J., Little, W., & Zorn, G. (1999). *Point-to-point tunneling protocol (PPTP)* (RFC 2637). Retrieved from http://www.ietf.org/rfc/rfc2637.txt

Ivkovic, I. (2001). *Improving Gnutella protocol: Protocol analysis and research proposals.* (Tech. Rep.). LimeWire LLC.

Microsoft. (2000). Understanding universal plug and play. White paper. Redmond WA: Author.

Reynolds, B., & Ghosal, D. (2002). STEM: Secure telephony enabled middlebox [Special issue]. *IEEE Communications.*

Yan, L., & Sere, K. (2003). Stepwise development of peer-to-peer systems. In *Proceedings of the 6th International Workshop in Formal Methods (IWFM'03),* Dublin, Ireland.

KEY TERMS

Action System: An action system is a notation for writing programs, due to Ralph Back. An action system is a collection of actions. It is executed by repeatedly choosing an action to execute. If it is the case that no action is able to be executed, then execution of the action system stops.

Firewall: A firewall is a piece of hardware and/or software which functions in a networked environment to prevent some communications forbidden by the security policy, analogous to the function of firewalls in building construction.

Peer-to-Peer (P2P): A peer-to-peer (P2P) computer network is a network that relies primarily on the computing power and bandwidth of the participants in the network rather than concentrating it in a relatively low number of servers. P2P networks are typically used for connecting nodes via largely ad hoc connections.

Chapter VIII
Identity Management for Wireless Service Access

Mohammad M. R. Chowdhury
University Graduate Center – UniK, Norway

Josef Noll
University Graduate Center – UniK, Norway

ABSTRACT

Ubiquitous access and pervasive computing concept is almost intrinsically tied to wireless communications. Emerging next-generation wireless networks enable innovative service access in every situation. Apart from many remote services, proximity services will also be widely available. People currently rely on numerous forms of identities to access these services. The inconvenience of possessing and using these identities creates significant security vulnerability, especially from network and device point of view in wireless service access. After explaining the current identity solutions scenarios, the chapter illustrates the on-going efforts by various organizations, the requirements and frameworks to develop an innovative, easy-to-use identity management mechanism to access the future diverse service worlds. The chapter also conveys various possibilities, challenges, and research questions evolving in these areas.

INTRODUCTION

Nowadays people are increasingly connected through wireless networks from public places to their office/home areas. The deployment of packet-based mobile networks has provided mobile users with the capability to access data services in every situation. The next-generation wireless network is expected to integrate various radio systems including third generation (3G), wireless LANs (WLANs), fourth generation (4G), and others. One motivation of this network is the pervasive computing abilities, which provide automatic handovers for any moving computing devices in a globally networked environment. Fast vertical handover is considered important for managing continued access to different types of network resources in next generation networks (Li et al., 2005). Such networks will provide ubiquitous service access taking the advantages of each of these forms of wireless communications. Service intake will be increased significantly through the availability and reach of innovative and easy-to-use services. Apart from the remote service access (Web services), the introduction of near field communication (NFC) in use with a mobile phone can enable many new proximity services.

User identity solutions and its hassle-free management will play a vital role in the future ubiquitous service access. Current identity solutions can no longer cope with the increasing expectations of both users and service providers in terms of their usability and manageability. Mobile and Internet service providers are increasingly facing the same identity management challenges as services in both domains continue to flourish. Real-time data communication capabilities of mobile networks will multiply the remote service accesses through mobile networks, if efficient identity management and security is ensured over the wireless access. Personalization through customized user profiles based on their preferences will become an important factor for success of future wireless service access. In more advanced service scenarios, open identity management architecture enables the use of standard user profile attributes, like age and gender, and authorizations for service, such as location, to bring a richer user experience. Users, network operators, and service providers can make use of an open standard technology for identity management to meet their own specific requirements through customizations. There is clearly a need for such a standard for identity management that can be applied to all ubiquitous service access scenarios. As user needs are at the center in the service world from business perspective, identity management mechanism should be user-centric.

The impressive capabilities and reach of emerging next-generation networks, the abundance of services, and on-going development in user device require proper address to the user identity management issues which have yet met the stakeholders' expectations. The main goal of this chapter is to discuss these concerns. The second section discusses the background of identity management. In the third section, requirements and framework of identity management mechanism for wireless service access are given mentioning the current efforts by various organizations. Security issues are also a part of this mechanism. The fourth section provides the future trends. The chapter concludes with the summary of all discussions.

BACKGROUND

In a broadest sense, identity management encompasses definitions and life-cycle management for user identities and profiles, as well as environments for exchanging and validating such information. A service provider issues identity to its users. Identity life-cycle management comprises establish/re-establishment of identity, description of identity attributes, and at the end revocation of identity. Attributes are a set of characteristics of an identity that are required by the service providers to identify a user during service interactions. User authenticates to the service providers as real owner of the identity for accessing services. Authentication is a key aspect of trust-based identity attribution, providing a codified assurance of the identity of one entity to another.

Next-generation wireless network includes state-of-the-art intelligent core network and various wireless access networks. It is expected to offer sufficient capacity, quality of service (QoS), and interoperability for seamless service access remotely. Currently the network and thereby the remote service access are often granted through numerous user identification and authentication mechanisms, such as, usernames/passwords/PIN codes/certificates. Users have to register prior to first usage and publish private information, often more than what is strictly necessary for service access. It hampers user's privacy. There is a growing consensus among the legislators across the world that individual's rights of privacy and the protection of personal data is equally applicable in the context of the Information Society as it is in the off-line world. To address this issue, a user-centric identity management framework is expected where users having complete control over the identity information transmission.

Some services happen in the proximity of users at local access points. These services are accessed through physical interactions with physical cards or devices, for example, payment and admittance. The use of NFC with mobile phones to transfer user information from one device to another boosts the intake of proximity services. The user personal device is often used to store his/her identity information. To protect unauthorized service access, users also need to be authenticated before accessing such devices. It is evident that a user is burdened with too many identities to access many remote and proximity services. An integrated approach is required to manage all those identities to access all these services.

Wireless service access results in more complexity to manage identities prior to accessing the services. Besides device authentications, users need to authenticate themselves before accessing the wireless networks. In addition to this, because of the size limitations, mobile devices are equipped with smaller screens and limited data entry capabilities using small keypads. For wireless services to succeed, it is critical that the mobile users are able to get convenient and immediate access to the information and services they need without going through long menus and having to enter various usernames and passwords.

In the future, one of the key issues of identity management in the wireless domain will be who the identity providers will be to the users and who will own/manage the subscriber identity module (SIM/USIM). It is because, currently, almost every service provider is also an identity provider for users to access that specific service. SIM card is in fact a smart card with processing and information storage capabilities. With the development of powerful, sophisticated as well as secure smart cards, it is now considered as the storage place for user's identity information. In current cellular models, the operator provides not only the wireless access but also owns and manages SIM/USIM. In this case, the user has little control over his/her identity. A user is having a SIM/USIM as his/her identity but is not allowed to modify or update it so that he/she cannot subscribe to new wireless providers or to whatever service providers he/she likes. A collaborative operator model has been thought where such identity module belongs to the user (Kuroda, Yoshida, Ono, Kiyomoto, & Tanaka, 2004, pp. 165-166). A third party can provide the infrastructure to manage such identity. This approach leads towards user-centric identity management and provides the user with flexibility in choosing wireless providers.

In general, common identity deployment architectures can be broadly classified into three types: Silo, Walled Garden, and Federation (Altmann & Sampath, 2006, p. 496). Current identity management in the service world is mostly silo-based. Silo is a simple architecture, which requires each service provider to maintain a unique ID for each user. This approach is simpler from a service provider's point of view but it is not only laborious but also problematic for the user. Moreover, it results in a huge waste of resources due to the possession of redundant identity information in the service world. As studies show, users who register with several service providers routinely forget their passwords for less frequently used accounts. This has a significant financial effect. On average, $45 is spent on password reset each time a user forgets a password (Altmann & Sampath, 2006, p. 496). Walled Garden is a centralized identity management approach where all service providers can typically rely on one singe identity provider to manage the user's identity. The user is benefited through managing only a single set of credentials. Its inherent weakness is, once the significant barrier of protection is compromised, a malicious user enjoys unbridled access to all resources. Lastly, in identity federation management a group of service providers forms a federation. Here, each service provider recognizes the identifiers of other service providers and thereby, consider a user who has been authenticated by another service provider to be authenticated as well. However, the real distinction between Walled Garden and Federation approach is that here service providers have their own unique identifiers and credentials. Though this approach is widely accepted considering the heterogeneity of service providers, many possible service interaction scenarios and the requirements of several levels of security make such a system far more complex.

IDENTITY MANAGEMENT FOR WIRELESS SERVICE ACCESS

Designing an identity management mechanism to access both remote and proximity services, without using numerous inconvenient identity solutions, is expected to be the main focus in the identity management for service access over wireless networks. This section also considers the selection of a user identity storage place, the role of identity provider, and various other requirements to develop such a mechanism from a wireless service access point of view.

Requirements of Identity Management Systems

Identity management system should be user-centric. It means such a system should reveal information identifying a user with user's consent. Security is one of the most important concerns of this system. The system should protect the user against deception, verifying the identity of any parties who ask for information to ensure that it goes to the right place. In the user-centric approach, the user will decide and control the extent of identifying information to be transmitted. The system must disclose the least identifying information possible. By following these practices, the least possible damage can be ensured in the event of a breach. These are some of the requirements to design a user-centric identity management system in The Laws of Identity (Cameron, 2005).

Identity management system requires an integrated and often complex infrastructure where all involved parties must be trusted for specific purposes depending on their role. Since there are costs associated with establishing trust, it will be an advantage to have identity management models with simple trust requirements (Jøsang, Fabre, Hay, Dalziel, & Pope, 2005). Success of an identity management system depends upon the ability to interoperate across a trusted network of businesses, partners, and services regardless of the platform, programming language, or application with which they are interacting. It should handle user identities for both remote (Web) and proximity service access. Above all, such a system should be user friendly.

Identity Management Solutions and Controversies

Various institutes and industries are working to develop the required identity management solutions. SXIP ("The SXIP 2.0 Overview," n.d.). identity has designed a solution to address the Internet-scalable and user-centric identity architecture. It provides user identification, authentication and Internet form fill solutions using Web interfaces for storing user identity, attribute profiles, and facilitating automatic exchange of identity data over the Internet. Windows CardSpace uses various virtual cards (mimic physical cards) issued by the identity providers for user identifications and authentications, each retrieving identity data from an identity provider in a secure manner (Chowdhury & Noll, 2007). In the Liberty Alliance Project (Miller et al., 2004), members are working to build open standard-based specifications for federated identity and interoperability in multiple federations, thereby fostering the usage of identity-based Web services. Within this, they are focusing on end-user privacy and confidentiality issues and solutions against identity theft. But these efforts are mainly focusing on identity management in the Internet domain.

Besides working for identity handling in a Web domain, Liberty Alliance (Miller et al., 2004) also provides solutions in identity management for mobile operators. It proposes single sign-on (SSO) to relieve the users from managing many usernames/passwords and for fast access to the resources. But in SSO, if a malicious attacker secures one of the user's accounts, he/she will enjoy an unbridled access to data pertaining not only to that account but also across all her accounts spread across domains. Therefore, some research approaches do not encourage such SSOs (Altmann & Sampath, 2006, p. 500). However, a current version of liberty, Shibboleth, reduces such risk by providing an attribute-based authorization system. But in wireless service access, especially for mobile devices seamless service sign-on solutions and one-click access to personalized services are key issues for successful identity management.

Apart from possessing numerous usernames/passwords/PIN codes for remote (Web) service access, the user is also carrying many physical identities for proximity service access. These include credit card, bank card, home/office access cards, and so forth. Many researchers working in these areas are proposing the smart cards, like SIM/USIM currently used in mobile phones, as the secure storage place for the user's identity information because it can be revoked, users nowadays can rarely be found without a mobile phone and there are possibilities of security enhancements. Custom made SIMs/USIMs having enough computational power and storage space can be used to manage users' identification information and multi-factor authentication mechanisms. Gemalto, a company providing digital security, is involved in developing sophisticated smart cards (e.g., SIM/USIM) based online or off-line identity management with associated software, middleware, and server-based solutions. NXP, a semiconductor company (formerly a division of Philips), is also offering identification products in areas like government, banking, access control, and so forth using secure innovative contactless smart cards and chips. Credit card companies are running various trials for providing user's payment identity handling solutions using mobile phones and NFC technology. *Tap N Go* is the name of a contactless payment trial powered by MasterCard *PayPass* (2007) in the U.S. started in 2006. In the same year, Visa completed contactless-based mobile pilots in Malaysia and the United States, using NFC-enabled phones, complementing existing programs in Japan and Korea. In February 2007, Visa International and SK Telecom of South Korea announced the world's first contactless payment application on a universal SIM card which is personalized over-the-air based on Visa's recently introduced mobile platform ("Visa's mobile platform initiative," 2007).

Identity providers issue identities to each user. They have a very important central role in the identity management business. The identity provider manages users' identities and their access rights to various services securely. It provides the authentication and authorization services to the users. Who can be the identity providers in future identity management systems, is a debatable issue. Liberty Alliance (Miller et al., 2004) believes that mobile operators are in a good position to become the most favored identity providers, because they possess valuable static and dynamic user information which can be transmitted to third parties in a controlled manner through open standard Web service interface. Mobile operators also have the ability to seamlessly authenticate users with the phone number on behalf of the service providers (SP). Many contradict such roles of mobile operators. Instead a more trusted third party, like financial institutes and governments are also well positioned to become preferred identity providers. They might provide identity services for their specific market and services that need stronger user identities. When a user wants to subscribe to a new wireless network, he/she asks the third party identity provider to add new identification data into his/her phone. In such a situation, it is possible that a third party can even manage SIM/USIM, which is currently done by cellular operators. It is expected that the next-generation wireless network will have such flexibility.

Components of User Identities

Identity management in wireless service access needs to address device-level security, network-level security, and service-level security (Kuroda et al., 2004, p. 169). Therefore, the over-all user identity comprises device, network, and service identities. The user's device is divided into two components, a personal smart card (e.g., SIM/USIM) and mobile devices with wireless access capabilities. The smart card includes user identification data that contains user's public or shared-secret keys, certificates for network operators, and service providers. The card and the device need to be mutually authenticated in the initial setup phase because both devices have built no relationship of trust to exchange security information from the very beginning. Afterwards, the user identifies him/herself to the card, since it stores sensitive personal information, which is used for network- and service-level authentication. The user can identify through PIN, password, or biometrics. After these authentication procedures,

Figure 1. User identifications to ensure device-, network-, and service-level security

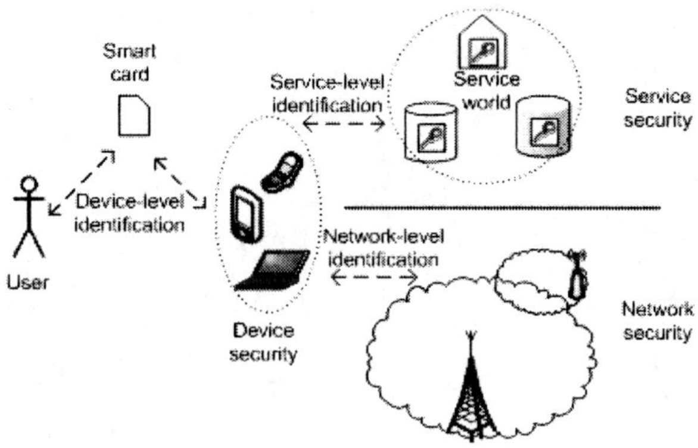

the card delegates user identity information to the mobile device to authenticate wireless access and thereby, service access. The user expects to use services without being concerned about the individual characteristics of each wireless access. Network-level authentication verifies that the user is a subscriber and has wireless access to the right network. Service-level authentication verifies that the user is a subscribed user to the right services. In each case, service or network providers and user device mutually authenticate each other. Figure 1 depicts the overview of device-, service-, and network-level identifications to ensure the security of user-device, network, and services for wireless service access.

Integrated Identity Management Mechanism

Every human being is playing numerous roles in life to live. To organize the user identities in a more structured way, all user identities can be broadly categorized into three areas based on the roles he/she exercises in real life (Chowdhury & Noll, 2006). These are personal identity (PID), corporate identity (CID), and social identity (SID). PIDs can be used to identify a user in his/her very personal and commercial service interactions. CIDs and SIDs can be used in professional and social, interpersonal interactions respectively. For example, PIDs include bank/credit card, home/office access card/code, and so forth. According to Dick Hardt (keynote speech at OSCON 2005 conference), founder and CEO of SXIP Identity, individual's interests, fondness, preferences, or tastes are also part of his/her identity. These roles can be dealt with by user's SIDs. Some of these identities are having very sensitive user information therefore very strict authentication requirements have to be met. Some others require less secure infrastructure as they possess not so sensitive user information.

Considering all these aspects, instead of storing user's vast identity information into a single place (a user device), these can be distributed into two places. The less sensitive user identity information, especially his/her SIDs can be stored in a secure network identity space. The most sensitive identity information like user's PIDs will be stored in user's personal device. The mobile phone (more correctly the SIM card) has been proposed as the user's personal device (Chowdhury & Noll, 2006). When the user subscribes to the identity services, the identity provider (IDP) issues a certificate to him/her. It will be stored in the device. At the same time, a secure identity space in the network will be allocated for the user too. The device identifies and authenticates the user to access his/her network identity space. When the user authenticates to the device and the network, he/she can also gain access to the network identity space (if it requires, an optional password can also protect such access). The user device holds the most sensitive user identity information. Depending on the security requirements of the services, the possession-based authentication (e.g., having a personal device) can be enhanced by a knowledge factor (e.g., PIN code). An additional knowledge-based authentication mechanism can be used here to grant access to sensitive PIDs stored in the device. This is how user identities can be stored in a distributed manner and multi-factor authentication mechanisms can protect the security of user's identities.

In future service access scenarios, the user expects a hassle-free use of identities. In this regard an approach is expected to integrate all user's identities to access every remote and proximity services into a single mechanism. The distributed identity infrastructure just being described can also

Figure 2. A generic diagram for integrated identity mechanism and service access

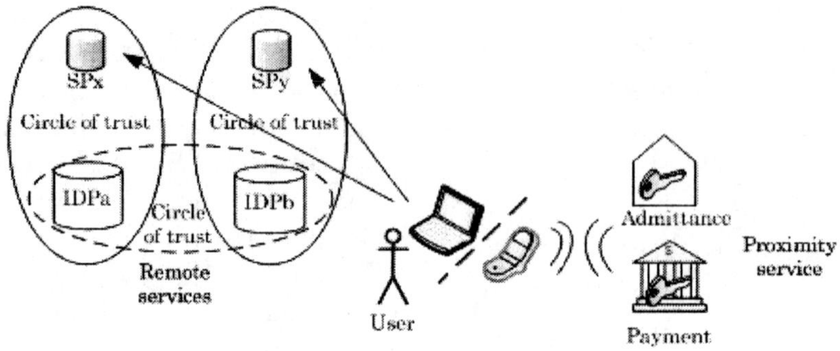

provide an integrated mechanism to handle the use of identities for every possible service access over the wireless network. For example, by using public key infrastructure (PKI) built in SIM card user can access services of banks remotely; with the NFC capable mobile phone user, can access home premises transferring the stored admittance key from user device. This is how a proximity service access can also be handled. Figure 2 shows a generic diagram for integrated identity mechanism to handle remote and proximity service access.

In a significant move towards providing secure ubiquitous digital credentials management; the banking industry of Norway with a partnership of a mobile operator initiated PKI-based BankID (Cybertrust Case Studies Library, 2005) for identification and signing agreement on the move. BankID for mobile phones will initially be used in four areas: (1) logging on to Internet banks, (2) mobile banking, (3) electronic service for business and the public sector, and (4) account-based payment service for the Internet and mobiles.

Security Infrastructure in Identity Management Systems

The wireless network along with the user device (mobile phone) can serve as the underlying secure infrastructure for exchanging user identity information and authentication messages. The next generation network will integrate 3G and WLAN to offer subscribers high-speed wireless data services as well as ubiquitous connectivity (Siddiqui et al., 2005). Users are expected to access countless services seamlessly over the wireless network. Hence, secure identity information handling is a crucial issue in wireless service access. In the mobile domain, the security of 3G mobile systems has already been strengthened by introducing longer cipher keys; mutual (network, user) authentication; signaling and data traffic integrity; and the extension of ciphering back into the network (Boman, Horn, Howard, & Niemi, 2002, pp. 192-200). WLAN security has been improved significantly with the adoption of IEEE 802.11i. It was created in response to several serious weaknesses researchers had found in the previous system, wired equivalent privacy (WEP). There have been discussions to accomodate both 3G and WLAN security frameworks on the IP layer or based on 3rd Generation Partnership Project (3GPP), but the resulting solution does not offer fast vertical handover which is critical for session continuity (Kuroda et al., 2004, p. 166). EAP-TLS and EAP-AKA have been proposed to provide strong end-to-end security and authentication to the user in such integrated network environment (Kambourakis, Rouskas, & Gritzalis, 2004, pp. 287-296). Due to the various weaknesses of Bluetooth/Infrared communications, NFC is considered as the secure technology to transfer user identity information between a user's mobile phone and the devices at service points. Its short range should mitigate the risk of eavesdropping by other reader devices. It is practically impossible to do man-in-the-middle attack on NFC link (Haselsteiner

& Breitfuss, 2006). Moreover, a secure channel can be established using cryptographic protocol between the two NFC devices.

Identity management and associated security infrastructure will play a vital role for seamless service interaction in the next generation wireless network. There are several important requirements of a successful identity management system. Such a system should be user centric rather than service centric. The user expects an integrated identity management mechanism that can handle identities for both remote and proximity service access. Security of the underlying infrastructure is also a crucial issue in wireless service access. This section has discussed all these aspects in brief.

FUTURE TRENDS

3G networks are covering a wide service area and providing ubiquitous connectivity to mobile users with low-speed data rates which is sufficient for most real-time communications. WLAN/Wi-Fi networks cover smaller areas and provide high data rates to static users. There exists a strong need for integrating WLANs/Wi-Fis with 3G networks to develop a hybrid of data networks capable of ubiquitous data services and very high data rates at strategic locations called "hotspots." The future wireless network is expected to attain this goal and thereby, seamless service access. Next generation wireless network architectures envisaged to constitute of an IP-based core network, whereas the access network can be based on a variety of heterogeneous wireless technologies depending on the nature of the access cell. In this environment, people can access many new innovative services from anywhere and anytime. Service intake will be increased significantly. Instead of current identity provisions, a new identity management mechanism is expected, especially in wireless service access.

People no longer use many usernames and passwords for remote service access. Instead SSO will become popular with the provisions of additional authentication levels to meet higher security requirements. In future identity management, the role of an identity provider will be very critical. The key issue will be who can be the identity provider? Whoever will be the identity provider, the user SIM/USIM card is in a good position to become a secure storage place for user identity information. Researchers are working to develop high capacity sophisticated smart cards to meet such demand in various service access scenarios. In this regard, it is very important to decide who will be the owner of such a user identity device (e.g., SIM/USIM). For acceptability of a SIM card as a secure identity storage place and to develop a user-centric identity management mechanism, the user should have rights to update or modify the SIM card. It is expected that the business model of the next generation network will have such flexibility. Mobile networks or other wireless access networks will play a vital role to ensure security for identity information exchange. Therefore, numerous efforts are going on to enhance the security infrastructure of access network's air interface and provide strong end-to-end protection for secure service access.

Introduction of NFC adds intelligence and networking capabilities to the phone and creates many new opportunities to add product and service capabilities to handset-like digital transactions in very good proximities. It can make the mobile phone an ideal device for payments and gaining access. Financial institutes like credit card companies and mobile manufacturers are running various trials with NFC-enabled mobile phoned in service access scenarios like admittance and payment. User identities for admittance and payment services are very sensitive in nature. Therefore, an integrated identity mechanism is expected to deal with these proximity services and as well as remote services.

Currently, the user expects and technology demands service personalization, including adaptation to personal preferences, terminal, and network capabilities. Rule-based personalization algorithms become too complex when handling user context and preferences, thus asking for new mechanisms allowing dynamic adaptability of services. Semantic descriptions of user preferences and user relations with the combination of current developments in security and privacy issues can create more dynamic service provisions and personalization. Semantic Web is seen as the next generation of the Internet where information has machine-readable and machine-understandable semantics.

CONCLUSION

Current reporting from the World Factbook states 1.5 to two times as many mobile users as Internet users for developed countries like UK, France, and Germany and roughly three times as many mobile users as Internet users in China (The World Factbook, 2006). Taking into account that mobile users are available 24/7 as compared to an average PC usage of 137.3 min/day for male (134.2 min/day for female) shows the importance of mobile service access. The emerging next generation network is expecting to integrate various access networks including 3G and WLAN/Wi-Fi networks. Ubiquitous access for seamless service interaction will be a reality soon.

The current identity provisions will not allow this to happen. Users possess many identities in various forms and identity information is stored in a scattered way in networks. Most of the recent developments are focused towards identity management in Internet domain to access remote services. However, some efforts also target service access located in the proximity of users. The success of future service access asks for an integrated identity mechanism to deal with both remote and proximity service access. The creation of a user's role-based identity in a dynamic way and use of Semantic Web technology will enhance user experience in service interaction. Such a dynamic and integrated identity mechanism together with mobile, sensor networks, and NFC-enabled mobile terminal can improve the healthcare system for better handling of patients, especially elderly and disabled people. Semantic descriptions of user preferences and relations can also improve user experience in social interactions.

The user personal wireless device along with a sophisticated smart card will play a key role for identity management for wireless service access in terms of user identity information storage and providing secure network and service authentication. With strong encryption, privacy, and data integrity mechanisms, mobile networks have the capability to provide the underlying security infrastructure for sensitive identity information exchange for mobile users. Mobile phones equipped with custom-made high capacity SIM cards can act as a secure user identity storage place. New developments in security mechanisms to protect mishandling of user identities over the air interphase as well as over the IP-based core network can make the identity management for wireless service access secure enough.

REFERENCES

Altmann, J., & Sampath, R. (2006, April). UNIQuE: A user-centric framework for network identity management. In *Proceedings of IEEE/IFIP Network Operations and Management Symposium, NOMS 2006* (pp. 495-506). Vancouver, Canada.

Boman, K., Horn, G., Howard, P., & Niemi, V. (2002, October). UMTS security. *Electronics and Communication Engineering Journal, 14*(5), 191-204.

Cameron, K. (2005). *The laws of identity.* Retrieved December 29, 2006, from http://identityblog.com/

Chowdhury, M. M. R., & Noll, J. (2006, November). *Service interaction through role based Identity.* Paper presented at Wireless World Research Forum Meeting 17, Heidelberg, Germany.

Chowdhury, M. M. R., & Noll, J. (2007, March). Distributed identity for secure service interaction. In *Proceedings of the Third International Conference on Wireless and Mobile Communications, ICWMC'07,* Guadeloupe, French Caribbean.

Cybertrust Case Studies Library. (2005). *BankID: Delivering bank-common trust for Web-based transactions.* Retrieved November 15, 2006, from https://www.cybertrust.com/intelligence/case_studies/

Damiani, E., De Capitani di Vimercati, S., & Samarati, P. (2003, November). Managing multiple and dependable identities. *IEEE Internet Computing, 7*(6), 29-37.

Haselsteiner, E. & Breitfuss, K. (2006). *Security in near field communication (NFC) strengths and weaknesses.* Paper presented at the Workshop on RFID Security—RFIDSec 06, Graz, Austria.

Jøsang, A., Fabre, J., Hay, B., Dalziel, J., & Pope, S. (2005). Trust requirements in identity management. In *Proceedings of the Australasian Information Security Workshop (AISW'05),* Newcastle, Australia.

Kambourakis, G., Rouskas, A., & Gritzalis, D. (2004). Performance evaluation of certificate based authentication in integrated emerging 3G and Wi-Fi network. In S. K. Katsikas et al. (Eds.), *EuroPKI 2004* (LNCS 3093, pp. 287-296). Berlin/Heidelberg, Germany: Springer.

Kuroda, M., Yoshida, M., Ono, R., Kiyomoto, S., & Tanaka, T. (2004). Secure service and network framework for mobile Ethernet. *Wireless Personal Communication, 29,* 161-190.

Li, M., Sandrasegaran, K., & Huang, X. (2005, July). Identity management in vertical handovers for UMTS-WLAN networks. In *Proceedings of the Internationl Conference on Mobile Business, ICMB'05* (pp. 479-484). Washington DC: IEEE Communication Society.

Mastercard PayPass. (n.d.). *The NYC mobile trial.* Retrieved February 09, 2007, from http://www.mastercard.com/us/paypass/mobile/

Miller, P. et al. (Eds.). (2004). *Tier 2 business guidelines: Mobile deployments.* Retrieved November 1, 2006, from http://www.projectliberty.org/liberty/resource_center/papers

Noll, J., Carlsen, U., & Kalman, G. (2006, August 7-10). *License transfer mechanisms through seamless SIM authentication.* Paper presented at the International Conference on Wireless Information Systems, Winsys 2006, Setubal, Portugal.

Noll, J., Lopez Calvet, J. C., & Myksvoll, K. (2006, July 29-31). *Admittance services through mobile phone short messages.* Paper presented at the International Conf. on Wireless and Mobile Communications ICWMC'06, Bucharest, Romania.

Park, D.-G., & Lee, Y.-R. (2003). The RBAC based privilege management for authorization of wireless networks. In G. Dong et al. (Eds.), *WAIM 2003,* Berlin/Heidelberg, Germany (LNCS 2762, pp. 314-326). Springer-Verlag.

Siddiqui, F., Zeadally, S., & Yaprak, E. (2005). Design architecture for 3G and IEEE802.11 WLAN Integration. In P. Lorenz & P. Dini (Eds.), *ICN 2005* (LNCS 3421, pp. 1047-1054). Berlin/Heidelberg, Germany: Springer.

The SXIP 2.0 Overview. In specifications of SXIP 2.0 protocol. (n.d.). Retrieved December 15, 2006, from http://sxip.net/Specs

Visa's mobile platform initiative. (2007). *Payment news.* Retrieved April 27, 2007, from http://www.paymentsnews.com/2007/02/visas_mobile_pl.html

KEY TERMS

Authentication: Authentication is to prove as genuine.

Biometrics: Biometrics is the biological identification of a person which may include characteristics of structure and of action such as iris and retinal patterns; hand geometry; fingerprints; voice response to challenges; the dynamics of hand-written signatures, and so forth.

Circle of Trust: Circle of trust is a trust relationship through agreement among various service providers.

EAP-TLS and EAP-AKA: EAP-TLS and EAP-AKA are authentication frameworks frequently used in wireless networks.

Federation: Federation is the joining together to form a union through agreement.

IDP: Identity providers.

Life Cycle: Life cycle is the progression through a series of different stages of development.

PIN: Personal identification number.

Personalization: Personalization is when something is customized or tailored for the user, taking into consideration that person's habits and preferences.

Pervasive Computing: Pervasive computing is the use of computing devices everywhere and these devices communicate with each other over wireless networks without any interactions required by the user.

Proximity Service: Proximity services are those available close to the users.

Revocation of Identity: Revocation of identity is the act of recalling or annulling the identity.

SP: Service providers.

Single Sign-On (SSO): SSO on is the ability for users to log on once to a network and be able to access all authorized resources within the domain.

Smart Card: Smart card is a card containing a computer chip that enables the holder to perform various operations requiring data stored on chip.

Ubiquitous: Ubiquitous is being or seeming to be everywhere at the same time.

Chapter IX
Privacy-Enhancing Technique: A Survey and Classification

Peter Langendörfer
IHP, Germany

Michael Maaser
IHP, Germany

Krzysztof Piotrowski
IHP, Germany

Steffen Peter
IHP, Germany

ABSTRACT

This chapter provides a survey of privacy-enhancing techniques and discusses their effect using a scenario in which a charged location-based service is used. We introduce four protection levels and discuss an assessment of privacy-enhancing techniques according to these protection levels.

INTRODUCTION

Privacy is a very complex topic that touches legal, social, and technical issues. In this chapter we are focussing on the technical aspect of how to preserve privacy on the Internet. Throughout this chapter we define privacy as users' capability to determine who may know, store, and compute their data.

Privacy is one of the major concerns of Internet users (Cranor, 2000). The combination of wireless technology and Internet provides a means to combine real-world and cyber-world behaviour. Thus, extending Internet use to mobile devices is going to aggravate privacy concerns. But, privacy concerns influence also the revenue of companies which are offering their service via the Internet (Federal Trade Commission [FTC], 1999). So there is an interest in proper preserving of privacy on both sides. Especially big enterprises may suffer a lot from loss of trust in case they cannot protect the privacy-relevant data or do not adhere to their own privacy policies (Anton, He, & Baumer, 2004; Barbaro & Zeller, 2006).

Privacy-enhancing technologies (PETs) have become a hot research topic in the last few years, leading to a plethora of approaches that intend to protect privacy. This chapter provides an overview of PETs and discusses their effect on information disclosed while using a location-based service from a mobile device. In addition, an assessment of the protection level that can be achieved by applying the introduced means is provided. Thus, this chapter helps scientists to understand what is going on in the privacy research area so they can identify new research topics more easily. In addition, it enables practitioners to find approaches that allow them to build a privacy-preserving system.

The rest of this chapter is structured as follows. We first discuss privacy protection goals and provide an example that outlines which information can be gathered while using a charged service. In the third section we explain privacy-enhancing technologies. A discussion of the protection level achieved by individual means is given in the fourth section. The chapter concludes with an investigation of the currently reached deployment of privacy-enhancing techniques and a discussion of new research challenges.

PRIVACY PROTECTION GOALS

While browsing the Web or doing e- or m-commerce every user exposes information about his/her interests, personal data, and so forth to one or several of the following service providers: network service provider, for example, telco company; Internet service provider, for example, online book store; context service provider, for example, location handling system; and payment service provider, for example, his/her bank. Perfect privacy can be achieved if and only if the user reveals no information at all. Since this excludes the user from all benefits online services provide it is not a reasonable choice. The most valuable alternative is to disclose as little information as possible and only to the service provider who essentially needs this information.

In order to achieve a reasonable good separation of information, personal data and communication habits have to be protected at network as well as at application level. The former is an essential prerequisite of the latter, that is, protection at the application level does not make any sense as long as no protection at the network level is used. Protection at application level is much more difficult to achieve than protection at the network level. Here some information has to be revealed in order to get a useful service, that is, data has to be given away and therefore it has to be protected somehow. At the application level two dimensions have to be considered to prevent detailed profiling: time and location (in the sense of data gathering entity). The time dimension hinders service providers to construct a relationship between different service uses executed by the same individual but at different points in time. The location dimension provides separation of information between several service providers so that each one of them knows only data of a specific type.

In the following subsection we discuss a service scenario in which the current position of the user is requested by the service provider, who is also charging for the service. We use this scenario to show which data is known by which party of the whole system. We will also refer to this scenario later on to illustrate the effect of the privacy-enhancing techniques discussed in the following section.

Example

In this section we present a charged location-based service scenario that shows privacy issues in detail. It shows the information flow between the involved parties and the resulting dependencies that may cause privacy flaws.

The service provides its mobile user with information that is dependent on the location of the user. Additionally, the user pays for the information using a payment protocol. As shown in Figure 1 there are five parties, besides the user, involved in this scenario.

1. Positioning system is used to sense the current location of the user. Depending on the kind of the system the location information is sent to the location handling subsystem either direct from the positioning system or is forwarded by the user. In the first case the role of the user in location information forwarding is passive, in the latter active.

Figure 1. The data exchange in the charged location-based service scenario

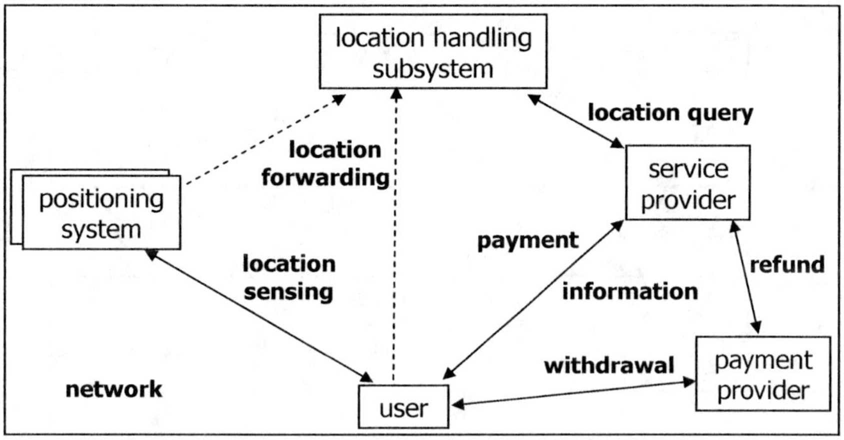

2. The location handling subsystem combines the location information together with the user identity. This subsystem may be a part of the service, or it may be a part of the positioning system with passive user role. Of course, it may also be a stand-alone infrastructure part that manages the location information for multiple users and provides it to multiple services.
3. The payment provider transfers money in order to allow the user to pay the service for the information.
4. The service provides the user with information based on the current location of the user received from the location handling subsystem.
5. An additional, but virtual party is the network. It may be a local network or the Internet. It may introduce parties that, generally, can be considered as eavesdroppers.

Each party of the scenario has some of the user data. In other words they have a kind of knowledge. Table 1 shows the distribution of knowledge between these parties. Additionally, if not protected by means of cryptography the user data is available to eavesdroppers.

The habits of the user are reflected in his/her location and purchase history. Even if the content of the purchased information is not known, the fact that the user communicated with or paid a specific service provider causes privacy flaws. Table 1 shows that in this scenario a detailed profiling of the user is possible if he/she always provides the same identifier to the individual service providers. So, almost every party in the scenario can create a kind of profile. The situation becomes even worse if the service providers are collaborating to get a more detailed profile of the user.

DISCUSSION OF PRIVACY-ENHANCING TECHNIQUES

In this section we provide an overview on privacy enhancing techniques but do not discuss basic technologies such as cipher means. Throughout this section we assume that all messages are encrypted in order to avoid eavesdropping and easy observation of user activities by third parties. We start with the discussion of network level protection means, before we describe application level approaches.

Network Level Privacy Protection

An often used approach intended to increase the security is the application of a proxy chain, which provides better security than use of a single proxy.

Table 1. Distribution of the knowledge about the user in the scenario. Information in brackets can also be available to the corresponding parties depending on the setup.

Party	Knowledge about the user
User	- Identity - Purchased information - Location
Positioning system	- (Location) - (Identity)
Location handling subsystem	- Location - Identity - Kind of purchased information
Service	- Identity - Purchased information - Location
Payment provider	- Identity - Kind of purchased information
network	- (Kind of purchased information) - (Identity) - (Purchased information) - (Location)

By this means the messages are forwarded from one proxy to another and eventually to the destination. Without additional security mechanisms this approach cannot be recommended since every proxy has access to data and at least the destination address, so that no protection improvement is achieved. An improvement of the proxy chain idea is the crowds approach (Reiter & Rubin, 1998). Each user runs a program called Jondo. This program is the local access of the user to the proxy network and also a proxy for other users in the network. If a Jondo proxy receives a packet it randomly decides whether to forward the packet to another Jondo or to send it to its destination. The receiver and also an external eavesdropper cannot decide whether the packet was originally sent by the direct peer or by another computer in the crowd, because the encrypted packets will be re-encrypted on every proxy. However, since every proxy still has access to content and destination address, the crowds approach still has security and privacy flaws.

In 1981 Chaum proposed mix networks as solution that solves the open issues. Mix networks are a combination of proxy chains and asymmetric cryptography. Instead of forwarding the plain message, every packet is encrypted using pubic-key cryptography (PKC). PKC allows every sender to encrypt the message with the publicly known public key of the proxy. Only the specific proxy is able to decrypt the message with the corresponding private key. The idea of mix networks is to encrypt the actual message with the public keys of a set of proxy servers (also called stages or mixes). Encryption is performed cascaded in reverse order of the mixes that will receive the packet. Additionally to the message, each encryption layer contains the address of the next mix in the chain or the final receiver. That is, first the sender encrypts the message together with the address of the receiver using the public key of the last mix in the chain, while this cipher text is encrypted together with the address of the last mix using the key of the second last mix, and so on. Every mix only knows the previous and the next computer in the chain. No mix but the first knows the sender, and no mix but the last has information about the receiver. A single proxy is not able to disclose sender or receiver. Only with the private keys of every mix in the network it is possible to reconstruct the path from the sender to the receiver. Such alliance is unlikely if individuals or organizations with different interests administrate the mixes on the route. As long as one mix on the route does not cooperate in order to reconstruct the route the anonymity is preserved. Indeed, it is required that many users use the mix network. In order to strengthen the privacy every mix can delay and reorder messages. Due to the successive decryption on every mix recognition of forwarded packets is prevented.

Though mix networks have been matured, are available, and very safe, they also have some problems. First, the effective transfer speed is limited. While for mail applications it is not a problem and for the Web mostly acceptable, for example, real-time video streams are hardly possible. A survey on mix networks available in Sampigethaya & Poovendran (2006) provides insight into both the design and weaknesses of existing solutions.

Similar approaches such as onion routing (Reed, Syverson, & Goldschlag, 1998), crowds (Reiter & Rubin, 1998), and Web mixes (Berthold & Köhntopp, 2000) have been reported in the past. The first two are in contrast to the original mix approach vulnerable to traffic analysis attacks, but they are more efficient. The Web mixes provide the same level of privacy as mix networks, but are optimized for real-time traffic such as browsing the Web. The comparison of these approaches discussed in Berthold and Köhntopp clearly shows that better protection of user privacy comes at the cost of less efficiency.

Several projects have realized implementations of mix networks. The Tor-network ("Tor: Overview," n.d.) is a freely available open peer-to-peer (P2P) solution of a mix network. Every Internet user may open a mix server that can be part of randomly selected routes through the network. Before transmitting a packet the sender selects a route and encrypts the message with the public keys of the corresponding mixes. Both input and output mix change from connection to connection. In contrast the Java Anon Proxy (JAP) (Project: AN.ON, n.d.) uses fixed routes, termed mix cascades. The mixes are administrated by independent, well-reputed partners. Though JAP is more reliable and more trustworthy than a P2P network, it shows a weakness with respect to privacy for the user. If the last mix detects illegal content, all mixes in the cascade work together and log the incident together with the subjects.

Mix networks are probably the best way to protect privacy on the network level. On this level they are a means that provide provable perfect security. However, the gain of privacy can turn useless if privacy is not additionally protected on the application layer.

Application Layer Privacy Protection

Location Protection

In Gruteser and Grunwald (2003) an approach is presented that reduces the accuracy of location information in order to prevent re-identification of objects using location-based services. Two dimensions, that is, space and time can be modified by the system. So, instead of a single position a region is reported to the location-based service, or instead of a single point in time an interval in which the user was at a certain position is reported. The fuzzification of the data is done at a trusted server which also extracts the user identity and network address. The communication between the user and the trusted server is protected by cryptographic means and use of mix networks. The major drawback of this approach is that the trusted server knows almost everything about the user, that is, his/her identity, network address, when he/she was where, as well as which services he/she used.

Another approach, which was not actually designed for privacy on the first hand, releases position information only in case they may be actually needed (Treu & Küpper, 2005). For proximity detection of two objects that actively report their location, location updates are not necessary as long as either of them remains in a certain circular surrounding. Consider two moving objects A and B that report their location to the infrastructure. There is a registration for a proximity event between A and B of equal or less than 1 km. Their current positions are at 101 km distance. Both objects are notified about a logical circular region with 50 km radius around their current position. These circles do not intersect and have a minimum distance of 1 km. That is, while either objects moves only within the given circle there is no chance that the objects are closer than 1 km. Hence neither of them needs to report its actual location to the infrastructure, so the location server only knows that a certain region within the user is moving and thus no detailed location tracking is possible. Since the system was not designed originally for privacy protection purposes, no means to withhold the user's identity from the server or protection of the communication between the location server and the user are investigated.

Use of Pseudonyms

The idea of pseudonyms is to hide the real identity of a user by using a bogus identity. Nicknames used in chat rooms are widely known pseudonyms. Pseudonyms prevent service providers from linking an isolated transaction to a certain user. There are several approaches that propose the use of pseudonyms in order to protect user privacy (Berthold & Köhntopp, 2000; Jendricke & Gerd tom Markotten, 2000; Jia, Brebner, & D'Uriage, 2004; Koch & Wörndl, 2001). The fundamental difference between those approaches is that Jendricke and Gerd tom Markotten and Berthold manage the different user pseudonyms at the user's own device, whereas Koch and Wörndl and Jia et al. propose the use of a centralized pseudonym service. The major drawback of systems relying on a centralized pseudonym management is that they still know the user's real identity, which services the user required and so forth, that is, they have a detailed user profile. Thus, such solutions provide only limited privacy to their users. The positive aspect of these systems is that information such as network addresses cannot be used by third parties to link pseudonyms, if the system acts as a proxy for its users as described in Jia et al.. In case of systems that still require direct interaction between their users and potential service provides such as Koch and Wörndl this benefit is no longer there. On the other hand, centralized systems provide means to identify individual users if necessary, that is, after incorrect behaviour was detected. This may help to increased acceptance of such systems on the service provider site. Decentralized approaches such as the one by Jendricke and Gerd tom Markotten allow each user to define pseudonyms himself/herself, so that there is no other entity, which has complete knowledge of real identity, and cyber world behaviour. The open issue that network addresses can be used to link pseudonyms together can be solved by additionally using mix networks.

Pseudonyms are also used to realize anonymous e-cash systems, which are explained in the next paragraph.

Anonymous Payment Systems

In an electronic payment scheme there are at least three parties. Let us describe the minimum setup. The shop or service provider delivers goods or services to the user in return of a monetary equivalent by means of payment provider. Figure 2 shows the flow of the virtual money in this setup. The payment provider issues some kind of electronic money the user uses to pay the service provider for its services. There can be multiple instances of aforementioned parties, but to simplify the description only the presented flow of virtual money is assumed.

But besides the flow of the money there is also flow of information about the user. If the payment provider can recognize the electronic money tokens it issued for the user it can create a history of the service providers the user prefers. On the other hand, if the service provider can recognize the user each time the user uses the service then the history of user purchases can be created. By combining the knowledge of these two parties a complete profile of the user can be created. To avoid the possibility of profiling the user, that is, to improve

Figure 2. The flow of the virtual money in an electronic payment scheme

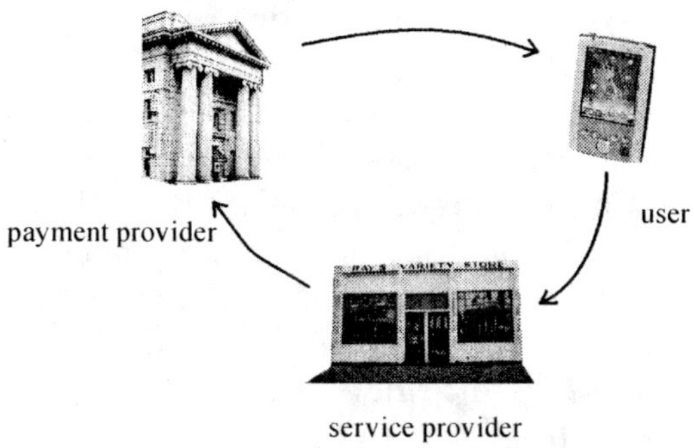

his/her privacy, there is a need for mechanisms that remove the link between the payments and the user identity. Thus, payment providers that use credit card like approaches, where the link to the user identity is strong, are to be avoided in privacy protecting systems.

The protection against profiling done on the payment provider side is especially important because usually during the token creation process the payment provider has access to the identity of the user. The remedy is to create the electronic money tokens in such a way that the payment provider cannot recognize them as they are returned by the service provider. The basic mechanisms that can be used for that purpose are blind signatures and anonymity as they were introduced in Chaum (1985). However, complete anonymity causes several problems, for example, the user can try to use one token twice without any consequences in case of success. To avoid this problem, improvements to the basic scheme were introduced. Revocable anonymity introduced in Brands (1993) allows linking the user identity with the token if the user used a single token twice. Generally, there is a trade-off between the security of the electronic payment scheme and the privacy level it provides to the user. As already stated, completely anonymous schemes usually suffer from security flaws. On the other hand, completely secure ones provide less privacy. Anyway, for security reasons there is a need for the inclusion of the identity of users in their tokens. This is usually done in an encrypted form that allows to reveal the user identity only in case of user misbehaviour.

To avoid the user profiling by the service provider, there is a need to remove any link between any two transactions or sets of transactions, depending on the desired granularity. To allow this, the electronic payment scheme tokens shall not contain any clue that might lead to linking any two tokens or sets of tokens that belong to the same user. Additionally, if any identification is needed, the user shall use pseudonyms while talking to the service provider. Changing the pseudonym frequent enough helps to remove the links between transactions.

Descriptive Approaches

There exist a number of technologies that do not actually protect privacy in a technical manner but rather in a descriptive way. That is, the parties involved in a data exchange agree on certain statements about the content and use of the data to be gathered, stored and /or processed. Most prominent representatives for such descriptive approaches are P3P (Cranor, Langheinrich, Marchiori, Presler-Marshall, & Reagle, 2002) and GEOPRIV (Peterson, 2005).

1. **P3P/APPEL**

The goal of P3P is to increase user trust and confidence in the Web. P3P provides a technical mechanism to inform users about the intended privacy policies of service providers and Web sites. The P3P specification defines the following:

- A standard schema for data a Web site may wish to collect, known as the "P3P base data schema."
- A standard set of uses, recipients, data categories, and other privacy disclosures.
- An XML format for expressing a privacy policy.
- A means of associating privacy policies with Web pages or sites, and cookies.

The P3P policy further states consequences in case of privacy breach. P3P complements legislature and self-regulatory programs in helping to enforce Web site policies.

APPEL (World Wide Web Consortium [W3C], 2002) can be used to express what a user expects to find in a privacy policy. P3P and APPEL merely provide a mechanism to describe the intentions of both sides than means to protect user data after agreeing to use the service.

There are several privacy-related tools that are based on P3P and APPEL specifications. AT&T's (n.d.) Privacy Bird is a free plug-in for Microsoft® Internet Explorer. It allows users to specify privacy preferences regarding how a Web site stores and collects data about them. If the user visits a Web site, the Privacy Bird analyzes the policy provided and indicates whether or not the policy fits to the users preferences. The Microsoft® Internet Explorer 6 (Microsoft, n.d.) and Netscape® 7 (Netscape, n.d.) embed a similar behaviour. They allow the user to set some options regarding cookies and are capable of displaying the privacy policy in human readable format. All these tools are a valuable step into the right direction, but they still lack means to personalize privacy policies. Steps towards personalized privacy policies are discussed by Maaser and Langendoerfer (2005) and Preibusch (2005). In Preibusch a fine-grained choice from a set of offered policies is proposed whereas a form of a bargaining in which neither party fully publishes all its options is proposed in Maaser and Langendoefer.

Privacy policies allow for "opting-out" of or "opting-in" to certain data or data uses. But they do not provide a technical protection means. The user has no control on the actual abidance of the policy but still has to trust that his/her personal data is processed in accordance to the stated P3P policy only. Enforcement of the policy abidance could be done by hippocratic databases or other means.

2. **IETFs GeoPriv**

GEOPRIV is a framework (Cuellar, Morris, Mulligan, Peterson, & Polk, 2004) that defines four primary network entities: (1) a location generator, (2) a location server, (3) a location recipient, and a (4) rule holder. For appropriate interaction between those three interfaces are defined, including a publication interface and a notification interface.

GEOPRIV specifies that a "using protocol" is employed to transport location objects from one place to another. Location recipients may request a location server to retrieve GEOPRIV location information concerning a particular target. The location generator publishes location information to a location server. Such information can then be distributed to location recipients in coordination with policies set by the rule maker, for example, the user whose position is stored.

A using protocol must provide some mechanism allowing location recipients to subscribe persistently in order to receive regular notification of the geographical location of the target as its location changes over time. Location generators must be enabled to publish location information to a location server that applies further policies for distribution.

One of the benefits of this architecture is that the privacy rules are stored as part of the location object (Cuellar et al., 2004). Thus, nobody can claim that he/she did not know that access to the location information was restricted. But misuse is still possible and it is still not hindered by technical means.

Server Side Means

In order to ensure privacy after agreeing to a certain privacy policy or privacy contract suitable means on the data gathering side are needed. Such could be hippocratic databases (Agrawal, Kiernan, Srikant, & Xu, 2002), HP Select Access (Casassa, Thyne, Chan, & Bramhall, 2005), Carnival (Arnesen, Danielsson, & Nordlund, 2004), PrivGuard (Lategan & Olivier, 2002). All these systems check whether an agreed individual privacy policy allows access to certain data for the stated purpose and by the requiring entity.

There are several approaches that try to protect privacy in location-aware middleware platforms (Bennicke & Langendörfer, 2003; Gruteser & Grunwald, 2003; Langendörfer & Kraemer, 2002; Synnes, Nord, & Parnes, 2003; Wagealla, Terzis,

& English, 2003). In Langendörfer and Kraemer; Bennicke and Langendörfer; and Wagealla et al. means are discussed that enable the user to declare how much information he/she is willing to reveal. In Synnes et al. the authors discuss a middleware that uses user-defined rules, which describe who may access the user's position information and under which circumstances. The approach investigated in Gruteser and Grunwald intentionally reduces the accuracy of the position information in order to protect privacy. All these approaches lack means to enforce access to user data according to the access policy defined by users. A combination of the location-aware middleware platforms with protection means sketched previously would clearly improve user privacy. A first step in this direction was reported in Langendörfer, Piotrowski, and Maaser (2006) where users are enabled to generate Kerberos tokens on their own device and where the platform checks these tokens before granting access to user data.

ASSESSMENT OF PRIVACY-ENHANCING TECHNIQUES

In this section we discuss the protection level that can be achieved by applying privacy-enhancing techniques. In order to clarify how different classes of approaches effect user privacy we resume our example from the *Privacy Protection Goals* section and show which data is protected by which means. Thereafter we identify the protection level achieved by each class of protection means.

Evaluation of Presented Techniques

For the evaluation of the privacy-enhancing techniques we resume our example. Table 2 shows that each class of privacy-enhancing techniques has its own merit and is applicable for a specific type of information. The fact that all techniques have been designed to protect specific information allows easy combination of several approaches to improve user privacy. In the case of e-cash with revocable anonymity the use of different pseudonyms is essential in order to prevent service providers from linking individual transactions by using un-altered pseudonyms. Along these lines, the use of identity management systems becomes essential in order to ensure that all pseudonyms are used correctly, when interacting with service providers. In addition, support for the generation of pseudonyms can be of help in order to guarantee a minimal level of pseudonym quality.

In Table 2 we have not included descriptive and server-side approaches. With the former data gathered depends on user preferences and the latter provides protection against misuse only after the fact, that is, it has no influence on the data accumulated in a certain service provider's database.

Protection Level

In order to asses the protection a certain PET can provide we use a classification with four protection levels:

- **High:** Technical means are given to ensure that the amount of data that can be gathered by a service provider is restricted to a minimum or matches the user's requirements. So, no detailed information can be deduced from gathered data. The downside is that no value-added services can be provided or a service may not be provided at all.
- **Medium:** The data that are gathered can not only be determined by the user, but he/she keeps somewhat control over them. This control might be either an active data control, that is, an obeyed request for deletion, or passive control that specifies certain rules on how these data shall be dealt with in the future or for certain purposes.
- **Low:** The user can determine which of his/her data is gathered. Especially if there is no proven technical means to protect the data, it is the task of the service provider to ensure the security of the gathered data. The drawbacks for service providers could be that users are hesitant to use their service if they cannot prove the security/privacy of the data.

Table 2. The sets of user data each party can link per transaction. The positing system can get information only if the user role is passive, that is, the system tracks the user.

Party	unprotected	pseudonyms	anonymous e-cash
User	1. Identity 2. Location 3. Service provider 4. Purchase details	1. Identity 1.1 Location system user pseudonym 1.2 Service user pseudonym 1.3 E-cash user pseudonym 2. Location 3. Service provider 4. Purchase details	1. Identity 1.1 Location system user pseudonym 1.2 Service user pseudonym 2. Location 3. Service provider 4. Purchase details
Positioning system	(1); (2)	(1.1); (2)	(1.1), (2)
Location handling subsystem	1; 2; 3	1.1; 2; 3	1.1; 2; 3
Service provider	1; 2; 3; 4	1.1; 1.2; 1.3; 2; 3; 4	1.1; 1.2; 2; 3; 4
Payment provider	1; 3	3	3
Network unencrypted	1; 2; 3; 4	1.1; 1.2; 1.3; 2; 3; 4	1.1; 1.2; 2; 3; 4
Network encrypted	3	3	3
Network with MIX	-	-	-

- **None:** The user, respectively, the owner of the data, has no influence on the kind of data that is gathered, which information gets inferred or derived. In addition, the service provider or data collector respectively applies no appropriate means to protect the information or privacy. In this case we cannot speak of privacy at all. Such an environment enables service providers or others to gather as much and almost any data they want. Besides the drawback for service users having no privacy at all is it most likely diminishes the trust of the users or potential customers respectively into such services.

In the classification of the PET according to protection levels we are focussing on the strength of the classes of mechanism and neglect the side effects. We are aware of the fact that real system properties such as the number of participants have significant impact on the protection level. For example, anonymous e-cash schemes provide a high level of protection since they prevent the user's bank from learning about the users online purchase habits as well as the service provider from revealing the users identity. But if the anonymous e-cash scheme is used by a single customer of the bank only, the protection provided by the anonymous e-cash scheme collapses to the protection against the service provider, since the bank can easily link the e-coins to the user's identity.

Table 3 shows the protection level of all presented classes of privacy-enhancing techniques such as mix networks and so forth. Here we did not consider individual differences in a class since weighting individual the drawbacks of similar approaches depends much on personal preferences and technical differences are already discussed in the *Discussion of Privacy-Enhancing Techniques* section.

Table 3. Protection level of the individual privacy-enhancing techniques at network and application level

	Mix networks	Pseudonyms	Anonymous e-cash	Descriptive approaches (DA)	DA + server side technologies	Location protection
Application level	none	medium	High	low	medium	low - medium
Network level	high	none	None	none	none	none

CONCLUSION

In this chapter we have presented privacy-enhancing techniques that have evolved during the last decades. If all these techniques are combined and used in the correct way, user privacy is reasonably protected. The sad point here is that despite the fact that some of these approaches are quite well understood, they are still not in place. So despite that privacy protection is theoretically possible in the real world it is hard to achieve. Only different versions of Chaum's (1981) mix network approach and P3P (Cranor et al., 2002)/APPEL (W3C, 2002) are currently in place to protect user privacy, and experienced Internet users are using different pseudonyms while browsing the Web or doing e- or m-commerce.

From our perspective, most of the privacy-enhancing techniques still suffer from acceptance issues. Anonymous e-cash lacks support from banks. Service providers might also be reluctant to accept fully anonymous e-cash due to the challenging fraud protection mechanisms involved. Even using mix networks is problematic nowadays. Many service providers block their access if they recognize usage of mix networks. Officially it is mostly justified with crime prevention, though it can be assumed that they do not want to lose valuable additional user information.

The paradigm shift in Internet use from wired to wireless also leads to new challenges. Resource consuming, privacy-enhancing techniques cannot be applied by mobile service users. This holds especially true for use of mix networks.

New technologies such as Web 2.0 allow completely new kinds of attacks. In Rao and Rohatgi (2000) and Novak, Raghavan, and Tomkins (2004) the individual way of writing was described as a means to link pseudonyms together. As long as service users are only entering a pseudonym and an e-mail address into Web forms they are still safe, but writing exhaustive comments in news groups or blogs provides sufficient material to link pseudonyms.

Pervasive computing is going to become a real challenge for privacy-enhancing techniques. A lot of information can be gathered by the environment and up to now it is still an open issue how such an environment can be adjusted to individual privacy preferences.

ADDITIONAL READING

Additional reading can be found on the Web pages of the EU-projects, Future of Identity in the Information Society (FIDIS), Privacy and Identity Management for Europe (PRIME), and Safeguards in a World of AMbient Intelligence (SWAMI). The first two projects are focusing on identity management issues whereas SWAMI deals with privacy issues in pervasive environments. The research agenda of FIDIS (http://www.fidis.net) includes virtual identities, embodying concepts such as pseudonymity and anonymity. PRIME (https://www.prime-project.eu) aims to develop a working prototype of a privacy-enhancing identity management system. In contrast to other research projects PRIME also aims at fostering market adoption of PETs. Privacy issues in pervasive environments have not been intensively investigated by the research community in recent years. A first attempt is made by the SWAMI project (http://swami.jrc.es), which focused on AMI projects, legal aspects, scenarios, and available PET.

The workshop series "Privacy Enhancing Technologies" published in Springer's LNCS series (2482, 2760, 3856, 3424, 4258) provides a great variety of publications dealing with technological, social, and legal aspects of privacy.

REFERENCES

Agrawal, R., Kiernan, J., Srikant, R., & Xu, Y. (2002, August 20-23). Hippocratic databases. In *Proceedings of the 28th International Conference on Very Large Data Bases.* Hong Kong, China.

Anton, A. I., He, Q., & Baumer, D. L. (2004). Inside JetBlue's privacy policy violations. *IEEE Security & Privacy.*

Arnesen, R. R., Danielsson, J., & Nordlund, B. (2004, November 4-5). *Carnival: An application framework for enforcement of privacy policies.* Paper presented at the 9th Nordic Workshop on Secure IT-systems. Helsinki, Finland.

AT&T Corporation. (n.d.). *AT&T privacy bird.* Retrieved January 1, 2007, from http://privacybird.com

Barbaro, M., & Zeller, Jr., T., (2006, August 9). A face is exposed for AOL searcher no. 4417749. *New York Times.* Retrieved from http://www.nytimes.com/2006/08/09/technology/09aol.html?ex=1167454800&en=f448108fbc40931e&ei=5070

Bennicke, M., & Langendörfer, P. (2003). Towards automatic negotiation of privacy contracts for Internet services. In *Proceeding of the 11th IEEE Conference on Networks (ICON 2003).* IEEE Society Press.

Berthold, O., & Köhntopp, M. (2000, July 25-26). Identity management based on P3P. In *Proceedings of the Workshop on Design Issues in Anonymity and Unobservability.* Berkeley, CA.

Brands, S. (1993). Untraceable off-line cash in wallets with observers. In *Proceedings of Crypto '93* (LNCS 773, pp. 302-318). Springer-Verlag.

Casassa Mont, M., Thyne, R., Chan, K., & Bramhall, P. (2005). *Extending HP identity management solutions to enforce privacy policies and obligations for regulatory compliance by enterprises. HPL-2005-110.* Retrieved January 1, 2007, from http://www.hpl.hp.com/techreports/2005/HPL-2005-110.html

Chaum, D. (1981). Untraceable electronic mail, return addresses, and digital pseudonyms. *Communications of the ACM, 24*(2).

Chaum, D. (1985). Security without identification: Transaction systems to make big brother obsolete. *Communications of the ACM, 28*(10), 1030-1044.

Cranor, L. F. (2000). Beyond concern: Understanding net users' attitudes about online privacy. In I. Vogelsang & B. M. Compaine (Eds.), The Internet upheaval: Raising questions, seeking answers in communications policy (pp. 47-70). Cambridge, MA: The MIT Press.

Cranor, L., Langheinrich, M., Marchiori, M., Presler-Marshall, M., & Reagle, J. (2002, April 16). *The platform for privacy preferences 1.0 (P3P1.0) Specification.* Retrieved January 1, 2007, from http://www.w3.org/TR/P3P/

Cuellar, J., Morris, J., Mulligan, D., Peterson, J., & Polk, J. (2004). *GEOPRIV requirements* (RFC 3693). Retrieved from http://www.rfc-archive.org/getrfc.php?rfc=3693

Federal Trade Commission (FTC). (1999). *The FTC's first five years: Protecting consumers online.* Retrieved from http://www.ftc.org

Gruteser, M., & Grunwald, D. (2003, May 5-8). *Anonymous usage of location-based services through spatial and temporal cloaking.* Paper presented at the ACM/USENIX International Conference on Mobile Systems, Applications, and Services (MobiSys). San Francisco, CA.

Jendricke, U., & Gerd tom Markotten, D. (2000). Usability meets security—The identity-manager as your personal security assistant for the Internet. In *Proceedings of the Computer Security Applications, 2000, ACSAC'00, 16th Annual Conference*, New Orleans, LA (pp. 344-353).

Jia, G., Brebner, G., & D'Uriage, M. (2004). *Privacy protection system and method.* U.S. Patent: US 2004/0181683 A1.

Koch, M., & Wörndl, W. (2001). Community support and identity management. In *Proceedings of the European Conference on Computer Supported Cooperative Work (ECSCW 2001)*, Bonn, Germany.

Langendörfer, P., & Kraemer, R. (2002). Towards user defined privacy in location-aware platforms. In *Proceeding of the 3rd international Conference on Internet computing.* CSREA Press.

Langendörfer, P., Piotrowski, K., & Maaser, M. (2006). A distributed privacy enforcement architecture based on Kerberos. *WSEAS Transactions on Communications, 5*(2), 231-238.

Lategan, F. A., & Olivier, M. S. (2002). PrivGuard: A model to protect private information based on its usage. *South African Computer Journal, 29*, 58-68.

Maaser, M., & Langendoerfer, P. (2005, July 26-28). *Automated negotiation of privacy contracts.* Paper presented at the Computer Software and Applications Conference, Edinburgh, Great Britain.

Microsoft. (n.d.). *Microsoft announces privacy enhancements for Windows, Internet Explorer.* Retrieved January 1, 2007, from http://www.microsoft.com/presspass/press/2000/Jun00/P3Ppr.asp

Netscape. (n.d.). *Netscape 7.0—7.2 release notes.* Retrieved January 1, 2007, from http://wp.netscape.com/eng/mozilla/ns7/relnotes/7.html#psm

Novak, J., Raghavan, P., & Tomkins, A. (2004). AntiAliasing on the Web. In *Proceedings of the 13th international conference on World Wide Web*, New York.

Peterson, J. (2005). *A presence architecture for the distribution of GEOPRIV location objects* (RFC 4079). Retrieved from http://www.ietf.org/rfc/rfc4079.txt

Preibusch, S. (2005, July 19-22). Implementing privacy negotiation techniques in e-commerce. In *Proceedings of the 7th IEEE International Conference on ECommerce Technology, IEEE CEC 2005*, Technische Universität München, Germany.

Project: AN.ON—Anonymity.Online. (n.d.). *Protection of privacy on the Internet.* Retrieved January 1, 2007, from http://anon.inf.tu-dresden.de/index_en.html

Rao, J. R., & Rohatgi, P. (2000). Can pseudonymity really guarantee privacy? In *Proceedings of the Ninth USENIX Security Symposium.*

Reed, M., Syverson, P., & Goldschlag, D. (1998). Anonymous connections and onion routing. *IEEE Journal on Selected Areas in Communications, 16*(4).

Reiter, M., & Rubin, A. (1998). Crowds: Anonymity for Web transactions. *ACM Transactions on Information and System Security, 1*(1), 66-92.

Sampigethaya, K., & Poovendran, R. (2006). A survey on mix networks and their secure applications. *Proceedings of the IEEE, 94*(12).

Synnes, K., Nord, J., & Parnes, P. (2003, January). Location privacy in the Alipes platform. In *Proceedings of the Hawaii International Conference on System Sciences (HICSS-36)*, Big Island, HI.

Tor: Overview. (n.d.). Retrieved January 1, 2007, from http://tor.eff.org/overview.html

Treu, G., & Küpper, A. (2005). Efficient proximity detection for location based services. In *Proceedings of the 2nd Workshop on Positioning, Navigation and Communication 2005 (WPNC05)*, Hannover, Germany: SHAKER-Publishing.

Wagealla, W., Terzis, S., & English, C. (2003). Trust-based model for privacy control in context-aware systems. In *Proceedings of the 2nd Workshop on Security in Ubiquitous Computing*, Ubicomp.

World Wide Web Consortium (W3C). (2002, April 15). *W3C: A P3P preference exchange language 1.0 (APPEL1.0).* Retrieved January 1, 2007, from http://www.w3.org/TR/P3P-preferences/

KEY TERMS

Anonymous E-Cash: Electronic payment system or protocol that provides anonymity to its users.

APPEL: A language specification to define rules for acceptance of certain P3P policies.

GeoPriv: An IETF working group, which assesses the authorization, integrity and privacy requirements of transfer, release or representation of geographic location information through an agent.

Mix Networks: Ccombination of proxy chains and asymmetric cryptography that enables hard-to-trace communication over unprotected networks.

P3P: The Platform for Privacy Preferences Project (P3P) enables Websites to express their privacy practices in a standard format that can be retrieved automatically and interpreted easily by user agents.

Privacy Enhancing Techniques: Technical means that provide anonymity, intractability in networks.

Pseudonyms: Bogus identity, possible temporary, used in order to hide the real identity.

Chapter X
Vulnerability Analysis and Defenses in Wireless Networks

Lawan A. Mohammed
King Fahd University of Petroleum and Minerals, Saudi Arabia

Biju Issac
Swinburne University of Technology – Sarawak Campus, Malaysia

ABSTRACT

This chapter shows that the security challenges posed by the 802.11 wireless networks are manifold and it is therefore important to explore the various vulnerabilities that are present with such networks. Along with other security vulnerabilities, defense against denial of service attacks is a critical component of any security system. Unlike wired networks where denial of service attacks has been extensively studied, there is a lack of research for preventing such attacks in wireless networks. In addition to various vulnerabilities, some factors leading to different types of denial of service (DoS) attacks and some defense mechanisms are discussed in this chapter. This can help to better understand the wireless network vulnerabilities and subsequently more techniques and procedures to combat these attacks may be developed by researchers.

INTRODUCTION

Due to the increasing advancement in wireless technologies, wireless communication is becoming more prevalent as it is gaining more popularity in both public and private sectors. Wireless networks are based on a technology that uses radio waves or radio frequencies (RF) to transmit or send data along a communication path. Companies and individuals are using wireless technology for important communications that they want to keep private. A recent report by a market research firm Cahners In-Stat (In-Stat, 2006) predicts sales of 802.15.4 devices (using low powered network standard) could grow by a compound annual growth rate (CAGR) of 200% from 2004 to 2009. In a similar survey, the Infornetics projected that 57% of small, 62% of medium, and 72% of large organizations in North America will be using wireless LANs (WLANs) by 2009 (Richard, 2005).

Wired networks requires a physical setup (i.e., cable wiring) for a user to get access and a misbehaved network card can be tracked down and its switch port can be disconnected remotely using network management tools. But wireless users are not connected to any physical socket, and being in an unknown location, network access can be obtained almost spontaneously. Generally speaking, typical wireless networks are defenseless against individuals who can find unsecured networks. The wireless server dutifully grants the unauthorized computer or mobile device an IP address, and the attacker is able to launch a variety of attacks such as breaking into specific servers, eavesdropping on network packets, unleashing a worm, and denial of service (DoS) or distributed denial of service (DDoS) attacks, and so forth. In this chapter, we discuss some security threats along with DoS attacks in a typical wireless networks and survey some counter measures.

OVERVIEW OF SECURITY CHALLENGES IN WIRELESS NETWORKS

Security has traditionally consisted of ensuring confidentiality of data, the complete integrity of the data, and the availability of the data when ever needed—where service is not denied. Generally speaking, both wired and wireless network environments are complicated. Security solutions are most effective when they can be customized to a specific installation. Unfortunately, a high percentage of individuals involved in building and maintaining inter-networks and infrastructures for these environments have little knowledge of security protocols. As a result, many of today's systems are vulnerable. Recent reports indicated that the wireless networks are becoming more popular. As these networks deployments increase, so does the challenge to provide these networks with security. Wireless networks face more security challenges than their wired counterparts. This is partly duc to the nature of the wireless medium as transmitted signals can travel through the walls, ceilings, and windows of buildings up to thousands of feet outside of the building walls. Moreover, since the wireless medium is airwaves, it is a shared medium that allows any one within certain distance or proximity to intrude into the network and sniff the traffic. Further, the risks of using a shared medium is increasing with the advent of available hacking tools that can be found freely from hacker's Web sites. Additionally, some default wireless access points (APs) come from the manufacturers in open access mode with all security features turned off by default. Therefore, insecure wireless devices such as APs and user stations, can seriously compromise wireless networks, making them popular targets for hackers.

Securing wireless networks requires at least three actions to be taken: first, authenticating users to ensure only legitimate users have access to the network; second, protecting the transmitted data by means of encryption; and third, preventing unauthorized connections by eliminating unauthorized transmitter or receiver. This emphasizes the need for a security framework with strong encryption and mutual authentication as explained later.

Specific Challenges and Key Issues

The security challenges in wireless networks can be roughly divided into two main categories, based on their scope and impact. The first category involves attacks targeting the entire network and its infrastructure. This may include the following:

- **Channel jamming:** This involves jamming the wireless channel in the physical layer thus denying network access to legitimate users. Typical example is the DoS attack.
- **Unauthorized access:** This involves gaining free access to the network and also using the AP to bypass the firewall and access the internal network. Once an attacker has access to the network, he/she can then launch additional attacks or just enjoy free network use. Although free network usage may not be a significant threat to many networks, however network access is a key step in address resolution protocol (ARP)-based man-in-the-middle (MITM) attacks.
- **Traffic analysis:** This attack enables gaining information about data transmission and network activity by monitoring and intercepting patterns of wireless communication. This involves analyzing the overhead wireless traffic to obtain useful information. There are three forms of information that an attacker can obtain. First, he/she can identify that there is

activity on the network. Secondly, he/she can find information about the location of APs in the surrounding area. This is because unless turned off, APs broadcast their service set identifiers (SSIDs) for identification. Thirdly, he/she may learn the type of protocols being used in the transmission.

The second category involves attacks against the communication between the stations and the AP. This may include the following:

- **Faking/replay attack:** This involves the ability to guess the structure of transmitted information (even if it is encrypted) and replace the legitimate message with one which has the correct structure but that has altered fields. This is known as faking. A simple form of faking, and one that absolutely must be protected against, is that of replay. In this, an attacker simply records and then replays a message from one legitimate party to another. A new form of replay attack is known as ***wormhole*** attack. In this attack, the attacker records packets or individual bits from a packet at one location in the network, tunnels them to another location, and replays it there as described in Yih-Chun (2006).
- **Eavesdropping:** This implies the interception of information/data being transmitted over the wireless network. When the wireless link is not encrypted, an attacker can eavesdrop the communication even from some few miles away. The attacker can gain two types of information from this attack; he/she can read the data transmitted in the session and can also gather information indirectly by examining the packets in the session, specifically their source, destination, size, number, and time of transmission. Eavesdropping can also be active; in this case the attacker not only listens to the wireless connection, but also actively injects messages into the communication medium.
- **Man-in-the-middle attack (MITM):** In this attack, the attacker resides between the station and the AP, and can intercept and modify the message, then release the modified message to the target destination. This can be done by setting a rogue AP as described in Lynn and Baird (2002).
- **Message forgery:** In this attack, as the wireless link is not protected for message integrity, an attacker can inject forged messages into both directions of the communication.
- **Session hijacking:** In this attack, an attacker causes the user to lose his/her connection, and he/she assumes his/her identity and privileges for a period. It is an attack against the integrity of a session. The target knows that it no longer has access to the session but may not be aware that the session has been taken over by an attacker. The target may attribute the session loss to a normal malfunction of the WLAN.

Analysis of Wired Equivalent Privacy (WEP) Protocol

Wired equivalent privacy (WEP) has been part of the 802.11 standard since its initial ratification in September 1999. It is designed for data privacy and encryption to protect messages form unauthorized viewing in case they are intercepted in the air. Its goals are to provide integrity, availability, and confidentiality to the wireless networks. However, notable security research findings have shown deficiencies and flaws in the design of WEP (Fluhrer, Mantin, & Shamir, 2001; Gast, 2002, pp. 93-96).

War Driving and Its variants

The process of identifying and categorizing the wireless networks by using pre-configured laptops from within a moving vehicle is called war driving. War drivers use laptops and some special software to identify wireless networks and let them understand the security associated with any particular wireless network that they have recorded. They also upload their war driving results to a Web site where others who have access will be able to see exactly where these unsecured wireless networks are located. The use of GPS has aided this objective even further.

War driving Web site http://www.worldwideward-rive.org has done the data collection during four rounds of war driving world wide from 2002 to 2004. Their first worldwide war driving started on August 31 and finished on September 7, 2002. During this time, 9,374 APs were located and in only 30.13% had WEP encryption enabled. The second drive lasted from October 26 to November 2, 2002 when they tracked 24,958 APs, with only 27.2% having WEP enabled. During the third drive which happened from June 28 to July 5, 2003, 88,122 APs were located with only 32.26% WEP enabled. The fourth drive started in June 2004 for some months, located 228,537 APs and the total number of wireless networks running WEP was found to be 38.3%.

Security Enhancements

In the context of the aforementioned deficiencies, an IEEE 802.11i or IEEE 802.11 Task Groupi (TGi) developed a new set of WLAN security protocols to form the future IEEE 802.11i standard. The new security standard, 802.11i, which was confirmed and ratified in June 2004, eliminates all the weaknesses of WEP. It is divided into three main categories (Strand, 2004) and these enhancements are described as follows:

1. **Temporary key integrity protocol (TKIP):** This is essentially a short term solution that fixes all WEP weaknesses. It would be compatible with old 802.11 devices and it provides integrity and confidentiality.
2. **Counter mode with cipher block chaining-message authentication code protocol (CCMP):** This is a new protocol designed with planning, based on RFC 2610 which uses Advanced Encryption Standard (AES) as cryptographic algorithm. Since this is more CPU intensive than RC4 (used in WEP and TKIP), new and improved 802.11 hardware may be required. It provides integrity and confidentiality.
3. **Extensible authentication protocol (EAP):** EAP is a general protocol for point-to-point (PPP) authentication that supports multiple authentication mechanisms.

Temporary Key Integrity Protocol (TKIP)

Wi-Fi protected access (WPA) was designed to replace WEP with the combination of the TKIP, which provides data confidentiality through encryption, and a new cryptographic message integrity code called MIC or Michael, which provides data integrity. TKIP comprises the same encryption engine and RC4 algorithm defined for WEP. However, unlike WEP the TKIP uses a 128 bits key for encryption and 64 bits key for authentication. This solves the problem of a shorter WEP key. TKIP also added a per-packet key mixing function to de-correlate the public initialization vectors (IVs) from weak keys. Furthermore, TKIP also provides a rekeying mechanism to provide fresh encryption and integrity keys by giving each user a unique shared key per session and by using IV as a counter. It discards any IV value received out of sequence. If the IV space is exhausted, a new key is negotiated. This makes TKIP protected networks more resistant to cryptanalytic attacks involving key reuse. TKIP provides better security than the WEP by adding four new algorithms:

- It provides a nonlinear hash function (Michael) that produces a 64 bit output. Unlike CRC used in WEP, Michael is keyed. Only those who know the secret key can compute a valid hash.
- It provides a new IV sequencing discipline to remove replay attacks from the attacker's arsenal.
- It also has a per-packet key mixing function to de-correlate the public IVs from weak keys.
- Finally, it provides a rekeying mechanism, to provide fresh encryption and integrity keys, undoing the threat of attacks stemming from key reuse.

Table 1 shows how WPA uses TKIP and Michael to address the cryptographic weaknesses of WEP (Cable, 2004).

Counter CBC-MAC Mode

Counter with cipher block chaining-message authentication code or simply (CCM) is a mode

Table 1. WPA vs. WEP

WEP weakness	How weakness is addressed by WPA
IV is too short	In TKIP, the IV has been doubled in size to 48 bits.
Weak data integrity	The WEP-encrypted CRC-32 checksum calculation has been replaced with Michael. The Michael algorithm calculates a 64-bit message integrity code (MIC) value, which is encrypted with TKIP
Uses the master key rather than derived key	TKIP and Michael use a set of temporal keys that are derived from a master key and other values. The master key is derived from the extensible authentication protocol-transport layer security (EAP-TLS) or Protected EAP (PEAP) 802.1X authentication process. Additionally, the secret portion of the input to the RC4 PRNG is changed with each frame through a packet mixing function.
No rekeying	WPA rekeys automatically to derive new sets of temporal keys.
No replay protection	TKIP uses the IV as a frame counter to provide replay protection.

of operation for a symmetric key block cipher algorithm. CCM may be used to provide assurance of the confidentiality and the authenticity of computer data by combining the techniques of the counter (CTR) mode and the cipher block chaining-message authentication code (CBC-MAC) algorithm. CCM is based on an approved symmetric key block cipher algorithm whose block size is 128 bits, such as the AES. CCM consists of two related processes: generation-encryption and decryption-verification, which combine two cryptographic primitives: counter mode encryption and cipher block chaining based authentication. Only the forward cipher function of the block cipher algorithm is used within these primitives. In generation-encryption, cipher block chaining is applied to the payload, the associated data, and a nonce to generate a message authentication code (MAC); then, counter mode encryption is applied to the MAC and the payload to transform them into a ciphertext. Thus, CCM generation-encryption expands the size of the payload by the size of the MAC. In decryption-verification, counter mode decryption is applied to the purported ciphertext to recover the MAC and the corresponding payload; then, cipher block chaining is applied to the payload, the received associated data, and the received nonce to verify the correctness of the MAC. Successful verification provides assurance that the payload and the associated data originated from a source with access to the key (Dworkin, 2004).

CCMP is the preferred encryption protocol in the 802.11i standard. CCMP is based upon the CCM mode of the AES encryption algorithm. Thus, CCMP utilizes 128-bit keys, with a 48-bit IV. As with the CCM, confidentiality and authentication are provided by the counter mode (CM) and the cipher block chaining message authentication code (CBC-MAC).

CCMP addresses all known WEP deficiencies, but without the restrictions of the already-deployed hardware. The protocol has many properties in common with TKIP (Cam-Winget et al., 2003). WEP, TKIP, and CCMP can be compared as in Table 2.

802.1x/EAP Authentication

IEEE 802.1x was created for authentication in PPP. It ties a protocol called EAP, which can be applied to both the wired and wireless networks. It also supports multiple authentication methods, such as EAP-Message Digest (EAP-MD5), EAP-One Time Password (EAP-OTP), EAP-Transport Layer Security (EAP-TLS), EAP-Tunneled TLS (EAP-TTLS), EAP-Generic Token Card (EAP-GTC), Microsoft CHAP version 2 (EAP-MSCHAPv2), and EAP-FAST (Blunk & Vollbrecht, 1998).

In 802.1x EAP authentication process, a client attempts to connect with an authenticator (AP). The AP responds by enabling a port for passing only EAP packets from the client to an authentication server located on the wired side of the AP. The AP blocks all other traffic, such as HTTP, DHCP, and POP3 packets, until the AP can verify the client's identity using an authentication server (e.g., RADIUS). Once authenticated, the AP opens the client's port for other types of traffic. The summary of the process is as shown in Figure 1.

Table 2. WEP, TKIP, and CCMP comparison (Cam-Winget, Housley, Wagner, & Walker, 2003)

	WEP	TKIP	CCMP
Cipher	RC4	RC4	AES
Key size	40 or 104 bits	128 bits encryption, 64 bits authentication	128 bits
Key lifetime	24-bit IV, wrap	48-bit IV	48-bit IV
Packet key integrity	Concatenating IV to base key	Mixing function	Not needed
Packet data	CRC-32	Michael	CCM
Packet header	None	Michael	CCM
Replay detection	None	Use IV sequencing	Use IV sequencing
Key management	None	EAP-based (802.1x)	EAP-based (802.1x)

- Client or supplicant sends an association request to the authenticator (AP)
- The authenticator or AP replies with associated response to the supplicant (client)
- Supplicant sends an EAP-start message to the authenticator
- The authenticator replies with an EAP-request identity message
- The supplicant sends an EAP-response packet containing the received identity to the authentication server
- The authentication server uses a specific authentication algorithm to verify the client's identity
- The authentication server will either send an acceptance or rejection message to the AP
- The AP sends an EAP-success packet (or reject packet) to the client

If the authentication server accepts the client, then the AP will transition the client's port to an authorized state and forward additional traffic. However, potential for MITM attacks in tunneling EAP protocols such as PEAP and EAP-TTL are documented in an IACR report that is available online (http://eprint.iacr.org/2002/163/).

Other Protocols

This section will briefly introduce other protocols that are being used or being developed in securing wireless networks.

- **TIK protocol:** TESLA with instant key disclosure protocol or simply TIK protocol was proposed in Yih-Chun (2006). It is an extension of the TESLA broadcast authentication protocol (Perrig, Canetti, Tyger, & Song, 2000). It implements temporal leashes and provides efficient instant authentication for broadcast communication in wireless networks. The intuition behind TIK is that the packet transmission time can be significantly longer than the time synchronization error. In these cases, a receiver can verify the TESLA security condition (that the corresponding key has not yet been disclosed) as it receives the packet; this fact allows the sender to disclose the key in the same packet. TIK implements a temporal leash and, thus, enables the receiver to detect a wormhole attack. It is based on efficient symmetric cryptographic primitives (a message authentication code is a symmetric cryptographic primitive). It requires

Figure 1. General EAP authentication process

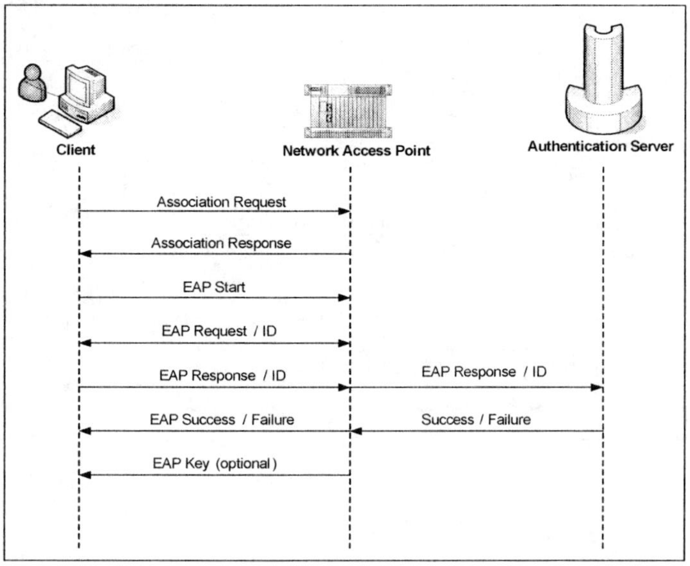

accurate time synchronization between all communicating parties, and requires each communicating node to know just one public value for each sender node, thus enabling scalable key distribution.

- **SSTP protocol:** Microsoft is working on a remote access tunneling protocol that allows client devices to securely access networks via a *virtual private network* (VPN) from anywhere on the Internet without any issues with typical port blocking problems. The secure socket tunnelingprotocol (SSTP) makes a VPN tunnel that goes over Secure-HTTP, eliminating issues associated with VPN connections based on the point-to-point tunneling protocol (PPTP) or layer 2 tunneling protocol (L2TP) that can be blocked by some Web proxies, firewalls, and network address translation (NAT) routers that sit between clients and servers. The protocol is only for remote access and will not support site-to-site VPN tunnels. (Fontana, 2007)

Other Attacks on Wireless Security

There are some other effective attacks that can be launched against 802.11 wireless networks and they are briefly explained next (Earle, 2006).

WPA Passive Dictionary Attack

This attack can be launched against a WPA pre-shared key (with four-way handshake) setup in 802.11g networks using a dictionary file of words (Takahashi, 2004). As a precaution, avoid dictionary words for the pass phrase during AP configuration and make pass phrase more than 20 characters.

The process of the four-way handshake shown in Figure 2 can be explained as follows. The AP and communicating station need an individual pairwise transient key (PTK) to shield the unicast conversation between them. To come out with a different PTK for each AP-station pair, a pairwise master key (PMK) is included in the algorithm, along with MAC address, ANonce, and SNonce (two random values). The first two messages manage to derive the same PTK without transmitting in the air. The AP also generates a group transient key (GTK) to shield all conversations, especially multicast and broadcast. As all stations on the wireless network needs that same GTK to decrypt broadcast or multicast frames, the AP sends the current GTK in the third message of the handshake.

To stop someone from hacking the communication, the GTK is encrypted with the PTK. To avoid forgery in these handshake messages,

Figure 2. WPA-PSK four-way handshake

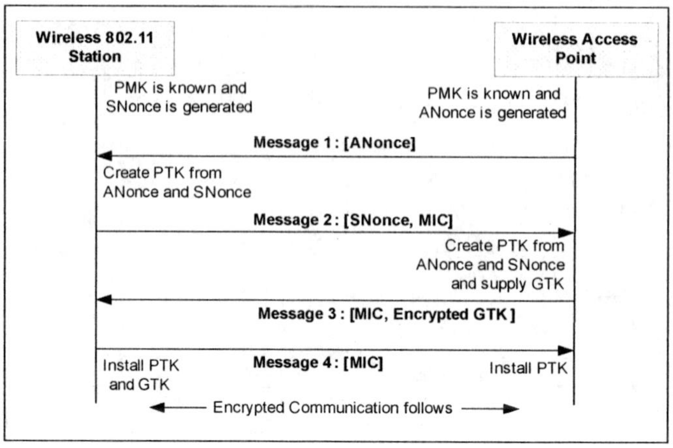

second, third and fourth messages have a message integrity code (MIC). The MIC is generated by hashing a specified portion of the message and then encrypting that hash with the PTK. This four-way handshake occurs whenever someone connects to a WLAN using WPA. It also occurs thereafter, whenever the AP decides to refresh the transient keys (Phifer, 2007).

Attack on Michael MIC

Michael MIC was introduced to prevent attacks through message modification. It uses a feature known as TKIP countermeasure procedure, which works by disabling the AP if it receives two MIC failures within one second. After exactly one minute, the AP comes back to life and would need all its past and current users to re-key to gain access to the network. An attacker could send corrupt packets to the AP which can pass the frame CRC check, but would trigger the TKIP countermeasure eventually shutting down the AP, especially after repeated corrupt traffic.

Encryption Attacks on Known Plaintext, Double Encryption, and Message Modification

For WEP encryption process, an XOR operation of message (or plaintext) with encryption key is done. For decryption process, the encrypted text is XOR-ed with the key to get the plaintext. Firstly, the known plaintext attack is done when the attacker knows two things: cleartext and the encrypted text of a message communication. Having both the encrypted and unencrypted form of the same information allows one to perform this attack and to retrieve the encryption key. The attacker needs to XOR cleartext and encrypted text to get the key. Secondly, to carryout the double encryption attack, a frame must be captured and the attacker must change the frame header destination MAC address to that of the attacker's wireless client. After this subtle change, the attacker must wait for the IV to reset to one minus the original IV (of the modified frame), so that he/she can replay the captured frame into the air. When the AP sees the frame with the expected IV, it will encrypt the frame, actually being fooled into decrypting the frame instead of encrypting it. After doing the unknowing decryption process, the AP will forward the cleartext frame across the air to the forged MAC address specified by the attacker. Thirdly, to achieve the message modification attack, the attacker must capture an encrypted packet that is going to another subnet, modify a single bit, and attempt to resend it. The modification will offset the IC and the packet will be rejected. After trying a number of times, the bits that are flipped will make the IC correct again, although the packet would be malformed. The attacker can do this numerous times without any logging or alerts from the AP. Once the packet passes the AP's IC check, it will reach the route. The router will observe that the packet is malformed and would send a response that contains the cleartext and associated encrypted text packet to the initial sender. This will give the attacker the ingredients to perform cleartext cryptanalysis. A solution is to encrypt the 802.11 frames within a layer 3 (network layer) wrapper, so that any tampering cannot go undetected.

General WLAN Security Measures

General security measures to minimize some of the mention flaws are listed as follows (Held, 2003; Hurton & Mugge, 2003; Issac. Jacob, & Mohammed, 2005):

1. Encrypt the network traffic. WPA with TKIP/AES options can be enabled. Upgrade the firmware on AP to prevent the use of weak IV WEP keys.
2. Ensuring mutual authentication through IEEE 802.1x protocol. Client and AP should both authenticate to each other. Implementing IEEE 802.1x port-based authentication with RADIUS server (with PEAP/MS-CHAPv2) would be a good choice.
3. Make the wireless network invisible by disabling identifier broadcasting. Turning off the SSID broadcast by AP and configure the AP not to respond to probe requests with SSID "any," by setting your own SSID. Meaning, rename the wireless network and change the default name.
4. Changing the default WEP key settings, if any. Changing the default IP address in the AP to a different one. Change administrator's password from the default password. If the wireless network does not have a default password, create one and use it to protect the network.
5. Enabling the MAC filtering in AP level or in RADIUS server or in both can tighten the security more, as there is a restriction in the use of MAC addresses (this step in itself, can be defeated through MAC spoofing).
6. Positioning and shielding of the antenna can help to direct the radio waves to a limited space.
7. Enabling of accounting and logging can help to locate and trace back some mischief that could be going on in the network. Preventive measures can then be taken after the preliminary analysis of the log file. Allow regular analysis of log files captured to trace any illegal access or network activity.
8. Using intrusion detection software to monitor the network activity in real time and to inform alerts.
9. Using honey pots or fake APs in the regular network to confuse the intruder so that he/she gets hooked to that fake AP without achieving anything.
10. Turn off the network during extended periods of non-use or inactivity.
11. Use file sharing with caution. If the user does not need to share directories and files over the network, file sharing should be disabled on his/her computers.
12. Do not auto-connect to open Wi-Fi (wireless fidelity) networks.
13. Connect using a VPN as it allows connecting securely. VPNs encrypt connections at the sending and receiving ends through secure tunnels.
14. Use firewalls in between wireless and wired network segments and implement filters.
15. Generally avoid dictionary words for pass phrase in any authentication. Also make the pass phrase more than 20 characters, especially if WPA-Pre Shared Key security is employed.

TYPES OF DENIAL OF SERVICE ATTACKS AND PREVENTIVE MEASURES

DoS simply means the inability of a user, process, or system to get the service that it needs or wants. Common DoS attacks on networks include direct attacks, remote controlled attacks, reflective attacks, and attacks with worms and viruses.

DoS attacks are quite effective against wireless networks. The wireless management frames which are transmitted in cleartext in a wireless network, informs the clients that they can connect or disconnect. The de-authentication frame will disassociate a wireless end device from an AP. Since they are sent in cleartext, they can easily be forged to force legitimate users out of the network. This can be accomplished by replaying a previous disassociation frame with a wireless sniffer. An attack on 802.11b with 802.11g mixed network mode can affect the clear channel assessment (CCA) process that brings down the probability that two wireless nodes will transmit on the same frequency simultaneously. This attack can cause all nodes in range to shut down until the attacker stops injecting the malicious frame. A layer 2 encryption would be the only solution to this. The EAP-DoS attack involves injecting a number of EAP stat frames to an AP and if the AP cannot properly process all these frames, there is the chance that it might become inoperable. Another attack against the AP involves sending malformed EAP messages. One of the recent attacks against the AP involves filling up the EAP identifier space that allows 255 ID tags to keep track of each client instance. If an attacker can flood the AP with a large number of client connection instances, using up this counter, a DoS attack can be achieved (Earle, 2006).

Different researchers have categorized DoS and DDoS from different perspectives. As documented in Christos and Aikaterini (2003), DoS attacks can be classified into five different categories, namely: (1) network device level attack, (2) operating system (OS), level attack, (3) application level attack, (4) data flood attack, and (4) protocol attack.

Network device level attack includes attacks that might be caused either by taking advantage of bugs or weaknesses in driver software or by trying to exhaust the hardware resources of network devices. Network level attacks may also involve compromising a series of computers and placing an application or agent on the computers. The computer then listens for commands from a central control computer. The compromise of computers can either be done manually or automatically through a worm or virus.

The OS level DoS attacks rely on the ways operating systems implement protocols. A typical example is the ping of death attack in which Internet control message protocol (ICMP) echo requests having total data sizes greater than the maximum IP standard size to be sent to the targeted victim. This attack often has the effect of crashing the victim's machine.

In application-based attacks, machine or a service are compromised and set out of order either by taking advantage of specific bugs in network applications that are running on the target host or by using such applications to drain the resources of their victim. It is also possible that the attacker may have found points of high algorithmic complexity and exploits them in order to consume all available resources on a remote host.

In data flooding attacks, an attacker uses all network bandwidth or any other device bandwidth by sending massive quantities of data and so causing it to process extremely large amounts of data. For instance, the attacker bombards the targeted victim with normal, but meaningless packets with spoofed source addresses.

DoS attacks based on protocol features take advantage of certain standard protocol features such as IP and MAC source addresses. Typically, the attacker spoofs these features. Several types of DoS attacks have focused on domain name systems (DNSs), and many of these involve attacking DNS cache on name servers. An attacker who owns a name server may coerce a victim name server into caching false records by querying the victim about the attackers own site. A vulnerable victim name server would then refer to the rogue server and cache the answer (Davidowicz, 1999).

Other researchers such as Papadimitratos and Hass (2002) and Marti, Giuli, Lai, and Baker (2001) describe DoS attacks in relation to routing layer and those at the link or MAC layer.

Attacks at the routing layer could consist of the following: (1) the attacker participates in routing and simply drops a certain number of the data packets. This causes the quality of the connections to deteriorate and further ramifications on the performance if TCP is the transport layer protocol that is used; (2) the attacker transmits falsified route updates. The effects could lead to frequent route failures thereby deteriorating performance; (3) the attacker could potentially replay stale updates. This might again lead to false routes and degradation in performance; and (4) reduce the time-to-live (TTL) field in the IP header so that the packet never reaches the destination. Routing attacks are usually directed at dynamic routing protocols such as border gateway protocol (BGP), open shortest path first (OSPF), and enhanced interior gateway routing protocol (EIGRP). Direct DoS or DDoS attacks against routing protocols can lead to regional outages. Another form of routing attack is called route injection, which can lead to traffic redirection, prefix hijacking, and so forth. Attacks at the MAC layer are described next.

Flooding and Spoofing Attacks

Flooding attack, as the name implies, involves the generation of spurious messages to increase traffic on the network. While spoofing attacks involves the creation of packets with spoofed (i.e., forged) source IP addresses and other credentials.

In smurf attack, an attacker sends a large amount of ICMP echo traffic to a set of IP broadcast addresses, multiplying the traffic by the number of hosts responding. ICMP flooding attack uses public sites that respond to ICMP echo request packets within an IP network to flood the victim's site. It involves flooding the buffer of the target computer with unwanted ICMP packets. SYN flood attack is also known as the transmission control protocol (TCP) SYN attack and is based on exploiting the standard TCP three-way handshake. In this case, an attacker sends SYN packet to initiate connection. The victim responds with the second packet back to the source address with SYN-ACK bit set. The attacker never responds to the reply packet. In this case, the victim's TCP receive queues would be filled up, denying new TCP connections. Another variant of this attack is called user datagram protocol (UDP) flooding attack (Craig, 2000). This attack is based on UDP echo and character generator services provided by most computers on a network. In MAC spoofing attack, an attacker spoofs his/her original MAC address to the MAC address he/she wants to spoof. An attacker can learn the MAC address of the valid user by capturing wireless packets using any packet capturing software by passively or actively observing the traffic. Web spoofing permits an attacker to observe and change all the Web traffic sent to the victim's machine and capture all data entered into the Web page forms (if any) by the victim. The attack can be done using Web plug-ins and JavaScript segments. The attack, once implemented, is started when the victim visits a malicious Web page through a Web link in a malicious e-mail message sent by the attacker. DNS spoofing is where the attacker makes a DNS entry to point to another IP address than it would be generally pointing to. It works through stealth by unknowingly forcing a victim to generate a request to the attacker's server, and then spoofing the response from that server. IP spoofing is a process used to gain unauthorized access to computers, whereby the attacker sends packets to a computer with spoofed IP address implying that the message is coming from a trusted and genuine host.

DDoS Attack

DDoS attacks usually refer to an attack by use of multiple sources that are distributed throughout the network. In this attack, an attacker installs the DDoS software controls on a network of computers, mostly through security compromise. This allows the attacker to remotely control compromised computers, thereby making it handlers and agents. From a "master" device, the attacker can control the slave devices and direct the attack on a particular victim. Thousands of machines can be controlled from a single point of contact as shown in Figure 3. There are several types of DDoS attacks, but their methods are very similar in that they rely on a large group of previously compromised systems to direct a coordinated distributed flood attack against a particular target.

Christos and Aikaterini (2003) classified DDoS based on the degree of the attack automation. These classifications are manual, semi-automatic, and automatic DDoS attacks. The manual attack involves manual scanning of remote machines for

Figure 3. DDoS attack scenario using agents/zombies to flood the victim

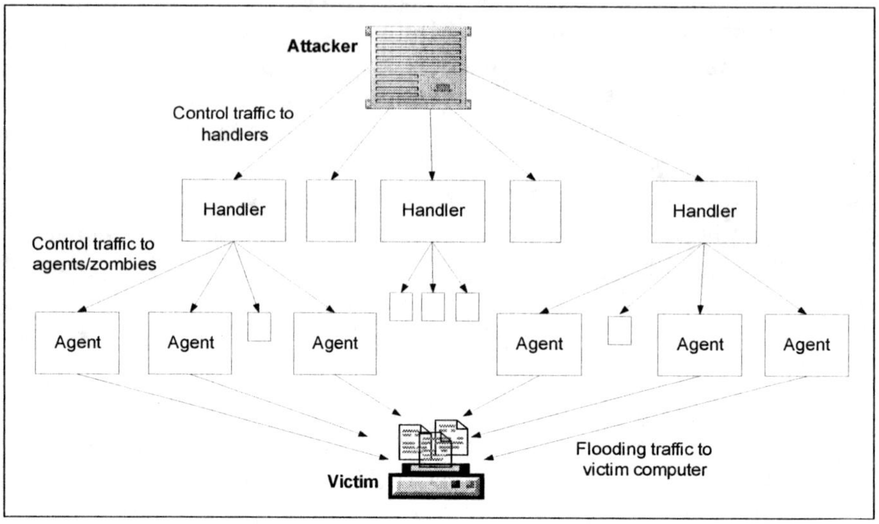

vulnerabilities, then the attacker breaks into anyone of them to install attack codes. Semi-automatic attacks are partially manual and partially automatic. In this case, the attacker scans and compromises handlers and agents by using automated scripts. He/she then types the victims address manually and the onset of the attack is specified by the handler machines. In automatic DDoS attacks the communication between attacker and agent machines is completely avoided. In most cases the attack phase is limited to a single command through the attack code file. All the features of the attack, for example the attack type, the duration, and the victims address are preprogrammed in the attack code. This way, the possibility of revealing the attacker's identity or source is very minimal. A number of DDoS tools that are available from the Internet have been identified by the Internet Security Systems (ISS) (www.iss.net).

Defense Mechanisms Against DoS Attacks

Several techniques to counter DoS and DDoS attacks have been proposed by researchers, and we briefly discuss some of these techniques. A challenge based mechanisms was proposed by Kandula, Katabi, Jacob, and Berger (2005). For example, an image-based challenge may be used to determine whether the client is a real human being or an automated script. A similar approach based on capabilities was proposed in Agarwal, Dawson, and Tryfonas (2003), and the method generally relies on clients having to ask the server for permission to send packets. If the server decides to allow the connection, it replies with a capability token, which the client includes in subsequent packets and which the network polices.

Greenhalgh, Handley, and Huici (2005) described an approach consisted of diverting traffic going to protected servers so that it traverses control points. These control points would encapsulate the traffic, sending it to a decapsulator near the server. The server could then tell which control point a malicious flow had traversed, and request it be shut down at this boundary. Signature-based and anomaly based detection techniques are proposed in Park and Lee (2001) and Shields (2002). Some solutions involve the use of strong digital signature based transport level authentication mechanisms as recently proposed in Dierks and Allen (2006).

Mechanisms Against Spoofing

Attackers launching spoofing usually hide the identity of machines they used to carry out an attack by falsifying the source address of the network communication. This makes it more difficult to identify the sources of attack traffic. It is therefore important to use network switches that have MAC binding features that store the first MAC address that appears on a port and do not allow this mapping to be altered without authentication. To prevent *IP spoofing*, disable source routing on all internal routers and use ingress filtering. *Web spoofing* depends mainly upon social engineering tricks and it is thus important to educate users and to be generally aware of the address window in a browser that displays the Web address that they are directed to. That can help if some suspicious Web site address comes up. *DNS spoofing* can be prevented by securing the DNS servers and by implementing anti-IP address spoofing measures (Paul, Ben, & Steven, 2003). Some vendors have added access control lists (ACL), implemented through MAC address filtering, to increase security. MAC address filtering amounts to allowing predetermined clients with specific MAC addresses to authenticate and associate. While the addition of MAC address filtering increases security, it is not a perfect solution given that MAC addresses can be spoofed. Also, the process of manually maintaining a list of all MAC addresses can be time consuming and error prone. Therefore MAC address filtering is probably best left for only small and fairly static networks (Mohammed & Issac, 2005).

Filtering Techniques

Filtering requires being able to filter the flood packets. This can be achieved with a signature-based packet filter. If one can create signatures for typical flood packets (TCP packets with zero data size for example, or unusually large ICMP packets), and filter out those packets, one can then filter the flood packets while allowing "normal" traffic to proceed.

Another filtering option is to reject the first IP packet from any IP address. This works with many current generations of attack tools because they tend to use a flat distribution random number generator to generate spoofed source addresses, and they only use each random address once. Another possibility is to divert traffic based on IP protocol to different servers or even route it differently. Thus, for a Web server it might be possible to route ICMP and UDP traffic bound for the Web server somewhere else entirely, or even block it at the router, so that only TCP-based floods will succeed. This at least narrows the scope of attacks that can be made.

Another filtering technique is called ingress filtering. This filtering prevents spoofed attacks from entering the network by putting rules on point-of-entry routers that restrict source addresses to a known valid range.

Filtering can also be based on channel control. This method is known as channel control filtering and can be achieved by filtering out DDoS control messages; this prevents the attacker from causing the attack servers to begin the attack. This can also be accomplished using a signature-based packet filter. If we can develop signatures for most control channel packets, we can simply reject them at the control channel packet filter, and they will disappear from the network.

FUTURE TRENDS

Due to the rapid changes in treat level and attacking techniques, existing defense mechanisms may not be adequate to counter the threats of the future attacks. Therefore, it is important for researchers to continue analyzing different threats as they emerge and develop more effective and efficient defense mechanisms. For instance, detecting distributed and automated attacks still remains a challenge. Due to the drawback of some of the exiting solutions or defense mechanisms as well as the emergence of new attack tools, further study is needed to combine well-known security drawbacks with defense techniques that are already mature and very effective. Moreover, it is also important to look into the developing of DoS management framework for protecting, detecting, and reacting to attacks when they occur. The following summarizes expected future trends in DoS and DDoS attacks—attacks on emerging technologies; attacks against anti-DoS infrastructure; attacks with the aid of malware, adware, or spyware; recursive DNS attacks or the use of DNS server for DoS attack; and attacks against OpenEdge WebSpeed platforms, and so forth.

CONCLUSION

This chapter explores some of the security vulnerabilities associated with 802.11 wireless networks. Here basic issues with WEP and better protocols like TKIP and CCMP were discussed with some advice on security precautions. Later emphasis was given on DoS and DDoS attacks to show how complicated and varied they are in nature. DoS attacks are done quite effectively against wired and wireless networks and it costs much in terms of the damages done. Defense mechanisms against such attacks are still not perfect and the chapter eventually reviews and explains some sets of defense mechanisms that could help against such attacks.

REFERENCES

Agarwal, S., Dawson, T., & Tryfonas, C. (2003). *DDoS mitigation via regional cleaning centers* (Tech. Rep. No. RR04-ATL-013177). Sprint ATL Research Report.

Blunk, L., & Vollbrecht, J. (1998). *PPP extensible authentication protocol (EAP)* (RFC 2284). Retrieved December 25, 2006, from http://www.ietf.org/rfc/rfc2284.txt

Cable, G. (2004). *Wi-Fi protected access data encryption and integrity.* Retrieved December 17, 2006, from http://www.microsoft.com/technet/community/columns /cableguy/cg1104.mspx

Cam-Winget, N., Housley, R., Wagner, D., & Walker, J. (2003). Security flaws in 802.11 data link protocols. *Communications of the ACM,* 35-39.

Christos, D., & Aikaterini, K. (2003). DoS attacks and defense mechanism: Classifications and state-of-the-art. *Computer Networks, 44,* 643-666.

Craig, A. H. (2000). *The latest in denial of service attacks: Smurfing description and information to minimize effects.* Retrieved May 17, 2006, from http://www.pentics.net/denial-of-service/white-papers/smurf.cgi

Davidowicz, D. (1999). *Domain name system (DNS) security.* Retrieved June 23, 2006, from http://compsec101.antibozo.net/ papers/dnssec/dnssec.html

Dierks, T., & Allen, C. (2006). *The TLS protocol* (RFC 2246). Retrieved December 7, 2006, from http://www.ietf.org/rfc/rfc2246.txt

Dworkin, M. (2004). *Recommendation for block cipher modes of operation: The CCM mode of authentication and confidentiality.* Retrieved November 17, 2006, from http://csrc.nist.gov/publications/nistpubs/800-38C/SP800-38C.pdf

Earle, A. E. (2006). *Wireless security handbook.* Auerbach Publications, Taylor & Francis Group.

Fluhrer, S., Mantin, I., & Shamir, A. (2001). *Weaknesses in the key scheduling algorithm of RC4.* Retrieved July 25, 2005, from http://downloads.securityfocus.com/ library/rc4_ksaproc.pdf

Fontana, J. (2007). *Network World.* Retrieved April 5, 2007, from http://www.networkworld.com/news/2007/011907-microsoft-secure-vpn-tunneling-protocol.html

Gast, M. (2002). *802.11 wireless networks—The definitive guide.* CA: O'Reilly Media.

Greenhalgh, A., Handley, M., & Huici, F. (2005). Using routing and tunneling to combat DoS attacks. In *Proceedings of the 2005 Workshop on Steps to Reducing Unwanted Traffic on the Internet.*

Held, G. (2003). *Securing wireless LAN.* Sussex, England: John Wiley & Sons.

Hurton, M., & Mugge, C. (2003). *Hacknotes—Network security portable reference.* CA: McGraw-Hill/Osborne.

In-Stat. (2006). *In-stat market survey.* Retrieved May 11, 2007, from http://www.in-stat.com

Issac, B., Jacob, S. M., & Mohammed, L. A. (2005). The art of war driving—A Malaysian case study. In *Proceedings of IEEE International Conference on Networks (ICON)* (pp. 124-129).

Kandula, S., Katabi, D., Jacob, M., & Berger, A. (2005). Botz-4-sale: Surviving organized DDoS attacks that mimic flash crowds. In *Proceedings of the 2nd Symposium on Networked Systems and Design and Implementation.*

Lynn, M., & Baird, R. (2002). *Advance 802.11 attack, Blackhat 2002.* Retrieved June 19, 2006, from http://www.blackhat.com/html/bh-usa-02/bh-usa-02-speakers.html#baird

Marti, S., Giuli, T., Lai, K., & Baker, M. (2001). Mitigating routing behavior in mobile ad hoc networks. In *Proceedings of Mobicom,* Rome.

Mohammed, L. A., & Issac, B. (2005). DoS attacks and defense mechanisms in wireless networks. In *Proceedings of the IEE Mobility Conference (Mobility 2005),* Guangzhou, China (pp. P2-1A).

Papadimitratos, P., & Haas, Z. J. (2002). Secure routing for mobile ad hoc networks. In *Proceedings of the SCS Communication Networks and Distributed Systems Modeling and Simulation Conference (CNDS 2002),* San Antonio, TX.

Park, K., & Lee, H. (2001). On the effectiveness of route-based packet filtering for distributed DoS attack prevention in powerless Internet. In *Proceedings of the ACM SIGCOMM_01 Conference on Applications, Technologies, Architectures, and Protocols for Computer Communications* (pp. 15-26) New York: ACM Press.

Paul, C., Ben, C., & Steven, B. (2003). *Security+ guide to network security fundamentals.* Thomson Course Technology (pp. 47-84).

Perrig, A., Canetti, D., Tyger, D., & Song, D. (2000). Efficient authentication and signature of multicast streams over lossy channels. In *Proceedings of the IEEE Symposium on Security and Privacy* (pp. 90-100).

Phifer, L. (2007). *WPA PSK crackers: Loose lips sink ships.* Retrieved April 2, 2007, from http://www.wi-fiplanet.com/tutorials/article.php/3667586

Richard, W. (2005). *Voice over wireless LAN adoption triples by 2007.* Retrieved January 05, 2007, from http://www.infonetics.com/resources/purple.shtml? upna05.wl.nr.shtm

Shields, C. (2002). What do we mean by network denial of service? In *Proceedings of the 2002 IEEE workshop on Information Assurance* (pp. 196-203). U.S. Military Academy.

Strand, L. (2004). *802.1x port-based authentication HOWTO.* Retrieved July 15, 2005, from http://www.tldp.org/HOWTO/8021X-HOWTO

Takahashi, T. (2004). *WPA passive dictionary attack overview* (White Paper).

Yih-Chun, H. (2006). Wormhole attacks in wireless networks. *IEEE Journal on Selected Areas in Communications, 24*(2), 370-380.

Additional Important Links/References:

CERT Coordination Center References

http://www.cert.org/advisories/CA -2000-11.html

http://www.cert.org/research/JHThesis/Chapter11.html

http://www.cert.org/incident_notes/IN-2000-04.html

http://www.cert.org/tech_tips/denial_of_service.html

http://www.cert.org/archive/pdf/DoS_trends.pdf

http://www.cert.org/research/isw/isw2000/papers/42.pdf

Other links

http://www.kb.cert.org/vuls/

http://www.usenix.org/publications/login/2000-7/apropos.html

http://www.iss.net

http://www-1.ibm.com/services/continuity/recover1.nsf/files/Downloads/$file/DOS.pdf

http://www.cymru.com/~robt/Docs/Articles/dos-and-vip.html

KEY TERMS

Denial of Service (DoS): Denial of service are attacks to prevent legitimate users from receiving services from the service provider.

Distributed Denial of Service (DDOS): DDOS is a type of DoS attack conducted by using multiple sources that are distributed throughout the network.

Flooding Attack: Flooding attack involves the generation of spurious messages to increase traffic on the network for consuming server's or network's resources.

Information Security: Information security is a mechanism dealing with providing confidentiality, integrity, authentication, and non-repudiation.

Network Security: Network security is a mechanism dealing with protection of the networking system as a whole and sustaining its capability to provide connectivity between the communicating entities.

Spoofing Attack: Spoofing attack involves the creation of packets with a forged or faked source IP addresses.

Wireless Networks: Wireless networks are based on a technology that uses radio waves or radio frequencies to transmit or send data.

Chapter XI
Key Distribution and Management for Mobile Applications

György Kálmán
University Graduate Center – UniK, Norway

Josef Noll
University Graduate Center – UniK, Norway

ABSTRACT

This chapter deals with challenges raised by securing transport, service access, user privacy, and accounting in wireless environments. Key generation, delivery, and revocation possibilities are discussed and recent solutions are shown. Special focus is on efficiency and adaptation to the mobile environment. Device domains in personal area networks and home networks are introduced to provide personal digital rights management (DRM) solutions. The value of smart cards and other security tokens are shown and a secure and convenient transmission method is recommended based on the mobile phone and near-field communication technology.

A PROBLEM OF MEDIA ACCESS

On the dawn of ubiquitous network access, data protection is becoming more and more important. While in the past network connectivity was mainly provided by wired connections, which is still considered the most secure access method, current and future users are moving towards wireless access and only the backbone stays connected by wires. In a wired environment, eavesdropping is existent, but not as spread and also not easy to implement. While methods exist to receive electromagnetic radiation from unshielded twisted pair (UTP) cables, a quite good protection can be achieved already by transport layer encryption or deploying shielded twisted pair (STP) or even fibre.

New technologies emerged in the wireless world, and especially the IEEE 802.11 family has drastically changed the way users connect to networks. The most basic requirements for new devices are the capability of supporting wireless service access. The mobile world introduced general packet radio service (GPRS) and third generation (3G) mobile systems provide permanent IP connectivity and provide together with Wi-Fi access points continuous wireless connectivity. Besides communications devices such as laptops, phones, also cars, machines, and home appliances nowadays come with wireless/mobile connectivity.

Protecting user data is of key importance for all communications, and especially for wireless communications, where eavesdropping, man-in-the-middle, and other attacks are much easier. With a simple wireless LAN (WLAN) card and corresponding software it is possible to catch, analyse, and potentially decrypt wireless traffic. The implementation of the first WLAN encryption standard wired equivalent privacy (WEP) had serious weaknesses. Encryption keys can be obtained through a laptop in promiscuous mode in less than a minute, and this can happen through a hidden attacker somewhere in the surrounding. Data protection is even worse in places with public access and on factory default WLAN access points without activated encryption. Standard Internet protocols as simple mail transport protocol (SMTP) messages are not encoded, thus all user data are transmitted in plaintext. Thus, sending an e-mail over an open access point has the same effect as broadcasting the content. With default firewall settings an intruder has access to local files, since the local subnet is usually placed inside the trusted zone. These examples emphasise that wireless links need some kind of traffic encryption.

When the first widespread digital cellular network was developed around 1985, standardisation of the global system for mobile communication (GSM) introduced the A5 cryptographic algorithms, which can nowadays be cracked in real-time (A5/2) or near real-time (A5/1). A further security threat is the lack of mutual authentication between the terminal and the network. Only the terminal is authenticated, the user has to trust the network unconditionally. In universal mobile telecommunications system (UMTS), strong encryption is applied on the radio part of the transmission and provides adequate security for current demands, but does not secure the transmission over the backbone. UTMS provides mutual authentication through an advanced mechanism for authentication and session key distribution, named authentication and key agreement (AKA).

Figure 1. A basic problem of broadcast environment

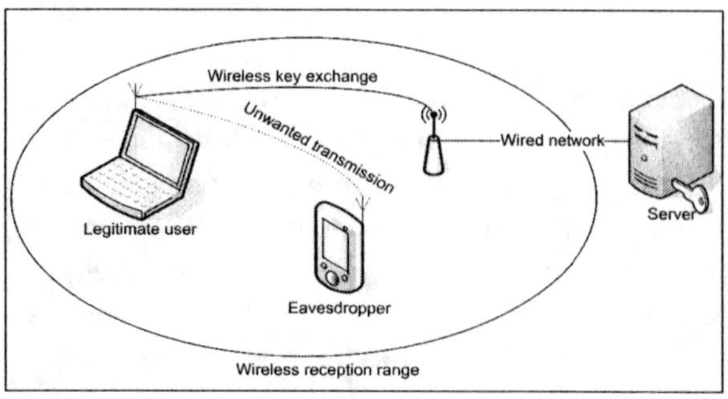

A LONG WAY TO SECURE COMMUNICATION

Applying some kind of cryptography does not imply a secured access. Communicating parties must negotiate the key used for encrypting the data. It should be obvious that the encryption key used for the communication session (session key) cannot be sent over the air in plaintext (see Figure 1).

In order to enable encryption even for the first message, several solutions exist. The simplest one, as used in cellular networks is a preshared key supplied to the mobile terminal on forehand. This key can be used later for initialising of the security infrastructure and can act as a master key in future authentications.

In more dynamic systems the use of preshared keys can be cumbersome. Most of WLAN encryption methods support this kind of key distribution. The key is taken to the new unit with some kind of out of band method, for example with an external unit, as indicated in Figure 2. Practically all private and many corporate WLANs use static keys, allowing an eavesdropper to catch huge amounts of traffic and thus enable easy decryption of the content. This implies that a system with just a secured access medium can be easily compromised. Non-aging keys can compromise even the strongest encryption, thus it is recommended to renew the keys from time to time.

Outside the telecom world it is harder to distribute keys on forehand, so key exchange protocols emerged, which offer protection from the first message and do not need any preshared secret. The most widespread protocol is the Diffie-Hell-

Figure 2. (a) Diffie-Hellmann key exchange and (b) out-of-band key delivery

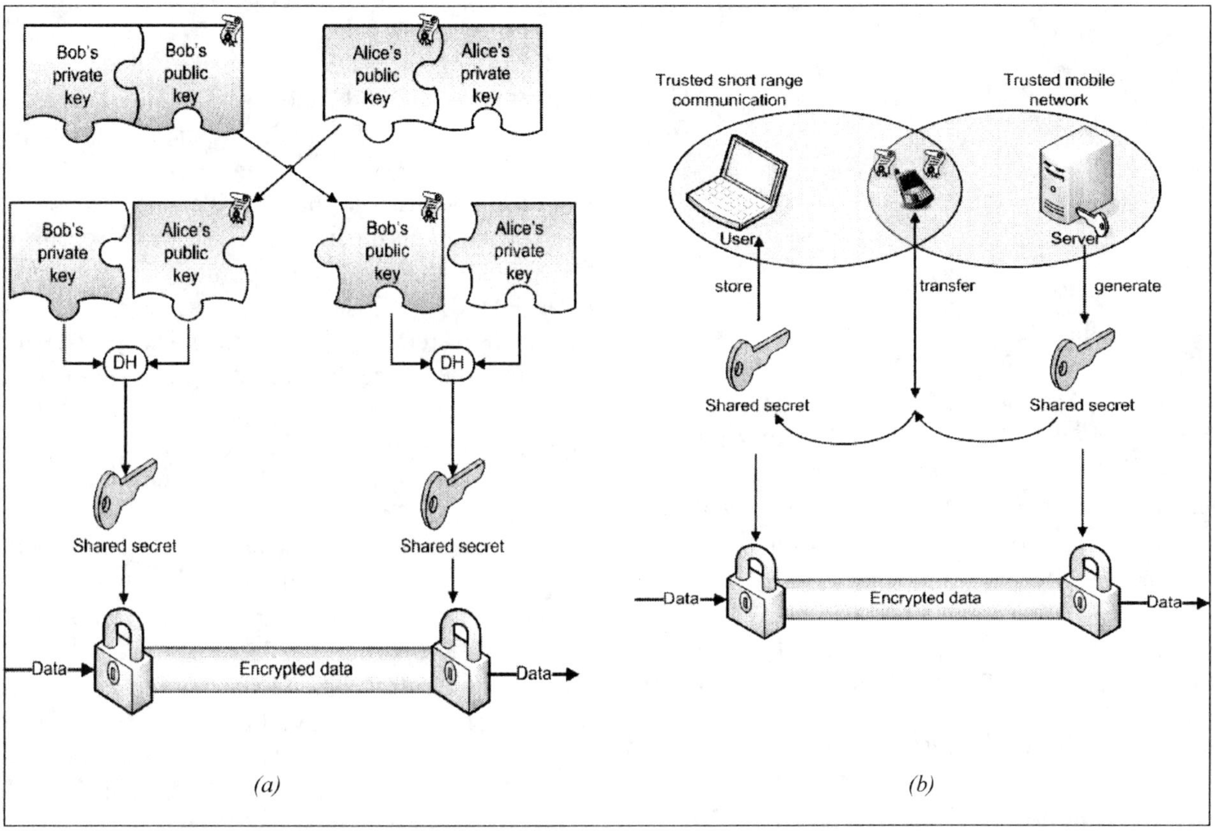

man (DH) key exchange of Figure 2, which allows two parties that have no prior knowledge of each other to jointly establish a shared secret key over an insecure communications channel.

This protocol does not authenticate the nodes to each other, but enables the exchange data, which can be decoded only by the two parties. Malicious attackers may start a man-in-the-middle attack (see Figure 4). Since this problem is well-known, several modifications enable identity based DH, for example Boneh, Goh, and Boyen (2005) showed a hierarchical identity based encryption method, which is operating in fact as a public key system, where the public key is a used chosen string.

Public key infrastructure (PKI) can help defending corresponding parties against man-in-the-middle attacks. Public key cryptography is based on the non polynomial (NP) time problems, for example of factorisation or elliptic curves.

Two keys, a public and a private are generated. The public key can be sent in plaintext, because messages encrypted with the public key can only be decoded by the private key and vice versa. The two way nature of public keys makes it possible to authenticate users to each other, since signatures generated with the public key can be checked with the public key. Message authenticity can be guaranteed. Still, the identity of the node is not proven. The signature proves only that the message was encoded by the node, which has a public key of the entity we may want to communicate with.

Identity can be ensured by using certificates. Certificate authorities (CA) store public keys and after checking the owner's identity out of band, prove their identity by signing the public key and user information with their own keys. This method is required for financial transactions and business and government operations. Without a

Figure 3. Principle of a man-in-the-middle attack

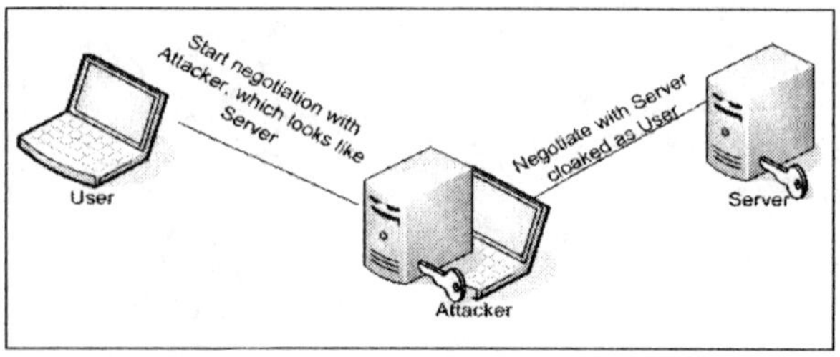

CA, the public keys can be gathered into a PKI, which provides an exchange service. Here, most commonly, a method called web of trust is used. A number of nodes, who think that the key is authentic, submit their opinion by creating a signature. The solution enables community or personal key management, with a considerable level of authenticity protection.

While public keys can be sent, private keys must be kept secret. Although they are protected usually with an additional password, this is the weakest point in the system. If the user saves a key in a program in order to enter the key automatically, security provided by the system is equal to the security of the program's agent application. Private firewalls and operating system policies usually will not stop a good equipped intruder.

Another security issue for terminals is the lack of tamper resistant storage. Usage of smart cards is a solution to this issue, but introduces additional hardware requirements. The lack of secure storage is getting much attention in DRM schemes. Most DRM schemes use a software-based method, but also hardware-assisted ones have lately been introduced.

All these authentication methods, secure storage and rights management support secure data exchange, but they do not protect the privacy of user credentials, preferences, and profiles. Ad hoc networks, like personal area networks (PANs), which move around and are dynamically configured open for intrusion attacks on the privacy.

Thus, protection of user credentials in wireless environments is one of the focal points of current research. Before addressing privacy, we will first summarise issues in key management protocols.

FROM KEY EXCHANGE TO ACCESS CONTROL INFRASTRUCTURE

Mobility and wireless access introduced new problems in network and user management, as compared to fixed network installations with, for example, port-based access restrictions. The network operators want to protect the network against malicious intruders, charge the correct user for the use, and provide easy and open access to their valued services.

The first step to get access to an encrypted network is to negotiate the first session key. This has been solved in coordinated networks like mobile networks through pre-shared keys. Authentication and access control is provided by central entities to ensure operations.

In computer networks, which are not controlled in such way and usually not backed-up by a central authorisation, authentication, and accounting (AAA), different methods have been created for connection control. The basic method is still to negotiate encryption keys based on a preshared secret. Typical preshared keys are a password for hash calculation, one time password sent via cell phone or keys given on an USB stick.

There are several solutions to protect the data transmitted over a wireless link. In private networks, security based on preshared keys is a working solution. In corporate or public networks, a more robust solution is needed. The most promising way is to integrate session key negotiation into the AAA process. Since providers or companies have to identify the connected user, they rely on an AAA infrastructure and have an encryption of user credentials as compulsory policy. A certificate-based medium access control and AAA system is advised, where AAA messages can carry also the certificates needed to secure the message exchange.

As public key operations induce a lot of network traffic, the negotiated session keys have to be used in the most efficient way. Encryption protocols designed for wired environments, like transport layer security (TLS) do not consider problems associated with the broadcast transmissions and limitations of mobile devices. In a wired, or at

Figure 4. TLS key negotiation

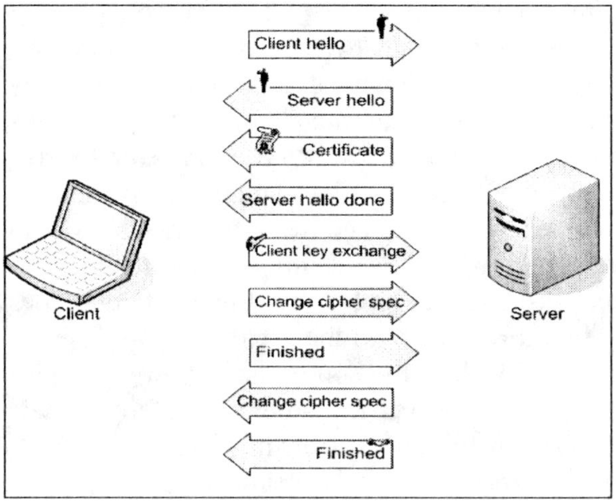

Figure 5. TLS-KEM key negotiation

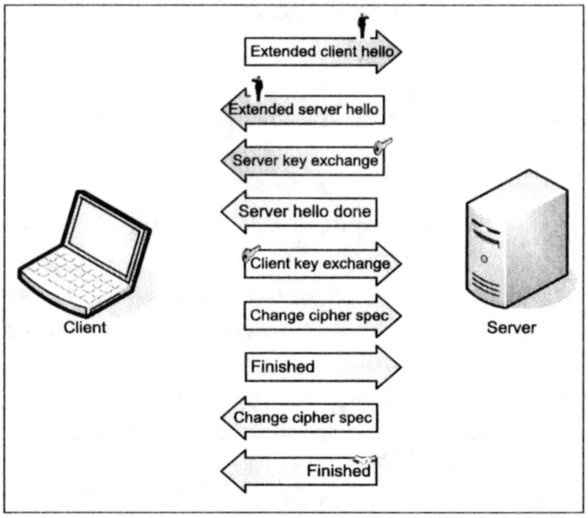

least fixed environment, computational cost of key negotiations is usually neglected. For example TLS is using several public key operations to negotiate a session key. This can be a problem for mobile devices, since computational cost is much higher in asymmetric encryption. The standard TLS suite uses lots of cryptographic operations and generates a too large message load on wireless links (see Figure 5).

If a mobile device wants to execute mutual authentication with a service provider, with certificate exchanges, it can lead to big amounts of data transferred over the radio interface beside the high computing power needs.

In environments with limited resources, authentication and identity management based on preshared keys is still the most effective solution. Badra and Hajjeh (2006) propose an extension to TLS, which enables the use of preshared secrets instead the use of asymmetric encryption. This is in line with the efforts to keep resource needs at the required minimum level in mobile devices. A preshared key solution was also proposed by the 3rd Generation Partnership Projects (3GPP, 2004) and (3GPP2, 2007) as an authentication method for wireless LAN interworking. The problem with the proposed solution is preshared keys does not provide adequate secrecy nor identity protection in Internet connections. To deal with this problem, the TLS-key exchange method (TLS-KEM) provides identity protection, minimal resource need, and full compatibility with the original protocol suite as seen in Figure 6.

In direct comparison, the public key based TLS needs a lot more computing, data traffic, and deployment effort.

In UMTS networks, an array of authentication keys is sent to the mobile in authentication vectors. In the computer world a good solution would be using hash functions to calculate new session keys, as these consume low power and require little computing.

A moving terminal can experience a communication problem, as the overhead caused by key negotiation might extend the connection time to a network node. A preserved session key for use in the new network is a potential solution in a mobile environment, as it speeds up the node's authentication. Lee and Chung (2006) recommend a scheme, which enables to reuse of session keys. Based on the AAA infrastructure, it is possible to forward the key to the new corresponding AAA server on a protected network and use it for authentication without compromising system security. This can reduce the delay for connecting, and also reduces the possibility of authentication failure. Since the old session key can be used for authenticating the node towards the new AAA server, connection to the home AAA is not needed any more. The messages are exchanged as follows (Lee & Chung, 2006): when sending the authorisation request to the new network, the node also includes the old network address it had. The foreign agent connects to the new local AAA server and sends an authentication request. The new AAA server connects to the old one sending a message to identify the user. The old AAA authenticates the message by checking the hash value included, and generates a nonce for the terminal and the foreign agent. The server composes an AAA-terminal answer, which is composed from a plain nonce, an encrypted nonce using the key shared between the old foreign agent and the terminal. Then the whole message is signed and encrypted with the key used between the two AAA servers. When the new AAA receives it, decrypts and sends the message to the new foreign agent. Based on the plain nonce, the agent generates the key and sends down the reply, which includes also the nonce encrypted by the old AAA. After the authentication of the user towards the network, the user can start using services.

Key distribution and efficiency in e-commerce applications is another important aspect. The network's AAA usually does not exchange information with third parties or can not use the authentication data of the network access because of privacy issues. Current security demands require mutual identification of communicating parties in an e-commerce application. This can easily lead to compromising the customer to companies (for example in a GSM network, the user has to trust the network unconditionally). If the user can also check the identity of the service provider, at least man-in-the-middle attacks are locked out.

When a user starts a new session with a service provider, this session should be based on a new key set. The session key has to be independent from the previous one in means of traceability and user identity should not be deductible from the session key, thus ensuring user privacy. For mutual identification, a key exchange method is proposed by Kwak, Oh, and Won (2006), which uses hash values to reduce resource need. The key calculation is based on random values generated by the parties, which ensures key freshness.

The use of hash functions is recommended in mobile environments, providing better performances for public key based mechanisms (Lim, Lim, & Chung, 2006). Mobile IPv4 uses symmetric keys and hashes by default. Since symmetric keys are hard to manage, a certificate-based key exchange was recommended, but this demands more resources. To lower the resource demand, a composite architecture was recommended (Sufatrio, 1999). The procedure uses certificates only in places where the terminal does not require processing of the public key algorithm and does not require storage of the certificate.

The result of the comparison shows that hash is by far the most efficient method in terms of key generation, but suffers from management difficulties. Lim et al. (2006) also demonstrates that a pure certificate-based authentication is unsuitable for mobile environments. Partial use of certificates and identity-based authentication with extensive use of hash functions can be a potential way ahead.

AUTHENTICATION OF DEVICE GROUPS

In a ubiquitous environment, moving networks appear. PANs and ad hoc connections based on various preferences emerge and fall apart. These devices communicate with each other and have usually very limited capabilities in terms of computing power and energy reserves. In order to provide secure communication between any part of the network, hierarchical key management methods emerged (Kim, Ahn, & Oh, 2006). Here a single trusted server is used to manage the group key. These entities are usually storing the keys in a binary tree, where nodes are the leaves.

Public key operations are usually required when a terminal wants to connect to a group for the first time. A group management system needs frequent key generation rounds, because it has to ensure forward and backward secrecy. Strict key management policies ensure that no new node is capable of decoding former traffic and none of the old nodes have the possibility to decrypt current traffic. To adjust resource usage to mobile environment, a management scheme which uses mainly simple operations like XOR and hash is advisable (Kim et al., 2006). As the key in the root of the

Figure 6. Keys in a binary tree

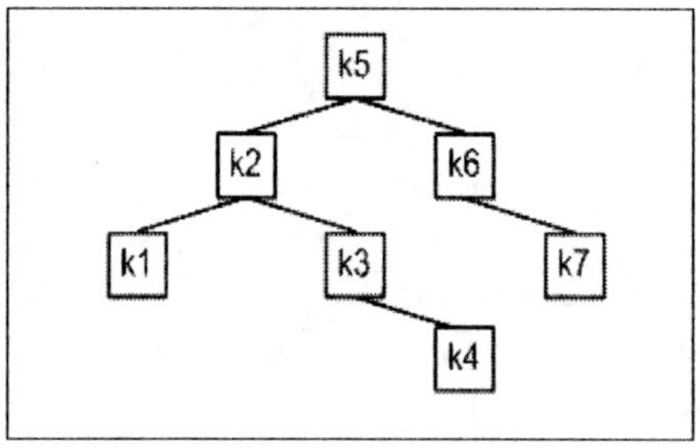

binary tree is used to authenticate the whole group, keys need to be regenerated when a node leaves the network. This procedure is starting from the parent of the former node and goes up to the root. Then the management unit sends out the new keys in one message. Building a tree from keys ensures fast searches and a simple, clean structure. In addition, all keys in the internal nodes are group keys for the leaves under them. So a subset of devices can be easily addressed.

The root unit has to compute these keys in acceptable time, requiring a more complex architecture. In PANs this is usually not a problem, but when a member of a larger subnet is leaving, calculations could be more demanding. A standard group key handling method is the Tree-based Group Diffie-Hellman (TGDH), where management steps assume that all nodes have the same processing capabilities. To ensure maximal efficiency, the highest performance unit shall be the one in the root of the tree (Hong & Lopez-Benitez, 2006). When node computing capabilities are showing big differences, the overhead caused by tree transformations does not represent a drawback.

Another significant group of devices that need encryption can be found in home networks, where the focus is on management of content and personal data.

SECURE HOME NETWORK AND RIGHTS MANAGEMENT

Deployment of wired or wireless home networks happens in roughly 80% of all households with broadband access (Noll, Ribeiro, & Thorsteinsson, 2005). Network-capable multimedia devices, media players, game consoles, and digital set-top boxes are widespread and part of the digital entertainment era. Content is stored within this network, and provided through the Internet to other users. Since the birth of peer-to-peer (P2P) networks, such technologies are in the crosshair of content providers. Recently, some software developers and a few musicians started using the torrent network for cost effective delivery of their content. A digital rights management method designed for such network is still missing.

Current right protection solutions are not compatible with each other and the user friendliness is also varying. The basic problem is, that just a very few devices are equipped with tamper resistant storage and integrated cryptographic capabilities. Beside software solutions, which are meant as weak solutions, hardware-based encryption can severely limit the lawful use of digital content. Recent lawsuits related to Sony's rootkit protection mechanism also reveals that customer rights of usage is considered to be more important than the legitimate wish of content providers to protect the content.

Trusted platform modules (TPM) are the most likely candidate for content protection in hardware-based solutions. While providing encryption capabilities, it is very likely that these components will be used to dispose the users' right to decide over the user's own resources.

The current discussions on DRM for audio content are regarded as minor when compared to high definition (HD) content protection. Even the connection to the screen has to use strong encryption, which has to exceed GSM/UMTS encryption in order to be acceptable for content providers. Enforcing a digital, end-to-end encrypted stream means that a HD-TV purchased at the end of 2006 may not work with the new encryption standards for HD. There is no current solution for computers to legally play full resolution HD. By the end of 2006 it was announced, that a workaround is arising to deal with the advanced content protection system of HD.

A more discrete, but not intrusive business model discussion for digital content management is presented in order to visualise the requirements of this market. Apple's FairPlay enables making backup copies of audio tracks, which is permitted by law in several European countries, and copy of content between the user's iPod players. This solution is considered being to open for some content providers, and the distribution is limited to a server-client infrastructure. For HD content with high bandwidth needs such a server-client infrastructure is not advisable, both from a server and network point of view. The ever growing size of P2P networks form a perfect infrastructure to deliver content with high bandwidth need practically without substantial transmission costs. P2P networks are usually run without any DRM support. An additional infrastructure supporting DRM in a P2P network used to transmit content will enable high volume distribution of digital content (Pfeifer, Savage, Brazil, & Downes, 2006). If seamless license delivery and user privacy could be guaranteed, such a network could be the foundation of a low cost content delivery scheme.

While the usage of P2P networks is an excellent idea, the recommended solution proposed by Nützel and Beyer (2006) is similar to the Sony's rootkit solution: It bypasses the user control and is thus not acceptable. While the primary goal is to secure content, the software used in such solutions acts like hidden Trojans and opens backdoors not only for the content providers, but also other hackers.

Content usage across platforms is not supported yet, as a common standard does not exist. Pfeifer et al. (2006) suggests a common management platform for DRM keys with an XML-based, standard MPEG-REL framework. Users will also produce content with digital protection, in order to ensure that personal pictures cannot be distributed electronically. Social networks and groups of interest, as well as distribution of content in PANs is a challenge for DRM development. Zou, Thukral, and Ramamurthy (2006) and Popescu, Crispo, Tanenbaum, and Kamperman (2004) propose a key delivery architecture for device groups, which could be extended by a local license manager. The central key management unit could distribute licenses seamlessly to the device, which wants to get access, without invading user experience.

Kálmán and Noll (2006) recommend a phone-based solution. This represents a good trade-off between user experience and content protection. The phone is practically always online, most of them have Bluetooth or other short range radio transmitters, so licenses can be transmitted on demand. Since the phone has a screen and a keyboard, it is possible to request authorisation from the user before every significant message exchange, so the user can control the way licenses are distributes.

If we look aside the issues related to business aspects, computational issues still remain. Highly secure DRM entities will use asymmetric encryption and certificates. Sur and Rhee (2006) recommend a device authentication architecture, which eliminates traditional public key operations except the ones on the coordinator device. This is achieved by using hash chains including the permission, for example, a device can get keys to play a designated audio track ten times or permission to use five daily permits on demand. Such schemes allow end devices to be simpler and lower network communication overhead.

If a central device is not appreciated, a composite key management scheme may be used. The parties in the PAN will form a web of trust like in a confidentiality scheme, for example, pretty good privacy (PGP). In this web, the main key is split between nodes and cooperation is needed for significant operations. This means that if the scheme is operating on a (k, n) basis, k-1 nodes can be lost before the system needs to be generate a new key. Fu, He, and Li (2006) mention the problem of the PAN's ad hoc nature as the biggest problem. Since this scheme selects n nodes randomly, the ones that are moving between networks fast can cause instability in the system. Also, the resource need of this proposal is quite high on all nodes present.

When a scheme is enabling off-line use of license keys, attention should be given to problems arising from leaving or compromised nodes. Identity-based schemes become popular recently because of their efficiency in key distribution. The main drawback is that these proposals do not provide a solution for revocation and key renewal. Hoeper and Gong (2006) propose a solution based on a heuristic (z, m) method. The solution is similar to the threshold scheme shown before, but enables key revocation. If z nodes are accusing one node to be compromised, based on their own opinion, the node is forced to negotiate a new key. If a node reaches a threshold in number of regenerations in a time period, it could be locked out, since most likely an intruder is trying to get into the system or the internal security of the node is not good enough. The assumptions about the system are strongly limiting the effectiveness of the solution. The most stringent assumption is that they require to nodes to be in promiscuous mode. This can lead to serious energy problems. Another requirement is that there has to be a unit for out-of-band key distribution. This unit could be the cellular phone.

SMART CARDS AND CELLULAR OPERATORS

The use of smart cards has its roots in the basic problem of security infrastructures: even the most well designed system is vulnerable to weak passwords. A card, which represents a physical entity, can be much easier protected compared to a theoretical possession of a password. Smart cards integrate tamper resistant storage and cryptographic functions. They are usually initialised with a preshared key and creating a hash chain, where values can be used as authentication tokens.

The remote authentication server is using the same function to calculate the next member. The encryption key is the selection of a collision resistant hash function. While the tokens they provide are quite secure, a problem with smart cards is that they represent a new unit that has to be present in order to enable secure communication, and user terminals must be equipped with suitable readers. The additional hardware does not only cause interoperability problems, but is usually slow, as a measurement conducted shows (Badra & Hajjeh, 2006). This becomes eminent when high traffic is associated with asymmetric encryption; sending a "hello" message with standard TLS to the smart card needed 10 seconds. In contrast, the modified TLS-KEM needed 1.5 s.

A user-friendly, seamless key delivery system can be created with the help of cellular operators and SIM cards with enhanced encryption capabilities. The SIM and USIM modules used in GSM/UMTS are quite capable smart cards. They offer protected storage with the possibility of over the air key management, good user interface, and standard architecture. Danzeisen, Braun, Rodellar, and Winiker (2006) shows the possible use of the mobile operator as trusted third party for exchanging encryption keys out of band for other networks.

Delivery of the mobile phone key to a different device can be problematic, since most devices do not have a SIM reader, or it is inconvenient to move the SIM card from the mobile phone to another device. New developments in near field communication may overcome this and enable short range secure key transfer.

BREAKING THE LAST CENTIMETRE BOUNDARY

Frequency of authentication request is a key factor in user acceptance. If a system asks permanently for new passwords or new values from the smart card hash chain, it will not be accepted by the user. On the other hand, if a device gets stolen and it asks for a password only when it is switched on, then a malicious person can impersonate the user for a long time. A potential solution is to create a wearable token with some kind of wireless transmission technology and define the device behaviour such that if the token is not accessible, it should disable itself in the very moment of notification.

Since the main challenge is not securing data transfer between the terminal and the network, but to authenticate the current user of the terminal, a personal token has to be presented. As proposed by Kálmán and Noll (2007), the mobile phone can be a perfect personal authentication token if it is extended by a wireless protocol for key distribution.

With the capabilities of user interaction, network control of the mobile phone, it can be ensured that critical operations will need user presence by requiring PINs or passwords. Possible candidates for key exchange are Bluetooth

(BT), radio frequency identification (RFID), and Near Field Communications (NFC). NFC is a successor of RFID technology in very short range transmissions. BT is close to the usability limit, since its transmit range reaches several meters. But the two later ones are promising candidates. Depending on the frequency, general RFID has a range of several meters while NFC operates in the 0-10 cm range. NFC is recommended, as the range alone limits the possibilities of eavesdroppers and intruders who want to impersonate the token while it is absent. The use of repeaters in the case of NFC, a so-called wormhole attack as described by Nicholson, Corner, and Noble (2006), looks not feasible because of the tight net of repeaters required. Also, the capability of user interaction provides an additional level of security.

Mobile phones with integrated NFC functionality are already available and serve as user authentication devices. To use these devices as tokens for other terminals, they have to be placed very close to each other. This prevents accidental use in most cases. To check presence of the token, heartbeat messages might be introduced. By design, this solution is very capable of distributing preshared keys for other devices out of band. Meaning, the phone can get the keys from the cellular network from an identity provider and send it down to the appropriate device by asking the user to put the devices close to each other for a second or two.

Transmission of the key must be done only when needed, so the programmable chip on the phones has to be in a secured state by default and only activated by the user's interaction. Protection of RFID tags is shown by Rieback, Gaydadjiev, Crispo, Hofman, and Tanenbaum (2006), where a proprietary hardware solution is presented. In case of a phone-based NFC key transmission, additional active devices might be unnecessary to use, but for general privacy protection, IDs with RFID extensions must be treated with care.

Transmission of certificates would not need additional encryption over the NFC interface, while other keys may require a preshared key between the phone and the terminals, which can be done via a wired method or by the phone provider. Most providers have at least one secret key stored on phones and a public key connected to that one. Based on this, DH key exchange would be possible between terminals and the phone using the cellular network as a gateway. An NFC-enabled phone could be the central element of a home DRM service, as it is online, capable of over the air downloads, and still able to ensure user control.

ON THE DAWN ON PERSONAL CONTENT MANAGEMENT

From the viewpoint of secure data transmission and user authentication, access and distribution of digital content can be ensured. Open issues remain for moving PANs and devices with limited capability. Focus nowadays is on protecting the user's privacy. As usage of digital devices with personal information was limited, user privacy was not of primary concern for a long time. Since PANs and home networks hold a large amount of critical personal data, this has to change (Jeong, Chung, & Choo, 2006; Ren, Lou, Kim, & Deng, 2006).

In a ubiquitous environment users want to access their content wherever they are. This has to be enabled in a secure manner. With upcoming social services, also fine grained access control methods have to be deployed inside the personal infrastructure. The focus of DRM research has to shift towards the end user, who will also require the right to protect himself/herself and his/her content with the same strength as companies do.

Extending the phone's functions may be problematic because of energy consumption and limited computing power. This could be easily solved by the technology itself, since a new generation of mobile terminals is arriving every half year. The capacity and functionalities of the SIM cards will be extended, the newest 3GPP proposals are predicting high capacity and extended cryptographic possibilities.

Regarding legal aspects, extending the SIM possibilities may cause some concern, since the SIM cards are currently owned by the network operators.

CONCLUSION

Transport encryption and authentication of devices has been the subject of research for a long time and resulted in sufficient secure solutions with current technologies. The focus in recent proposals is on the limited possibilities of mobile terminals and adoption of encryption technologies for mobile and wireless links.

Distributing keys between nodes is solved, except for the first step, which usually requires out-of-band transmissions. A solution for this initial key distribution might be the mobile phone with its integrated smart card and already existing communication possibility. As phones come with NFC, they may act as contact-less cards to distribute keys between devices.

While device authentication is handled sufficiently, user identity is hard to prove. A knowledge-based password or PIN request is not a user-friendly solution. Current proposals tend to be insecure when performing the trade-off between user experience and security.

Focus on research should be paid towards personal area and home networks. These networks hold most of the user's personal private data and content, either purchased or created by the user. Currently no standard solution exists for managing content rights or for access control of own content.

REFERENCES

3rd Generation Partnership Projects (3GPP). (2004, July). *Technical standardization groups-system and architecture (TSG-SA) working group 3 (Security) meeting, 3GPP2 security—Report to 3GPP, S3-040588.* Retrieved December 20, 2006, from www.3gpp.org/ftp/TSG_SA/WG3_Security/TSGS3_34_Acapulco/Docs/PDF/S3-040588.pdf

3rd Generation Partnership Projects (3GPP)2. (2007). *TSG-X/TIA TR-45.6, 3GPP2 system to wireless local area network interworking to be published as 3GPP2 X.S0028.* Retrieved December 22, 2006

Badra, M., & Hajjeh, I. (2006). Key-exchange authentication using shared secrets. *IEEE Computer Magazine, 39*(3), 58-66.

Boneh, D., Goh, E.-J., & Boyen, X. (2005). Hierarchical identity based encryption with constant size ciphertext. In *Proceedings of Eurocrypt '05.*

Danzeisen, M., Braun, T., Rodellar, D., & Winiker, S. (2006). Heterogeneous communications enabled by cellular operators. *IEEE Vehicular Technology Magazine, 1*(1), 23-30.

Fathi, H., Shin, S., Kobara, K., Chakraborty, S. S., Imai, H., & Prasad, R. (2006). LR-AKE-based AAA for network mobility (NEMO) over wireless links. *IEEE Selected Areas in Communications, 24*(9), 1725-1737.

Fu, Y., He, J., & Li, G. (2006). A composite key management scheme for mobile ad hoc networks. In *On the move to meaningful Internet systems, OTM 2006 Workshops* (LNCS 4277).

Hoeper, K., & Gong, G. (2006). Key revocation for identity-based schemes in mobile ad hoc networks, ad-hoc, mobile, and wireless networks (LNCS 4104).

Hong, S., & Lopez-Benitez, N. (2006). Enhanced group key generation algorithm. In *Network 10th IEEE/IFIP Operations and Management Symposium, NOMS 2006* (pp 1-4).

Jeong, J., Chung, M. Y., Choo, H. (2006). Secure user authentication mechanism in digital home network environments. In *Embedded and Ubiquitous Computing* (LNCS 4096).

Kálmán, Gy., & Noll, J. (2006). *SIM as a key of user identification: Enabling seamless user identity management in communication networks.* Paper presented at the WWRF meeting #17.

Kálmán, Gy., & Noll, J. (2007). SIM as secure key storage in communication networks. In *The International Conference on Wireless and Mobile Communications ICWMC'07.*

Kim, S., Ahn, T., & Oh, H. (2006). An efficient hierarchical group key management protocol for a ubiquitous computing environment. In *Computational Science and Its Applications—ICCSA 2006* (LNCS 3983).

Kwak, J., Oh, S., & Won, D. (2006). Efficient key distribution protocol for electronic commerce in mobile communications. In *Applied Parallel Computing* (LNCS 3732).

Lee, J.-H., & Chung, T.-M. (2006). Session key forwarding scheme based on AAA architecture in wireless networks. In *Parallel and Distributed Processing and Applications* (LNCS 4330).

Lim, J.-M., Lim, H.-J., & Chung, T.-M. (2006). Performance evaluation of public key based mechanisms for mobile IPv4 authentication in AAA environments. In *Information Networking. Advances in Data Communications and Wireless Networks* (LNCS 3961).

Nicholson, A. J., Corner, M. D., & Noble, B. D. (2006). Mobile device security using transient authentication. *IEEE Transactions on Mobile Computing, 5*(11), 1489-1502.

Noll, J., Ribeiro, V., & Thorsteinsson, S. E. (2005). Telecom perspective on scenarios and business in home services. In *Proceedings of the Eurescom Summit 2005* (pp 249-257).

Nützel, J., & Beyer, A. (2006). How to increase the security of digital rights management systems without affecting consumer's security, In *Emerging Trends in Information and Communication Security* (LNCS 3995).

Pfeifer, T., Savage, P., Brazil, J., & Downes, B. (2006). VidShare: A management platform for peer-to-peer multimedia asset distribution across heterogeneous access networks with intellectual property management. In *Autonomic Management of Mobile Multimedia Services* (LNCS 4267).

Phillips, T., Karygiannis, T., & Kuhn, R. (2005). Security standards for the RFID market. *IEEE Security & Privacy Magazine, 3*(6), 85-89.

Popescu, B. C., Crispo, B., Tanenbaum, A. S., & Kamperman, F. L. A. J. (2004). A DRM security architecture for home networks. In *Proceedings of the 4th ACM workshop on Digital rights management,* Washington, DC.

Ren, K., Lou, W., Kim, K., & Deng, R. (2006). A novel privacy preserving authentication and access control scheme for pervasive computing environments. *IEEE Transactions on Vehicular Technology, 55*(4), 1373-1384.

Rieback, M. R., Gaydadjiev, G. N., Crispo, B., Hofman, R. F. H., & Tanenbaum, A. S. (2006, December 3-8). *A platform for RFID security and privacy administration.* Paper presented at the 20th USENIX/SAGE Large Installation System Administration Conference—LISA 2006, Washington, DC.

Sufatrio, K. Y. L. (1999, June 23-25). *Registration protocol: A security attack and new secure mini-mal public-key based authentication.* Paper presented at the International Symposium on Parallel Architectures, Algorithms and Networks, ISPAN'99. Fremantle, Australia.

Sur, C., & Rhee, K. H. (2006). An efficient authentication and simplified certificate status management for personal area networks. In *Management of Convergence Networks and Services* (LNCS 4238).

Zou, X., Thukral, A., & Ramamurthy, B. (2006). An authenticated key agreement protocol for mobile ad hoc networks. In *Mobile Ad-hoc and Sensor Networks* (LNCS 4325).

KEY TERMS

Diffie-Hellman Key Exchange: Diffie-Hellman key exchange is a procedure, which allows negotiating a secure session key between parties, who do not have any former information about each other. The negotiation messages are in band, but because of the non-polynomial (NP) problem used in the procedure, adversaries are not able to compromise it.

Mutual Authentication: Mutual authentication occurs when the communicating parties can mutually check each others identity, thus reducing the possibility of a man-in-the-middle attack or other integrity attacks.

Out of Band Key Delivery: Out of band key delivery occurs when an encryption key is delivered with a mean, which is inaccessible from inside the network it will be used in. An example is to carry a key on an USB stick between parties, where the key will never be transmitted over the network.

Rootkit: Rootkit is a kind of software to hide other programs. Mainly used by Trojans, they enable hidden applications to access local resources without user knowledge.

Seamless Authentication: Seamless authentication is a method where the user is authenticated towards an entity without the burden of credential requests. For high security requirements, transparent methods are not applicable, but can provide additional security in traditional username/password or PIN-based sessions.

Session Key: Session key is a short life, randomly generated encryption key to protect one or a group of messages. The main purpose is to use expensive encryption operations only when starting a session and use a simpler to manage cipher in the later part.

Chapter XII
Architecture and Protocols for Authentication, Authorization, and Accounting in the Future Wireless Communications Networks

Said Zaghloul
Technical University Carolo-Wilhelmina – Braunschweig, Germany

Admela Jukan
Technical University Carolo-Wilhelmina – Braunschweig, Germany

ABSTRACT

The architecture, and protocols for authentication, authorization, and accounting (AAA) are one of the most important design considerations in third generation (3G)/fourth generation (4G) telecommunication networks. Many advances have been made to exploit the benefits of the current systems based on the protocol remote authentication dial in user service (RADIUS)protocol, and the evolution to migrate into the more secure, robust, and scalable protocol Diameter. Diameter is the protocol of choice for the IP multimedia subsystem (IMS) architecture, the core technology for the next generation networks. It is envisioned that Diameter will be widely used in various wired and wireless systems to facilitate robust and seamless AAA. In this chapter, we provide an overview of the major AAA protocols RADIUS and Diameter, and we discuss their roles in practical 1xEV-DO network architectures in the three major network tiers: access, distribution, and core. We conclude the chapter with a short summary of the current and future trends related to the Diameter-based AAA systems.

INTRODUCTION

Many 3G cellular providers consider the architecture for the authentication, authorization, and accounting (AAA) system as one of the most important functional blocks for the success of service delivery. Typically, users are authenticated when requesting a service and only after successful authentication they are authorized to use the service. Once the user is granted access to the service, the network generates accounting messages based on the user's activity. Currently, the remote authentication dial in user service (RADIUS) protocol is the most widely deployed protocol in cellular networks to perform subscriber AAA. Since RADIUS is susceptible to various security threats, a standard developed by the Internet Engineering Task Force (IETF), called Diameter, was proposed to substitute RADIUS in the future. Unlike its predecessor RADIUS, Diameter offers reliable and secure communication enabling seamless roaming among operators and support of auditability, capability negotiation, and peer discovery and configuration. Diameter augments its reliable transmission capabilities by defining failover mechanisms and thus embraces two crucial elements for the robust communication of sensitive billing and authentication messages. Since most of the current equipment and radio standards only support RADIUS for authentication, it is evident that cellular network operators will be running both protocols in the near future. Therefore, it can not be sufficiently emphasized that prudent decisions need to be made when designing AAA systems with multiple protocols in mind at the three major tiers: access, distribution, and core.

The purpose of this chapter is to address the specific aspects of the AAA system architecture of these three major tiers. Given the broadness of the scope and the myriad of the existing AAA standards, we sharpen our focus on a reference 3G cellular network architecture which we define and show in Figure 1. As can be seen from Figure 1, a typical AAA system in 3G architectures is characterized by three distinctive architectural elements: (1) radio access network (RAN), (2) distribution network based on mobile IP (MIP), and (3) a multimedia domain (MMD)[1] based core including both IP multimedia system (IMS) networks and Internet access deployments. The RAN, based on one of the 1x carrier evolution data only (1xEV-DO) standards/revisions for wireless transmission, consists of various base stations (BSs) and radio network controllers (RNCs). The distribution network consists of the MIP elements, that is, the packet data serving node (PDSN) playing the foreign agent's (FA) role and the home agent (HA). It is worth observing that this architecture has a hierarchical nature, where multiple BTSs are governed by a single RNC and multiple RNCs are covered by a single PDSN region. Finally, at the core, we have the IMS elements, including it standardized elements such as the call session control functions (CSCF) and home subscriber servers (HSS) enabling robust applications and services such as gaming, presence, voice over IP (VoIP), and so forth.

Upon receiving a mobile subscriber call, the RNC authenticates the subscriber's request by communicating with the access network AAA (AN-AAA) over the RADIUS-based A12 interface. Once authenticated, the RNC contacts the PDSNs through the A10/A11 interface (3rd Generation Partnership Project 2 [3GPP2] A.S0008-B, 2006). Note that since the A12 interface is RADIUS based, a translation agent (TA) needs to be used to translate the RADIUS requests to Diameter for authentication. In Figure 1, we illustrate that the AAA contacts an Oracle-based users' database to authenticate the incoming calls. We assume that the TA, AAA, and the AN-AAA are collocated in the same physical platform for simplicity. For higher reliability, RNCs usually connect to multiple AAAs (one primary and another secondary AAAs) to allow redundancy to admit users into the system in case of AN-AAA connectivity problems.

Once admitted, the mobile node (MN) starts a point-to-point (PPP) session with the PDSN. During the process of PPP establishment, the PDSN advertises itself as a MobileIP FA and challenges the user. The user then replies with a Mobile IP registration request that answers the PDSN's challenge. The PDSN forwards this information to the AAA. The AAA validates the user's response

Figure 1. A 1xEV-DO reference network architecture

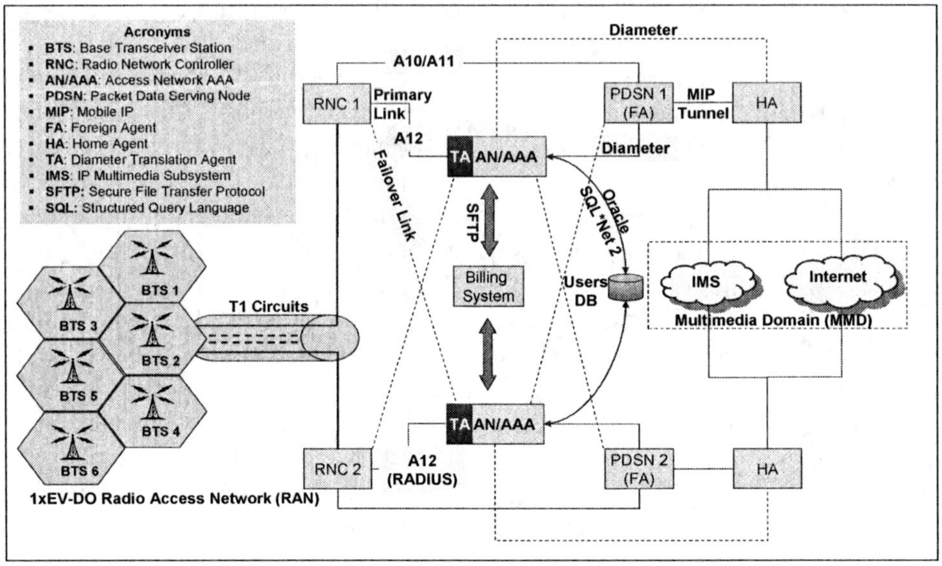

based on the MN-AAA shared secret and responds to the PDSN. In case of successful authentication, the PDSN proceeds with the MobileIP registration process with the HA and establishes a MobileIP tunnel to serve the user's traffic. At this point, the PDSN starts to generate accounting towards the AAA server to reflect the subscriber's usage. Accounting data is reformatted and is communicated to the upstream billing systems for further processing. Here, we assume simple secure FTP (SFTP) communication. Note that the PDSN also connects to multiple AAA's for redundancy purposes. In our illustrative architecture, the PDSN implements the Diameter MobileIP application and thus needs no translation functionality.

In this illustrative reference architecture, RADIUS is deployed in the access tier and translation agents were utilized to convert between RADIUS and Diameter, while Diameter applications at the distribution and network tiers were natively supported. Following this example, we organize the chapter as follows. First, we present the AAA concept and quickly survey RADIUS and its current deployment features. Then, we discuss the evolution from RADIUS to Diameter and shortly review the current Diameter standard. Afterwards, we illustrate a prospective end-to-end application of Diameter at all the three major network tiers in the wireless network, including access, distribution, and core. Finally, we summarize the chapter and discuss open issues and future work.

BACKGROUND

The RADIUS Protocol

AAA systems received significant attention from network service providers throughout the past decade. The need for a standardized, simple, and scalable protocol that accomplishes the required AAA functionality was the main motivation for the introduction of the (RADIUS) protocol (Rigney, 2000; Rigney Willens, Rubens, & Simpson, 2000; RFC2866). In 1998, RADIUS was the only protocol that seemed to satisfy the IETF NASREQ working group's requirements for authentication and authorization (Rigney, 1998). Due to its wide implementation by many networking equipment vendors, its simplicity and scalability, it became the protocol of choice for many service providers. RADIUS was quickly extended to support various networking protocols such as MobileIP (Perkins, 2002), IP security (IPsec) (Kent & Seo, 2005), and the IEEE 802.1x authentication.

Figure 2. A simple service provider's architecture with AAA functionality

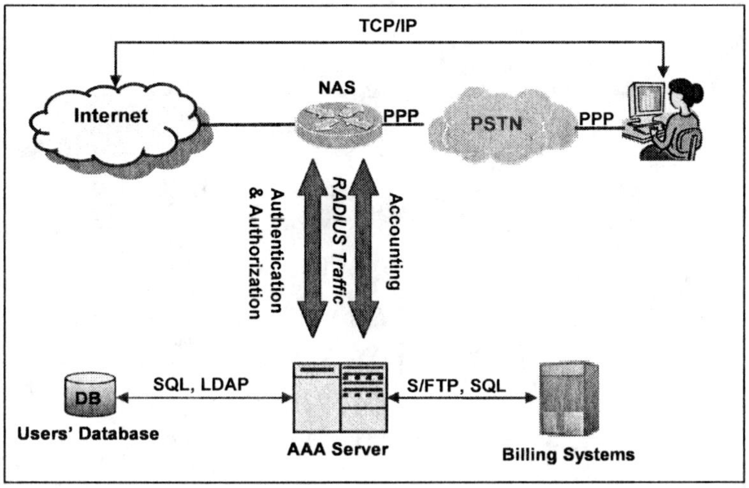

In RADIUS, after the user is granted access, the network access server (NAS) generates accounting messages based on the user's activity. The NAS is usually the gateway to the IP network. Routers, WiFi access points (APs), PDSNs, and gateway general packet radio service (GPRS) support nodes (GGSN) in GPRS networks, are typical examples of NASs in telecom networks. As shown in Figure 2, a user tries to access the Internet through a dialup modem connection. The PPP protocol is mainly used to establish the communication between the user and the NAS, that is, the router in this example. The NAS attempts to authenticate the user either through the password authentication protocol (PAP) or the challenge handshake authentication protocol (CHAP). Upon obtaining the responses from the client, the NAS generates an Access-Request and sends it to the RADIUS server in order to validate the user's responses. Typically, the RADIUS server is connected to an external database that contains the user's credentials and authorized services. Thus, the RADIUS server returns an Access-Accept message if the user credentials are valid, otherwise it returns an Access-Reject. The Access-Accept message may contain authorization information. For example, an Access-Accept message may contain: filters to grant the user access to internal networks, specific routing instructions to the NAS, quality of service (QoS) settings, and so forth. This authorization set is returned as a group of attribute value pairs (AVP) in the Access-Accept message.[2] Once the user is granted access, the NAS generates accounting messages based on user's activity (connection time, total bytes used, etc).

The RADIUS message format is shown in Figure 3. It consists of a 20 octet header followed by multiple AVPs. AVPs include standardized types and values. For example, the username is passed to the AAA server using the *User-Name* attribute. To allow expandability, the AVP type 26 is reserved for vendor-specific AVPs (VSAs). Thus, a vendor requests a *Vendor ID* from the Internet Assigned Numbers Authority (IANA) to be able to define specific attributes for his equipment. The following are sample vendor ID values: Cisco (9), Nortel (2637), 3GPP (10415), and 3GPP2 (5535). Usually, AAA implementations include dictionary files that define the AVP type and the expected values, for example refer to Braunöder (2003). RADIUS accounting is composed of three primary message types: (1) *Accounting-Start*, (2) *Accounting-Interim*, and (3) *Accounting-Stop*. Accounting messages usually carry the user's session information. For example, in CDMA2000-based systems accounting messages may contain the user's assigned IP address; user's sent and received byte counts; user's electronic serial number; calling and called station numbers; accounting session ID; BS ID; and so forth, (3GPP2 A.S0008-B, 2006; 3GPP2 X.S0011-005-C, 2006). Note that the electronic serial number and the BS ID attributes are 3GPP2 VSAs augmented to the standard RADIUS AVPs.

Figure 3. RADIUS message format

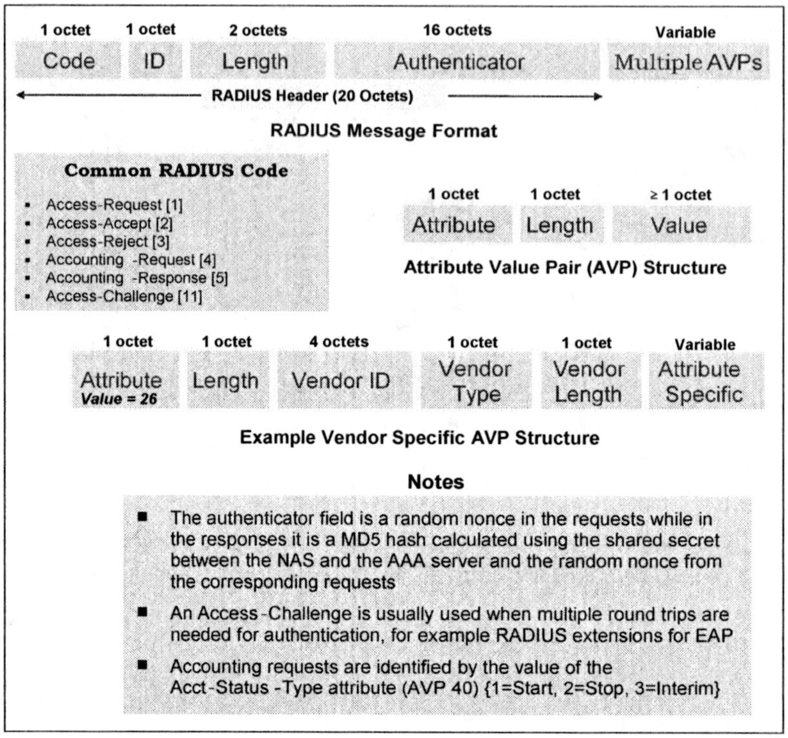

RADIUS offers reliability over the intrinsically unreliable user datagram protocol (UDP)[3] by requiring a response for each request. If a response is not received within a predefined time period (TO), the request times out. It is then up to the requestor (RADIUS client) to either retry the same server, another RADIUS server, or even drop the request. The timeout value and the maximum number of allowed retransmissions are configurable parameters at the client. It is noteworthy to mention that the failover mechanism was not standardized in RADIUS and often raised interoperability issues due to the inherent differences in the AAA implementations (Calhoun, Loughney, Guttman, Zorn, & Arkko, 2003).

RADIUS follows a client/server model where clients maybe NASs or other RADIUS servers. RADIUS clients and servers share a common secret to secure their communications. This method is weak and is only intended to secure communication within a trusted network.[4] Sometimes an AAA server serves as a RADIUS client/proxy when it is provisioned with a policy instructing it to forward the request to another RADIUS server. Such policies are occasionally based on the domain in the user's network access identifier (NAI). Standards (Aboba & Vollbrecht, 1999) refer to this setup as the proxy-chain configuration. For instance, in a roaming scenario the host AAA is usually configured to forward AAA requests from the hosting NAS to the home AAA. Note that multiple proxies maybe traversed along the path to the home AAA server as shown in Figure 4.

Evolution from RADIUS to Diameter

Diameter Protocol Overview

As network architectures evolved and with the tremendous growth in the wireless data infrastructures, secure inter-domain communication among various AAA servers to exchange subscribers' credentials, profiles, and accounting information became an absolute necessity. Despite its tremendous success, RADIUS inherent security vulnerabilities, its questionable transport reli-

Figure 4. Proxy chain configuration

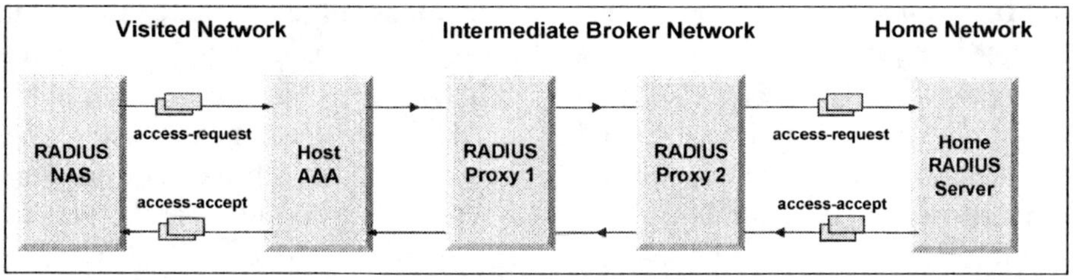

Figure 5. Diameter protocol

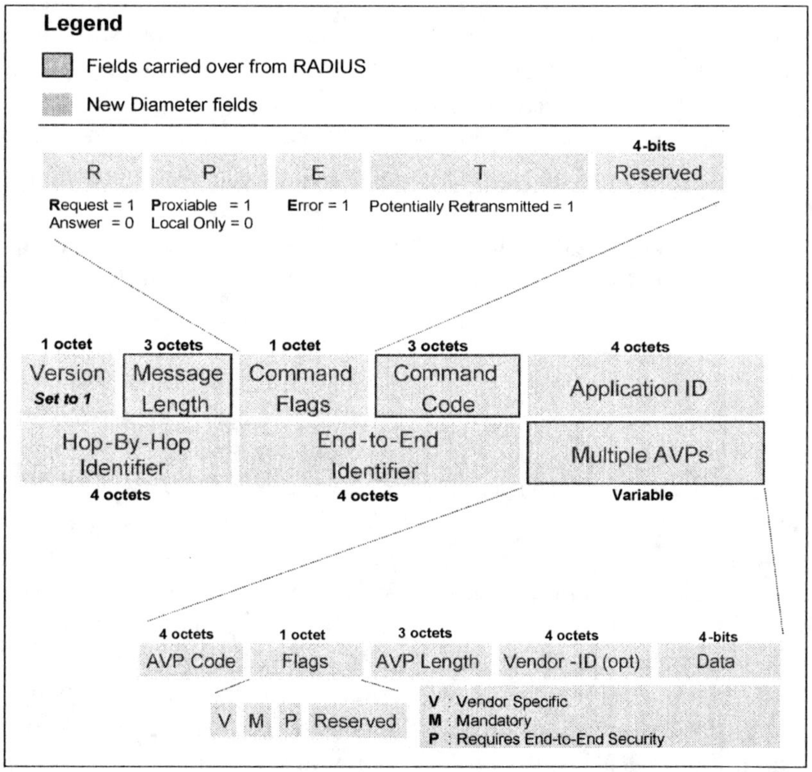

ability, and its limited redundancy support were the primary reasons for the introduction of the Diameter protocol (Calhoun et al., 2003) as a substitute protocol. Diameter was carefully designed to address security and reliability while thoroughly exploiting the benefits of RADIUS. Thus, secure transmission mechanisms using a choice of IPsec or transport layer security (TLS) protocols were integrated into Diameter, while reliable transport was enhanced by designing Diameter to run over either stream control transmission protocol (SCTP) or transmission control protocol (TCP) supported by standardized failover and failback (recovery) mechanisms.

Diameter RFC reused many of the RADIUS message codes and attributes and extended them. Figure 5 shows Diameter's header format. The framed fields in Figure 5 are those carried over from RADIUS. In contrast to RADIUS, note the introduction of the Version, Command Flags, Application ID, Hop-By-Hop ID, and End-to-End ID fields in Diameter. Also note the increase in size of the message length field (from 2 octets in RADIUS to 3 in Diameter). Note also that the authenticator field is no longer present as security is guaranteed by the integrated IPsec and TLS protocols. Command codes in Diameter start from 257 to maintain compatibility with RADIUS. Unlike in RADIUS, the requests and answers have the same command codes in Diameter, for example, the accounting request (ACR) and answer (ACA) commands have the command-code of 271. Diameter nodes can recognize message types (e.g., whether it is ACA or ACR) based on the "R" flag in the command flags shown in Figure 5. The "P" flag instructs nodes whether a message must be processed locally and should not be forwarded. The "E" flag along with the result-code AVP is used to indicate errors (and possibly redirection as we will see later). Finally, the "T" flag is used to indicate a possible duplication in case of retransmissions after a failover. Figure 5 also shows Diameter's AVP structure. The most significant addition is the inclusion of the flags field.

Diameter Agents

To facilitate migration from the current RADIUS infrastructure, Diameter offers indirect backward compatibility by introducing translation agents to convert RADIUS messages into Diameter messages and vice versa. Besides the main incentive of reusing as much of the RADIUS codes and attributes as possible for simpler migration, such reuse is also beneficial in reducing the amount of processing on the translation agents. Diameter supports a broader definition of scalability to suite roaming scenarios by including relay and redirect agents while still maintaining the RADIUS proxy agent model, therefore allowing the deployment of different architectures.

A proxy agent is used to forward Diameter traffic to another Diameter peer in order to handle the request. The decision to forward requests is policy based as in RADIUS. Proxy agents may modify packets and may originate rejection messages in case of policy violation, for example, in case of receiving requests from unknown realms. On the other hand, relay agents only forward requests without modifying any of the non-routing attributes. Relays and proxies are required to append the route-record AVP with the identity of the peer it received the request from prior to forwarding it. The reader is encouraged to refer to Calhoun et al. (2003) for more information on routing AVPs and their usage. Finally redirect agents, as their name implies, are used to refer clients to alternative AAAs. Redirect agents may act as proxies or end servers for other requests. For example, an AAA server may handle the Diameter base accounting messages while redirecting requests that require Diameter server support for MobileIP. Figure 6 summarizes the functionality of Diameter agents and illustrates the message flow in order of transmission. In Figure 6a, the relay agent only forwards the Diameter *Authentication and Authorization Request* (AAR) to Provider's B Diameter server. In Figure 6b, the proxy agent has an outbound policy for AAR to add or override the session Idle-Timeout attribute to 4,000 seconds and maximum link MTU to 1,300 bytes. It is also configured with an inbound policy for the *Authentication and Authorization Answers* (AAA) to remove any instructions for compression. Figure 6c shows the translation agent's role. Note that a translation agent may at the same time act as a proxy, that is, add, modify, or remove AVPs while converting between RADIUS and Diameter. Finally, as shown in Figure 6d, the Diameter client issues an AAR towards the redirect agent. Once received, the redirect agent sends back an AAA with the "E" flag set with the result-code AVP set to DIAMETER_REDIRECT_INDICATION instructing the Diameter client to contact *dest.com* by using the Redirect-Host AVP. A redirect agent may also provide indication on the usage of the redirect instruction, that is, whether its response is meant for all realms or simply restricted for the request's realm, whether the redirection policy should be cached at the requestor (Client), and for how long, and so forth.

Server Initiated Messages in Diameter

Unlike RADIUS, Diameter is a peer-to-peer protocol where any Diameter node may act as a client or server at any time. Peers are simply the next hop nodes that a Diameter node communicates with. A significant improvement over RADIUS is that Diameter has mandatory support of server-initiated messages to allow operations like re-authentication

Figure 6. Diameter agents' operation

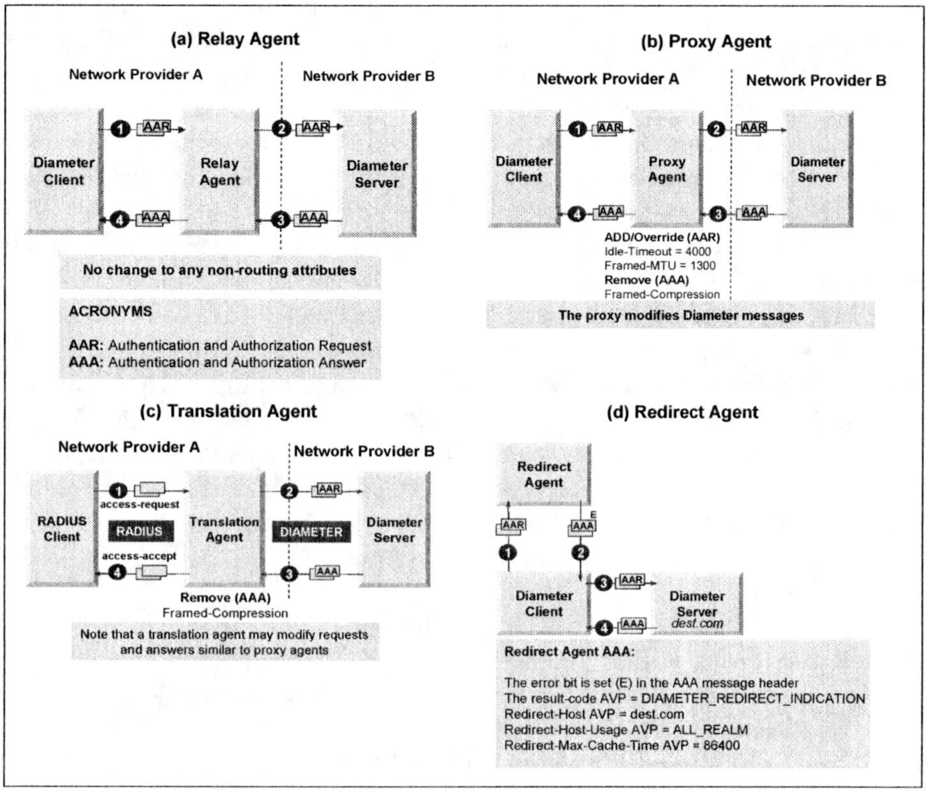

and network triggered session abortion. Diameter outlines a policy based framework for end-to-end security[5] and establishes auditability and proof of agreement by mandating message path authorization in case a message traverses multiple Diameter agents between two providers. This is accomplished by mandating authentication and authorization for each Diameter node along the path between two Diameter end-nodes. For instance, a service level agreement (SLA) among providers A, B, and C prevents the intermediary provider C from passing A and B's accounting traffic through the untrusted network U. Here, Diameter offers path authorization by requiring that each Diameter agent (provider C in this example) append the identity of the peer the request is received from prior to forwarding it. The Diameter servers must validate the conformance of the route-record attributes with the service policy. Thus, if servers on A or B detect entries for any untrusted servers, an AUTHORIZATION_REJECT error message is sent.

Diameter Applications

So far, we have only presented a summary of the so-called Diameter-based protocol, which must be implemented by every Diameter node. One of the most powerful features in Diameter is the introduction of the so-called "Diameter Applications." A node's capability to support certain applications is exchanged upon connection setup in the so-called Diameter capability exchange request and answer (CER, CEA) messages (Calhoun et al., 2003). Note that although many extensions were also added to RADIUS to support different applications (e.g., the Extensible Authentication Protocol [EAP], for WiFi in Rigney, 2000), RADIUS does not include any mechanisms to inform clients whether servers support such extensions. In other words, there is no standardized method to allow clients to discover whether the EAP extension is supported on an arbitrary server. This problem was solved in

Diameter by introducing the concept of Diameter applications.

It is important to understand that RFC 3588 defines the minimum prerequisites for a Diameter node implementation and maybe used by itself only for accounting. In case of authentication and authorization, a Diameter node must implement a specific application. The most common applications are Diameter NAS (Calhoun, Zorn, Spence, & Mitton, 2005) and MobileIPv4 (Calhoun, Johansson, Perkins, Hiller, & McCann, 2005). The Diameter NAS application defines NAS-related requirements where PPP-based authentication/authorization is needed. Diameter MobileIPv4 application defines AAA functionality in scenarios where users roam into foreign provider networks. The concept of Diameter applications was employed in many areas, and the following is a summary of three major Diameter applications,

- Diameter credit control application (Hakala, Mattila, Koskinen, Stura, & Loughney, 2005) is proposed to handle online billing for prepaid solutions. Prepaid billing implies real-time rating for the requested service, user's balance validation, and service suspension once the user's account is exhausted. Debiting and crediting are also supported for some applications such as gaming. Note that Diameter accounting defined in Calhoun et al. (2003) is mostly suitable for postpaid services where off-line processing of accounting records is performed.
- Diameter EAP (Eronen, Hiller, & Zorn, 2005) is used to support end-to-end authentication in dial-up, 802.1x, 802.11i, and in IPsec IKEv2. It eliminates the possibility of man-in-the-middle attacks if node is compromised within a proxy chain.
- The Diameter Session Initiation Protocol (SIP) application (Garcia-Martin, Belinchon, Pallares-Lopez, Canales-Valenzuela, & Tammi, 2006) supports HTTP digest authentication (RFC2617) mandated by SIP (Rosenberg et al., 2002) to allow SIP user agents and proxies to authenticate and authorize user's requests to access certain resources. This application does not depend on the Diameter NAS nor MobileIPv4 applications, where as it supports the Diameter credit control application but does not depend on it. Moreover, Diameter SIP allows locating SIP servers when a SIP agent requests routing information. Finally, it provides a mechanism for pushing updated user profiles to the serving SIP server in case the profile is (administratively) updated.

Finally, it is extremely important to understand that Diameter applications need to be defined only when none of the existing Diameter applications can support the required message flow without major modifications. Such major changes include adding new mandatory AVPs, commands requiring different message flows from any of the currently defined applications, or requiring support for new authentication methods with new AVPs (Fajardo & Ohba, 2006).

Protocol Mechanisms

Diameter Peer Discovery

Diameter offers three primary means to discover Diameter peers: static, Service Location Protocol Version 2 (SLPv2) queries, and domain name system. Thus, a *peer table* entry is created after peer discovery is executed. Note that peer discovery maybe triggered upon the reception of a CER. In some cases, policies may allow establishing connections with unknown peers. In this case, the peer table entry is built from the peer's identity in the CER and expires as soon as the connection is closed. In most of the cases, peer table entries for known peers are created along with their advertised applications. Thus, only requests for advertised applications are forwarded to these peers.

Diameter Policies

Routing tables provide guidance to the Diameter node on how to process a received request. Figure 7 illustrates an example realm routing table for Relay/Proxy Agent. Note that a policy includes a realm, an application identifier, and an action. When forwarding is needed, the next hop server is given and whether the route entry was statically or dynamically discovered (through a redirect, for example), along with its expiration time. The

Figure 7. Sample routing policy

```
RealmName=ourrealm.com AND destination=ourid,
ApplicationID=any, Action=LOCAL

RealmName=myMIPdomain.com, ApplicationID=MobileIPv4,
Action=REDIRECT, Next-Hop=ServerMIP.com,
Dynamic:ExpirationTime=900

RealmName=myMIPdomain.com, ApplicationID=DiameterNAS,
Action=PROXY, Next-Hop=ServerACT.com, Static, Proxy_Policy =
{outbound[Idle-Timeout=400],inbound[remove framed-compression]}

DefaultPolicy -Answer result-code=DIAMETER_UNABLE_TO_DELIVER
```

default policy in case no route is available is to return an error message with the DIAMETER_UNABLE_TO_DELIVER result code.

Diameter Request Routing

Diameter request routing refers to the process needed when originating, sending, and receiving requests. When originating a request, the Diameter node sets the Application-ID, the Origin-Host and Origin-Realm AVPs along with the Destination-Host and/or realm. When receiving a message, the node checks the route-record AVP to make sure that there are no routing loops.[6] It also checks whether it is the ultimate destination of the message. If not, the node acts as an agent and according to its policy it relays, proxies, or redirects the message. Each forwarded (i.e., proxied[7] or relayed) message is updated with a locally generated hop-by-hop identifier. This field is used to match requests and answers. Answers are routed opposite to how requests are routed and using the hop-by-hop identifiers the expected answers at each hop are recognized. Using the hop-by-hop identifier and the saved sender's information, the answer is forwarded back to the previous node with the hop-by-hop identifier restored to its original value. This process ends once a node finds its identity in the origin-host.

Diameter's Failover and Failback Algorithms

Diameter implements the so-called watchdog algorithm to detect communication trouble and initiate the failover mechanism. A Diameter node may have a primary server and multiple secondary servers for redundancy. When a communication problem is detected, a secondary server is promoted to primary and the primary is suspended. Notice that this is important to guarantee consistent failover for all requests.

The link is considered responsive as long as acknowledgements arrive. If the link is idle for "tw" seconds then a device watchdog request (DWR) is sent. If no device watchdog answer (DWA) arrives in "tw" seconds, the primary is suspended, the secondary server is promoted, and all subsequent communication is sent to the promoted server. Note that outstanding messages maybe sent on the failover link and in this case the "T" flag is set in each message to indicate (to the end server) that such messages maybe duplicates. If another "tw" seconds pass without receiving the DWA on the suspended primary link, then the transport connection is closed. The connection may be retried periodically, but for reopened connections, a connection validation procedure must be initiated. In this case, three watch-dog messages must be answered before failing back to the original primary link (Aboba & Wood, 2003; Calhoun et al., 2003).

A Summary of Diameter's Session Management and Accounting

A *session* is defined as "a related progression of events devoted to a particular activity" (Fajardo & Ohba, 2006). When a Diameter node is required to keep track of sessions for later use the node is considered stateful, otherwise it is stateless. For example, in the case where a server needs to trigger re-authentication, it needs to maintain the session state. This implies that session management is application specific. For example, a Diameter accounting server maybe configured to keep track of accounting messages such that it is able to eliminate duplicates and fraudulent messages (e.g., a unique Accounting-Start message should not arrive before an Accounting Stop message for an opened session). In cases where the server is stateful, a Diameter client must always send a session-termination-request (STR) to the server so that the server frees its allocated resources for the session.

RFC3588 (Calhoun et al., 2003) and RFC4005 (Calhoun, Zorn, et al., 2005) outline the accounting process. Similar to RADIUS, Diameter accounting requests (ACR) are sent and answers (ACA) are received from servers. A new accounting type, Event record, has been introduced to be used for short connections where accounting Start and Stop records may arrive during very short time periods (e.g., for push-to-talk services). Accounting Event records are also used to indicate accounting problems. For long connections (e.g., VoIP conferencing and file downloads), Start, Interim, and Stop records are used. It is noteworthy to state that in case of reauthorization, an accounting Interim may be sent to summarize the pervious state. In case connection details are modified considerably, an accounting Stop followed by an accounting Start message are sent. The later is case is widely used in practice.

DIAMETER-BASED ARCHITECTURES

As we have seen in the introduction section, there are three network tiers: access, distribution, and core (see Figure 1). In this section, we analyze a selected Diameter application in each tier.

At the Access Layer: 1xEV-DO with a translation agent

Figure 1 shows a simplified 1xEV-DO network where radio network controllers (RNCs) authenticate the mobile call through the RADIUS based A12 interface (3GPP2 A.S0008-B, 2006). The AN-AAA returns the subscriber's International Mobile Subscriber Identity (IMSI) in the Callback-ID AVP to the RNC in the RADIUS access-accept message. Note that since the 1xEV-DO standard does not support Diameter yet, operators may utilize Diameter TAs to convert between RADIUS and Diameter queries. The TA maybe collocated with the AN-AAA as shown in Figure 1. Note that RNCs maybe configured to failover to another AN-AAA for redundancy. Here, the reader should be aware that such failover is RADIUS based and is not based on the Diameter failover mechanisms.

At the Distribution Layer: Diameter MobileIPv4

The PDSN is considered the first IP gateway in 1xEV-DO networks. In MobileIPv4 architectures, MNs are expected to move from one PDSN region into another resulting in MobileIP handovers (HO). The HA represents the home network to which the MN's IP address (Home Address) belongs. Here, we assume that the PDSN/FA and the HA natively support the Diameter MobileIPv4 application (i.e., no translation is involved). When the MN moves into a foreign network, it attaches through a FA that tunnels its traffic back to its home agent enabling it to maintain its IP address while moving (Perkins, 2002; Perkins & Calhoun, 2005).

In 1xEV-DO architectures, the PDSN normally plays the FA role (as well as the NAS role for Diameter) and tunnels the MN's traffic to its HA. The MN establishes a PPP tunnel to the PDSN and broadcasts a registration request (RRQ). Upon receiving the RRQ, the PDSN forwards it towards the AAA for authentication in a Diameter AA-mobile-node-request (AMR) which includes: Session-ID, MN Home Address, Home Agent identity, and MN NAI (Calhoun, Johansson, et al., 2005). Note that such authentication is needed as the RAN may be operated by a different entity from the Internet Service Provider (ISP) who owns the PDSN, HA, and so forth. Thus, upon receiving the Diameter AA-mobile-node-answer (AMA) from the AAA server, the PDSN/FA establishes a MobileIP tunnel with the HA to serve the MN's traffic and starts sending accounting requests to the AAA server.

We know when a mobile node roams into a foreign network, the foreign network's AAA usually acts as a proxy and forwards the Diameter requests pertaining to the roaming mobile node to its home AAA (HAAA) server. Foreign mobile nodes are simply recognized by the domains in their NAIs. In these cases, the mobile node needs to establish security associations with HA and/or FA. The HAAA is an attractive element to assist a key distribution mechanism. The Diameter MobileIPv4 application focuses on the role of the AAA as the key distribution element. As shown in Figure 8, the MN-AAA shared secret[8] is used to generate the MN-HA and MN-FA secrets. The FA advertises itself and includes a random challenge and the mobile node replies to the challenge using its MN-AAA shared secret and formulates a registration request (Steps 1, 2). The registration request triggers an AMR at the PDSN to be eventually forwarded to the HAAA (Steps 3, 4). The HAAA validates the request and derives the session keys based on a combination of nonces and the MN-AAA shared secret, then forwards the keys in a home-agent-MIP-request (HAR) to the HA. The HA extracts the MN-HA session key and reformats the nonces generated by the HAAA according to the MobileIP standard and encapsulates them in the home-agent-MIP-answer (HAA) (Steps 5, 6). The HAAA then creates an AMA which includes the MN-FA session key as well as the reformatted nonces from the HAA and forwards it towards the PDSN. The PDSN eventually extracts the session key and sends a registration reply towards the MN (Steps 7-9). The mobile node derives the session keys using the provided nonces and the MN-AAA shared key. Afterwards, the PDSN generates accounting requests (ACRs) reflecting the user's activity (Steps 10-13). The HAAA may be further used to maintain session information such that the same session-ID is used after handovers (Calhoun, Johansson, et al., 2005).

At the Core: IP Multimedia Subsystem (IMS) Interfaces

In the last few years, convergent networking architectures were widely discussed. The IMS was proposed as a radio access agnostic core infrastructure that allows heterogeneous radio networks (e.g., WiMAX, 1xEV-DO, UMTS, WiFi)

Figure 8. AAA role in mobileIPv4 key distribution

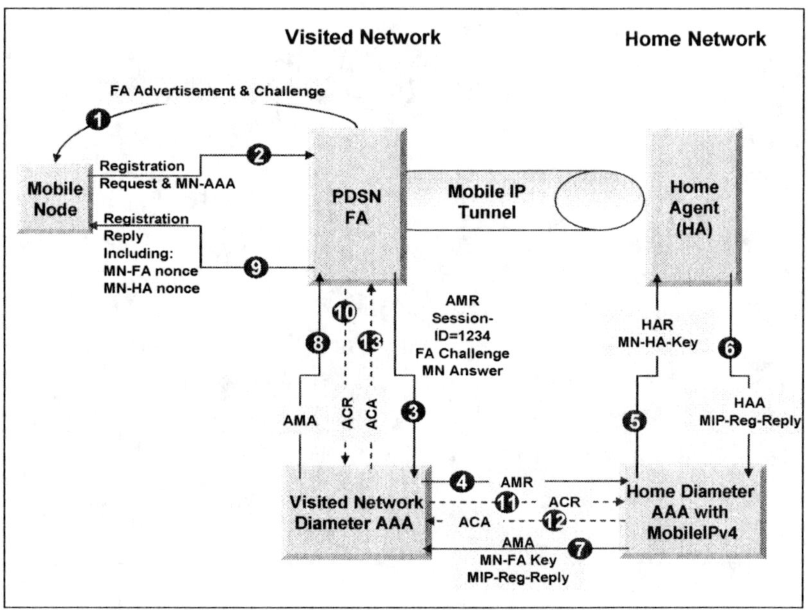

to communicate. As such, IMS offers unified services and enables seamless connectivity to the application servers (AS). In this section, we outline the role of Diameter in an IMS-based network. In IMS-based architectures users are granted a private identifier like (nai@operatorA.com) and multiple public identifiers (e.g., john.smith@corporate.com, smith_family@home.com), offering users the capability of sharing business and personal contact information, for instance. The users' profiles are stored in the Home Subscriber Server (HSS). Note that the HSS here plays an authentication and authorization role (AA) and this immediately implies the use of Diameter interfaces.

Let us assume that user 1 roams into provider Y's network and wishes to access a game service located in his/her home network. For that, user 1 first needs to register with the home network through operator Y's infrastructure. As shown in Figure 9, the first point of entry to the IMS network is the so-called Proxy Call Session Control Function (P-CSCF). The P-CSCF is responsible for SIP message processing and may perform various functions in security, compression, and policy enforcement over the SIP messages. The Interrogating CSCF (I-CSCF) is used to facilitate the communication among different operators. Operators have the I-CSCF addresses listed in their DNS servers to allow their I-CSCF to communicate with their peer I-CSCF in the other operator's networks. The I-CSCF normally proxies all SIP messages to the user's Serving CSCF (S-CSCF). The S-CSCF is the element that inspects all user's requests and confirms that they abide by access rights specified for that user. It also acts as a SIP router where it determines whether the SIP message needs to be sent to one or more ASs before granting service (Camarillo & García-Martín, 2004). Note that CSCFs communicate over the SIP-based Mw interface and that only I-CSCFs and S-CSCFs communicate with the HSS over the Cx interface (see Table 1 for the Cx Diameter commands). The Cx interface enables the S-CSCF to download users' profiles from the HSS.

Figure 9. Diameter role in IMS network environments

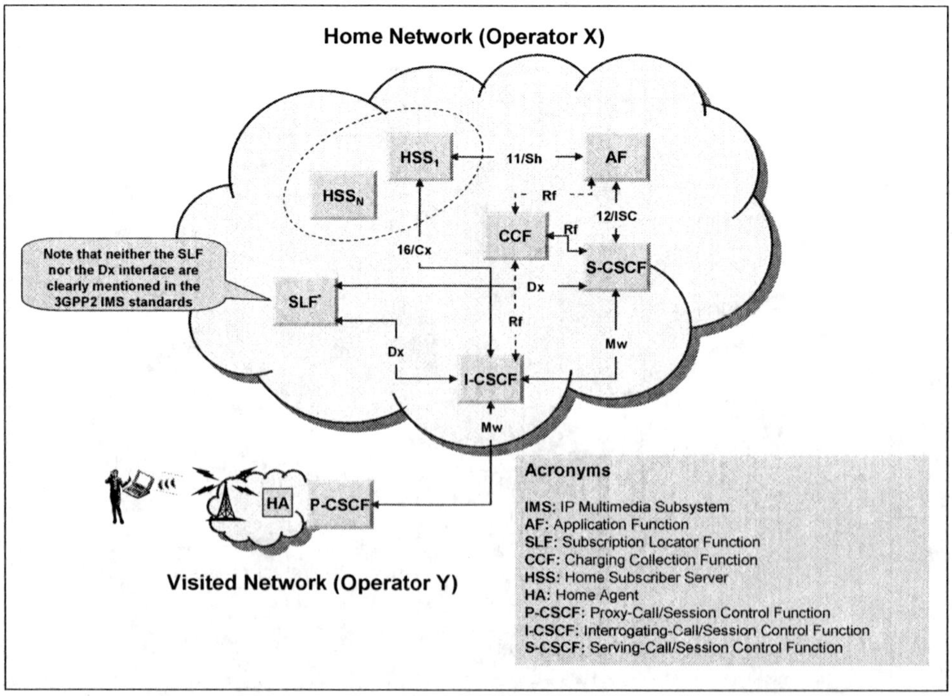

Table 1. The Cx interface commands

Source	Destination	Command-Name[9]	Abbreviation
I-CSCF	HSS	User-Authorization-Request	UAR
HSS	I-CSCF	User-Authorization-Answer	UAA
S-CSCF	HSS	Server-Assignment-Request	SAR
HSS	S-CSCF	Server-Assignment-Answer	SAA
I-CSCF	HSS	Location-Info-Request	LIR
HSS	I-CSCF	Location-Info-Answer	LIA
S-CSCF	HSS	Multimedia-Authentication-Request	MAR
HSS	S-CSCF	Multimedia-Authentication-Answer	MAA
HSS	S-CSCF	Registration-Termination-Request	RTR
S-CSCF	HSS	Registration-Termination-Answer	RTA
HSS	S-CSCF	Push-Profile-Request	PPR
S-CSCF	HSS	Push-Profile-Answer	PPA

The Dx interface, shown in Figure 9, is essentially the same as the Cx interface. When an I-CSCF wishes to locate the appropriate HSS that holds the user's profile (in order to contact the right S-CSCF for the user's request), it communicates with the subscription location function (SLF). The SLF is simply a Diameter redirect agent, which refers the I-CSCF to the right HSS. Although this interface is not clearly mentioned in (3GPP2 X.S0013-000-A, 2005), it can be simply viewed as a Diameter redirect for a Cx request.

The Sh interface between the HSS and the AS servers facilitates retrieving the application specific user's data, updating it, and receiving notifications when it is changed on the HSS. S-CSCF and AF may generate accounting records and in this case such accounting records are sent over the Diameter-based Rf interface towards the charging collection function (CCF). The CCF may reformat the billing records in the charging data record (CDR) format for further processing in the upstream billing system. It is noteworthy to mention that 3GPP2 X.S0013-000-A (2005) includes a 3GPP2 assigned interface name or number of each interface, for example, 16/Cx, along with the original IMS interface names. However, the use of interface numbering seems to be inconsistent in 3GPP2 standards as in most of the cases original names are only used (e.g., Cx not 16/Cx).

In Figure 10, we utilize a subset of the Cx interface commands to illustrate the IMS registration process for a roaming user. Once IP connectivity is established through the MobileIP procedures, the MN commonly referred to as the user agent (UA) in IMS initiates a registration request towards the P-CSCF. The P-CSCF recognizes that the user belongs to operator X, performs a DNS lookup for Operator X's I-CSCF, and forwards the request to the I-CSCF (Steps 1, 2). When the corresponding I-CSCF receives the registration request, it contacts the HSS over Diameter using the UAR command. Since, the REGISTER request usually carries both the user's public and private identities, the HSS validates that a roaming agreement exists with Operator Y and that the requestor is a valid user and returns a UAA to the I-CSCF (Steps 3, 4). The I-CSCF uses the information in the UAA to locate an S-CSCF and forwards the registration request to it (Step 5). Upon receiving the request, the S-CSCF issues a MAR towards the HSS to obtain appropriate authentication vectors to authenticate the user. The S-CSCF formats the response into a SIP response (401 Unauthorized) that carries a challenge (Steps 6, 7). Once the UA receives the response including a challenge (Step 10), it immediately responds with another registration message carrying a response for the

Figure 10. Initial registration with IMS over the Cx interface

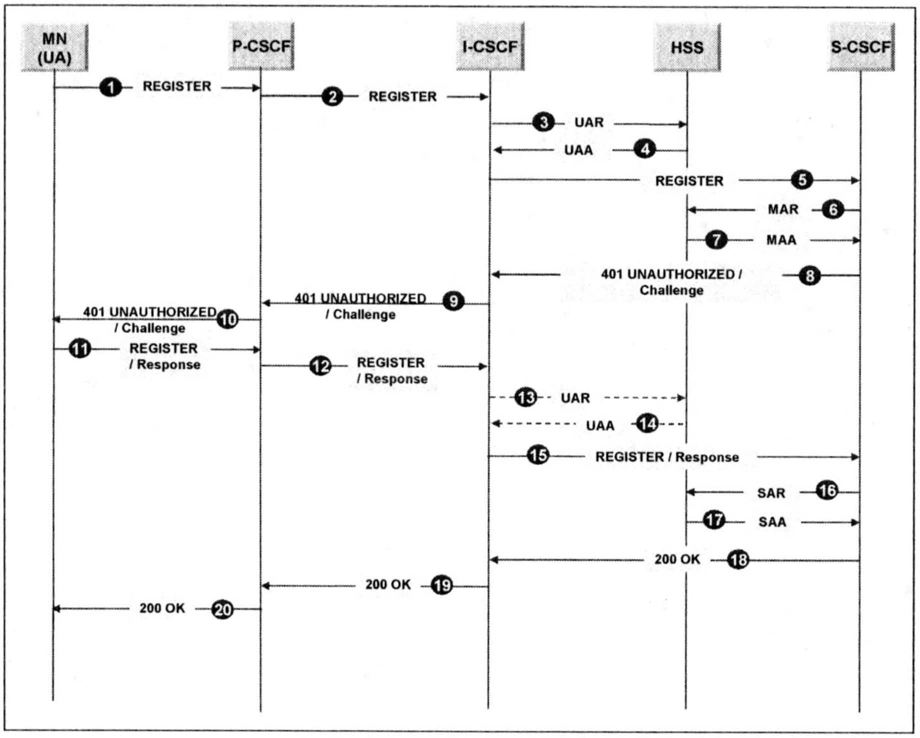

supplied challenge (Step 12). Note that the I-CSCF may perform another UAR to obtain the assigned S-CSCF (Steps 13, 14) either because it is stateless or it is another I-CSCF selected due to DNS load balancing. When S-CSCF receives the second registration request, it validates the user's response (Step 15) and if successful, it issues SAR to HSS requesting its assignment for the user's session and requesting the user's profile. The HSS assigns the S-CSCF for the user's session and sends the user's profile back to it (step 16, 17). At this point (step 18), the S-CSCF issues a SIP 200 OK message to the UA and once received (Step 20), the registration process is complete.

For registered users, when the I-CSCF receives a SIP INVITE request, it queries the HSS for the assigned S-CSCF using the LIR command. If the user's profile is updated, the HSS informs the serving S-CSCF of this change by sending a PPR. The HSS may terminate the user's session by issuing a RTR message towards the S-CSCF (3GPP2 X.S0013-005-A, 2005). As we can see from this short IMS registration walkthrough as well as from the previous sections, Diameter is envisioned to be one of the fundamental protocols used in the future 3G/4G telecommunication infrastructures.

ISSUES AND FUTURE TRENDS

Many standardization and research efforts are underway to upgrade and enhance the current AAA architectures to exploit the security and the scalability benefits of Diameter in the areas of session management, mobility support, distributed online and off-line accounting, and QoS assurance for user services over heterogeneous wireless networks. For instance, Eyermann, Racz, Stiller, Schaefer, and Walter (2006) discuss possible enhancements to maintain consistent accounting reporting in heterogeneous multi-operator environments by introducing a new Diameter accounting application including new commands and AVPs to allow sharing session context information. Moreover, efforts to attain seamless translation between RADIUS and Diameter are ongoing especially in the areas of matching requirements between RADIUS and Diameter and in the translation of VSAs (Mitton, 2006).

As the future telecommunication networks are expected to be based on IPv6, Diameter implementations over IPv6 were tested and some issues were identified (Lopez, Perez, & Skarmeta, 2005). The tests were conducted based on the Open Diameter[10] implementation. Integrating Diameter with MobileIPv6 is also an active area in both IETF and research. For example 3GPP2 X.P0047-0 (2006) discusses possible enhancements for MobileIPv6 to exploit the security features of the Diameter applications for MobileIPv6 tunnel setup. It also proposes enhancing MobileIPv6 by using Diameter for dynamic selection of home agents.[11]

Finally, continuous efforts are being made to establish a standardized framework for end-to-end QoS for services starting from the calling user at the RAN and ending at the called party whether it is located on the Internet or on another cellular network. 3GPP2 addresses such architectures in the service based bearer control draft document (3GPP2 X.S0013-012-0, 2006). It is noteworthy to mention that Diameter is quickly being considered to support many services. For instance Kim and Afifi (2003) discuss the integration of GSM SIM-based authentication with the AAA over Diameter-EAP application. Moreover, 3GPP2 has adopted Diameter architectures to support simple and multimedia messaging services (SMS and MMS) in (3GPP2 X.S0016-101-0, 2006).

SUMMARY

In this chapter we presented and discussed architecture and protocols for AAA as one of the most important design considerations in 3G/4G telecommunication networks. While many advances have been made to exploit the benefits of the current systems based on the RADIUS protocol, we illustrated its inherent security vulnerabilities. We then surveyed the details of the Diameter protocol and some of its applications. We showed that the Diameter protocol is not only the protocol of choice for the IMS architecture, but it also plays an increasingly important role in the three major network tiers, that is, access, distribution, and core. We demonstrated the role of Diameter in each tier by means of sample call flows in practical 1xEV-DO network architectures. We concluded the chapter with a short summary of the current and future trends related to the Diameter-based AAA systems.

REFERENCES

3rd Generation Partnership Project 2 (3GPP2) X.S0013-000-A. (2005). *All-IP core network multimedia domain—Overview (Ver. 1)* (3GPP2: TSG X Series). Retrieved from http://www.3gpp2.com/Public_html/specs/X.S0013-000-A_v1.0_051103.pdf

3rd Generation Partnership Project 2 (3GPP2) X.S0013-005-A. (2005). *All-IP core network multimedia domain—IP multimedia subsystem Cx interface signaling flows and message contents (Ver. 1)* (3GPP2: TSG X Series). Retrieved from http://www.3gpp2.com/Public_html/specs/X.S0013-005-A_v1.0_051103.pdf

3rd Generation Partnership Project 2 (3GPP2) A.S0008-B v1.0. (2006). *Interoperability specification (IOS) for high rate packet data (HRPD) radio access network interfaces with session control in the access network* (3GPP2: TSG A Series). Retrieved from http://www.3gpp2.org/Public_html/specs/A.S0008-B_v1.0_061019.pdf

3rd Generation Partnership Project 2 (3GPP2). X.P0047-0 v1.0. (2006). *Mobile IPv6 enhancement.* (3GPP2:Draft). Retrieved from http://www.3gpp2.org/Public_html/Misc/X.P0047-0v0.5_VV_Due_08_January-2007.pdf

3rd Generation Partnership Project 2 (3GPP2) X.S0011-005-C. (2006). *cdma2000 wireless IP network standard; Accounting services and 3GPP2 RADIUS VSAs* (3GPP2: TSG X Series). Retrieved from http://www.3gpp2.org/public_html/specs/X.S0011-005-C_v3.0_061030.pdf

3rd Generation Partnership Project 2 (3GPP2) X.S0013-012-0. (2006). *All-IP core network multimedia domain—Service based bearer control—Stage 2* (3GPP2:Draft). Retrieved from http://www.3gpp2.org/Public_html/Misc/X.P0013-012_SBBC_Stage-2_VV_Due_11_Sept-2006.pdf

3rd Generation Partnership Project 2 (3GPP2) X.S0016-101-0. (2006). *Multimedia messaging service; MM10 interface based on diameter protocol* (3GPP2:Draft). Retrieved from http://www.3gpp2.org/Public_html/SC/X.S0016-101-0_v1.0_060124.pdf

Aboba, B. (2005). *Re: End-to-end security in RFC 3588.* IETF Mail Archive, Message#01185. Retrieved from http://www1.ietf.org/mail-archive/web/aaa/current/msg01185.html

Aboba, B., & Vollbrecht, J. (1999). *Proxy chaining and policy implementation in roaming* (RFC 2607). Retrieved from http://www.ietf.org/rfc/rfc2607.txt

Aboba, B., & Wood, J. (2003). *Authentication, authorization and accounting (AAA) transport profile* (RFC 3539). Retrieved from http://www.ietf.org/rfc/rfc3539.txt

Braunöder, M. (2003). *Plug and phone software.* Retrieved from http://samuel.labs.nic.at/at43/dictionary

Calhoun, P., Bulley, W., & Farrell, S. (2002). *Diameter CMS security application.* IETF: DRAFT. Retrieved from http://www3.ietf.org/proceedings/02mar/I-D/draft-ietf-aaa-diameter-cms-sec-04.txt

Calhoun, P., Johansson, T., Perkins, C., Hiller, T., & McCann, P. (2005). *Diameter mobile IPv4 application* (RFC 4004). Retrieved from http://www.ietf.org/rfc/rfc4004.txt

Calhoun, P., Loughney, J., Guttman, E., Zorn, G., & Arkko, J. (2003). *Diameter base protocol* (RFC 3588). Retrieved from http://www.ietf.org/rfc/rfc3588.txt

Calhoun, P., Zorn, G., Spence, D., & Mitton, D. (2005). *Diameter network access server application* (RFC 4005). Retrieved from http://www.ietf.org/rfc/rfc4005.txt

Camarillo, G., & García-Martín, M. (2004). *The 3G IP multimedia subsystem (IMS): Merging the Internet and the cellular worlds.* John Wiley & Sons.

Eronen, P., Hiller, T., & Zorn, G. (2005). *Diameter extensible authentication protocol (EAP) application* (RFC 4072). Retrieved from http://www.ietf.org/rfc/rfc4072.txt

Eyermann, F., Racz, P., Stiller, B., Schaefer, C., & Walter, T. (2006). Diameter-based accounting management for wireless services. In *IEEE Wireless Communications and Networking Conference (WCNC'06)* (Vol. 4, pp. 2305-2311).

Fajardo, V., & Ohba, Y. (2006). Diameter base protocol details. In *The 67th IETF meeting.* San Diego, CA. Retrieved from http://www3.ietf.org/proceedings/06nov/slides/dime-3/dime-3.ppt

Garcia-Martin, M., Ed., Belinchon, M., Pallares-Lopez, M., Canales-Valenzuela, C., & Tammi, K. (2006). *Diameter session initiation protocol (SIP) application* (RFC:4740). Retrieved from http://www.ietf.org/rfc/rfc4740.txt

Hakala, H., Mattila, L., Koskinen, J.-P., Stura, M., & Loughney, J. (2005). *Diameter credit-control application* (RFC 4006). Retrieved from http://www.ietf.org/rfc/rfc4006.txt

Kent, S., & Seo, K. (2005). *Security architecture for the Internet protocol* (RFC 4301). Retrieved from http://www.ietf.org/rfc/rfc4301.txt

Kim, H., & Afifi, H. (2003). Improving mobile authentication with new AAA protocols. In *IEEE International Conference on Communications (ICC '03)* (Vol. 1, pp. 497-501).

Lopez, M., Perez, G., & Skarmeta, A. (2005). Implementing RADIUS and diameter AAA systems in IPv6-based scenarios. In *IEEE Proceedings of the 19th International Conference on Advanced Networking and Applications (AINA'05)* (Vol. 2, pp. 851-855).

Mitton, D. (2006). *Diameter/RADIUS vendor specific AVP translation.* IETF:DRAFT. Retrieved from http://internet-drafts.osmirror.nl/draft-mitton-diameter-radius-vsas-01.txt

Perkins, C. (2002). *IP mobility support for IPv4* (RFC 3344). Retrieved from http://www.ietf.org/rfc/rfc3344.txt

Perkins, C., & Calhoun, P. (2005). *Authentication, authorization, and accounting (AAA) registration keys for mobile IP* (RFC 3957). Retrieved from http://www.ietf.org/rfc/rfc4301.txt

Rigney, C. (1998). *2.4.10 Remote authentication dial-in user service (radius).* Snapshot of the 41st IETF meeting. In *Proceedings of the IETF March 1998.* Retrieved from http://www3.ietf.org/proceedings/98mar/98mar-edited-79.htm

Rigney, C. (2000). *RADIUS accounting* (RFC 2866). Retrieved from http://www.ietf.org/rfc/rfc2865.txt

Rigney, C., Willats, W., & Calhoun, P. (2000). *RADIUS extensions* (RFC 2869). Retrieved from http://www.ietf.org/rfc/rfc2869.txt

Rigney, C., Willens, S., Rubens, A., & Simpson, W. (2000). *Remote authentication dial in user service (RADIUS)* (RFC 2865). Retrieved from http://www.ietf.org/rfc/rfc2865.txt

Rosenberg, J., Schulzrinne, H., Camarillo, G., Johnston, A., Peterson, J., Sparks, R., et al. (2002). *SIP: Session initiation protocol* (RFC 3261). Retrieved from http://www.ietf.org/rfc/rfc3261.txt

Wikipedia. (n.d.). *RADIUS.* Retrieved from http://en.wikipedia.org/wiki/RADIUS

KEY TERMS

Diameter: Diameter is a new AAA protocol presented in RFC 3588 to replace RADIUS.

IP Multimedia Subsystem (IMS): IP multimedia subsystem is an access agnostic architecture proposed as a core technology for the next generation services.

One Carrier Evolution Data Only (1xEV-DO): 1xEV-DO is a CDMA2000 based cellular access technology proposed to support high rate data services.

Remote Access Dial In User Service (RADIUS): RADIUS is an AAA protocol defined in RFCs 2865 and 2866.

ENDNOTES

1 MMD is defined in all-IP core network standards (TSG X series) found at http://www.3gpp2.org/.

2 Notice that the authentication and the authorization operations are not separated in RADIUS. In other words, to obtain a user's authorization set, user must be successfully authenticated.

3 UDP ports 1812 and 1813 are the standard ports assigned for authentication and accounting respectively.

4 Inter-domain AAA traffic crossing untrusted networks such as in roaming scenarios is usually secured by dedicated VPNs.

5 According to (Aboba, 2005, message 01185) end-to-end security through Diameter CMS (Calhoun, 2002) mentioned in the standard (Calhoun, 2003, RFC 3588) has been abandoned and resolved by the introduction of the Diameter EAP application defined in (Eronen, 2005, RFC4702).

6 If a loop exists, the message is rejected with a DIAMETER_LOOP_DETECTED error message

7 More complex procedures may apply in case of translation.

8 Loosely speaking the user's password

9 These commands are based on the Diameter Cx Application (Application-ID = 16777216), more details can be found in (3GPP2 X.S0013-005-A, 2005; 3GPP2 X.S0013-006-A, 2005).

10 The Open Diameter project, located at [http://www.opendiameter.org/], offers open source C++ implementation of the Diameter base protocol.

11 Dynamic Home Agent (DHA) selection is a method used to dynamically select home agent based on the geographic location of the user such that the network backhaul delay is minimized.

Chapter XIII
Authentication, Authorisation, and Access Control in Mobile Systems

Josef Noll
University Graduate Center – UniK, Norway

György Kálmán
University Graduate Center – UniK, Norway

ABSTRACT

Converging networks and mobility raise new challenges towards the existing authentication, authorisation, and accounting (AAA) systems. Focus of the research is towards integrated solutions for seamless service access of mobile users. Interworking issues between mobile and wireless networks are the basis for detailed research on handover delay, multi-device roaming, mobile networks, security, ease-of-use, and anonymity of the user. This chapter provides an overview over the state of the art in authentication for mobile systems and suggests extending AAA mechanisms to home and community networks, taking into account security and privacy of the users.

INTRODUCTION

Today's pervasive computing environments raise new challenges against mobile services. In future visions, a converged user access network is projected. This means, that one network will be used to deliver different services, for example, broadcast TV, telephony, and Internet. Composed from mobile (e.g., Universal Mobile Telecommunications System [UMTS]), wireless (IEEE 802.11, IEEE 802.16, IEEE 802.20), and wired (cable, Asymmetric Digital Subscriber Line [ADSL]), these networks hide the border between the telecom, broadcast, and computer networks. The common service enables roaming terminals, which can access services independently of the currently used networking technology. Market players in both areas transform into wireless service providers across access networks. Telecom provide packet switched data and mobile services over the fixed network, while Internet service providers run voice over IP (VoIP) and video on demand (VoD) over mobile networks.

The changing environment also changes the management plane of the underlying networks. Providers on converged networks have to change their accounting and billing methods and need to redefine their business models. While commercial players demonstrate early examples, research in the AAA area focuses on providing a backplane for the upcoming ubiquitous services run over converged networks.

BACKGROUND

The AAA methods employed in current networks were developed for a single type of network, resulting in two different systems, one for telecommunication services and one for computer networks. This chapter addresses AAA in global system for mobil communications (GSM) and UMTS and computer network solutions based on Internet Engineering Task Force (IETF) standards.

The computer networks provide a unified AAA access, and research focuses on extending the existing methods to be suitable for telecommunication services. Extensions for Remote Authentication Dial In User Service (RADIUS) and Diameter are proposed. RADIUS is the current de facto standard for remote user authentication. It uses Universal Datagram Protocol (UDP) as transport. Authentication requests are protected by a shared secret between the server and the client, and the client uses hash values calculated from this secret. The requests are sent in plaintext except for the user password attribute. The Diameter protocol provides an upgrade possibility as compared to RADIUS. While enhancing the security through supervised packet transmission using the transmission control protocol (TCP) and transport layer encryption for reducing man-in-the-middle attacks, it lacks backward compatibility.

Both methods have a different background. The computer networks targeted the person using a computer in a fixed network environment, while mobile systems addressed a personal device in a mobile network. Thus a challenge for telcos is to enhance seamless network authentication towards user authentication for service access. Most companies are also Internet service providers (ISPs), this would be a natural unification of their AAA systems.

A generic approach is taken by extension of the Extensible Authentication Protocol (EAP) family. Development efforts of the Internet and telecommunication world were united on EAP. This protocol family has the potential for becoming the future common platform for user authentication over converged networks. EAP is a universal authentication framework standardised by IETF, which includes the authentication and key agreement (AKA) and Subscriber Identity Module (SIM) methods. EAP-AKA is the standard authentication method of UMTS networks.

Beside the fundamental differences of communication and computer networks, mobility is the key issue for both. Network services should not only be accessible from mobile terminals, but they should be adapted to the quality of service (QoS) requirements of a mobile/wireless link. Improvements of AAA methods are of fundamental importance for mobility, providing fast handover, reliable and secure communications on a user-friendly and privacy protecting basis.

Subscriber Authentication in Current Networks

In GSM networks, the integrated AAA is used for any type of user traffic. The authentication is just one way the user has to authenticate himself/herself towards the network.

To be more precise, the user is authenticated with a PIN code towards the SIM in the mobile phone, then the device authenticates itself towards the network. Device authentication instead of user authentication can hinder the upcoming personalised services because it is hiding the user behind the device. In UMTS, the authentication of the device is two-way. A device can also check the authenticity of the network with the help of keys stored on the SIM.

Integration of the mobile authentication with different external services is not widespread. The telecom providers have some internal services, which can authenticate the subscriber based on the data coming from the network. Credentials could be basically the CallerID, the Temporary International Mobile Subscriber Identity (TIMSI) or other data transformed with a hash function. Access control and authorisation is more an internal network task. Without considerable extension, the current mobile networks are more islands than connecting networks in the area of AAA. Equipment manufacturers are now recommending various IP multimedia subsystem (IMS) solutions for mobile providers in order to enable integrated and third party service convergence and to enable multimedia content over today's networks.

AAA protocols employed in computer networks are meant to provide services for authenticated users. Current single sign-on (SSO) protocols, like RADIUS, Diameter, or Kerberos provide the identity of the user to third parties. SSOs can use digital certificates, public key infrastructure (PKI) and other strong encryption methods. But, none of them is able to provide such a complete solution like the integrated AAA of the mobile network. Computer network protocols lack the support for fast mobility of moving clients and optimise resource usage for low bandwidth connections.

With incorporating seamless authentication used in network internal services in telecom world and SSO solutions provided by various protocols from computer networks, a unified AAA system will achieve a enhanced user acceptance and service security. In such a system, secure key storage and tamper resistant handling is crucial. Smart cards for key storage and generation will fulfil the security requirements, but usage and distribution of the smart cards is cumbersome. As most users have a mobile phone, the SIM card is a candidate to be a primary smart card used for AAA in a ubiquitous environment (Kálmán & Noll, 2006).

AAA IN CONVERGED NETWORKS

A converged network carries several types of traffic and enables seamless information exchange between different terminals, regardless of transport medium. To enable converged AAA, research work is going on in different areas: enabling wireless LAN (WLAN)-mobile network interworking, enhancing network mobility in wireless computer networks, and reducing resource requirements in cryptography.

Interworking Between Mobile and Wireless Networks

Network convergence is most significant in the wireless environment, having to face varying QoS measures on the radio interface, for example, propagation delay, variation of delay, bit error rate, error free seconds, distortion, signal to noise ratio, duration of interruption, interruption probability, time between interruption, bit rate, and throughput. These parameters will depend on the user and terminal environment and underline that an optimum access will have to use all available wireless and mobile connections. Leu, Lai, Lin, and Shih (2006) have provided the fundamental differences of these networks, summarised in Table 1.

Increased demand for security has improved the security on wireless links, resulting in Wi-Fi protected access (WPA) and WPA2 as draft implementations of the IEEE 802.11i standard. This standard aims at incorporating protocols of

Table 1. Comparison of cellular and WLAN networks

	Cellular	WLAN
Coverage	Country-wide	Local
Security	Strong	Depends on setup
Transmission rate	Low	High
Deployment cost	High	Low
License fee	Very high	No need
Construction	Difficult	Easy
Mobility support	High	Poor

Figure 1. Authentication in GSM and UMTS

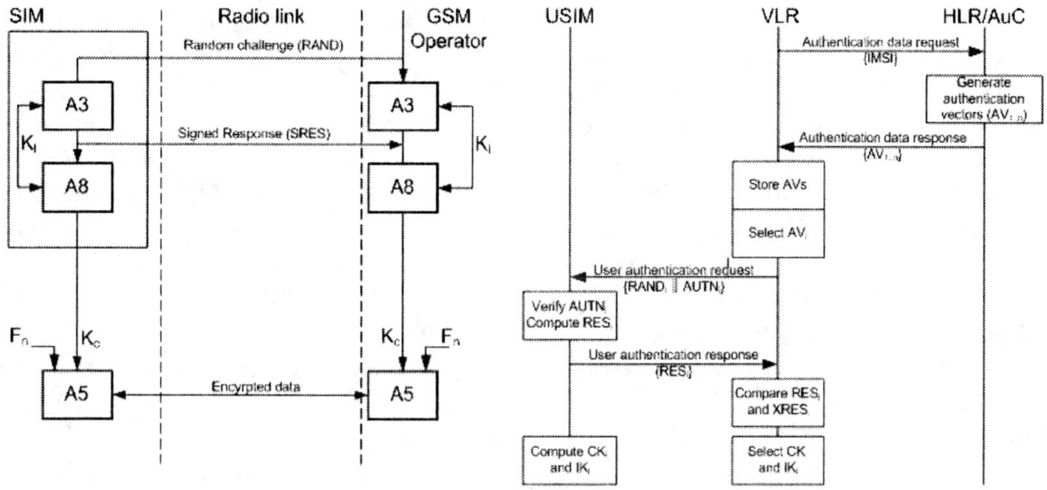

the EAP family, especially transport layer security (TLS) and SIM.

Most cellular operators are now providing WLAN services using the Universal Access Method (UAM) for authentication. UAM uses a layer 3 authentication method, typically a Web browser to identify the client for access to the WLAN. This raises the problem of mutual authentication, which has been a problem also in GSM networks. By extending to EAP-SIM it would be possible to enable SIM-based authentication in these environments for SIM-enabled devices.

Roaming between access providers is a second issue. Since data between access points are carried over an IP backbone, it is natural to use a network-based protocol such as Radius, suggested by Leu et al. (2006). Transport encryption inside the backbone is indifferent from normal wired practice, hence out of scope for this chapter. In a converged network, where users can switch between mobile networks and WLAN services, a common AAA system has to be operational to ensure correct operation. A unified billing scheme is proposed by Janevski et al. (2006), suggesting to use 802.1x on the WLAN side as shown on Figure 2. The mobile networks WLAN connection is suggested through the RADIUS server used also for access control in 802.1x.

The use of the IEEE 802.1x standard allows seamless authentication, since preshared certificates and key negotiation are provided to the cellular network, where the user is already authenticated. With the use of digital certificates, the system is getting closer to the preferred view of pervasive systems, where the user and the service providers are mutually identified. Since these systems authenticate the user towards several services, privacy is a primary concern. A possible solution, recommended by Ren, Lou, Kim, and Deng (2006) has a secure authentication scheme while preserving user privacy.

In pervasive environments a user connected will experience seamless authentication to all services when connected through a SSO service. Malicious tracking of his/her behaviour or eavesdropping of authentication messages can compromise the user credentials. The SSO service has to be extremely prudent when sending user-related information. Keeping a reasonable level of privacy, the system should deal with questions in location privacy, connection anonymity, and confidentiality (Ren et al., 2006). The recommendations are based on blind signatures and hash chains. Using hash is highly recommended, since a good hash function can provide good foundation for anonymous access and its resource needs are not too high for the current mobile devices, as sometimes blind signatures based on Rivest-Shamir-Adleman (RSA) scheme may be. In certain environments, the GSM integrated functions may also be used.

The user retains full control over authentication credentials when composing and generating authentication tokens like the identities suggested by Chowdhury and Noll (2007). Initial service access can be achieved showing one of these tokens after mutual identification between the service and the user. Based on these tokens, no user data can be retrieved nor traced back. If all of the initial identification steps succeed, the exchange of the required credentials can proceed using a freshly negotiated session key.

The base of most authentication techniques is a preshared key, delivered to the user device out-of-band. Authentication can be done for example in mobile phones by inserting a master private key on the SIM at the activation of the card (Kálmán & Noll, 2006).

A different approach is to extend the current mobile network with additional elements to enable network integrated AAA also in an Internet environment. Khara, Mistra, and Saha (2006) suggest including a new node, called Serving GPRS Access Router. This entity acts as a gateway for the WLAN traffic to enter the general packet radio services (GPRS) backbone and enable GPRS signalling to control WLAN. The new protocol set eliminates the need of Signalling System 7 (SS7) in addition to the IP backbone. Khara et al. claim that this solution is superior in terms of speed and overhead compared to the RADIUS-based methods suggested previously. The main drawback is the need of special dual mode devices with a split IP layer, a solution which might not be practical having in mind the basis of 2.5 billion mobile phones available in the market.

For mobile devices limited computational resources and battery power require an effective AAA mechanism. Extension of the GPRS/UMTS network could be potentially more expensive than deploying RADIUS authentication. Handover delay caused by terminal mobility is an issue which might favour GPRS/UMTS protocols.

Figure 2. Integration of radius and mobile network authentication

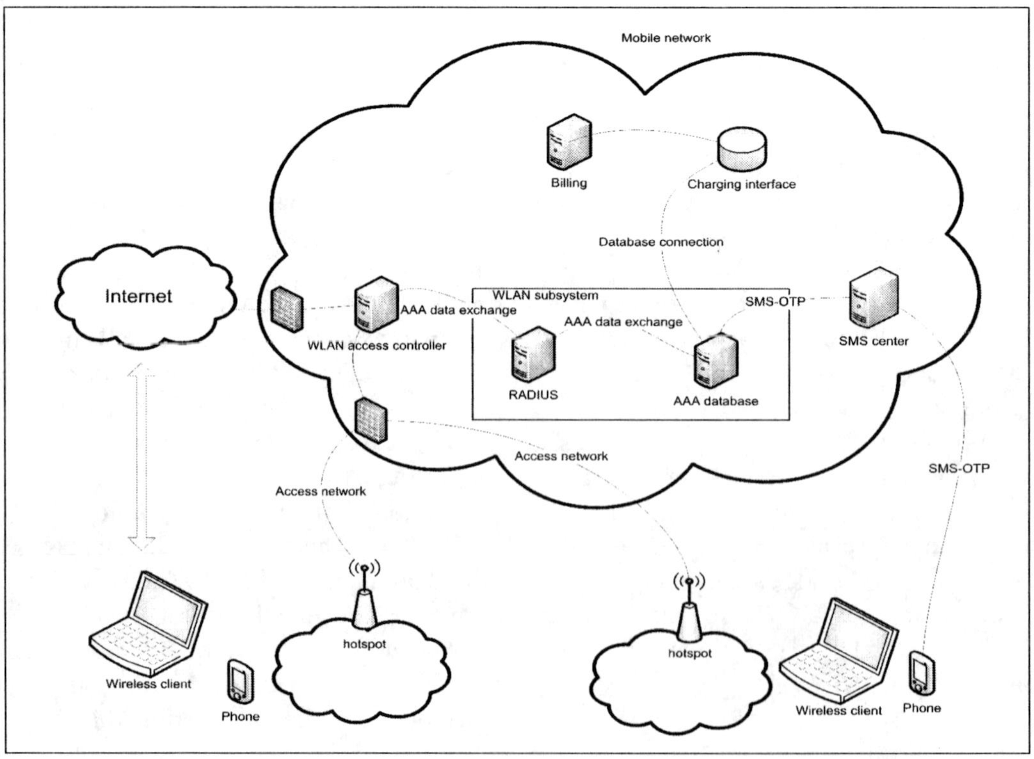

Authentication in Converged Networks

From the data traffic's point of view, the speed of the network's internal routines does not play a primary role, in VoIP and other sensitive services, QoS is a key parameter. Delay reduction is currently the topic having intensive focus. Interconnecting mobile and IP networks for data traffic is not a challenge, since GPRS has an IP backbone, and UMTS is practically an IP network. Most of the problems begin when the network has to provide a certain QoS in order to support service with time-critical transmission, that is, voice or video calls. Delay in the wired network can be reduced by additional bandwidth to reduce collisions, alternate routing paths, or other methods. But in wireless environments, where terminals move around and connect to different networks, which may be "far" away in terms of network topology, switching the data transfer path is a challenging task.

In the IP world, Mobile IPv6 (MIPv6) was introduced to deal with mobility problems. This protocol works flawlessly for clients that are changing networks with quite low frequency and are connected to a wired network, where additional signalling and other overheads are not causing bandwidth problems. The convergence time of the routing in MIPv6 is quite slow. In a wireless environment every additional message exchange or signalling overhead has a direct influence on usability. When the terminal is moving fast between these distant networks, it may reach a speed, where the routing of MIPv6 can not keep the connection in a correct state. This means that while data traffic could be able to transmit with low average speed, QoS cannot be kept on an adequate level to support VoIP or VoD services, for example. To fight this problem, several micromobility (local area) protocols were developed to support fast moving nodes. Different approaches are used, for example in hierarchical MIPv6 with fast handover adds a *local home agent* into the network. Seamless handoff for MIPv6 tries to lower the handover time with instructing the nodes to change networks based on precalculated patterns.

Handoff between neighbouring IP networks could be done in reasonable time if they are cooperating, but with introducing converged network access, it is likely that the terminal moves between WLAN and UMTS networks and back in less than a minute. Session mobility, for example a VoIP call without interruption, cannot be achieved using current protocols. The key is to reduce the handoff delay in interworking networks. To reduce delay inside the UMTS network, Zhang and Fujise (2006) show a possible improvement for the integrated authentication protocol. One cause of the long delay is getting an authentication vector (AV) if the Serving GPRS Support Node (SGSN) and Home Location Register (HLR) are far away. While roaming, the AV consumption is higher, if the terminal is moving frequently or it is producing significant traffic. The specifications allow a high blocking ratio of 20% for the UTMS network in case of requesting new AVs. The proposal claims to lower this rate to 2%. For each authentication instance, the SGSN consumes one AV from a first in first out (FIFO) storage.

A fundamental question is to allow the size of the AV vector to be customised based on the terminal's behaviour. In the default way, the SGSN executes a distribution of authentication vector (DAV) procedure if all AVs are consumed. Communication between the terminal and the SGSN cannot proceed until the reply is received from the HLR, inserting a potentially high delay into the system. This can lead to call failure, errors in location update, or unacceptable delays in services running on GPRS. The proposed protocol from Zhang and Fujise (2006) implies no change in case of the first authentication to the SGSN, but keeps track of the number of available AVs and sends out a new request when hitting a predefined level. This level can be customised for a network, to reduce or even remove the possible delay of waiting for an AV. The proposal also changes the basic behaviour, asking for new AVs when they are consumed. The original 3rd Generation Partnership Project (3GPP) system asks for them when a new event comes in and no AVs are available.

While reducing delay inside the GPRS network can reduce block probability in reaching network services, also handover functions in IP have to be revised in order to achieve reasonably fast mobility support. The basic challenge is that currently AAA and MIPv6 are operated independently. This means that the terminal has to negotiate with two different entities in order to get access to the new network.

In MobileIPv6, the terminal is allowed to keep connections to a home agent (HA) and a correspondent node (CN), even when the terminal changes point of attachment to that network. The terminal has two addresses, the home address (HoA) and the care-of address (CoA). The HoA is fixed, but the CoA is generated by the visited network. The mobile IP protocol binds these two addresses together. To ensure an optimal routing in the network, the terminals switch to *route optimalisation* mode after joining a new network. Then it executes a *return routability* procedure and a *binding update* (BU) to communicate to the correspondent node directly. The return routability procedure consists of several messages, which together induce a long delay.

The handover between networks implies even more steps and consumes more time: movement detection, address configuration, home BU, return routability procedure, and a BU to the correspondent node. The terminal cannot communicate with the CN before the end of the procedure.

Fast handover capability is a major research item in IETF for MIPv6, including the standards FMIPv6 and HMIPv6. In addition to these schemes, Ryu and Mun (2006) introduce an optimisation in order to lower the amount of signalling required and thus lower the handover delay between domains. In an IPv6 system, the IP mobility and AAA are handled by different entities. This architecture implies unnecessary delays. Several solutions are proposed to enable the mobile terminal to build a security association between the mobile node and the HA. This enables home BU during the AAA procedure. Route optimisation is a key topic in efficient mobility service provision. MIPv6 optimises the route with the use of the return routability procedure. In wireless environments, the generated signalling messages represent a considerable part of the whole overhead. Moving route optimisation into the AAA procedure can reduce the delay by nearly 50% (Ryu & Mun, 2006). This was enabled by embedding the BU message into the AAA request message and so optimising the route while authenticating. This solution can solve MIPv6's basic problem of supporting different administrative domains and enable scalable large scale deployment.

Lee, Huh, Kim, and Lee (2006) define a novel communication approach to enable communication between the visited AAA servers for a faster and more efficient authentication mechanism. If a terminal visits a remote network, the AAA must be done by the remote system. IETF recommends integrating Diameter-based authentication into the MIPv6 system. But, when the user is using services on the remote network, the remote AAA has to keep a connection with the home AAA. The proposed new approach of Lee, Huh, et al. suggests enabling faster authentication when the terminal moves between subnets inside a domain by exchanging authentication data between visited AAA servers without the need of renegotiation with the HA. Connection to the HA is needed only after the authentication when the terminal executes a BU.

One other aspect is shown by Li, Ye, and Tian (2006) suggesting a topology-aware AAA overlay network. This additional network could help MIPv6 to make more effective decisions and to prepare for handovers and other changes in network configuration. Based on the AAA servers and connections between, a logical AAA backbone can be created, which can serve as administration backbone for the whole network. Signals delivered over this network are topologically aware, so the optimal route can easily be selected and signalling messages can be transmitted over the best route. In exchange to the build cost of this backbone network and some additional bandwidth consumed, MIPv6's security and performance can be enhanced.

As the route of the service access is secured, optimised and delay reduced, one basic problem still remains: how to ensure that the user is the one, the network thinks he/she is. Lee, Park, and Jun (2006) suggest using smart cards to support interdomain roaming. The use of the SIM might be preferable because of its widespread use and cryptographic capabilities (Kálmán & Noll, 2006). The problem of having multiple devices is also raised here, since a system based on the SIM as smart card will require SIM readers in every device—if a secure key exchange method between the devices is not in place.

Lee, Park, et al. (2006) suggest an entity called *roaming coordinator* ensuring seamless roaming services in the converged network. This additional node provides context management services and enables seamless movement between the third generation (3G) network and WLAN to enforce security in converged networks. In order to provide good user experience in a pervasive environment, additional intelligence needs to be added to the traditional AAA systems to ensure that the terminal selects the most appropriate connection method. This method has to be based on the context and has to be supported in all networks. A smart-card-based secure roaming management framework enables the transfer of the terminals context without renegotiating the whole security protocol set. When the terminal moves into a new network, the roaming coordinator, AAA servers, and proxies take charge of the authentication process. The coordinator, having received a roaming request, evaluates the available networks and chooses the best available one, and then triggers the context transfer between the corresponding AAA servers. When transferring whole user contexts, the system has to consider privacy requirements of the user's identity and his/her profile.

Anonymity and Identity

In pervasive environments, privacy is of key importance. With computers all around, gathering information about traffic, movements, service access, or physical environment, customer privacy must be protected. Køien (in press) suggests a protocol, which is able to provide better protection for the user's privacy than the normal 3G network. Changes in the EAP-AKA protocol are suggested to use only random generated user authentication values. He defines three user contexts implying different key management and authentication schemes, like existing keys for short-term and fresh keys for medium-term access. Identity-based encryption is recommended to enable a flexible binding of the security context to protect the permanent subscriber identity and location data, which will only be discoverable by the home register. The main drawback of the suggested protocol is its higher computing requirements as compared to EAP-AKA, potentially limiting the applicability.

Security and Computing Power

A security protocol in a wireless environment should be fast and secure, and it has to be effective in terms of computing power and low data transfer need. In low power environments an authentication scheme with high security and low computing power is advised. One solution is based on hash functions and smart cards, allowing minimised network traffic and short message rounds used for authentication. Anonymity can be ensured through one-time passwords. While accepting the advantages of a system with smart cards, the use of extra hardware like a card reader is not advisable, due to compatibility issues and power requirements.

Software-based solutions have an advantage, as they only require computing power. Showing the importance of power consumption, a comparison of cryptographic protocols is presented by Lee, Hwang, and Liao (2006) and Potlapally, Ravi, Raghunathan, and Jha (2006) showing, that twice of the transmit energy of one bit is needed to run asymmetric encryption on that piece of information. Symmetric encryption needs, in contrast, around one half of the transmit energy. Most overhead is generated by session initialisation, meaning longer sessions induce lower overhead. There is a trade-off between security and session length. While negotiation overhead is getting lower with long sessions, security risks are getting higher.

This overhead can be lowered by special hardware or software solutions. Hardware needs some power and bigger silicon, while software requires a faster CPU. Hash functions have an energy requirement of around half a percent compared to PKI in generating session keys (Potlapally et al., 2006). Key exchange protocols using elliptic curve Diffie-Hellman (DH) come out much more energy efficient as compared to the same traditional strength DH. The DH calculations demonstrate the trade-off between power consumption and security. In order to have an efficient operation, the security protocol needs to have the possibility to adapt encryption to the needs of the current application. Authentication token generation can be problematic for devices with limited computing capabilities. Personal area networks (PAN) with multiple devices raise this problem by their very nature.

Security in Personal Area and Home Networks

Efficient authentication and certificate management ensures better usability of PAN devices. By using efficient security protocols, content-adaptive encryption, efficient key and certificate management, considerably longer battery operation is achievable. To enable key management in a PAN a personal certificate authority (CA) entity is suggested (Sur & Rhee, 2006; Sur, Yang, and Rhee, 2006), which will be responsible for generating certificates for all mobile devices within the PAN or home device domain (Popescu, Crispo, Tanenbaum, & Kamperman, 2004). Because of the context of use, the authentication protocol is focused on efficiency by reducing computational overheads for generating and verifying signatures.

Main focus is on reducing PKI operations, which have been proven to be energy consuming. Instead, it proposes to use hash chains to lower communication and computational costs for checking certificates. Former research suggested hash trees in order to authenticate a large number of one-time signatures. By extending these with fractal-based traversal, it has been proven that these trees provide fast signature times with low signature sizes and storage requirements. The personal CA has to be a unique trusted third party in the PAN. It needs to have a screen, a simple input device, and has to always be available for the members of the network. A cell phone with the SIM is a perfect candidate to be a personal CA (Kálmán & Noll, 2006).

In home environments, basically two types of authentication are distinguished: (1) user authentication, and (2) device authentication (Jeong, 2006). Mutual authentication has to be used in order to prevent impersonation attacks (*identity theft*). This requires an SSO infrastructure, which can be for example Kerberos or RADIUS. A special aspect of resource access over the home LAN is that specific privileges are given to selected programs. The AAA server maintains an access control list to ensure correct privilege distribution.

To build the initial trust relationships some kind of user interaction is needed. The key should initially be distributed out-of-band, for example on an USB stick, or by using short range wireless technology, Near Field Communication (NFC), for example (Noll, Lopez Calvet, & Myksvoll, 2006). On home networks, where power consumption is not a problem, PKI may be used for negotiating session keys between devices, since key management in a PKI is simpler than in symmetric encryption and the delay caused by checking certificates and so forth will not be noticeable in this environment. Users authenticated towards the AAA infrastructure can access the resources seamlessly. Initial authentication is done with PKI. In case of mobile devices, also the home AAA can use previously calculated hash values in chain to lower computational cost. These AAA infrastructures can be connected to a providers AAA, for example to use in digital rights management (DRM) or home service access from a remote network (Popescu et al., 2004).

A user moving with his/her devices to the home raises another AAA challenge, the mobile nodes.

Mobile Nodes (Network Mobility)

Movement of whole networks like PANs or networks deployed on a vehicle, introduce a new level of AAA issues. In a conventional network a standard mobility support does not describe route optimisation. Several procedures are suggested to provide this functionality for mobile nodes, like Recursive Binding Update Plus (RBU+), where route optimisation is operated by MIPv6 instead of the network mobility (NEMO) architecture. This means, that every node has to execute its own BU with the corresponding HAs. To solve problems with pinball routing, it uses the binding cache in the CN. When a new BU message arrives, the RBU+ has to execute a recursive search, which leads to serious delays with a growing cache size. One potential route optimisation is presented by Jeong (2006).

A designated member of the network, called a mobile router is elected to deal with mobility tasks to reduce network overhead. The AAA protocol for this environment defines a handover scheme and tree-based accounting to enable efficient optimisation. They recommend using dual BU (DBU) procedure instead of the existing procedures like RBU+ as a solution for the reverse routing problem raised by mobility. DBU operates with additional information placed into the messages sent in a BU process. This is the CoA of the top level mobile router (TLMR). By monitoring the messages, the CNs in the subnet can keep optimal route towards the TLMR.

Moving subnets are the subject of eavesdropping and possible leakage of the stored secrets. A secure AAA is proposed for network mobility over wireless links, which deals with these problems (Fathi et al., 2006). Secret leakage can be caused by malicious eavesdroppers, viruses, or Trojans. A possibility is to store the keys in tamper resistant modules, like smart cards, the SIM, or trusted hardware modules. Deploying additional modules can be problematic and expensive. Fathi et al. propose a protocol based on a short secret, which can be remembered by humans and used in a secure protocol called Leakage-resilient authenticated key exchange protocol (LR-AKE). This protocol is used for AAA to reduce NEMO latency under 300 ms in order to provide session continuity, for example in VoIP applications, which is important in keeping a good user experience. However, short passwords as proposed with LR-AKE are not advisable. If complex, they will be noted down by the user, and if weak, they are easy to guess.

As network mobility has considerable security issues, it may be not the way to go. Functionality of a mobile network might be achieved by using a dedicated device as a gateway of the PAN. Only this device will show up in the wireless network, and all traffic originating and arriving to the PAN will go through this device and its HA.

After these technical issues of authentication the next chapter will deal with authentication from the user viewpoint.

Customer Ergonomics

There is always a trade-off between user security and ease of use. If the system is prompting for a password for every transaction, it can assume with quite high probability, that the access is enabled just for the correct user. But, that is unacceptable for most of the users in private environments, where convenience is more valued than security. In corporate networks, policies are just enforced and users have to accept it. It would however be problematic if the credentials were only asked once at start-up or connecting to the network, since mobile devices are threatened by theft, loss, and other dangers by their nature of use.

Smart cards could be a solution to have a good trade-off between the usability and security. Since the user will have a token, which he/she has to care of, and exchange keys generated by it, at least it could be secured that the user who is accessing a specified service holds the authentication token. The mobile phone with the integrated smart card, the SIM, is a potential tool for this purpose. As indicated by Leu et al. (2006) the requirement of carrying a SIM reader or equipping all the equipment with SIM cards is neither convenient nor cost effective. The possibility of secure key exchange between user equipment shall be provided.

The cell phone can act as a key negotiator, with its tamper resistant cryptographic functions integrated into the SIM and then exchange the session keys with other terminals with the use of a short range wireless solution. Currently, most of the security problems, besides the user behaviour, are coming from security holes in the software. Having the capability to download new software over the air to the phone ensures the use of recent updates and eliminates this type of security threat (Kálmán & Noll, 2006). Compared to a security token, it may be better to use the phone, since the SIM card can be locked by the provider, so if the device gets lost, the authentication credentials can be withdrawn within short time.

OUTLOOK

Current research is focused on merging basic network functions to enable pervasive computing and network access. The result of these efforts is a converged infrastructure, which is able to handle most of user needs in high quality. The problem of QoS control in wireless systems remains an open one, but experiences of VoIP and VoD services in wireless networks show the adaptability of the user to the current environment.

Mobility of packet data is still to be enhanced, with the challenge of reducing the handover delay. Remote access to home content is just beginning to be spread between early adopters. MIPv6 will address most of the issues sometime in the future, and with the promising extensions, the protocol will be able to handle sessions together with the AAA infrastructure without service interruption. Mobile networks will use WLAN as a high capacity data service, although upcoming solutions and MIPv6 extensions may be able to threaten their use inside dense populated areas, assuming global Wi-Fi roaming mechanisms are in place.

Efforts are being made towards an easy deployable home AAA infrastructure, which can later bear the tasks associated with inner (user management, remote access, user content DRM, purchased media DRM) and outer (authentication towards corporate, provider- or public-based AAA) authentication and access control.

Educating the user might be the biggest challenge, as mobile phone users represent the whole population, and not just the *educated* computer community. The enforcement of the use of smart cards is advisable, where the possible use of the mobile phone shall be investigated.

Now, we can experience the dawn of new social and community services over the Internet. This raises the problem of privacy protection as never before. AAA services must take care of user credentials, and even must ensure that data collected from different AAA providers cannot be merged. So, research in the area of one-way functions, blind signatures, and different PKI methods is recommended.

Finally, current market players also have to change their business plans. Research in the economical area has to point out new objectives to ensure a good working, open, and secure AAA infrastructure which can be used by every service provider while keeping information exchange on the required minimal level.

CONCLUSION

The biggest effort in AAA systems is on extending the capabilities of the existing solutions in telecommunication and in computer networks to an integrated network approach enabling seamless service access of mobile users.

While telecom solutions are usually more secure, user privacy is not a primary concern here. In computer networks AAA solutions are more open and flexible, while the widespread model of "web of trust" methods is not acceptable for commercial service exchange. Ongoing research indicates the potential for a common mobile/Internet authentication suite, potentially based on the EAP.

Interworking issues between mobile and wireless networks are the basis for detailed research on handover delay, multi-device roaming, mobile networks, security, ease-of-use, and anonymity of the user. This chapter provided an overview of the state of the art in authentication for mobile systems.

Extended AAA mechanisms are suggested for home and community networks, taking into account security and privacy of the users. These networks will keep a high amount of personal data, and thus need stronger privacy protection mechanisms. By using link layer encryption, smart cards, and secure key transfer methods the security and privacy protection can be greatly enhanced.

REFERENCES

Chowdhury, M. M. R., & Noll, J. (2007). Service interaction through role based identity. In *Proceedings of the The International Conference on Wireless and Mobile Communications (ICWMC2007).*

Fathi, H., Shin, S., Kobara, K., Chakraborty, S. S., Imai, H., & Prasad, R. (2006). LR-AKE-based AAA for network mobility (NEMO) over wireless links. *IEEE Journal on Selected Areas in Communications, 24*(9), 1725-1737.

Janevski, T., Tudzarov, A., Janevska, M., Stojanovski, P., Temkov, D., Kantardziev, D., et al. (2006). Unified billing system solution for interworking of mobile networks and wireless LANs. In *Proceedings of the IEEE Electrotechnical Conference MELECON 2006* (pp. 717-720).

Jeong, J., Chung, M. Y., & Choo, H. (2006). Secure user authentication mechanism in digital home network environments. In *Embedded and Ubiquitous Computing* (LNCS 4096).

Jeong, K. C., Lee, T.-J., Lee, S., & Choo, H. (2006). Route optimization with AAA in network mobility. In *Computational Science and Its Applications—ICCSA 2006* (LNCS 3981).

Kálmán, Gy., Chowdhury, M. M. R., & Noll, J. (2007). Security for ambient wireless services. In *Proceedings of the 65th IEEE Vehicular Technology Conference (VTC2007).*

Kálmán, Gy., & Noll, J. (2006). SIM as a key of user identification: Enabling seamless user identity management in communication networks. In *Proceedings of the WWRF meeting #17.*

Khara, S., Mistra, I. S., & Saha, D. (2006). An alternative architecture for WLAN/GPRS integration. In *Proceedings of the IEEE Vehicular Technology Conference, 2006, VTC 2006* (pp. 37-41).

Køien, G. M. (in press). Privacy enhanced mobile authentication. *Wireless Personal Communications.*

Lee, C.-C., Hwang, M.-S., & Liao, I.-E. (2006). Security enhancement on a new authentication scheme with anonymity for wireless environments. *IEEE Transactions on Industrial Electronics, 53*(5), 1683-1687.

Lee, M., Park, S., & Jun, S. (2006). A security management framework with roaming coordinator for pervasive services. In *Autonomic and Trusted Computing* (LNCS 4158).

Lee, S.-Y., Huh, E.-N., Kim, Y.-W., & Lee, K. (2006). An efficient authentication mechanism for fast mobility service in MIPv6. In *Computational Science and Its Applications—ICCSA 2006* (LNCS 3981).

Leu, J.-S., Lai, R.-H., Lin, H.-I., & Shih, W.-K. (2006). Running cellular/PWLAN services: Practical considerations for cellular/PWLAN architecture supporting interoperator roaming. *IEEE Communications Magazine, 44*(2), 73-84.

Li, J., Ye, X.-M., & Tian, Y. (2006). Topologically-aware AAA overlay network in mobile IPv6 environment. In *Networking 2006* (LNCS 3976).

Long, M., & Wu, C.-H. (2006). Energy-efficient and intrusion-resilient authentication for ubiquitous access to factory floor information. *IEEE Transactions on Industrial Informatics, 2*(1), 40-47.

Noll, J., Lopez Calvet, J. C., & Myksvoll, K. (2006). Admittance services through mobile phone short messages. In *Proceedings of the International Conference on Wireless and Mobile Communications ICWMC'06.*

Popescu, B. C., Crispo, B., Tanenbaum, A. S., & Kamperman, F. L. A. J. (2004). A DRM security architecture for home networks. In *Proceedings of the 4th ACM Workshop on Digital Rights Management.*

Potlapally, N. R., Ravi, S., Raghunathan, A., & Jha, N. K. (2006). A study of the energy consumption characteristics of cryptographic algorithms and security protocols. *IEEE Transactions on Mobile Computing, 5*(2), 128-143.

Ren, K., Lou, W., Kim, K., & Deng, R. (2006). A novel privacy preserving authentication and access control scheme for pervasive computing environments. *IEEE Transactions on Vehicular Technology, 55*(4), 1373-1384.

Ryu, S., & Mun, Y. (2006). An optimized scheme for mobile IPv6 handover between domains based on AAA. In *Embedded and Ubiquitous Computing* (LNCS 4096).

Sur, C., & Rhee, K.-H. (2006). An efficient authentication and simplified certificate status management for personal area networks. In *Management of Convergence Networks and Services* (LNCS 4238).

Sur, C., Yang, J.-P., & Rhee, K.-H. (2006). A new efficient protocol for authentication and certificate status management in personal area networks. In *Computer and Information Sciences—ISCIS 2006* (LNCS 4263).

Zhang, Y., & Fujise, M. (2006). An improvement for authentication protocol in third-generation wireless networks. *IEEE Transactions on Wireless Communications, 5*(9), 2348-2352.

KEY TERMS

Authentication, Authorisation, and Accounting (AAA): AAA is a system that handles all users of the system to ensure appropriate right management and billing.

Converged Network: Converged network is a network carrying various types of traffic. Such a network is providing services to different terminals, which can access and exchange content regardless of the current networking technology they are using.

Diameter: Diameter is a proposed successor of RADIUS. It uses TCP as a transport method and provides the possibility to secure transmissions with TLS. It is not backward compatible with RADIUS.

Digital Rights Management (DRM): DRM is a software solution that gives the power for the content creator to keep control over use and redistribution of the material. Used mostly in connection with digital media provider companies, but in pervasive environments, users may also require a way to have a fine-grained security infrastructure in order to control access to own content.

Extensible Authentication Protocol (EAP): EAP, a flexible protocol family, which includes TLS, IKE protocols, and also the default authentication method of UMTS, EAP-AKA.

International Mobile Subscriber Identity (IMSI), Temporary-IMSI (TMSI): IMSI and TIMSI is the unique identity number used in UMTS to indentify a subscriber. The temporary one is renewed from time to time, and that is the only one that is used over the air interface.

Public Key Infrastructure (PKI): PKI is a service that acts as a trusted third party, manages public keys, and binds users to a public key.

Remote Authentication Dial in User Service (RADIUS): RADIUS is the de facto remote authentication standard over the Internet. It uses UDP as a transport method and is supported by software and hardware manufacturers. Privacy problems may arise when used on wireless links, since only the user password is protected by an MD5 hash.

Rivest-Shamir-Adleman (RSA): RSA is the de facto standard of public key encryption.

Smart Card: Smart card is a tamper resistant pocket sized card, which contains tamper resistant non-volatile storage and security logic.

Subscriber Identity Module (SIM): SIM is the smart card used in GSM and UMTS (as USIM) networks to identify the subscribers. It has integrated secure storage and cryptographic functions.

Chapter XIV
Trustworthy Networks, Authentication, Privacy, and Security Models

Yacine Djemaiel
University of the 7th of November at Carthage, Tunisia

Slim Rekhis
University of the 7th of November at Carthage, Tunisia

Noureddine Boudriga
University of the 7th of November at Carthage, Tunisia

ABSTRACT

Wireless networks are gaining popularity that comes with the occurrence of several networking technologies raising from personal to wide area, from centralized to distributed, and from infrastructure-based to infrastructure-less. Wireless data link characteristics such as openness of transmission media, makes these networks vulnerable to a novel set of security attacks, despite those that they inherit from wired networks. In order to ensure the protection of mobile nodes that are interconnected using wireless protocols and standards, it is essential to provide a depth study of a set of mechanisms and security models. In this chapter, we present the research studies and proposed solutions related to the authentication, privacy, trust establishment, and management in wireless networks. Moreover, we introduce and discuss the major security models used in a wireless environment.

INTRODUCTION

Wireless networks are gaining popularity. Such popularity comes with the occurrence of several networking technologies raising from personal to wide area, from centralized to distributed, and from infrastructure-based to infrastructure-less. However wireless data link characteristics such as openness of transmission media, make these networks vulnerable to a novel set of security attacks. In order to protect such networks, multiple security solutions were proposed for the authenticating of users, ensuring privacy, and establishing trust. Deploying wireless networks without considering the threats associated to this technology may lead to the compromise of the interconnected resources and also the loss of security.

To ensure the protection of mobile nodes that are interconnected using wireless protocols, several security mechanisms and security models have been provided. The solutions were made to cope with the features of the wireless environment and the mobile nodes. In this chapter, we present the research work and security solutions related to authentication, privacy, and trust management. Moreover, we introduce and discuss the major security models used in a wireless environment.

The first section of this chapter takes interest to the concept of trust, which can be defined as the firm belief in the competence of an entity to act dependably, securely and reliably within a specified context. Starting from this definition, it is significant that trust implies a level of uncertainty and judgment. This may depend on many factors due to risks associated to wireless networks. In this section, we define the trust in wireless context and discuss its models.

The second section discusses the authentication, which is a crucial mechanism that ensures that a resource is used by the appropriate entities. Actors, architecture, and issues related to authentication in wireless environment are discussed.

The third section discusses authentication models and protocols in wireless LAN (WLAN), cellular, ad hoc, wireless mobile access networks (WMAN) networks. As Mobile IP is becoming a unifying technology for wireless networks, allowing mobile nodes to change their point of attachment without loosing their connections, a particular interest is also given to authentication in Mobile IP.

The fourth section of this chapter discusses privacy regarding location and transaction in wireless environment. The fifth section presents two aspects regarding security modeling in wireless environments. The first is related to the specification of trust, modeling, and verification. The second addresses the specification and verification of security policies that take into consideration wireless threats.

TRUST MANAGEMENT

Trust management represents the skeleton of any network security framework. The absence of a centralized entity, for example, in ad hoc networks makes trust management a challenging problem to address.

Trust Establishment Basis

Trust describes a set of relations among entities engaged in various protocols, which are established based on a body of assurance evidence. A trust is established between two different entities further to the application of an evaluation metric to trust evidence. The established relations may be composed with other trust relations to generate new relations. Trust may influence decisions including access control. To clarify the process of trust establishment, we consider the following example. Assume two trust relations *A* and *B*. Relation *A* states that "*a certification authority CA1 accepts entity X's authentication evidences*" and is established off-line upon delivery of some evidences (e.g., identity, employment card) by X to B. Upon the establishment of A, the certification authority *CA1* issues a certificate binding a public key to X. Then, it stores the relation in its trust database registering X with its certificate. Relation *B* states that "*a certification authority CA2 accepts CA1's authentication of any entity registered by CA1*". To establish *B*, certification authority *CA2* may ask *CA1* to deliver some evidences such as: (1) *CA1*'s authentication of entities is done using satisfactory mechanism and policy; and (2) certification authority *CA1*'s trust database is protected using satisfactory security mechanisms and policies. The establishment of such trust relation leads to the publication of a certificate signed by *CA2*, associating *CA1*'s public key. The relation is then stored in *CA2*'s trust database. The composition of the two trust relations leads to the acceptance of *CA1*'s authentication of *X* by *CA2*.

One of the main properties that need to be handled during trust establishment techniques is transitivity. To decide whether a trust relation is transitive or not, evidences used to establish trust should ensure (1) availability, meaning that evidences can be evaluated at any time by the entities wishing to establish trust; (2) uniformity, meaning that evidences satisfy the same global metrics of adequacy, (3) stability, which means that authentication mechanism cannot change accidentally or intentionally, and (4) log-term existence, meaning that evidences last as long as the time used to gather and evaluate it.

Need for Trust Management in Mobile Networks

While there are extensive research works that contributed to the management of trust in complex systems, the great majority of them was set up for fixed infrastructures; assumed long-term availability and validation of evidences; and generated lengthy validation process. Several characteristics of wireless networks including unreliable transmission range and topology changes made trust management a challenging task. The focus on ad hoc networks was based on the fact that these networks are self-organized and barely suppose the existence of trustworthy nodes. In infrastructure-based wireless networks such as cellular networks and WLAN, the base stations (BSs) (or access points [APs]) are considered trustworthy. Three main requirements need to be fulfilled by trust establishment process in wireless ad hoc networks. First, trust should be established in a distributed manner without a pre-established trust infrastructure. In fact, connectivity to certification authorities' directory servers in the node's home domain cannot be guaranteed in mobile ad hoc network (MANET) when needed. As a consequence, trust establishment in MANET must support peer-to-peer trust relations.

Second, trust establishment should be performed online and trust relations should have short-life period. This is mainly due to the fact that in MANET, when a node moves randomly from a location to another, its security context may change. For instance, when a node moves to a location in which its compromise becomes possible, any trust relation that involves such node should be withdrawn. Such behavior should not affect network connectivity and new trust evidences should be gathered as a consequence. Third, trust establishment should be tolerant to incomplete evidence or unavailable trust relations. In fact, in MANET, it becomes unfair to suppose that all evidences are available to all nodes when they are required to establish trust. Therefore, trust relations should be established using incomplete or uncertain trust evidences, based on the incomplete amount of information that each node holds.

When nodes plan to communicate, they must initially interact with each other and establish a certain level of trust. The change of such level may be triggered further to interaction between neighboring nodes or further to a recommendation from a third party. As a node in the MANET has only a partial view of the whole network, additional mechanisms should be designed to allow these nodes identifying valid trust evidences and prevent intruders from altering them or modifying the trust value of other nodes. To clarify this issue, Figure 1 depicts two networks. In the first network, users *User1x* need to communicate very often with server *Server1*. In the second network, users *User2x* need to communicate very often with server *Server2*. Different trust relations can be established. Nodes in network 1 and 2 trust each other based on identity certificates which are registered by certification authority *CA1* and *CA2*, respectively. In this scenario, User12 has lost communication with server 1 and User11, because it moved out of the coverage. Some among User2x can be found under the communication range of User12. To reach Server1, User12 has to authenticate itself to any User2x and get access to the second ad hoc network. To do so, User12 provides its certificate (as signed by CA1) to User21. User21 has to decide whether to accept such trust evidence. Assume now that the access policy requires that any node that wants to access the ad hoc network should provide a valid identity certificate from a trusted authority. Thus, User21 should contact its trusted certification authority (CA2) and get the CA1 certificate signed by CA2. After that, User21 will be able to valid the certificate of User12. Transitivity of the trust relation is thus established.

Recent Advances in Trust Management

Former trust establishment solutions focused mainly on procedures to locate the communicating peer's certificate in order to determine the cryp-

Figure 1. Trust establishment in ad hoc network

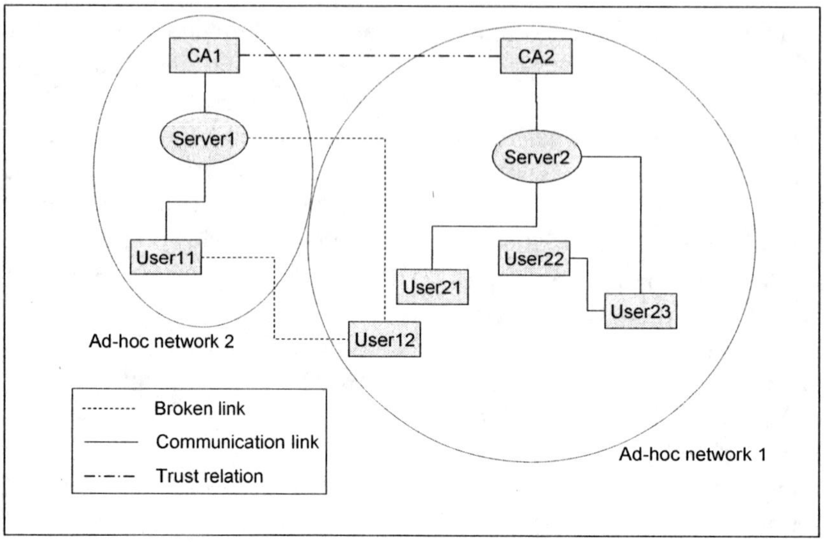

tographic key. In this context, Balfanz, Smetters, Stewart, and Wong (2002) base its solution on using a location-limited channel to allow nodes performing pre-authentication of each other. As the propagation of the channel is limited, intruders have an outside chance to mount a successful passive attack. While pre-authentication does not require a heavy bandwidth, the existence of location-limited channel represents a very restrictive assumption. The approach proposed in Ren et al. (2004) assumes a minimum storage requirement to establish trust in mobile ad hoc networks. A centralized secret dealer is introduced into the network during the system bootstrapping phase and is supposed to be trusted by all nodes. Every node is assumed to have a pair of public/private key where the public key is known by the secret dealer.

In the first part of bootstrapping, every network node receives a pre-computed short list, say SL, from the secret dealer. SL represents k tuples binding node identifiers to related public keys. These bindings are distributed symmetrically, meaning that if node j receives the node identifier of i and its corresponding public key, then node i will also receives node i identifier and its public key. In the second part of the bootstrapping phase, each node generates k certificates, one certificate for every received binding, assuming that every certificate contains the signature on the selected binding from the received secret list. These certificates will therefore be stored locally. The value of k is chosen so that there is a sufficient trust relationship in the network and the distribution scheme should ensure the certainty of being able to establish a trust chain between any two nodes.

After the system bootstrapping phase is finished, there is no need for the secret dealer to continue existing. To accommodate the dynamic changing of the network structure, every node is assumed to be able to establish independent trust relationship with at least two nodes.

When a node leaves the network properly, it broadcasts information about its departure and signs them. Consequently, the receiving nodes revoke the certificate that was issued to that leaving node. One major advantage of this solution lies in the fact that (1) it decreases the length of the trust path, and (2) it is slightly affected by the dynamic nature of the ad hoc network. However, guaranteeing that sufficient trust relationships exist in the network requires a large care during the selection of value of k.

On one hand, the work in Baras and Jiang (2004) proposed to investigate the stability of trust establishment by modeling a MANET as an indirect graph where edges represent pre-trust relations. The two authors cast the problem of trust computation and evaluation by every individual node as a cooperative game and base it on elementary voting methods. In Theodorakopoulos and Baras (2004), the process of trust relation establishment is formulated as a path problem on a weighted directed graph. The vertices in the graph represent the entities and a weighted edge (i, j) represents the opinion that entity i has about entity j. Such opinion consists of two numbers: the trust value and the confidence value. The trust value is an estimate of the trustworthiness of the target, while the confidence value corresponds to the accuracy related to the assignment of the trust value. Using the formal theory of semirings, one can show how two nodes can establish an indirect trust relation without previous direct interaction. For that case, two operators were developed allowing to combine trust opinions along different paths and compute the trust-confidence value between pair of nodes.

GENERAL MODELS FOR AUTHENTICATION IN WIRELESS NETWORKS

In addition to authentication solutions applied to specific wireless technologies, some general models are introduced in wireless networks. This section discusses these models.

Actors in an Authentication System

Basically, an authentication system is composed of three actors: (1) a supplicant, (2) an authenticator, and (3) a trusted third party (TTP). The supplicant is an entity that requests access to network resources. It may be a person, or an application running on a mobile node. The access to protected resources is gained only if the credentials provided by the supplicant are validated by the authenticator. In an authentication system, a credential is an identifier that is used by an authenticator to check whether the supplicant is authorized. It may be symmetric key, a public/private key pair, a generated hash, or some contextual information such as physical characteristic that uniquely identifies a supplicant (e.g., GPS location, signal to noise ratio, etc.). Finally, a TTP is an entity that is mutually trusted by the supplicant and the authenticator and facilitating mutual authentication between the two parties (Aboudagga, Refaei, Eltoweissy, DaSilva, & Quisquater, 2005). An authentication process is made up of a set of messages that are exchanged between these actors (as e illustrated by Figure 2). Authentication includes four components as follows: (1) "S" denotes the supplicant; (2) "D" denotes the destination mobile node; (3) "As" denotes the authenticator; and (4) "Ad" denotes the destination authentication server. Adding a TTP to this model introduces additional exchanged messages in order to establish trust between the different interacting nodes.

Authentication Management Architecture

An authentication system is based on an authentication protocol that fixes the interaction between the different components described previously. The interaction is made using a set of messages between system components. In a wireless environment, node mobility offers many advantages, but at the same time it may affect the overall system efficiency. Consequently, deploying an authentication system in a wireless environment needs to consider several aspects including authenticators' number and placements. The choice made on the placement of these servers has an effect on the time spent to authenticate a mobile node and the packet loss ratio. Typically, two strategies may be adopted concerning the authentication servers placement. The former aims at placing authentication servers on the same network within mobile nodes. This solution leads to route the two traffics (exchanged data and authentication traffic) within the same network. Consequently, the contention and the packet loss ratio are increased. However, the time spent during authentication is reduced compared with the second solution that aims to place authentication servers outside of the network and thus forwarding authentication traffic outwards. The latter solution reduces the packet loss ratio and liberates the network bandwidth for useful traffic.

Figure 2. Exchanged messages in a hierarchical authentication model

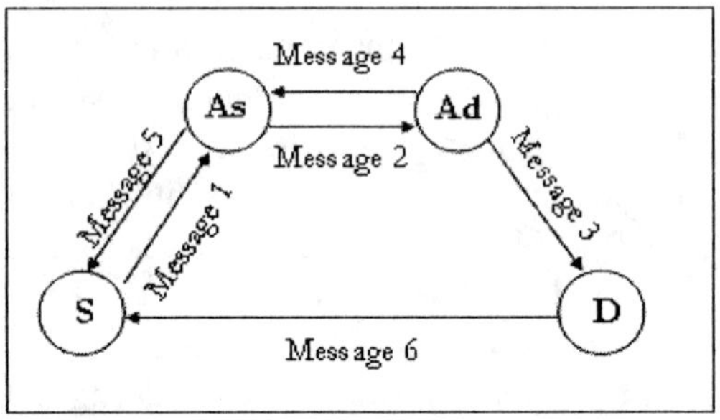

Figure 3. Exchanged messages in a flat authentication model

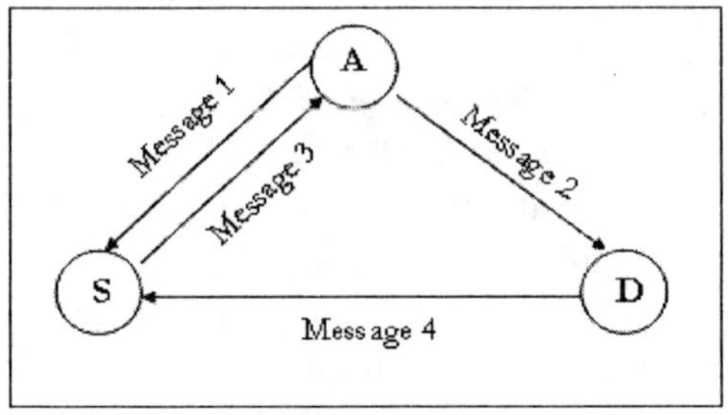

When deploying an authentication system, a choice should be made concerning how authentication servers manage credentials associated to authorized users. Basically, two architectures may be adopted. The former is an architecture where all authentication servers share the authentication status of all nodes in the network. This model reduces the amount of messages exchanged between interacting nodes, as shown by Figure 3, but holding all authentication information on each authentication server may introduce a processing overload since the verification process is performed on all stored credentials. The second architecture that is deployed is hierarchical where the authentication status of each node is known to a single authentication server. According to this approach, the number of exchanged messages during the authentication process is increased compared with the flat architecture.

Authentication Issues in Wireless Networks

Authentication is among the security services that should be considered in wireless networks. They allow the identification and the validation of credentials provided by users to access services. Deploying authentication in wireless networks needs to consider several issues including:

- **Mobility:** Routes used between communicating nodes change with time and links are unreliable. As a result, exchanged packets may be lost. In ad hoc networks, a set of packets can be exchanged between nodes that belong to the same group. These packets hold information to update a group key. For example, losing these packets is similar to the node that has not received the updated key from the group.
- **Large key size:** Large keys are desirable for their high entropy, but at the same time they would introduce processing overload and also require additional storage space at the mobile node.
- **Reconnection activities:** In wireless networks, connections are often reset and re-established by communicating devices. In this case, every reconnection requires fresh authentication parameters associated to the established sessions. This adds significant overhead.
- **Continuous authentication checks:** To prevent attacker attempts against authentication systems, authentication checks should be continuously performed. These checks introduce additional processing overhead and storage requirement.

In Wei and Wenye (2005), the impact of authentication on the security and quality of service (QoS) was quantitatively illustrated. A possible solution to key length could be sending a cryptographic hash instead of the whole key, but finding a generic cryptographic hash could be difficult for two reasons. First, in a wireless environment, devices are in most cases mobile. The conditions, in which two samples (one sent by the user and the one held by the server) are taken, vary considerably over time since they are taken under different working conditions. Therefore, two samples of the same object, such as a fingerprint, generated by two different sensors are most likely not identical. Cryptographic hash functions however, do not usually preserve distances and hence two samples of the same object may result in different digests at different conditions.

Introducing sampling in the authentication process may be a solution to reduce bandwidth and power requirements in a wireless environment. As an example of schemes that follow this principle, LAWN is a remote authentication protocol that enables repetitive remote authentication with large keys (Arnab, Rajnish, & Umakishore, 2005). This approach is motivated by the concept of a holographic proof (Polishchuk & Spielman, 1994; Spielman, 1995). A holographic proof is a proof of some fact, so constructed. To verify the proof, one does not need to scan through its entire length (Arnab et al., 2005). The verification process is limited to the examination of small parts randomly selected. According to this technique, a small sample of the authentication token is prepared, which can be used at the remote end to perform authentication with high probability of correctness. This technique allows saving bandwidth and power, since if the length of the original authentication token is n, then the selected sample is only $O(\log n)$. As samples may be different, a function that computes the difference between the patterns is needed. This may be achieved by computing the Hamming distance that gives as a result the number of bit positions where two strings differ in.

In the following, several general authentication techniques are detailed.

Password Authentication

Password authentication is among the solutions that are frequently required in wireless networks. Implementations according to this principle are vulnerable to multiple attacks that generally have targeted stored passwords or passwords sent across the network. As a solution to these threats, a proposed scheme should include cryptography techniques to protect stored credentials. Some security protocols are used to secure credentials. As an example of techniques that fulfil these needs, we mention the proposed scheme in I-En, Cheng-Chi, and Min-Shiang (2006) that supports the Diffie-Hellman key agreement protocol over insecure networks and function according to three phases: registration, phase, and authentication.

This scheme employs basic concepts, such as one-way hash function and discrete logarithm problem. During the registration phase, the server assigns smart cards to the users requesting registration. The registration phase is performed only when a new user needs to join the system. However, the login and authentication phases are performed at each user login attempt. During registration, the user chooses an identifier (ID) and a password (PW), it computes $h(PW)$ using a one-way function h. ID and $h(PW)$ are sent to the server through a secure channel. After receiving the registration message, the server calculates $B = g^{h(x\|ID)+h(PW)} \bmod p$, where p is a large prime number initially selected by the server, and g is a primitive number in $GF(p)$. After computing B, the server issues a smart card holding ID, B, p, g and delivers it to the user securely. During login, the user inserts the smart card to a terminal and introduces his/her ID and PW. Then, the terminal generates a login request message based on introduced information then it sends it to the server. At the server side, $B'' = g^{h(x\|ID)R} \bmod p$ is computed, where x is the server's secret key, ID is the user's identity, and R is a random number generated by the server. After that, the server calculates $h(B'')$ and sends it in addition to R to the user. When received at the user side, the user's smart card computes $B' = (Bg^{-h(PW)})^R \bmod p$ and the validity of the server is checked by comparing $h(B')$ and $h(B'')$.

If the server is considered valid, the user's module computes $C = h(T\|B')$, where T is the timestamp associated to the current login, otherwise the server is considered invalid and the user moves again to the login phase. At this step, the user's module sends (ID, C, T) to the server. After receiving the request, the server performs the checks to determine whether the user is allowed to login. It starts by checking if the format of ID is correct. If that is the case, the server compares T with T'; otherwise, it rejects the login request. T' is the time when the server receives the login request and ΔT is an acceptable time interval with respect to the transmission delay (needed for protection against replaying attack). If $T' - T \geq \Delta T$, the user's request is rejected; otherwise, it computes $C' = h(T \| B'')$ and compares C to C'. If the two values are equal, the login request is considered valid; otherwise it is rejected.

The RADIUS Protocol

The remote authentication dial-in user service (RADIUS) is an authentication protocol that has wide deployment in dial-up Internet services (Rigney, Rubens, Simpson, & Willens, 1997). RADIUS is a stateless transaction-based protocol; it runs over user diagram protocol (UDP). Using this protocol implies that at the end point RADIUS entities must integrate their own reliability mechanisms to handle lost packets. A simple reliability mechanism that is used by the most RADIUS implementations is the retransmission of lost packets.

RADIUS protocol is a client-server model, where authentication messages are exchanged between the RADIUS client and the RADIUS server, through one or more RADIUS proxies. This protocol is used for dial-up services; therefore the client is typically the network access server (NAS) that is in general connected to the remote access server (RAS), which represents the end point of the dial-up connection. The authenticator according to the RADIUS protocol is the RADIUS server that is responsible for receiving user connection requests and authentication. Moreover, RADIUS offers functionalities such as the support for authentication, authorization, and accounting (AAA). The use of plaintext passwords, the hash of passwords, and the keyed hash mechanism based on a shared secret are the basic authentication approaches that are used by RADIUS. These approaches are integrated in the two authentication protocols most used by RADIUS—they are PAP (Lloyd & Simpson, 1992) and challenge-handshake authentication protocol (CHAP) (Simpson, 1996). However, some vulnerabilities and weaknesses can be found including:

- **Lack of per-packet authentication for access-request packets.** A request authenticator (RA) that is a 128-bit pseudorandom number is included in the access request message. This value is used to hide the user password and does not really provide authentication of access-request messages. As a solution to this problem, it is possible to use RADIUS over IP security (IPsec).
- **Off-line dictionary attacks on the shared secret.** Several implementations of RADIUS allow only the use of shared secrets that are ASCII characters and having lengths that do not exceed 16 characters. Consequently, these secrets have low entropy. Based on this weakness, an attacker may collect access-requests and access-response packets and launch off-line dictionary attacks.

Authentication Between Heterogeneous Wireless Environments

When integrating heterogeneous networks, such as WLAN access networks and mobile cellular networks, many issues should be handled especially for authentication and roaming. For global system for mobile communications (GSM)/general packet radio service (GPRS) networks, the subscriber identity module (SIM) card is used for user identification, authentication, and message encryption. Therefore, it is feasible to authenticate the subscribers in WLAN via exchanging the authentication information between mobile cellular networks and subscribers' SIM cards (Yuh-Ren & Cheng-Ju, 2006). It is assumed that the WLAN access networks and the GSM/GPRS networks can interoperate and exchange system information via the GSM-MAP (Mobile Application Part) interface based on Signaling System 7 (SS7). In this context, a proposed protocol in Yuh-Ren and Cheng-Ju is used to authenticate a GSM/GPRS subscriber in a WLAN access network via the GSM/GPRS SIM card. This goal is achieved by exchanging some information to verify that the client and the GSM/GPRS have the same secret. The SIM-based authentication mechanism is divided into two phases: Temporary IP Address Acquisition Phase and Subscriber Identity Verification Phase. In Temporary IP Address Acquisition Phase, the mobile station (MS) attaches to the WLAN access network and discovers the DHCP Server to acquire the IP network configuration parameters. Subsequently, in the Subscriber Identity Verification Phase, the MS exchanges the authentication information with the WLAN Authentication Server to manifest the subscriber's identity (Yuh-Ren & Cheng-Ju). For Universal Mobile Telecommunications System (UMTS), the UMTS subscriber identity module is used for authentication.

AUTHENTICATION APPROACHES FOR CELLULAR AND MESH NETWORKS

In this section, we introduce authentication solutions for cellular and mesh networks.

Key Management in Wireless Sensor Networks

The wireless sensor network (WSN) represents a promising technology whose key idea lies in the scattering of tiny devices which are endowed with sensing, processing power, and wireless communication capability. These devices are intended to be deployed in a specific geographic area to sense change of some parameters (e.g., temperature, object movement, and noise) for several purposes including target tracking, environmental monitoring, and surveillance.

A large set of security issues and challenges in WSN networks are described in Pathan, Lee, and Hong (2006) and Karlof and Wanger (2003). Due to the known resources constraints in sensor nodes, it is not feasible to use the traditional pairwise key establishment techniques such as public key cryptography, which are too computationally intensive to ensure authentication and privacy. On the other hand, symmetric cryptography in WSN is not appropriate since the key may be easily eavesdropped during distribution. To address the issues, recently, several key distribution schemes were proposed which are based on pre-distributed keys or keying materials for key generation. The issue comes down to finding an efficient way for distributing key segments and materials before deployment of nodes.

In sensor networks, key distribution solutions can be classified into random, deterministic, and hybrid ones. For additional information, the reader is refereed to Camtepe and Yener (2004) where a survey on key distribution in sensor networks is provided. For peer-to-peer wireless sensor networks, where no infrastructure exists and nodes are randomly deployed, Eschenauer and Gligor (2002) proposed a random approach that is based on probabilistic key sharing among sensors and relies on a shared key discovery protocol for distribution and revocation of keys. The key distribution scheme has three phases: (1) the key pre-distribution, (2) shared-key discovery, and (3) path-key establishment. The key pre-distribution is preformed off-line before sensor nodes deployment. It consists of distributing a set of keys (key ring) from a large key pool. Shared-key discovery takes place during the initialization of the network. Each node broadcasts the list of identifiers of keys on their key ring. Consequently, every node in the network will discover the list of neighbors with which it shares a key. A routing link will exist between two nodes only if they share a key. As nodes give trust to each other, the same key can be shared by more than a pair of nodes. During the last phase, a path-key is assigned between a pair of sensor nodes that do not share the same key by relying on the set of intermediate secured links established at the second phase.

Du, Deng, Han, Chen, and Varshney (2003) exploit the knowledge of the node deployment to pre-distribute keys. A pair-wise key is distributed between each pair of neighboring nodes by exploiting the knowledge about the nodes that are likely to be neighbors of a sensor node. However, as the number of neighbors can be huge, a sensor may not be able to store all the related secret keys. To alleviate the problem, the use the random key pre-distribution scheme proposed in Eschenauer and Gligor (2002) reduced the amount of memory required.

In hierarchical wireless sensor networks, a hierarchy among nodes exists. BSs play the role of cluster supervisor and are more powerful that other sensor nodes in terms of transmission range and processing and storage capability. These BSs are considered as tamper resistant nodes. In this case of sensor networks, the key distribution becomes easier where BSs will handle the distribution of keys. A BS may share distinct pairwise keys with each sensor in the cluster that it handles. These keys can be used to establish other secure links between two sensor nodes where a BS will intermediate the establishment of a pairwise key between two different nodes.

Authentication Models for WLANs

Several models have been proposed for authentication in WLANs. These models are classified into categories according the techniques used to authenticate mobile nodes.

Web-Based Authentication Model

The Web-based approach for authentication is adopted due to the simplicity of the approach and the possible use of this technique without the need, at the user side, of special software or hardware. The simplicity of this technique is based on the use of secure socket layer (SSL). A Web server intercepts the user's HTTP traffic and redirects the user to the authenticator Web interface. The user provides his/her identity and password. The transmission of these credentials is protected by the SSL session, which encrypts the traffic between the mobile node browser's and the Web server. This method is simple to implement but does not result in a negotiated encryption key at the WLAN frame layer that is used by algorithms such as temporal key integrity protocol (TKIP). Consequently, after the accomplishment of the login phase, traffic at the MAC layer may remain unencrypted.

802.1X Authentication Framework

This framework is adopted for WLAN authentication and integrated with various key agreement protocols between the supplicant and the 802.11 AP for deriving the layer-2 cryptographic keys.

There are many authentication protocols in 802.1X that are certificate-based and can be used for devices as well as users. A digital certificate issued to a device and integrated physically into that device provides a strong mean to identify the device during the authentication process. A device may be of two kinds: (1) a user device (e.g., laptop, PDA, etc.), or (2) network equipment such as APs, routers, and so forth. A unique identity of the device may be deduced from the combination of the media access control (MAC) address, the product serial number, and others parameters, by using a hash function, for example. This unique identity may be used as the subject identity in the device certificate. A mutual authentication is needed in this case to ensure strong authentication. During this phase, the device uses its private key in to perform some verification tasks such as signing a nonce.

The Point-to-Point VPN Model

The use of IPsec virtual private networks (VPNs) provides an interesting mean to ensure confidentiality for data across the air interface. Thus, it is possible to use it to perform authentication for mobile nodes. In this model, the supplicant software establishes an IPsec VPN with a VPN server, which may be the AP or another device behind the AP. All data traffic generated by the client is tunnelled through the established VPN. According to this scheme, failure in the authentication process means that the IPsec VPN fails to be established and the supplicant's IP address is de-allocated. In this case, a threshold associated to the number of allowable failures by the supplicant may be fixed.

Authentication Protocols for GSM

Authentication protocols are considered among the main components of the GSM architecture. Several protocols proposed in this context have presented several drawbacks such as bandwidth consumption between the Visitor Location Register (VLR) and the Home Location Register (HLR), storage overhead in VLR and other insufficiencies such as bandwidth consumption if the MS moves frequently and requests several VLRs in a short period. The Authentication Center (AuC) is in charge of performing authentication in the GSM. It keeps the secret key K_i shared with the subscriber and generates the set of security parameters for requests associated to the authentication protocol of HLR. Subscriber's secret key is hold in the SIM card of the MS. When the subscriber registers for the first time, it gets a unique identity and an International mobile subscriber identity (IMSI) from the AuC. Among the solutions to improve such protocols, the authentication protocol proposed in Chang, Lee, and Chang (2005) provides mutual authentication between VLR and the MS. According to this protocol, the HLR makes the visiting VLR and MS share a temporary secret key K_T, which is computed by HLR using the algorithm A3and having as input both K_i (the secret key shared between MS and HLR) and R (a random number generated by HLR).

Moreover, HLR computes the certificate $CERT\_VLR = A3(T, K_i)$ for the visiting VLR of MS, where T is the timestamp sent by MS. This certificate is used to authenticate the validity of VLR. According to this authentication process, an authentication request including the temporary mobile subscriber identity (TMSI), the location area identity (LAI) and T is sent to VLR when the MS enters a new visiting area and asks for new communication services. After receiving the request, the new VLR uses the received TMSI to get the IMSI from the old VLR that will be sent with its identification D_v and T to the HLR through a secure channel. The HLR checks the validity of the D_v and T. If they are valid, it computes $CERT\_VLR = A3(T, K_i)$ and $K_T = A3(R, K_i)$ then it transmits the computed results and R to the visiting VLR. Otherwise, the HLR will terminate the authentication process. Receiving this information, the VLR computes $SRES = A5(R_1, K_T)$ and stores it in its database, where R_1 is a random number generated by the same component for the current communication. After that, the set of parameters composed of R, R_1, T and $CERT\_VLR$ is passed to the MS. Then the MS checks first the validity of T. If it is valid, it computes $CERT\_VLR'$ and compares it with the received $CERT\_VLR$. The authentication process will be halted if the two values are not equivalent; otherwise, MS computes $K_T = A3(R, K_i)$ and $SRES' = A5(R_1, K_T)$. Then, the computed $SRES'$ is sent back by MS to VLR. The latter compares it with the $SRES$ stored in its database to decide whether the request is considered valid; otherwise, it is rejected.

As long as the MS stays in the service area of the same visiting VLR, the latter does not need to request HLR for another authentication parameters but it generates only the random number, called R_j, at each jth communication (where $j > 1$ and $j \in 1$). This random number is used to compute $SRES_j = A5(R_j, K_T)$ that will be stored in the VLR database then R_j will be sent to the MS. The latter will compute and send $SRES'_j = A5(R_j, K_T)$ to VLR that checks whether the two values are equal or not. In the case of repetitive communications that are performed by the MS at the same area of the visiting VLR, mutual authentication is not ensured since only MS is authenticated, not the VLR.

To overcome this drawback, an improvement providing mutual authentication to the previous authentication scheme, is proposed in Chang et al. (2005). According to this scheme and while MS asks for the jth communication, VLR uses both K_T and T_j as inputs for the A3 algorithm to generate the certificate $CERT\_VLR_j$, where T_j is the timestamp generated by MS and included in the authentication request. This certificate is the means that will be used by the MS to authenticate the VLR. This check will be performed by the MS when it receives the $CERT\_VLR_j$ and proceeds to the computation of $CERT\_VLR'_j = A3(T_j, K_T)$. If the two certificates are equivalent then the VLR is authenticated successfully.

To enhance authentication in addition to ensuring mutual authentication, some modifications were proposed in Chang et al. (2005) including basically two phases: (1) the first authentication in the visiting VLR; and (2) the jth authentication between the same visiting VLR and the MS. The main idea of the first phase is to use $(R\|T_1)$ instead of R_1 to compute. The second phase handles repetitive communication between the visiting VLR and MS, the signed result $SRES_j$ is computed by using $T_{j-1}\|T_j$ and K_T as the inputs of A5 algorithm, where T_{j-1} is the timestamp associated to the $(j-1)th$ authentication and T_j is the timestamp generated by the MS for the jth authentication. Moreover, the certificate $CERT\_VLR_j$ of VLR is computed by using K_T and T_j using A3.

Kerberos Based Authentication Schemes for Ad Hoc Networks

Several authentication solutions that have been proposed at first for wired networks may be used in wireless environment by introducing some enhancements. Among these schemes, we mention Kerberos that has been implemented and tested in multiple production environments. In this context, a Kerberos assisted authentication in MANETs, known as *Kaman,* is proposed in Mcdonald and Pirzada (2004). This scheme adapts the Kerberos standard version to the wireless environment constraints. According to this scheme, secret keys or passwords are only known by users whereas the servers know a cryptographic hash of these passwords. Moreover, All Kaman servers share a secret key. These servers periodically, or on-demand, replicate their databases with each other in order to avoid a single-point-of-failure issue. Kaman uses a modified version of the Kerberos 5 protocol for authentication in ad hoc networks that eliminates the use of a ticket granting server (TGS). The adopted authentication protocol is described in Mcdonald and Pirzada . In addition to the basic authentication steps, the proposed protocol details key revocation operations, server elections and the replication of repository strategy.

Authentication in Mobile Ad Hoc Networks (MANETs)

According to the various constraints that characterize a mobile ad hoc network, an authentication mechanism should present low computational complexity and low bandwidth consumption. In Tsai and Wang (2007), two authentication mechanisms have been proposed to ensure cluster and individual authentication. The first authentication mechanism aims to verify whether the mobile node belongs to the same group or whether the message came from a node in this group. During the authentication process, the originator sends the original message and the cluster signature. The cluster signature is generated using the timestamp and the original message. At the destination node, the cluster signature is verified to determine if the received packet is valid or belongs to an attack traffic. The output packet, PKT_M, sent by the source is the following:

$$PKT_M = \{MAC_T, T_{stamp}, E_{Ks}(MAC_M, T_{stamp}, M)\},$$

where $MAC_T = H(K_C, T_{stamp})$ is generated using the timestamp and the common secret key, K_C that is used if there is not an available session key, denoted K_S. This key will be generated only if the individual authentication is performed successfully and is used to calculate MAC_M using the expression: $MAC_M = H(K_S, T_{stamp}, M)$. If the output packet is received by an intermediate node, the latter performs the following checks:

1. It computes $MAC_T = H(K_C, T_{stamp})$
2. It checks if T_{stamp} is within a reasonable time delay range.

If these two conditions are satisfied, the intermediate node forwards the packet to the next node; else it will be discarded. If the next node is the destination, it performs the two tasks in a similar way to an intermediate node, then it decrypts $E_{Ks}(MAC_M, T_{stamp}, M)$ and checks MAC_M by computing $H(K_S, T_{stamp}, M)$. After that, it verifies if the decrypted T_{stamp} is the same as the one that is added to the packet without encryption. If all these checks are performed successfully and all conditions are satisfied, the packet is considered valid; otherwise, it is discarded.

The second authentication process, called individual authentication for unicast, allows the verification of the identity of a user or a node in a group or the verification of the message originator. Since public-key cryptosystems are unsuitable for mobile ad hoc networks due to their high computational complexity, a low-power-consumption authentication procedure, based on secret sharing, is proposed. As mentioned in cluster authentication, individual authentication is required before generating the session key, which is shared between the source and the destination contrary to the common key that is shared between all the group nodes. The individual authentication mechanism is performed as follows: the source node S initiates a route discovery process to deduce a routing path from S to the destination node D. Next, S generates a random number, called a_0 and a random challenge number $RAND_S$.

S uses the function $f_1(x) = a_1 x + a_0 (\bmod\ p - 1)$. Based on this function, it generates a secret shadow associated with S and D, using an identity number of D (ID_S) that has the following expression: $K_{S,W} = f_1(D_S)$. The parameter a_1 is computed by S by satisfying the following expression:

$$K_{S,W} = f_1(ID_S)\bmod(p-1) = (a_1 \cdot ID_S + a_0)\bmod(p-1).$$

After that, S generates the session key $g^{a_0} \bmod p$ that is recovered at D. Γ is computed by S using $f_1(x = 1)$ as follows: $\Gamma = g^{f_1(1)} \bmod p$. Since these operations are performed, S sends the authentication packet that includes Γ and $RAND_S$ to D through the routing path deduced at the starting of the authentication process. At destination node, Z_D is computed that is the inverse of $(D_D - 1)$ on modulo $p - 1$ which satisfies the following relation: $(D_D - 1) \times Z_i \equiv 1 (\bmod\ p - 1)$. D uses the received Γ and a set of secret parameters called, $\Lambda_{D,S}$ ($\Lambda_{D,S} = (g^{K_{D,S}}) \bmod p$) to compute K_S according to the following expression:

$$K_S = (\Gamma^{ID_D x Z_D} / \Lambda_{D,S}) \bmod p.$$

Using this key, the destination encrypts the received challenge, $RAND_S$, and obtains the authentication reply code $AUTHR_S$ defined by $E_{K_S}(RAND_S)$. The generated $AUTHR_S$ and a regenerated $RAND_D$ are included into a confirmation packet that is forwarded to S via the routing path. Upon the reception of this packet by S, it computes the session key V as follows: $V = g^{a_0} \bmod p$. Then, it compares the received $AUTHR_S$ with $E_V(RAND_S)$. If the two parameters are equal, a common session key is obtained $K_S = V = g^{a_0} \bmod p$. After performing all these steps, S transmits the data packets via the secure routing path to D. The first packet should contain $AUTHR_D$ that is needed by D to verify the identity of S.

Authentication in 802.16 WMANs

As described by Figure 4, the 802.16 protocol layer consists of the physical layer and the MAC layer that is divided into three sublayers that are: (1) the security sublayer (or MAC privacy sublayer), (2) the MAC common-part sublayer, and (3) the MAC convergence sublayer. The MAC security sublayer supports security mechanisms such as authentication, key exchange, and also encryption. An 802.16 network or cell consists of one (or more) BSs and multiple subscriber stations (SSs). In addition to these components, it may be also additional entities within the network, such as repeater stations (RSs) and routers, ensuring connectivity of the network to core or backbone networks. An SS must perform a number of tasks that include authentication before gaining access to the network. During this phase, the SS must be authenticated by the BS, using the privacy key management (PKM) protocol that is detailed in Hardjono and Dondeti (2006). In such networks, each SS has an X.509 digital certificate (Housley, Polk, Ford, & Solo, 2002), which is integrated into the device hardware at manufacturing. The private key that is assigned to the certificate is embedded in the hardware in a way that is unfeasible for an intruder to read or modify it.

According to the IEEE 802.16 standard, only the SS is assigned a certificate and not the BS. Consequently, the mutual authentication is not ensured since the BS does not authenticate itself to the SS.

The PKM protocol first establishes a secret symmetric shared key between the SS and the BS that is called authorization key (AK).This key is used to ensure the exchange of the traffic encryp-

Figure 4. The 802.16 protocol layers

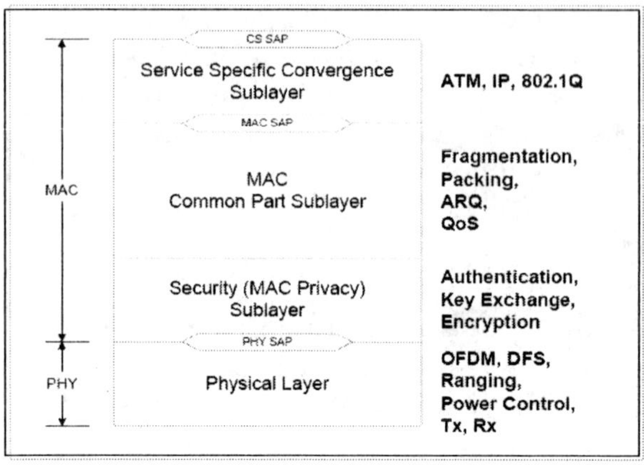

tion keys (TEK), in a secure manner. According to the PKM protocol, the authentication is performed through the exchange of three messages: (1) the authentication information; (2) the authentication request; and (3) the authorization reply sent by the BS to the SS. The first message sent by the SS during the authentication process contains the following information related to the SS: the MAC address, Rivest, Shamir, & Adleman (RSA) public key, X.509 certificate, list of the cryptographic capabilities, identifier of the primary SA (SAID), and the X.509 CA certificate of the manufacturer of the SS device. This message is facultative and allows the BS to verify if the SS has a valid primary SA and both certificates (for SS and manufacturer) are valid.

The second message sent by the SS to the BS in order to request an AK and the set of SAIDs of any static SA where the SS is authorized to participate. This message contains the same parameters as the first message (without the CA certificate of the manufacturer), only the added parameter is the SS device serial number and the manufacturer ID. The third message is sent by the BS to the SS after verifying its certificate and checking the set of cryptographic capabilities associated to it. The authorization reply message contains: A unique AK, encrypted with the RSA public key of the SS; a 4-bit key sequence number, used to identify different generated AKs; a key lifetime value associated to the AK and the SAIDs and properties of the primary SA and the set of static SAs, if there exist, for which the SS is authorized to obtain keying information.

Authentication in Mobile IPv6

In a Mobile IPv6, a mobile node (MN) exchanges with the home agent (HA) a set of messages that include signalling messages, binding update (BU) and binding acknowledgment (BA) (Johnson, Perkins, & Arkko, 2004). During the registration process, BU and BA messages are used. Moreover, BU messages may be used by an MN to inform other IPv6 nodes about its current care-of address that is registered at the home agent. To secure MIPv6 signaling messages between MNs and HAs, especially for BU messages, an alternate solution to IPsec, has been proposed in Patel, Leung, Khalil, Akhtar, and Chowdhury (2006). This method is a lightweight mechanism to authenticate the MN at the HA. According to this mechanism, authentication is ensured using a MIPv6-specific mobility message authentication option that can be added to MIPv6 signaling messages. Moreover, authentication is based on a shared-key-based mobility security association between the MN and the respective authenticating entity. As defined in Patel et al. (2006), the *shared-key-based mobility security association* is a security relation between the MN and its HA, used to authenticate the MN. The shared-key-based mobility security association consists of a mobility Security Parameter Index (SPI), a shared key, an authentication algorithm, and the replay protection mechanism in use.

MN must use the Mobile Node Identifier option, specifically the MN-NAI mobility option as defined in Patel, Leung, Khalil, Akhtar, and Chowdhury (2005) to identify itself while authenticating with the HA. This option is used to authenticate BU and BA messages. When a BU and BA is received without this option field and the entity receiving it is configured to use it, the received messages will be discarded by the HA. The structure of the mobility message authentication option is detailed in Patel et al. (2006).

PRIVACY PROTECTION

In this section, we detail basically a set of techniques that allow location privacy, and also ensure transaction based privacy.

Location Privacy

When designing wireless communication systems, location privacy should be considered to protect mobile nodes against a set of attacks. Location may be determined by an attacker by decoding packet contents and addresses or even by correlating different transmissions using a model of the user's movement. Using localization information, an attacker is able to track users especially where anonymous users are communicating. In this way, solutions ensuring location privacy are needed in addition to the set of techniques used to ensure confidentiality of transmitted data over the wireless network. The aim of these techniques is to render the intruder unable to correlate two locations.

For ad hoc networks, the privacy problem appears when adopting geographic routing that utilizes only location information to perform packet forwarding. While geographic routing improves routing performance, it introduces a problem related to location privacy. The information may be used by an intruder to reveal the target location. To preserve location privacy, an anonymous geographic routing algorithm that is based on the principle of dissociating user's location information with its identity, is proposed in Zhou and Yow (2005).

For wireless personal area networks, the location privacy problem is faced with some available standards. For example, packets exchanged between Bluetooth devices always contain the Bluetooth hardware address of the sender and the destination or an identifier which is directly mapped to this address. Consequently, an intruder is able to collect the Bluetooth hardware addresses of these devices. The eavesdropping of such kinds of information may be done remotely, by a device having a stronger antenna. Then the intruder can keep track of the place and time mobile devices. In Singelee and Preneel (2006), techniques that aim to solve the location privacy problem have been proposed. Since Bluetooth operates on the media access control layer, proposed solutions are integrated at this level, focusing the following four possible scenarios: (1) the mobile devices share a symmetric key, (2) the mobile devices know each others' addresses, (3) a secure extra communication channel is used, and (4) nothing shared between mobile devices.

The technique proposed for the last scenario is interesting since the two communicating mobile nodes have nothing shared and do not know each other's addresses. In this case, the mobile device that offers service (identified as A) generates a random identifier RA and broadcasts it during the initialization phase. When a mobile device (identified as B) needs to use these services, it generates a random identifier RB then puts the RA in the address field of the destination and its identifier in the source address field. According to this technique, the device does not know with whom it is exchanging messages. The reuse of RA and RB allows an attacker to link all messages in one communication round. Another technique may be adopted when the two mobile devices do not share information and that ensures location privacy, is to broadcast any message without specifying identifiers for communicating nodes.

Transaction Based Privacy

One goal of privacy is to ensure the concept of unlinkable unit of communications. According to this concept, an attacker is tolerated to link any communications within one unit but it is prevented from making the link between two different units. This unlinkability may be provided on a transaction granularity, as proposed in Yih-Chun and Helen (2005). A transaction may be defined as a stream of packets sent from the source to the destination, and another stream of packets sent as response by the destination. Session-based services such as SSH or Telnet may be considered as a sequence of transactions. Linkability between transactions may be decreased with time; after a sufficiently long time, the correlation between two transactions is unfeasible. Based on this principle, the introduction of random silent period for each node may be a solution to ensure privacy. During this period, the node does not forward or transmits any packets. This technique is not supported by all kind of applications. Voice over IP is among applications where latency is not tolerated.

In addition, information about the occurrence of a transaction should be protected to prevent attacks that are executed at the completion of a given transaction. For example, an intruder may need to determine the occurrence time of a transaction in order to capture the transaction output that may hold a set of information that helps compromising the system or collecting information about it. Ensuring time privacy may be achieved using several techniques that are mostly integrated in transaction-based systems. A random behavior may be another solution that prevents intruder from learning the transaction occurrence time. For example, it is possible that the execution of a transaction delivers only a partial output and the remaining results are provided when needed by requesting them or by informing the service provider node about this need. This technique ensures time privacy since the completion time associated to the execution of a transaction is based in some cases on the reception of the complete output that may be variable in time. Another technique that may be adopted is a dynamic scheme that order transactions following a priority order that takes into consideration multiple parameters such as outputs needed for next transactions, available resources (storage, processing) for processing nodes, and so forth.

SECURITY MODELS

This section takes interest into three aspects in modeling security. The first aspect is related to theories of trust representation, modeling, and verification. The second aspect is related to security policy specification and verification.

Trust Representation, Modeling, and Verification

Trust management systems represent a generalization of the traditional security mechanisms such as authentication and privacy by adding heterogeneity and distribution. Several trust management systems were proposed including Policymaker, Keynote trust management system, and Relational-based formalism of trust.

A number of trust management systems were developed including X.509 (Arsenault & Turner, 2002) model, SPKI (Ellison et al., 1999) and PGP trust model. The X509 is centralized bringing a hierarchical certificate management model with a tree structure and a root. In the context of wireless networks where the trust management should be distributed, the X509 model cannot fit. SPKI provides flexibility for trust management by allowing delegation. However, there is no restriction to control the delegated signature chain or allow the issuer to update the trust value of each delegated certificate. The PGP trust model supports trust management in distributed networks. It uses a trust model that has no centralized or hierarchical relationship between certification authorities as in X.509. The underlying assumption is that trustees may validate digital certificates from other entities or trust a third party to validate certificates. However, entities are fully trusted and there is a need for a mechanism to compute trustworthiness of every trusted or signed key. A degree of trust such as completely trusted, marginally trusted, or untrusted, should be supported.

Following the previous works, a number of automated trust management systems were proposed including PolicyMaker (Blaze, Feigengaum, & Strauss, 1998a) and KeyNote (Blaze, Feigenbaum, & Keromytis, 1998b). The PolicyMaker provides an application independent framework for constructing and validating the security policies. In fact, the security policy is specified by the system, whereby the applications can create their own actions and policies but are not required to do the security verification themselves. The system is then used to verify whether an action is consistent with a local policy. This system takes as input a set of local policy statements, a collection of credentials, and a string describing the actions to be performed. It evaluates these policies and credentials by verifying whether an action is consistent with the local policy. The output, which is made by a compliance checker, is a positive or a negative response or even a set of additional requirements to be fulfilled to let the actions be permissible.

KeyNote, which is the successor of PolicyMaker, uses the same design principle of assertions and queries. However, it brings an additional enhancement in the trust management engine and supports special language for assertion which allows a simple integration with the compliance checker. Note that the signature verification is performed in the keynote engine making it well-suited for trust management in public-key infrastructure. KeyNote evaluation is performed after receiving a set of local security assertions, a collection of credential assertions, which may represent certificates signed by entities delegating the trust, and a collection of attributes defining an action environment and containing all information relevant to the request.

Guemara-ElFatmi, Boudriga, and Obaidat (2004) proposed a trust management scheme based on the use of the relational calculus, which allows proving and verifying compliance of communication protocols with security policy in the case where the system is based on of use of public key certificates (PKC). The proposed trust management scheme integrates (1) a relational language for modeling entities (e.g., certificates, security policy) and actions (e.g., delegations); (2) a relational calculus for performing proofs; (3) a mechanism for identifying users; and (4) a compliance engine which allows to decide whether an action, which is performed by a principal can be granted or not. The model of trust management uses (1) a set of certificates/request, say X, denoting requests sent for online checks and operations; (2) a set of response returned by online checks, say Y, to represent the acceptance and rejections of requests; and (3) a binary relation from X^+ (the set of words in X) to Y, called the conformance checking relation, where X^+ denote the set of non-empty chains in X. This relation denotes all pairs of the form (C,y) where C is an input chain and y is the related response.

For complex situations, Guemara-ElFatmi et al. (2004) proposed a deductive system denoted by Δ=(Axiom, Rule, W) where W include formula of the first order logic interpreted over X+ and Y and formula of the form "$(C, y) \in R$". An element in Axiom denotes a formula in the form of $(C, y) \in R \wedge \pi(C) \wedge p(y)$ where π and p represent predicates interpreted over X+ and Y, respectively. y stands for the response at time t for an input chain denoted by C. Elements of Rule have equivalence between certificate chains and have the following form:

$$\frac{(C', y) \in R \qquad \pi(C, C')}{(C, y) \in R}$$

The previous rule states that if y is a response to chain C' and if C is related to C' through π then y is the response to chain C. As a consequence, a pair (C, y) is in R, at time t, if and only if the formula $(C, y) \in R$ is a theorem in Δ. The form of the axiom denoted previously allows the characterization and computation of R as a least fixpoint of a function. The trust management model takes into consideration several issues including security policy representation, compliance correctness, and system state determination. The system state determination issue, for instance, shows how to reduce the complexity of online checks by letting the server not to remember its past inputs but rather to just remember a summary of past input history. The aforementioned trust management model was validated using three case studies which are anonymous payment system, clinical information system, and distributed firewall system.

Security Policy Specification

To effectively protect themselves from security threats, organizations should define a security policy, which according to the ISO 17799 represents a document that provides management direction and support for information security. A security policy can be seen as a specification for security solutions and form a conceptual model of them as the specifications do for software. Improving correctness of security policies is thus essential in order to guarantee the security level required by the secured system. Several formal methods can be used to validate whether a security requirement for systems or components are complete, correct, and can be met by the security policy. Examples include model checking, theorem proving, and executable specifications.

In model checking technique, the security policy may be specified using temporal logic formulas which describe how the permissible system transitions may occur and what modification do they introduce to the system. By adding a description of the initial system state, and the set of security properties that should be followed by the security policy (e.g., all users are authenticated within the system), the model checking technique can be used to verify whether the model satisfies the expressed properties. To do so a graph is generated where nodes represent the system states through which the system progresses while edge represent possible transitions which may alter these states. The main drawback of the model checking technique lies in the state explosion problem that must be addressed to cope with non-finite sate specification. Several approaches can be used to defeat such a problem including the use of binary decision diagrams, partial order reduction to reduce the number of interleaving of non-concurrent processes, and abstraction to prove properties on a system after simplifying it. In theorem proving, specification is based on using formalisms such as first-order logic and higher-order logic. To prove that properties are met by the specification, proof techniques such as induction, rewriting, simplification, and decision procedure using can be followed.

The executable security policy (ESP) specification, which is defined by: "a specification whereby security policy is executed on a computer simulating its behavior when interacting with its environment," is a recent technique for security policy validation. In fact, a security policy is validated by comparing the behavior of the related ESP to what is specified in it. The methodology is based on defining the security policy in natural language then translating it to algebraic specification. After, the algebraic specification is then analyzed for syntactic verification purpose, it is translated to an executable language (e.g., S-TLA+ [Rekhis & Boudriga, 2005]) following a set of rules that have to present some properties including completeness and termination. At this level, the executable security policy is run to detect vulnerabilities.

To validate the security policy, the behavior of the executable security policy can be checked using the model checking technique. The satisfaction of the set of invariants in the security policy can be verified during the generation of potential scenarios by the model checker.

CONCLUSION

Ensuring security for wireless networks signifies several goals to achieve including the security of connected mobiles nodes, the security of services provided by wireless nodes and that are accessible from public networks, and the security of exchanged data between wireless nodes. All these goals may be achieved through the use of a set of mechanisms and models that have been the subject of this chapter. Authentication, privacy protection, trust, and security models are the concepts that are detailed in this chapter showing the possible solutions that may be adopted to prevent a set of attacks that represent a potential threat for wireless networks.

REFERENCES

Aboudagga, N., Refaei, M. T., Eltoweissy, M., DaSilva, L. A., & Quisquater, J.-J. (2005). Authentication protocols for ad hoc networks: Taxonomy and research issues. In *Proceedings of the 1st ACM International Workshop on QoS & Security in Wireless and Mobile Networks* (pp. 96-104).

Arnab, P., Rajnish, K., & Umakishore, R. (2005). LAWN: A protocol for remote authentication introduced over wireless networks. In *the Fourth IEEE International Symposium on Network Computing and Applications (NCA),* Cambridge, MA.

Arsenault, A., & Turner, S. (2002, July). *Internet X.509 public key infrastructure: Roadmap.* Retrieved January 2, 2007, from http://www1.tools.ietf.org/html/draft-ietf-pkix-roadmap-09.txt

Balfanz, D., Smetters, D. K., Stewart, P., & Wong, H. C. (2002, February). Talking to strangers: Authentication in ad-hoc wireless networks. In *Symposium on Network and Distributed Systems Security (NDSS '02),* San Diego, CA.

Baras, J., & Jiang, T. (2004). Cooperative games. Phase transitions on graphs and distributed trust in MANET. In *Proceedings of the 43rd IEEE Conference on Decision and Control,* Nassau, Bahamas.

Blaze, M., Feigenbaum, J., & Strauss, M. (1998a). Compliance checking in the policymaker trust management system. In *2nd International Conference on Financial Cryptography,* London, (LNCS 1465, pp. 254-274).

Blaze, M., Feigenbaum, J., & Keromytis, A. (1998b, April). Keynote: Trust management for public-key infrastructures. In *the 1998 Security Protocols International Workshop,* Cambridge, England (LNCS 1550, pp. 59-63).

Camtepe, S. A., & Yener, B. (2004). Key distribution mechanisms for wireless sensor networks; A survey. (LNCS 3193).

Chang, C., Lee, J., & Chang, Y. (2005). Efficient authentication protocols of GSM. *Computer Communications, 28*(8), 921-928.

Du, W., Deng, J., Han, Y. S., Chen, S., & Varshney, P. K. (2003, July). *A key management scheme for wireless sensor networks using deployment knowledge* (Tech. Rep.), New York: Syracuse University. Retrieved from http://www.cis.syr.edu/#wedu/Research/paper/ ddhcv03.pdf

Ellison, C., Frantz, B., Lampson, B., Rivest, R., Thomas, B., & Ylonen, T. (1999, September). *SPKI certificate theory* (RFC 2693). Retrieved January 2, 2007, from http://www.ietf.org/rfc/rfc2693.txt

Eschenauer, L., & Gligor, V. D. (2002, November 18-22). A key management scheme for distributed sensor networks. In *Proceedings of the 9th ACM Conference on Computer and Communications Security,* Washington, DC.

Guemara-ElFatmi, S., Boudriga, N., & Obaidat, M. (2004, July). Relational-based calculus for trust management in networked services. *Journal of Computer Communications. 27*(12), 1206-1219.

Hardjono, T., & Dondeti, L. R. (Eds.). (2006). *Security in wireless LANs and MANs.* Norwood, MA: Artech Press.

Housley, R., Polk, W., Ford, W., & Solo, D. (2002). *Internet X.509 public key infrastructure certificate and certificate revocation list (CRL) profile* (RFC 3280). Retrieved from http://www.ietf.org/rfc/rfc3280.txt

I-En, L., Cheng-Chi, L., & Min-Shiang, H. (2006). A password authentication scheme over insecure networks. *Journal of Computer and System Sciences, 72,* 727-740.

Johnson, D., Perkins, C., & Arkko, J. (2004). *Mobility support in IPv6* (RFC 3775). Retrieved from http://www.ietf.org/rfc/rfc3775.txt

Karlof, C., & Wagner, D. (2003, May). Secure routing in wireless sensor networks: Attacks and countermeasures. In *Proceedings of the First IEEE International Workshop on Sensor Network Protocols and Applications* (pp. 113-127).

Lloyd, B., & Simpson, W. (1992). *PPP authentication protocols* (RFC 1334). Retrieved from http://www.ietf.org/rfc/rfc1334.txt

Mcdonald, C. S., & Pirzada, A. A. (2004). Kerberos assisted authentication in mobile ad-hoc networks. In *Proceedings of the 27th Australasian Computer Science Conference (ACSC'04)* (Vol. 26, pp. 41-46).

Patel, A., Leung, K., Khalil, M., Akhtar, H., & Chowdhury, K. (2005). *Mobile node identifier option for mobile IPv6* (RFC 4283). Retrieved from http://www.ietf.org/rfc/rfc4283.txt

Patel, A., Leung, K., Khalil, M., Akhtar, H., & Chowdhury, K. (2006). *Authentication protocol for mobile IPv6* (RFC 4285). Retrieved from http://tools.ietf.org/html/rfc4285

Pathan, A. K., Lee, H. W., & Hong, C. S. (2006, February 20-22). Security in wireless sensor networks: Issues and challenges. In *Proceedings of the 8th International Conference on Advanced Communication Technology (ICACT 2006)* (pp. 1043-1048).

Polishchuk, A., & Spielman, D. A. (1994). Nearly-linear size holographic proofs. In *Proceedings of the 26th ACM Symposium on Theory of Computation (STOC)* (pp. 194-203).

Rekhis, S., & Boudriga, N. (2005, September). A temporal logic-based model for forensic investigation in networked system security. In the *3rd International Workshop on Mathematical Methods, Models and Architectures for Computer Networks Security (MMM-ACNS-05),* St. Petersburg, Russia (LNCS 3685, pp. 325-338).

Ren, K., Li, T., Wan, Z., Bao, F., Deng, R., & Kim, K. (2004). Highly reliable trust establishment scheme in ad-hoc networks. *Computer Networks: The International Journal of Computer and Telecommunications Networking, 45*(6), 687-699.

Rigney, C., Rubens, A., Simpson, W., & Willens, S. (1997). *Remote authentication dial in user service (RADIUS)* (RFC 2058). Retrieved from http://tools.ietf.org/html/rfc2058

Simpson, W. (1996). *PPP challenge handshake authentication protocol (CHAP)* (RFC 1994). Retrieved from http://www.ietf.org/rfc/rfc1994.txt

Singelee, D., & Preneel, B. (2006). Location privacy in wireless personal area networks. In *Proceedings of The 7th International Conference on Web Information Systems Engineering (WiSe'06).*

Spielman, D. (1995). *Computationally efficient error correcting codes and holographic proofs.* Unpublished doctoral thesis. Retrieved from http://www-math.mit.edu/ spielman/PAPERS/thesis.pdf

Theodorakopoulos, G., & Baras, J. (2004). Trust evaluation in adhoc networks. In *Proceedings of the 2004 ACM workshop on Wireless security,* Philadelphia (pp. 1-10).

Tsai, Y., & Wang, S. (2007). Two-tier authentication for cluster and individual sets in mobile ad hoc networks. *Computer Networks, 51*(3), 883-900.

Wei, L., & Wenye, W. (2005). On performance analysis of challenge/response based authentication in wireless networks. *Computer Networks, 48,* 267-288.

YihChun, H., & Helen, J. W. (2005). 2A framework for location privacy in wireless networks. In *Proceedings of the ACM Special Interest Group on Data Communication (SIGCOMM) Applications, Technologies, Architectures, and Protocols for Computer Communications,* Asia Workshop.

Yuh-Ren, T., & Cheng-Ju, C. (2006). SIM-based subscriber authentication mechanism for wireless local area networks. *Computer Communications: Monitoring and Measurements of IP Networks, 29*(10), 1744-1753.

Zhou, Z., & Yow, K. C. Anonymizing geographic ad hoc routing for preserving location privacy. In *Proceedings of the 3rd International Workshop on Mobile Distributed Computing (MDC'05),* IEEE ICDCS.

KEY TERMS

Authentication: Authentication is the process of attempting to verify the digital identity of the sender of a communication such as a request to log in. The sender being authenticated may be a person using a computer, a computer itself, or a computer program.

Digital Certificate: Digital certificate is an electronic document which incorporates a digital signature to bind together a public key with an identity—information such as the name of a person or an organization, their address, and so forth. The certificate can be used to verify that a public key belongs to an individual.

Digital Signature: Digital signature is a type of asymmetric cryptography used to simulate the security properties of a signature in digital rather than written form. Digital signature schemes normally give two algorithms, one for signing which involves the user's secret or private key, and one for verifying signatures which involves the user's public key. The output of the signature process is called the "digital signature."

Hash Function: Hash function is a function that takes a long string (or "message") of any length as input and produces a fixed length string as output, sometimes termed a message digest or a digital fingerprint.

Privacy: Privacy is the fact of protecting personal data and information related to a communication entity to be collected from other entities that are not authorized. Privacy is sometimes related to anonymity and can be seen as an aspect of security.

Wireless Network: Wireless network refers to any type of network that is wireless, the term is most commonly used to refer to a telecommunications network whose interconnections between nodes is implemented without the use of wires.

Wireless Sensor Network (WSN): WSN is a wireless network consisting of spatially distributed autonomous devices using sensors to cooperatively monitor physical or environmental conditions, such as temperature, sound, vibration, pressure, motion, or pollutants at different locations. The development of wireless sensor networks was originally motivated by military applications. However, wireless sensor networks are now used in many civilian application areas.

Chapter XV
The Provably Secure Formal Methods for Authentication and Key Agreement Protocols

Jianfeng Ma
Xidian University, China

Xinghua Li
Xidian University, China

ABSTRACT

In the design and analysis of authentication and key agreement protocols, provably secure formal methods play a very important role, among which the Canetti-Krawczyk (CK) model and universal composable (UC) security model are very popular at present. This chapter focuses on these two models and consists mainly of three parts: (1) an introduction to CK model and UC models; (2) A study of these two models, which includes an analysis of CK model and an extension of UC security model. The analysis of CK model presents its security analysis, advantages, and disadvantages, and a bridge between this formal method and the informal method (heuristic method) is established; an extension of UC security model gives a universally composable anonymous hash certification model. (3) The applications of these two models. With these two models, the four-way handshake protocols in 802.11i and Chinese wireless LAN (WLAN) security standard WLAN authentication and privacy infrastructure (WAPI) are analyzed.

INTRODUCTION

Key agreement protocols are mechanisms by which two parties that communicate over an adversarially controlled network can generate a common secret key. Key agreement protocols are essential for enabling the use of shared-key cryptography to protect transmitted data over insecure networks. As such they are a central piece for building secure communications and are among the most commonly used cryptographic protocols.

The design and analysis of secure key agreements protocols has proved to be a non-trivial task, with a large body of work written on the topic. Among the methods for the design and analysis of key agreement protocols, formal methods have always been a focused problem in the international investigation of cryptography. Over the years, two distinct views of formal methods, symbolic logic method and computational complexity method, have developed in two mostly separate communities (Martin & Phillip, 2002). The symbolic logic method relies on a simple but effective symbolic formal expression approach, in which cryptographic operations are seen as functions on a space of symbolic formal expressions (e.g., BAN, communicating sequential processes [CSP], NRL) (Wenbo, 2004). The other one, computational complexity method, relies on a detailed computational model that considers issues of complexity and probability of successful attacks, in which cryptographic operations are seen as functions on strings of bits.

Provably secure formal method, which is based on the computational complexity method, is a very hot research point at present. Its salient property is that the security protocols designed by them are provably secure. Among the provably secure formal methods, CK model and UC security model are very popular.

In 2001, Canetti and Krawczyk presented the CK model for the formal analysis of key-exchange (KE) protocols. A session-key security definition and a simple modular methodology to prove a KE protocol with this definition are introduced in this model. One central goal of the CK model is to simplify the usability of the definition via a modular approach to the design and analysis of KE protocols. It adopts the indistinguishability approach (Bellare, Canetti, & Krawczyk, 1998) to define security: A KE protocol is called secure if under the allowed adversarial actions it is infeasible for the attacker to distinguish the value of a key generated by the protocol from an independent random value. The security guarantees that result from the proof by the CK model are substantial as they capture many of the security concerns in the real communications setting.

Concurrent composition is a fact of life of real network settings. Protocols that are proven secure in the stand-alone model are not necessarily secure under composition. Therefore, it does not suffice to prove that a protocol is secure in the stand-alone model. UC security model proposed by Canetti in 2001 (Birgit & Michael, 2001) is for representing and analyzing cryptographic protocols under concurrent circumstance (Yeluda, 2003). The salient property of definitions of security in this framework is that they guarantee security even when the given protocol is running in an arbitrary and unknown multi-party environment. An approach taken in this framework is to use definitions that treat the protocol as stand-alone but guarantee secure composition. Security in complex settings (where a protocol instance may run concurrently with many other protocol instances, or arbitrary inputs and in an adversary controlled way) is guaranteed via a general composition theorem. On top of simplifying the process of formulating a definition and analyzing protocols, this approach guarantees security in arbitrary protocol environments, even unpredictable ones that have not been explicitly considered. The abstract level of UC security goes far beyond other security models, therefore, it tends to be more restrictive than other definitions of security. The most outstanding nature of UC framework is its modular design concept: may alone design a protocol, so long as the protocol satisfies the UC security, it can be guaranteed secure while runs concurrently with other protocols.

This chapter focuses mainly on the introduction, analysis, and applications of these two provably secure formal methods. The rest of this chapter is organized as follows. The next section, the CK model and the UC security model are introduced. In the third section, we analyze the security of the CK model. A bridge between this formal method and the informal method (heuristic method) is established. What is more, the advantages and disadvantages of the CK model are given. In the *Universally Composable Anonymous Hash Certification Model* section, an extension of the UC security model is presented. The UC security model fails to characterize the special security requirements of anonymous authentication with other kind of certificates. Therefore the UC security model is extended, and a new model—Universally Composable anonymous hash certification model is presented. In this model, an anonymous hash certification ideal function is introduced, which fulfills the identity authentication by binding the identity to special hash values. In addition, a more universal certificate CA model is presented, which can issue the certificate with specific form (for example hash value). In the fifth section, we analyze the four-way handshake protocol in 802.11i with the CK model and UC security model. In sixth section, first, the authentication modules in the Chinese WLAN national standard WAPI and its implementation plan are analyzed with the CK model. Then we point out that how the implementation plan overcomes the security weaknesses in the original WAPI. The last two sectionscontain the future trends and conclusions.

BACKGROUND OVERVIEW

Definition 1: Key-agreement protocol (Menezes, Van Oorschot, & Vanstone, 1996). A key-agreement protocol or mechanism is a key establishment technique in which a shared secret is derived by two (or more) parties as a function of information contributed by, or associated with, each of these, (ideally) such that no party can predetermine the resulting value.

The CK model and UC security model are very popular provably secure formal methods for key-agreement protocols at present. In this section, these two security models are introduced respectively, and the relationship between the security definitions in these two models is also given.

The Canetti-Krawczyk Model

A KE protocol is run in a network of interconnected parties where each party can be activated to run an instance of the protocol called a session. A KE session is a quadruple (*A, B, X, Y*) where *A* is the identity of the holder of the session, *B* the peer, *X* the outgoing messages in the session, and *Y* the incoming messages. The session (*B, A, Y, X*) (if it exists) is said to be matching to the session (*A, B, X, Y*). Matching sessions play a fundamental role in the definition of security (Canetti & Krawczyk, 2001).

Attacker Model

The attacker is modeled to capture realistic attack capabilities in open networks, including the control of communication links and the access to some of the secret information used or generated in the protocol. The attacker, denoted M, is an active "man-in-the-middle" adversary with full control of the communication links between parties. M can intercept and modify messages sent over these links, it can delay or prevent their delivery, inject its own messages, interleave messages from different sessions, and so forth. (Formally, it is M to whom parties hand their outgoing messages for delivery.) M also schedules all session activations and session-message delivery. In addition, in order to model potential disclosure of secret information, the attacker is allowed access to secret information via session exposure attacks (a.k.a. known-key attacks) of three types: state-reveal queries, session-key queries, and party corruption.

- **State-reveal query:** A state-reveal query is directed at a single session while still incomplete (i.e., before outputting the session key) and its result is that the attacker learns the session state for that particular session (which may include, for example, the secret exponent of an ephemeral Diffie-Hellman algorithm (DH) value but not the long-term private key used across all sessions at the party).
- **Session-key query:** A session-key query can be performed against an individual session after completion and the result is that the attacker learns the corresponding session key.
- **Party corruption:** Party corruption means that the attacker learns all information in the memory of that party (including the long-term private key of the party as well all session states and session keys stored

at the party); in addition, from the moment a party is corrupted all its actions may be controlled by the attacker. Indeed, note that the knowledge of the private key allows the attacker to impersonate the party at will.

Three Components in CK Model

- **The unauthenticated–links adversarial model (UM):** UM is the real network environment, the attacker in this model is an active one. It has all the attack ability mentioned previously.
- **The authenticated-links models (AM):** The adversarial model called AM is defined in a way that is identical to the UM with one fundamental difference: The attacker is restricted to only delivering messages truly generated by the parties without any change or addition to them.
- **Authenticators:** Authenticators are special algorithms which act as automatic "compliers" that translate protocols in the AM into equivalent (or "as secure as") protocols in the UM. Now there are two kinds of authenticators, one is based on the public key digital signature, the other one is based on the message authentication code (Bellare et al., 1998).

With the CK model, one can firstly design and analyze a protocol in AM, then transforms these protocols and their security assurance to the realistic UM by using an authenticator.

Definition of Session-Key Security

In addition to the regular actions of the attacker $\mathcal{M}$ against a KE protocol π, he/she can perform a *test session query.* That is, at any time during its run, $\mathcal{M}$ is able to choose, a *test-session* among the sessions that are completed, unexpired, and unexposed at the time. Let *k* be the value of the corresponding session key. We toss a coin b, b $\xleftarrow{R} \{0,1\}$. If b = 0 we provide $\mathcal{M}$ with the value *k*. Otherwise we provide $\mathcal{M}$ with a value *r* randomly chosen from the probability distribution of keys generated by protocol π. The attacker $\mathcal{M}$ is not allowed state-reveal queries, session-key queries, or party corruption on the test-session or its matching session. At the end of its run, $\mathcal{M}$ outputs a bit b' (as its guess for b).

An attacker that is allowed test-session queries is referred to as a KE-adversary.

Definition 2: Session-key security. A KE protocol π is called session-key secure (or SK-secure) if the following properties hold for any KE-adversary $\mathcal{M}$:

1. Protocol π satisfies the property that if two uncorrupted parties complete matching sessions then they both output the same key; and
2. The probability that $\mathcal{M}$ guesses correctly the bit b (i.e., outputs b' = b) is no more than 1/2 plus a negligible fraction ε in the security parameter. ε is called "advantage."

The Universal Composable Model

Universally composable security is a framework for defining the security of cryptographic protocols (Canetti, 2001). In this framework, an uncorruptable ideal functionality $\mathcal{F}$ which can provide a certain service, a set of dummy parties $\tilde{P}$ and an ideal adversary $\mathcal{S}$ are defined respectively. Only the dummy parties $\tilde{P}$ and ideal adversary $\mathcal{S}$ can access ideal functionality $\mathcal{F}$, each dummy party can not communicate directly with the others, and the ideal adversary can corrupt any dummy party at any time. The ideal adversary $\mathcal{S}$ is informed of when a message is sent, but not of the content, it is allowed to delay the delivery of such a message, but not change its content. On the other hand, an actual protocol π that can achieve the special service, a set of real parties *P*, and a real-world adversary $\mathcal{A}$ are correspondingly defined. Each real party can communicate with the others directly and the real-world adversary $\mathcal{A}$ can control all communications among them, meaning that $\mathcal{A}$ can read or alter all messages among the real parties, what is more, $\mathcal{A}$ can also corrupt any real party at any time. An environment Z is defined in the UC framework that can simulate the whole external environment;

Figure 1. The framework of universally composable security

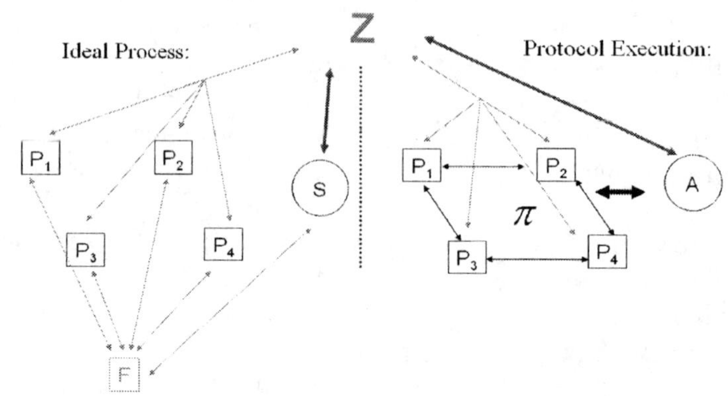

it can generate the inputs to all parties ($\tilde{P}$ or P), read all outputs, and in addition interact with the adversary ($\mathcal{A}$ or $\mathcal{S}$) in an arbitrary way throughout the computation. The environment 9 pt is forbidden to directly access the ideal functionality $\mathcal{F}$.The framework of universally composable security is shown in Figure.1.

Definition 3: Universally composable security (Canetti, 2001). In UC framework, real protocol π securely realizes ideal functionality $\mathcal{F}$ if, for $\forall\mathcal{A}$ and $\forall Z$, it has the same "action" as $\mathcal{F}$. Formally, a protocol π securely realizes an ideal functionality $\mathcal{F}$ is for any real-life adversary $\mathcal{A}$ there exists an ideal-process adversary $\mathcal{S}$ such that, for any environment Z and on any input, the probability that Z outputs "1" after interacting with $\mathcal{A}$ and parties running π in the real-life model differs by at most a negligible fraction from the probability that Z outputs "1" after interacting with $\mathcal{S}$ and $\mathcal{F}$ in the ideal process.

Definition 4: Composition theorem (Canetti, 2001). The key advantage of UC security is that we can create a complex protocol from already designed sub-protocols that securely achieves the given local tasks. This is very important since complex systems are usually divided into several sub-systems, each one performing a specific task securely. Canetti presented this feature as the composition theorem. This theorem assures that we can generally construct a large size "UC-secure" cryptographic protocol by using sub-protocols which is proven as secure in UC-secure manner.

Definition 5: Hybrid model (Canetti, 2001). In order to state the aformentioned definition and to formalize the notion of an actual protocol with access to multiple copies of an ideal functionality, Canetti also introduced the hybrid model which is identical to the actual model with the following: On top of sending messages to each other, the parties may send messages to and receive messages from an unbounded number of copies of an ideal functionality $\mathcal{F}$. The copies of $\mathcal{F}$ are differentiated using their session identifier SIDs. All messages addressed to each copy and all messages sent by each copy carry the corresponding SID.

The Relationship Between the SK-Security and UC-Security

UC-security is strictly stronger than SK-security. That is, for a KE protocol, UC-security is stronger than SK-security. Any UC-secure key-agreement protocol is SK-secure, but an SK-secure key agreement is not necessarily UC-secure.

Claim 1: Any protocol that is UC-secure is SK-secure. This holds in the UM and in the AM.

Definition 6: Acknowledgment (ACK) property. Let $\mathcal{F}$ an ideal functionality and let π be an SK-secure key protocol in the $\mathcal{F}$-hybrid model. An algorithm I is said to be an internal state simulator for π if for any environment machine Z and any adversary $\mathcal{A}$ we have:

$$\text{HYB}^{\mathcal{F}}_{\pi,\mathcal{A},Z} \approx \text{HYB}^{\mathcal{F}}_{\pi\ \mathcal{A},Z,I}$$

Protocol π is said to have the ACK property if there exists a good internal state simulator for π.

- **Theorem 1:** Let π be a KE protocol that has the ACK property and is SK-secure; then π is UC-secure (Canetti & Krawczyk, 2002).

SECURITY ANALYSIS OF THE CANETTI-KRAWCZYK MODEL

In the past 20 years, researchers have made a lot of efforts in designing and analyzing KE protocols (Diffie & Hellman, 1976; Diffie, Van Oorschot, & Wiener, 1992; Krawczyk, 1996; Shoup, 1999), they realize that the potential impact of the compromise of various types of keying material in a key-agreement protocol should be considered, even if such compromise is not normally expected (Menezes et al., 1996). So some desirable security properties that a key-agreement protocol should have are identified. Such security properties include perfect forward security (PFS), loss of information, known-key security, key-compromise impersonation, unknown-key share, key control, and so on.

The main goal of the CK model is to design and analyze key-agreement protocols. Then what is the relationship between the CK model and the desirable security attributes for a key-agreement protocol? This is the main motivation of this section.

Properties of Key-Agreement Protocols

Definition 7: (Implicit) key authentication (Menezes et al., 1996). Key authentication is the property whereby one party is assured that no other party aside from a specifically identified second party (and possibly additional identified trusted parties) may gain access to a particular secret key.

A key-agreement protocol, which provides implicit key authentication to both participating entities, is called an *authentication and key-agreement (AKA)* protocol.

Definition 8: Key confirmation (Menezes et al., 1996). Key confirmation is the property whereby one party is assured that a second (possibly unidentified) party actually has possession of a particular secret key.

Definition 9: Explicit key authentication (Menezes et al., 1996). Explicit key authentication is the property obtained when both (implicit) key authentication and key confirmation hold.

A key-agreement protocol which provides explicit key authentication to both participating entities is called *authenticated key agreement with key confirmation (AKC)* protocol (Menezes et al., 1996).

A secure key-agreement protocol should be able to withstand both passive attacks and active attacks. In addition to implicit key authentication and key confirmation, a number of desirable security attributes of key-agreement protocols have been identified (Law, Menezes, Qu, Solinas, & Vanstone, 1998).

1. **(Perfect) forward secrecy:** If long-term private keys of one or more entities are compromised, the secrecy of previous session keys established by honest entities is not affected (Menezes et al., 1996).
2. **Loss of information:** Compromise of other information that would not ordinarily be available to an adversary does not affect the security of the protocol. For example, in Diffie-Hellman type protocols, security is not comprised by loss of $\alpha^{S_i S_j}$ (where S_i represents entity *i*'s long-term secret value) (Blake-Wilson, Johnson, & Menezes, 1997).
3. **Known-key security:** A protocol is said to be vulnerable to a known-key attack if compromise of past session keys allows either a passive adversary to compromise future session keys, or impersonation by an active adversary in the future (Law et al., 1998).
4. **Key compromise impersonation:** Suppose *A*'s long-term private key is disclosed. Clearly an adversary that knows this value can now impersonate *A*, since it is precisely this value

that identifies A. However, it may be desirable that this loss does not enable an adversary to impersonate other entities to A (Law et al., 1998).

5. **Unknown key-share:** Entity A cannot be coerced into sharing a key with entity B without A's knowledge, that is, when A believes the key is shared with some entity $C \neq B$, and B (correctly) believes the key is shared with A (Law et al., 1998).
6. **Key control:** Neither entity should be able to force the session key to a preselected value (Law et al., 1998).

The Relationship Between the CK Model and the Desirable Secure Attributes

- **Theorem 2:** A key-agreement protocol designed and proved secure by the CK model offers almost all the desirable security properties mentioned above except key control (Li, Ma, & Moon, 2005).

The Relationship Between the Security Attributes and the Two Requirements of SK-Security

In the CK model, some security attributes can be ensured by the first requirement of SK-security, while others by the second requirement. In the following, Theorem 3 and Theorem 4 are presented for a detailed explanation:

- **Theorem 3.** The first requirement of SK-security guarantees a protocol to resist impersonation attacks and unknown key-share attacks (Li et al., 2005).

- **Theorem 4.** The second requirement of SK-security guarantees a protocol to offer PFS, known-key security (Li et al., 2005).

It should be noticed that the first requirement is the precondition of SK-security. Only under the consistency condition, does it make sense to investigate the security properties of PFS and known-key security.

Advantages and Disadvantages of the CK Mode

Advantages of the CK Model

Why is the CK Model Applicable for Designing and Analyzing Key-Agreement Protocols?
First, the indistinguishability between the session key and a random number is used to achieve the SK-security of a key-agreement protocol in the AM. If an attacker can distinguish the session key from a random number with a non-negligible advantage, a mathematics hard problem will be resolved. According to the reduction to absurdity, a conclusion can be gotten: no matter what methods are used by the attacker (except party corruption, session state reveal and session key query), he/she cannot distinguish the session key from a random number with a non-negligible advantage. So the protocol designed and proved secure by the CK model can resist known and even unknown attacks.

Second, the CK model employs authenticators to achieve the indistinguishability between the protocol in the AM and the corresponding one in the UM. Through this method, the consistency requirement of SK-security is satisfied.

From the previous analysis, it can be seen that this model is a modular approach to provably secure protocols. With this model, we can easily get a provably secure protocol which can offer almost all the desirable security attributes. And the CK model has the composable characteristic and can be used as an engineering approach (Bellare & Rogaway, 1993; Mitchell, Ward, & Wilson, 1998). Therefore, it is possible to use this approach without a detailed knowledge of the formal models and proofs, and is very efficient and suitable for applications by practitioners.

Disadvantages of the CK Model

Though the CK model is suitable for the design and analysis of key-agreement protocols, it still has some weaknesses as follows:

1. The CK model cannot detect security weaknesses that exist in key-agreement protocols, however some other formal methods have this ability, such as the method based on logic (Burrows, Abadi, & Needham, 1990) and the method based on state machines (Tin, Boyd, & Nieto, 2003). But the CK model can confirm the known attacks, that is, this model can prove that a protocol that has been found flaws is not SK-secure.
2. In the aspect of the forward secrecy, the CK model cannot guarantee that a key-agreement protocol offers forward secrecy with respect to compromise of both parties' private keys; it can only guarantee the forward secrecy of a protocol with respect to one party. In addition, in ID-based systems this model lacks the ability to guarantee the key generation center (KGC) forward secrecy because it does not fully consider the attacker's capabilities (Canetti & Krawczyk, 2002).
3. From Theorem 2, we know that protocols which are designed and proved secure by the CK model cannot resist key control, which is not fully consistent with the definition of key agreement (Blake-Wilson et al., 1997).
4. A key-agreement protocol designed and proved secure by the CK model cannot be guaranteed to resist denial-of-service (DoS) attacks. However DoS attacks have become a common threat in the present Internet, which have brought researchers' attention (Burrows et al., 1990; Meadows, 1996).
5. Some proofs of the protocols with the CK model are not very credible because of the subtleness of this model. For example, the Bellare-Rogaway three-party key-distribution (3PKD) protocol (Bellare & Rogaway, 1995) claimed proofs of security, but it is subsequently found flaws (Choo & Hitchcock, 2005).

We know that a protocol designed and proved secure by the CK model can offer almost all the security attributes, and this model has the modular and composable characteristics, so it is very practical and efficient for the design of a key-agreement protocol. But this model still has weaknesses. So when the CK model is employed to design a key-agreement protocol, we should pay attention to the possible flaws in the protocol that may result from the weaknesses of CK model.

A UNIVERSALLY COMPOSABLE ANONYMOUS HASH CERTIFICATION MODEL

The essence and difficulty of UC security protocol design lays in the formalization and abstraction of a perfect ideal functionality which can be realized securely. We consider the special security requirements for ideal anonymous authentication, define the security notions for them, and realize an anonymous hash certification ideal functionality F_{Cred} in a universally composable security sense, and present a more universal certificate CA model F_{HCA} (Canetti, 2004), which can issue anonymous hash certificates.

Anonymous Hash Certification Ideal Functionality F_{Cred}

We use Merkle tree to build the hash chain, which is constructed from each leaf up to the root of the tree. For each unit of the chain, it contains a value and an order bit which identities whether the given value should be concatenated from the left or the right.

A hash chain is said to be valid under a collision-free hash function H if $h_0 = h_o'$ and $h_{d-1}' = v$, $h_{i-1}' = H(h_i \| h_i')/H(h_i' \| h_i)$ for $o_i = l/r$, where $i = d-1, d-2, \ldots, 1$. It is written as $isvalid(h) = 1$. We also define several other functions, for instance, $root(h)$ is to choose the root of a hash chain, $leaf(h)$ is to return the value of a leaf node of path h, $buildtree_H(C)$ is to build a Merkle tree with the values of set C, and $getchain_T(e)$ is to capture the path of node e.

Security Requirements of F_{Cred}

Definition 10. Let k be a security parameter and $\varepsilon(k)$ be a negligible function on k. Let s be a signature key, v be a verification key. We say that an anonymous hash certification protocol satisfies the security requirements if the following properties hold:

Completeness. For any valid credential (c, p_i, k, h),
$Prob[(s,v) \leftarrow gen(1k); 0 \leftarrow Verify\ Credential(c,z,k,p_j,h,\sigma,v)] < \delta(k)$,
where σ is the signature of $root(h)$.

Consistency. For any valid credential (c, p_i, k, h), the probability that

$$Verify\ Credential(c,z,k,p_j,h,\sigma,v)$$

generates two different outputs in two independent invocations is smaller than $\varepsilon(k)$.

Unforgeability. For any PPT forger F,

$\text{Prob}[(s,v) \leftarrow gen(1^k); (c,p_j,k,h) \leftarrow F^{\pi_{Cred}}(v), 1 \leftarrow Verify\ Credential(c,z,k,p_j,h,\sigma,v)] < \delta(k)$

and F never as the signature functionality F_{SIG} to sign $root(h)$.

The Construction of Anonymous Hash Certification Ideal Functionality F_{Cred}

The functionality is realized by using a signature scheme $SS = (Kg, Sig, V_f)$, a symmetric encryption scheme, a pseudorandom functions R and a collision-free, one-way hash functions H. we assume that SS is CMA secure for the simplified purpose.

In the anonymous hash certification ideal functionality, the entities are denoted as ASU for the authentication server, which is also denoted by P_0 for the simplified purpose, and $P_1, \ldots, P_m$ for the subscribers or authenticator respectively.

Two security parameters, k_1 and k_2, are used in this ideal functionality. The parameter k_1 is the key length of the symmetric cipher, and k_2 is the length of string used to identify the authenticator. A special function ℓ is used to map the identity of authenticator to $[k_2]$ such that $\ell(p_i)$ has cardinality $k_2/2$, and $\ell(p_i) \neq \ell(p_j)$ for $p_i \neq p_j$.

The credential is denoted as $c_i = (c, p_i, k, h)$, where c is the encrypted real identity of the subscriber, P_i is the owner identity of the credential, k is the secret information, that is, the hash pre-images, with its length is k_2, h is the Merkle hash chain path of this credential. The value of this credential is defined as $val_H(c_i) = c \| H(k_1) \| H(k_2) \| \cdots \| H(k_{k_2})$.

A counter t that is initialized to 0 and is used to index credential c_i that has been issued in period i. A set of credential $C = \cup c_i$ and a set of credential to be used $T_{prepared}$ are initialized to ϕ.

1. *Present Credential*

Upon receiving a message

$$(Present\ Credential, p_i, c, z, p_j)$$

from some party p_i, send

$$(Present\ Credential, p_i, c, z, p_j)$$

to the adversary. Return the message from the adversary to p_i.

2. *Verify Credential*

Upon receiving a message

$$(Verify\ Credential, p_i, c, z, \tilde{k}, p_j, h', \sigma, v)$$

from some party p_i,

$$(Verify\ Credential, p_i, c, z, \tilde{k}, p_j, h', \sigma, v)$$

send to the adversary. Return the message from the adversary to p_i.

3. Check Reuse

Upon reception of

$$(Check\ Reuse, p_s, c, z, \tilde{k}_1, \tilde{k}_2, p_{j_1}, p_{j_2}, h, \sigma, v)$$

from some party p_s, execute

$$(Verify\ Credential, p_i, c, z, \tilde{k}, p_j, h', \sigma, v)$$

for $i = 1, 2$.

- If at least one execution returns,

$$(Verify\ Credential, p_s, c, p_{j_i}, invalid)$$

then return

$(Verify\ Credential, p_s, c, p_{j_i}, invalid)$

to p_s.

- If $P_{j_1} = P_{j_2}$ then return,

$(Check\ Reuse, p_s, c, no)$
otherwise return

$(Check\ Reuse, p_s, c, yes)$
to p_s. (end)

Construction of UC-Secure Anonymous Hash Certification Protocol

In this section, we present a simple protocol that realizes F_{Cred} given F_{SIG}, with the aid of ideally authenticated communication with a "trusted anonymous hash certificate authority." This set-up assumption is formalized as an ideal functionality F_{HCA}.

Firstly we modify the definition of F_{SIG} (Canetti, 2004; Michael & Dennis, 2004) as follows.

1. *Key generation*

Upon reception of (*KeyGen, P*) from *P*:

- Sends (*KeyGen, P*) to the adversary S.
- After receiving the message (*VerificationKey, P,* θ) from S, records (*P,* θ) and sends (*VerificationKey, P,* θ) to *P*.

2. *Signature generation*

Upon reception of (*Sign, P, m*) from *P*:

- Sends (*Sign, P, m*) to S.
- After receiving the message (*Signature, P, m,* σ) from S, looks for the record $(m, \sigma, \theta, 0)$. If it is found, sends an error message to *P* and halts. Else, sends $(Signature, P, m, \sigma)$ to *P* and then records $(m, \sigma, \theta, 0)$.

3. *Signature verification*

Upon reception of $(Verify, P, m, \sigma, \theta')$ from a verifier *V*:

- Sends $(Verify, P, m, \sigma, \theta')$ to S.
- After receiving the message $(Verify, P, m, \sigma, \theta')$ from S, works as follows.

1. If $\theta' = \theta$ and there exists the record $(m, \sigma, \theta, 1)$, set $f = 1$.
2. If $\theta' = \theta$, *P* has not yet been corrupted by S, and there exists no record such that $(m, \sigma', \theta, 1)$ for $\forall \sigma'$, set $f = 0$.
3. If $\theta' \neq \theta$ and there exists the record (m, σ, θ', f'), set $f = f'$.
4. Else, set $f = \phi$, then records $(m, \sigma, \theta', \phi)$.

- Hands (*Verified, P, m, f*) to *V*. (end)

Then the anonymous hash certificate authority Functionality F_{HCA} is presented as follows.

1. *Key generation*
Upon reception of the message (*GenerateKey*) from ASU, send (*KeyGen, ASU*) to the adversary *S*, upon receiving (*Verification Key, ASU, encryption key, k*) from *S*, records (*ASU, v, k*) and return (*Verification Key, ASU, v*).
2. *Identity Encryption*
Upon reception of the message (*Identity encryption, p_i*) from p_i, proceed as follows:
   1. Verify that p_i is in the member list. If not, return (*Not A Member, p_i*) and quit.
   2. Else, send (*Identity encryption, p_i*) to the adversary *S*, receive the encryption identity *c* of p_i, return (*Encrypted identity, p_i, c*).
3. *Credential generation*
Upon reception of the message (*Credential generation, p_i, (c, p_i, k, z)*) from p_i, send this message to the adversary, and wait for an *OK* from the adversary. Then, Store credential *e* = (*c, p_i, k,* ϕ) into set C_i, return (*S, New Credential, p_i*) and (*p_i, New Credential, c, z*) to *S*.

4. *Build tree*
Upon reception of the message (*Buildtree, ASU*) from *ASU*, set $T \leftarrow buildtree_H(val_H(C_t))$ and modify each credential $e = (c, p_i, k, \phi)$ of C_t into $(c, p_i, k, getchain_T(val_H(e)))$, and send (*Sign, ASU, root(T)*) to the adversary *S*. Upon receiving the message (*Signature, ASU, root(T)*, σ) from *S*, verify that no entry (*root(T)*, σ, 1) is recorded, if it is, then output an error message to *S* and halt, else record the entry (*root(T)*, σ, 1), return (*Build Tree, ASU, T*, σ) to *S*, and set $t \leftarrow t + 1$.
5. *Add prepared credential*
Upon reception of the message (*Add prepared credential*, p_i, (c, p_j)) from p_i, send this message to the adversary, and wait for an *OK* from the adversary. Then, add (c, p_j) in the set $T_{prepared}$ and return *OK*.
6. *Check prepared credential*
Upon reception of the message (*Check prepared credential*, p_i, (c, p_j)) from p_i, send this message to the adversary, and wait for an *OK* from the adversary. Then, find (c, p_j) in the set $T_{prepared}$, return *OK* if this entry exists.
7. *Check exist of credential*
Upon reception of the message (*Check exist of credential*, p_i, (c, p_j, k, h)) from p_i, send this message to the adversary, and wait for an *OK* from the adversary. Then, find (c, p_j, k, h) in the set *C*, return *OK* if this entry exists.
8. *Reveal ID*
Upon reception of the message (*Reveal ID, ASU, c*) from ASU, find a credential $(c, p, \cdot, \cdot)$ in set *C*. If no such entry exists, then send (*Reveal ID, ASU, c*) to the adversary *S*. Once the message (c, p) is received from *S*, returning(*Reveal ID, ASU, c, p*).

Finally, we present a protocol π_{Cred} that realizes F_{Cred} in the (F_{SIG}, F_{HCA})-hybrid model in a straightforward way as follows.

1. *Present Credential*
   1. p_i receives a message (*Present Credential, c, z*, p_j),
   2. If p_i has not owned a credential, it creates a symmetric key $R^i \xleftarrow{R} R^{k_1}$ with a pseudorandom function and sends (*Identity encryption*, p_i) to F_{HCA}. Upon receiving the message (*Encrypted identity*, p_i, *c*) from F_{HCA}, it calculates secret information $k_j \leftarrow (R^i(c \| j))_{j=1}^{k_2}$, $z \leftarrow H(k)$, sent (*Credential generation*, p_i, (c, p_i, k, z)) to F_{HCA}.
   3. Else p_i sets $\tilde{k} \leftarrow k_{\ell(P_j)}$ and outputs $(Present\ Credential, c, \tilde{k})$, if $k_l \leftarrow (R^i(c \| l))_{l=1}^{k_2}$ and $z = H(k)$.
   4. Otherwise, it outputs the message (*Reject Present Credential, c*) and quits.
2. *Verify Credential*
   1. p_i verifies the validity of $(c, z, \tilde{k}, p_j, h, \sigma, v)$
   2. P_i sends (*Verify* p_i, *root(h)*, σ, *v*) to F_{SIG} and then executes the signature verification process of F_{SIG}.
   3. P_i sends (*Check prepared credential*, p_i, (c, p_i, p_j)) to F_{HCA}, and wait for an *OK* from F_{HCA}.
   4. P_i verifies $isvalid_H(h) = 1$ and $H(\tilde{k}) = z_{\ell(P_j)}$.
   5. If F_{SIG} returns 0 or any condition 3 or 4 is not satisfied, it returns (*Verify Credential*, p_i, *c*, p_j, *invalid*) and quits.
   6. else P_i returns (*Verify Credential*, p_i, *c*, p_j, *valid*).
3. *Check Reuse*
   1. P_i checks the reuse of
   $(c, z, \tilde{k}_1, \tilde{k}_2, h, \sigma, p_{j_1}, p_{j_2})$,
   2. It executes
   $(Verify\ Credential, c, z, \tilde{k}_i, h, \sigma, p_{j_i})$
   for $i = 1, 2$.
   3. If at least one execution returns (*Verify Credential, c*, p_{ji}, *invalid*), then p_i returns (*Check Reuse, c, invalid*) and quits.
   4. If $p_{j_1} = p_{j_2}$ then p_i returns (*Check Reuse, c, no*), otherwise it returns (*Check Reuse, c, yes*).

Proof of π_{Cred} Securely Realizes F_{Cred} in the (F_{SIG}, F_{HCA})-Hybrid Model

Theorem 5. Protocol π_{Cred} securely realizes F_{Cred} in the (F_{SIG}, F_{HCA})-hybrid model.

Figure 2. The construction of an adversary S

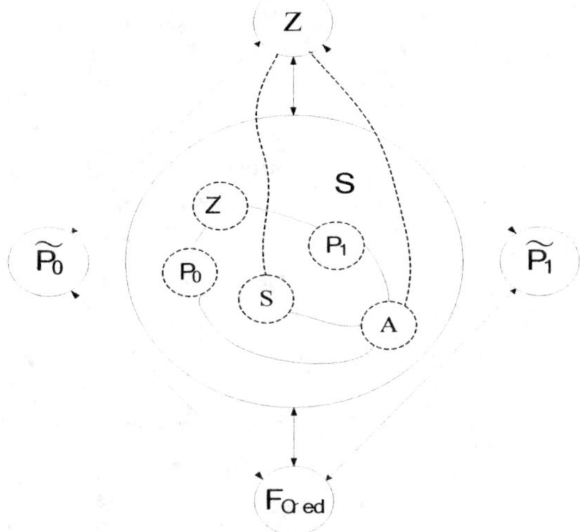

Proof. Let A be an adversary that interacts with entities running π_{Cred} in the (F_{SIG}, F_{HCA})-hybrid model. We construct an ideal-process adversary S such that the view of any environment Z of an interaction with A and π_{Cred} is distributed identically to its view of an interaction with S in the ideal process for F_{Cred}.

1. *The construction of adversary* S

The adversary S runs an internal copy of environment Z, adversary A and each of the involved parties p_i. All messages from Z to A are written to A's input tape. In addition, S does the following: For each player p_i that the real-world adversary A corrupts, the ideal adversary S corrupts the corresponding dummy player p_i. When a corrupted dummy player p_i receives a message m from Z, the adversary S lets Z' send m to p_i. When a corrupted p_i outputs a message m to Z', then S instructs the corrupted p_i to output m to Z. This corresponds to p_i being linked directly to Z. The construction of the adversary S is shown in Figure 2.

2. *The operations of adversary* S

Simulating Present Credential

When S receives in the ideal process F_{Cred} a message (*Present Credential*, p_i, *c*, *z*, p_j), it proceeds as follows:

1. If p_i has not owned a credential, then simulate for A the process of credential generation. That is, send to A (in the name of F_{HCA}) the message (*Identity encryption*, p_i), obtain the response from A, then it set a random number u^i as the key of P_i, i.e., $u^i \xleftarrow{R} \{0,1\}^{k_1}$, record ($P_i$, u^i) in the member list and then calculates secret information $k_j \leftarrow (U^i(c \| j))_{j=1}^{k_2}$, $z \leftarrow H(k)$, where $k = (k_1, k_2, ..., k_{k_2})$, and send to A the message (*Credential generation*, p_i, (*c*, p_i, *k*, *z*)) from F_{HCA}.
2. Simulate for A the process of present credential. That is, set $m \xleftarrow{R} \{0,1\}^{k_2}$, make sure the number of "1" is exactly $k_2/2$ and m never been produced before, construct the challenge information $k \leftarrow k_m$ by providing the pre-images of secret information k that corresponding to the bit "1" of m, send the message (*Add prepared credential*, p_i, (*c*, p_j)) to A from F_{HCA}, and send $(Present\ Credential, p_i, c, k)$ to F_{Cred}.

Simulating Verify Credential

1. If a message $(Verify\ Credential, p_i, c, z, k, p_j, h', \sigma, v)$ arrives from F_{Cred}, it proceed as follows.
2. Send the message $(Check\ exist\ of\ credential, p_i, (c, p_i, k, h))$ to A from F_{HCA}. If the message from F_{HCA} is not *OK*, send $(Verify\ Credential, p_i, c, p_j, invalid)$ to F_{Cred} and quit.
3. Else check the path, if $h' \neq h$, then send $(Verify\ Credential, p_i, c, p_j, invalid)$ to F_{Cred} and quit.
4. Else, verify the signature, send $(Verify\ p_i, root(h), \sigma, v)$ to A (in the name of)F_{SIG}, upon receiving the message $(Verified\ p_i, root(h), \phi)$ from A,

(1) If the entity $(root(h), \sigma, 1)$ is recorded, set $f = 1$.
(2) Else, if the signer is not corrupted, and no entry $(root(h), \sigma', 1)$ for any σ' is recorded, then set $f = 0$ and record the entry $(root(h), \sigma, 0)$.
(3) Else, if there is an entry $(root(h), \sigma, f')$ recorded , then let $f = f'$.
(4) Else, let $f = 0$ and record the entry $(root(h), \sigma, \varphi)$.

If $f = 0$, send $(Verify\ Credential, p_i, c, p_j, invalid)$ to F_{Cred} and quit.

5. Else, verify the validity of the credential,

(1) If p_i is not corrupted,
(a) Send message $(Check\ prepared\ credential, p_i, (c, p_j))$ to A from F_{HCA}.
(b)If the F_{HCA} message fromis not *OK* or $k \neq \tilde{k}_m$, send to F_{Cred} the message $(Verify\ Credential, p_i, c, p_j, invalid)$ and quit.
(2) Else if $H(\tilde{k}) \neq z_m$, send $(Verify\ Credential, P_i, C, P_j, invalid)$ to F_{Cred} and quit.

Otherwise return $(Verify\ Credential, p_i, c, p_j, valid)$ to F_{Cred}.

Simulating party corruptions

If A corrupts a patty p_i, then S corrupts the same party $\tilde{P}_i$ in the ideal process and hands A the internal data of that party $\tilde{P}_i$.

As for the other operations, like ***Check Reuse***, because their definitions are identical in the ideal functionality and real protocol, it is no use for them to be simulated for A.

As the simulation is perfect and the proof is direct, the proof procedure can be referred to Fan, JianFeng, & Moon, 2007).

THE SECURITY ANALYSIS OF FOUR-WAY HANDSHAKE IN 802.11I WITH THE CK MODEL AND UC MODEL

WLAN can provide great flexibility for the users. However, security is always a serious concern because of the openness of wireless medium for public access within a certain range. To solve the security problems of WLAN, the IEEE 802.11 has designed a new security standard, which is called IEEE 802.11i (IEEE P802.11i D3.0, 2002). In this standard, a concept of robust security network has been proposed. In addition, an authentication mechanism based on EAP/802.1X/RADIUS (Aboba & Simon, 1999; 802.1X-2001, 2001; Rigney, Willens, Rubens, & Simpson, 2000) has been developed to replace the poor open system authentication and shared-key authentication in WEP (Borisov, Goldberg, & Wagner, 2001). As a long-term solution to secure wireless links, the latest IEEE standard 802.11i has been ratified on June 24, 2004.

The four-way handshake (in short, 4WHS) protocol in 802.11i plays a very important role in the authentication and key-agreement process. Some works have been done on its security analysis. In Changhua and Mitchell (2004) the authors analyzed the four-way handshake protocol using a finite-state verification tool and find a DoS attack. The attack involves forging initial messages from the authenticator to the supplicant to produce inconsistent keys in peers. However the repair proposed by the authors involves only a minor change in the algorithm used by the supplicant and not involves the protocol itself.

In this section, we give a formal analysis of the four-way handshake. The results show that four-way handshake protocol is secure not only in the CK model, but also in the UC security model. So it can be securely used as the basic model of the authentication and key agreement of WLAN.

The Four-Way Handshake Protocol in 802.11i

In 802.11i, once a shared pairwise master key (PMK) is agreed upon between the authenticator and the supplicant, the authenticator may begin a four-way handshake by itself or upon request from the supplicant. The message exchange is shown, at an abstract level, in Figure 3. S represents the Supplicant and A represents the Authenticator; SPA and AA, SNonce and ANonce, represent the message authentication code (MAC) address and nonces of the supplicant and authenticator, respectively; sn is the sequence number; msg1, 2, 3, 4 are indicators of different message types; MICPTK{} represents the message integrity code (MIC) calculated for the contents inside the bracket with the fresh pairwise transient key (PTK). While MAC is commonly used in cryptography to refer to a MAC, the term MIC is used instead in connection with 802.11i because MAC has another standard meaning, medium access control, in networking.

The fresh PTK is derived from the shared PMK through a pseudo random function with output length X (PRF-X), say, PTK = PRF-X(PMK, "Pairwise key expansion" || Min{AA, SPA} || Max{AA, SPA} || Min{ANonce, SNonce} || Max{ANonce, SNonce}), and divided into Key Confirmation Key (KCK), Key Encryption Key (KEK), and Temporary Key (TK). Note that the MIC is actually calculated with KCK, which is only part of PTK.

The Security Analysis of Four-Way Handshake Protocol

According to the thought of CK model, we extract the protocols $4WHS_{AM}$ in AM and the authenticator λ_{prf}. We can further analyze that whether the protocol $4WHS_{AM}$ is SK-secure in the AM λ_{prf} or the protocolis an effective MT-authenticator or not, thus we can draw the conclusion that whether four-way handshake protocol can satisfy the definition of SK-security in the UM or not.

Protocol $4WHS_{AM}$

This protocol is described as follows:

1. Both players pre-share a key k_{ij}.
2. The initiator p_i, on input (p_i, p_j, s), chooses $r_i \xleftarrow{R} \{0,1\}^k$ and sends (p_i, s, r_i) to p_j;
3. Upon receipt of (p_i, s, r_i), the responder p_j chooses $r_j \xleftarrow{R} \{0,1\}^k$, where $r_j \neq r_i$, and sends (p_j, s, r_j, t_j) to p_i. Then p_j outputs session key $prf_{k_{ij}}(r_i, r_j)$.
4. Upon receipt of (p_j, s, r_j) player p_j outputs session key $prf_{k_{ij}}(r_i, r_j)$.

Figure 3. The idealized 4-way handshake protocol

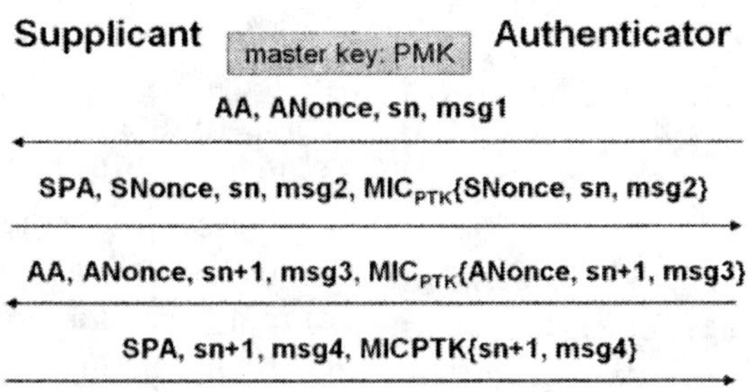

The Security Analysis of Protocol $4WHS_{AM}$

Theorem 6. If the pseudorandom function f is secure against chosen message attacks, the protocol $4WHS_{AM}$ is SK-secure without PFS in the AM.

Proof. The protocol $4WHS_{AM}$ is based on a pre-shared key, from which the session key k_{ij} is generated, thus it cannot provide the security attribute of perfect forward security (PFS). According to the model mentioned previously, let the session never expire.

To see that the first requirement of Definition SK-security is satisfied, according to the definition of AM, note that if both p_i and p_j are uncorrupted during the exchange of the key and both complete the protocol, then they both get the uncorrupted r_i and r_j, thus establish the same key, which is $prf_{k_{ij}}$. So the protocol $4WHS_{AM}$ satisfies the property 1of Definition SK-security.

Then we prove that the second property of Definition SK-security is also satisfied by protocol $4WHS_{AM}$. Assume there is a KE-adversary A in the AM against protocol $4WHS_{AM}$ that has a non-negligible advantage δ in guessing correctly b. We can construct an algorithm D that distinguishes pseudorandom and random function with non-negligible probability δ.

Let $Q_0=\{r,t,prf_k(r,t)\}$, and $Q_1=.\{r,t,random\ (\ )\}$ The input to D is denoted by $\{r,t,\gamma\}$ and is chosen from Q_0 or Q_1 each with probability $\frac{1}{2}$. Let L be an upper bound on the number of sessions invoked by A in any interaction. Algorithm D uses adversary A as a subroutine and is described as follows.

1. Choose $m\xleftarrow{R}.\{1,\ldots,l\}$
2. Invoke A, on a simulated interaction in the AM with parties $p_1,\ldots,p_n$ running $4WHS_{AM}$. Each of the parties shares $prf_{k_{ij}}(\ )$with the other one, except for those two in the m-th session, who share $prf_k(\ )$.
3. Whenever A activates a party to establish a new session (except for the m-th session) or to receive a message, D follows the instructions of $4WHS_{AM}$ on behalf of that party. When a party is corrupted or a session (other than the m-th session) is exposed, hands A all the information corresponding to that party or session as in a real interaction.
4. When the m-th session, say (p_i, p_j, s_m), is invoked within p_i, let p_i send the message (p_i, s_m, r) to p_j.
5. When p_j is invoked to receive (p_i, s_m, r), let p_j send the message (p_j, s_m, t) to p_i.
6. If session (p_i, p_j, s_m) is chosen by A as the test–session, then provide A with γ as the answer to this query.
7. If the m-th session (p_i, p_j, s_m) is ever exposed, or if a session different than the m-th session is chosen as the test–session, or if A halts without choosing a test-session then D outputs $b'\xleftarrow{R}\{0,1\}$ and halts.
8. If A halts and outputs a bit b', then D halts and outputs b' too.

The run of A by D (up to the point where A stops or D aborts A's run) is identical to a normal run of A against protocol $4WHS_{AM}$.

Consider the first case in which the m-th session is chosen by A to be tested and A get the response of γ. Thus, if the input to D came from Q_0 then the response was the actual value of the key. On the other hand, if the input to D came from Q_1 then the response to the test query was a random value. AS mentioned above, the input to D was chosen with probability 1/2 from Q_0 and Q_1. Then the distribution of responses provided by D to the test query of A is the same as specified by Definition SK-security. In this case, the probability that A guesses correctly whether the test value was "real" or "random" is $1/2+\delta$ for a non-negligible value δ. This is equivalent to guessing whether the input to the distinguisher D came from Q_0 or, Q_1 respectively. Thus, by outputting the same bit b' as A, we get that the distinguisher D guesses correctly the input distribution Q_0 or Q_1 with the same probability $1/2+\delta$ as A did.

Now consider the second case in which (p_i, p_j, s_m) is not chosen by A. In this case, D always halts and outputs a random bit, thus its probability to guess correctly the input distribution Q_0 or Q_1 is 1/2.

Since the first case happens with probability $\frac{1}{L}$,

while the second case happens with probability $1-\frac{1}{L}$,

the overall probability of D to guess correctly is

$$PR= (0.5+\delta) + \frac{1}{L} + 0.5 \times (1-\frac{1}{L}) = 0.5+\frac{\delta}{L}$$

Thus D succeeds in distinguishingfromwith non-negligible advantage, which is conflict to the Assumption that the *pseudorandom function is secure*. So the protocol $4WHS_{AM}$ satisfies the property 2 of Definition SK-security.

Thus the protocol $4WHS_{AM}$ is SK-secure without PFS in the AM. #

Authenticator λ_{prf}

Theorem 7. Assume that the pseudorandom function and MAC in use are secure against chosen message attacks. Then protocol λ_{prf} emulates protocol MT in unauthenticated networks. (Fan et al., 2007).

The Security Analysis of Four-Way Handshake Protocol in the UM

We have proved that the protocol $4WHS_{AM}$ is SK-Secure without PFS in the AM, and the protocol λ_{prf} is a MT-Authenticator, thus we get the result of security analysis of 4WHS in the UM.

Theorem 8. If the pseudorandom function and MAC function in use are secure against chosen message attacks, protocol four-way handshake is SK-Secure in the UM.

The Four-Way Handshake Protocol is UC-Secure

We have proved that 4WHS is SK-secure. According to Definition 6, now we prove that it has the ACK property, thus also satisfies the definition of UC-secure.

Theorem 9. The protocol 4WHS has the ACK property.

Proof. To prove the ACK property for 4WHS we construct the following internal state simulator *I*. Recall that before 4WHS actually generates output, the local state of the first party (p_i in the aforementioned description) consists of (k_1, k_2, s, p_i, p_j). The internal state of the other party (p_j in the aforementioned description) is identical (its internal state, like k_0, has been erased). The output of *I*, given (k_1, s, p_i, p_j) will be $l p_i = l p_j = (k_1, r_1, s, p_i, p_j)$, where r_1 is a random value of the same length as k_2. (Consequently, when the internal states of p_i and p_j are replaced with $l p_i$ and $l p_j$ respectively, the added protocol message will be computed and verified as $MAC_{R_1}(s, r_j)$ rather than $MAC_{K_2}(s, r_j)$. Next we proof that *I* is a good internal state simulator.

Le *F* be an ideal functionality which can securely realize key exchange and *A* be an adversary. If *I* is not a good internal state simulator, then the environment Z can distinguish between an interaction with *A* and 4WHS and an interaction with *A* and the above transformed protocol(replace the internal states of p_i and p_j with the outputs of *I*) with a non-negligible advantage . The only difference between the protocol resultant from the aforementioned transformation and 4WHS is the replacement of k_2 with r_1. So if *I* is not a good internal state simulator, then Z can distinguish between r_1 and k_2 with a non-negligible advantage. If the adversary can distinguish between k_2 and a random value with a non-negligible advantage, where $k_2 = second_{n_2}(k_0)$, then he/she can distinguish between k_0 and a random value with a non-negligible advantage. As we have proved that 4WHS is SK-secure, thus the adversary cannot distinguish between k_1 ($k_1 = first_{n_1}(k_0)$) and a random value with a non-negligible advantage, well then he/she cannot distinguish between k_0 and a random value with a non-negligible advantage, which reaches a contradiction. So the environment Z cannot distinguish between an interaction with (*A*, 4WHS) and (*A*, the transformed protocol) with a non-negligible advantage, thus we have $HYB^{F}_{\pi,A,Z} \approx HYB^{F}_{\pi,A,Z,I}$ and *I* is a good internal state simulator for 4WHS. According to Definition 6 and theorem 1, we know that 4WHS has the ACK property and is UC-secure. #

According to Theorems 8, 9, and 1, we get Theorem 10.

- **Theorem 10:** If the pseudorandom function and MAC function in use are secure against chosen message attacks, protocol four-way handshake is UC-Secure. #

THE SECURITY ANALYSIS OF CHINESE WLAN SECURITY STANDARD WAPI WITH THE CK MODEL

The Chinese WLAN standard WAPI (GB 15629.11-2003) (National Standard of the People's Republic of China, 2003), the first issued Chinese standard in the field of WLAN, has been formally implemented since November 1, 2003. WAPI is composed of two parts: WAI and wireless privacy infrastructure (WPI). They realize the identity authentication and data encryption, respectively. In March of 2004, China IT Standardization Technical Committee drafted out a new version, WAPI implementation plan (National Standard of the People's Republic of China, 2004), which improves the original standard WAPI. Compared with the original standard, the greatest change the implementation plan made lies in the WAI module.

As a national standard which is about to be deployed and implemented on a large scale, its security is undoubtedly the focus. But as far as we know, up to now, there are no articles that systemically analyze the security of WAPI and its implementation plan, which is imperfect for a national standard. This contribution discusses the security of WAPI and its implementation plan with the CK model. It has three contributions: (1) the security weaknesses of WAI in WAPI are given; (2) the WAI module in the implementation plan is proved secure in the CK model; and (3) how the implementation plan overcomes the security weaknesses of the original WAPI is pointed out. The analysis results can help us understand the necessity of the implementation plan and enhance the confidence of it. At the same time, as a case study, their analysis is helpful for the design of a secure key-agreement protocol.

WAIs in WAPI and its Implementation Plan

WAI adopts port-based authentication architecture that is identical with IEEE 802.1X. The whole system is composed of mobile guest STA, access point (AP), and authentication service unit (ASU).

WAI in WAPI

The interaction procedure of WAI in the original national standard WAPI is shown in Figure 4. From this figure, we can see that WAI is composed of two parts: certificate authentication and key agreement.

Figure 4. WAI in WAPI

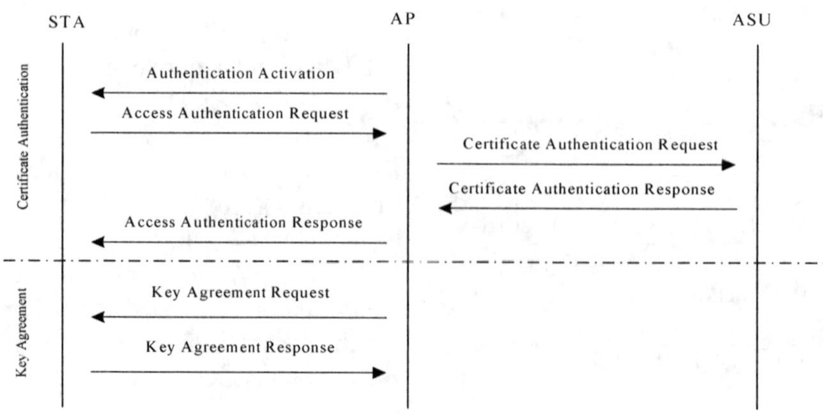

1. **Certificate authentication.** In this process, station (STA) sends its public key certificate and access request time to the access point (AP) in the access authentication request. AP sends its certificate, STA's certificate, STA's access request time, and its signature on them to authentication service unit (ASU) in certificate authentication request. After ASU validates AP's signature and the two certificates, it sends the certificates validation result, STA's access request time, and ASU's signature on them to STA and AP.
2. **Key agreement.**

Figure 5. The key agreement in the WAI of WAPI

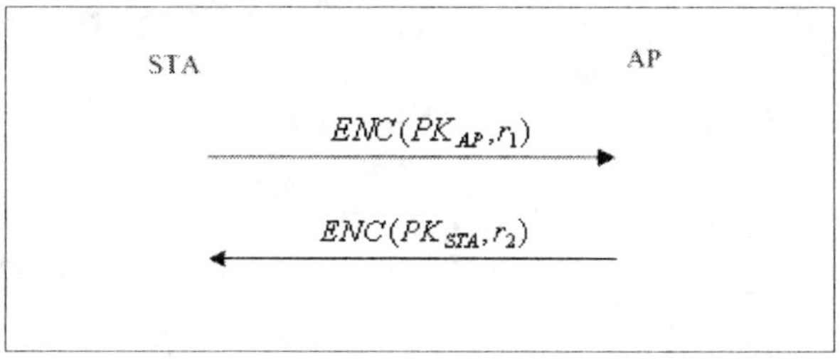

First, STA and AP negotiate the cryptography algorithms. Then, they respectively generate one random value r_1 and r_2. These random values are encrypted with the peer's public key and sent to each other. Both parties decrypt the encrypted random values and derive the session key $K=r_1 \oplus r_2$. The key agreement process is shown in Figure.5, where *ENC*( )is the encryption function, PK_{AP} and PK_{STA} are AP and STA's public key respectively.

WAI in the Implementation Plan

In the framework, WAI in the implementation plan is the same as that of the original WAPI, and it is also composed of certificate authentication and key agreement. Compared with the original standard WAPI, the implementation plan remains unchanged in the certificate authentication, but makes rather big improvement in the key agreement. The new key-agreement protocol is shown in Figure.6. It is different from the original one in the following points:

1. In the implementation plan, the key agreement request has to be initiated by AP. At the same time, the secure parameter index SPI, AP's signature on the encrypted random value and SPI are included in this request. The signature algorithm is ECDSA.
2. In the key agreement response, SPI and the STA's MAC on encrypted random and SPI are included. The MAC is computed through HMAC-SHA256 algorithm.
3. The keys derivation method is different. STA and AP first calculate the host key $k= r_{11} \oplus r_2$, then extend k with KD-HMAC-SHA256 algorithm to get the session key k_d, the authentication key k_a and integration check key.

The Security Weaknesses of WAI in WAPI

The WAI module in the original WAPI has several security weaknesses as follows:

1. **Its key-agreement protocol cannot resist the unknown key-share (UKS) attack (Burton & Kaliski, 2001).**

We assume that an attacker *E* gets a certificate where his/her public key PK_E is same as PK_{STA}. (In many practical settings, the certificate authority [CA] does not require a proof-of-possession of the corresponding private key from a registrant of a public key (Krawczyk, 2005), so an attacker *E* can get a certificate from the CA in which his/her public key is same as STA's.) In addition, in the certificate authentication process, ASU just verifies the authenticity and validity of a certificate, so *E* also can pass the certification authentication. Then he/she can launch the unknown-key share attack in the key agreement. When STA sends the first message $ENC(PK_{AP}, r_1)$, *E* forwards this message to AP and claims that this message is from *E*. Then AP replies with $ENC(PK_E, r_2)$. *E* forwards this message to STA. When the protocol completes, STA thinks that he/she agreed upon a key with AP, while AP thinks that he/she negotiated a key with *E*. And these two keys are same. So, the attacker *E* succeeds in the UKS attack.

Figure 6. The key-agreement protocol in WAI of the implementation plan

STA AP

$\longleftarrow$ $SPI, ENC(PK_{STA}, r_1), Sig_{AP}(SPI, ENC(PK_{STA}, r_1))$

$\longrightarrow$ $SPI, ENC(PK_{AP}, r_2), HMAC-SHA256_{K_a}(SPI, ENC(PK_{AP}, r_2))$

SPI=the MAC of the STA||the BSSID of the AP||the time of authentication request

Let us analyze this attack in the CK model. In the previous attack, the KE-adversary chooses the session in STA as the test session and expose the session in AP (because these two sessions are not matching sessions, the session in AP can be exposed). Because STA and AP get a same session key, the KE-adversary can completely get the session key of the test session. According to Definition 2, this protocol is not SK-secure. And Diffie et al. (1992) can be referred to for the consequences of this attack.

2. **Its key agreement protocol cannot resist key-compromise impersonation (KCI) attack.**

Let us analyze this attack in the CK model. First, we assume that STA's private key is compromised and the attacker chooses the session in STA as the test session after STA complete the matching sessions with AP. The attacker can first corrupt another mobile guest STA' and impersonates him/her to send message $ENC(PK_{AP}, r_1)$ to AP. We denote the session between STA' and AP as SID'. When AP receives this message from STA', he/she chooses another random value r_3 and responds with $ENC(PK_{STA'}, r_3)$. AP computes its session key of SID' $k' = r_1 \oplus r_3$. The attacker can expose this session and get k' (this session is not the matching session of the test session). In addition, the attacker can decrypt $ENC(PK_{STA'}, r_3)$ to get r_3. Thus he/she can get $r_1 = k' \oplus r_3$. In addition, the attacker can also decrypt $ENC(PK_{STA}, r_2)$ to get r_2. Then he/she can get the session key of the test session: $k = r_1 \oplus r_2$. Thus the attacker can impersonate AP to STA. According to Definition 2, this protocol is not SK-secure.

3. **It does not realize the explicit identity authentication of STA and perhaps lead to the faulty charge.**

From the WAI process, we can see that it does not realize the explicit identity authentication of STA to AP. An attacker can pass the certificate authentication and access the networks only if he/she gets a legal user's certificate, which will lead to the faulty charge if the networks charge the fee according to the access time.

The Security Analysis of WAI in the Implementation Plan

In the certificate authentication, AP makes signature in the certificate authentication request, and ASU makes signature in the certificate authentication response. Both these signatures include STA's access request time which ensures the freshness of the signatures. Therefore ASU can authenticate AP's identity and STA can authenticate ASU's identity. In addition, STA trusts ASU. So STA can authenticate the identity of AP after the certificate authentication. At the same time, AP authenticates the certificate provided by STA.

The key-agreement protocol in WAI of implementation plan is denoted by π. In the following, we will prove that π is SK-secure without PFS (Güther, 1990). That is, the protocol is SK-secure, but does not provide perfect forward secrecy of the session keys. In order to prove that π is SK-secure, we define a "game" as follows.

The Design of an Encryption Game

Let (*G, ENC, DEC*) be a key-generation, encryption and decryption algorithm, respectively, of a public-key encryption scheme that is secure against CCA2 attack (Wenbo, 2004). Let *K* be the security parameter. STA and AP have invoked *G*(*K*) to get their public and private key pairs.

This game integrates the CCA2-security of *ENC* with the key-agreement protocol (Canetti & Krawczyk, 2001; Wenbo, 2004). We will proceed to show that if an attacker can break the SK-security of p , then he/she can win the game, that is, he/she can break the CCA2-security of *ENC*.

The two participants in the game are $\mathcal{G}$ and $\mathcal{B}$ (for good and bad). $\mathcal{G}$ is the party against which $\mathcal{B}$ plays the game. $\mathcal{G}$ acts as a decryption oracle. $\mathcal{G}$ possesses a pair of public and private keys, PK_{STA} and SK_{STA} (generated via the key generation algorithm $\mathcal{G}$). $\mathcal{B}$ is the attacker of protocol π, he/she knows PK_{STA} but not SK_{STA}. He/she leverages the abilities he/she gets in the attack of π to take part in this game. The game process is shown in the following:

Phase 0: $\mathcal{G}$ provides $\mathcal{B}$ with a challenge ciphertext $c^*=ENC(PK_{STA},r_1)$ for $r_1 \xleftarrow{R} \{0,1\}^K$.

Phase 1: $\mathcal{B}$ sends a triple (*c, r, t*) to $\mathcal{G}$ who responds with HMAC- $SHA256_{k_a}(t)$.

($k_a' = last$(KD-HMAC-SHA256(k')), $k' = r \oplus r', r' = DEC(SK_{STA}, c)$. The *last*( ) is a function that extract out the last 16 bytes from a bit string.) This is repeated a polynomial number of times with each triple being chosen adaptively by $\mathcal{B}$ (i.e., after seeing $\mathcal{G}$'s response to previous triple), but he/she keeps *r* unchanged in every triple.

Phase 2: $\mathcal{B}$ sends a test string $t^*=(\text{SPI}\|PK_{AP}(r))$ to $\mathcal{G}$. Then $\mathcal{G}$ chooses a random bit

$b \xleftarrow{R} \{0,1\}$. If b=0 then $\mathcal{G}$ responds with HMAC-SHA256$_{k_a''}(t^*)$where $k_a'' = last$

(KD-HMAC-SHA256(k'') , $k'' = r_1 \oplus r$, r_1 is the value encrypted by $\mathcal{G}$ in phase 0. If b=1 then $\mathcal{G}$ responds with a random string s^* of the same length as HMAC-SHA256$_{k_a''}(t^*)$.

Phase 3: Same as Phase 1.

Phase 4: $\mathcal{B}$ outputs a bit b' as the guess of *b*.

And the winner is…$\mathcal{B}$ if and only if $b=b'$.

The following notes are made about the game:

1. The challenging ciphertext c^* in the phase 0 is also the ciphertext sent by AP in the key agreement request of π.
2. In Phase 1, $\mathcal{B}$ randomly chooses a test ciphertext *c*, random value *r* and string *t*, and sends them to $\mathcal{G}$ for process. It should be noticed that $\mathcal{B}$ cannot simultaneously chooses c^* and t^* as the input of $\mathcal{G}$.
3. $\mathcal{B}$ keeps *r* unchanged in every triple in order to reduce the difficulty of the attack.

Security Analysis of Key-Agreement Protocol in WAI

According to Definition 2, in order to prove that π is SK-secure, we have to argue that it can meet two requirements. The first one is that STA and AP can get a same session key after they complete matching sessions. The second one is that $\mathcal{B}$ cannot distinguish the session key k_d from a random value with a non-negligible advantage. In the following, we will prove that π can meet these two requirements.

Lemma 1. *If the encryption scheme ENC is secure against the CCA2 attack, then at the end of protocol π, STA and AP will complete matching sessions and get a same session key.*

Proof. Since the signature algorithm ECDSA is secure against existential forgery by adaptive chosen-message attack (Brown, 2001), in addition, SPI in the key agreement request can guarantee the freshness of this message and bind this message with the two communication parties, the attacker cannot forge or modify the request message.

In addition, the attacker $\mathcal{B}$ cannot forge a key agreement acknowledgment message. Let us prove this with the reduction to absurdity. It is assumed that the attacker can forge an acknowledgment message with a non-negligible probability during the run of the protocol π. That is, he/she can choose a random value (say r_3) and forge a message authentication code that AP can validate. Then $\mathcal{B}$ takes advantage of this ability to run the game above. In Phase 1, he/she also chooses r_3 as the random value r in the triple, while selects c and t randomly. Then, in Phase 2, he/she can work out HMAC-SHA256$_{k_a^*}(t^*)$ because this value is same as the forged message authentication code in the key agreement acknowledgment. Therefore the attacker can distinguish HMAC-SHA256$_{k_a^*}(t^*)$ from s^* and guess correctly b in Phase 4, thus wins the game, which indicates that the encryption scheme is not CCA2-secure. This contradicts with the presupposition. So during the run of the protocol π, the attacker cannot forge a key agreement acknowledgment with a non-negligible probability.

Therefore STA and AP will complete matching sessions and get a same session key at the end of protocol π, if ENC is CCA2-secure. #

Lemma 2. *If the encryption scheme ENC is secure against the CCA2 attack, the attacker cannot distinguish the session key k_d from a random value with a non-negligible advantage.*

Proof. It is assumed that the attacker $\mathcal{B}$ can distinguish the session key k_d from a random value with a non-negligible advantage η_1. In the CK model, the KE-attacker is not permitted to corrupt the test session or its matching session, so the attacker $\mathcal{B}$ cannot directly get the session key k_d from the attack of π. While k_d=*first*(KD-HMAC-SHA256(k)) (The *first*( ) is a function that extracts out the first sixteen bytes from a bit string), so the attacker $\mathcal{B}$ has only two possible methods to distinguish k_d from a random value. The first one: $\mathcal{B}$ learns k. The second one: $\mathcal{B}$ succeeds in forcing the establishment of a session (other than the test session or its matching session) that has the same key as the test session. In this case $\mathcal{B}$ can learn the test session key by simply querying the session with the same key, and without having to learn the value k. In the following, we prove that neither of these two methods is feasible.

The first method means that, from the attack of π, the attacker can distinguish $k'' = r_1 \oplus r$ from a random value with a non-negligible advantage. Based on this ability, $\mathcal{B}$ also can distinguish $k'' = r_1 \oplus r$ from a random value with a non-negligible advantage. This is because r in the k'' is selected by the attacker himself, which makes the difficulty that he/she distinguishes k'' from a random value no bigger than that he/she distinguishes k from a random value. It is assumed that the advantage that $\mathcal{B}$ distinguishes k'' from a random value is η_2, then $\eta_2 \geq \eta_1$. And because $k_a'' = last$(KD-HMAC-SHA256(k'')), $\mathcal{B}$ can get k_a''. Further, he/she can work out HMAC-SHA256$_{k_a^*}(t^*)$ with a non-negligible probability, which enables the attacker to win the encryption game. That means the encryption scheme is not secure against CCA2 attack. This contradicts the presupposition. So the attacker $\mathcal{B}$ can not get k with a non-negligible probability. Then this method is not practical.

As for the second method, there are two strategies that the attacker can take. (1) After STA and AP complete the matching sessions, the attacker $\mathcal{B}$ establishes a new session with AP or STA. But the session key of this session will not be k_d, because the encrypted random value is chosen randomly by AP or STA. (2) When AP and STA perform the key agreement, $\mathcal{B}$ intervenes this negations and makes them get a same session key without the completion of the matching sessions. That is, STA and AP get a same session key but they do not complete matching sessions. Then the attacker can get the test session key by breaking the unmatching session that has the same session key. But from Lemma 1, we know that if the encryption scheme ENC is secure against the CCA2 attack, $\mathcal{B}$ cannot succeed in this intervention. So this method is not feasible either.

Let us sum up the previous analysis. The attacker $\mathcal{B}$ neither can get the host key k, nor can he/she force to establish a new session with STA or AP that has the same session key as the test session. So the attacker cannot distinguish the session key k_d from the random value with a non-negligible advantage. #

Theorem 11. *If the encryption scheme ENC adopted is secure against CCA2 attack, then π is SK-secure without PFS.*

Proof. According to Lemma 1 and Lemma 2, we know that STA and AP will get a same session key after the key agreement and the attacker cannot distinguish the session key from a random value with a non-negligible advantage. Then in accordance with Definition 2, the protocol π is SK-secure.

In addition, if the private keys of STA and AP are compromised, the attacker can get the random values exchanged and can work out all the session keys that have been agreed about. Thus this protocol cannot provide PFS. So we can get that the key-agreement protocol is SK-secure without PFS. #

The Implementation Plan Overcomes the Weaknesses of the Original WAPI

We know that WAI in the original WAPI has some security weaknesses. But WAI in the implementation plan is secure in the CK model, and according to Li et al. (2005), we get that the WAI module of the implementation plan can resist KCI attack and UKS attack. In the following, we will analyze how the implementation plan overcomes the security weaknesses in the original WAPI.

1. The key-agreement protocol in the implementation plan can resist UKS attack. In the implementation plan, even though the attacker $\mathcal{B}$ gets a certificate in which his/her public key is the same as STA's or AP's, he/she cannot launch the UKS attack. Because the implementation plan requires that the key agreement request be sent by AP, STA just accepts the request from AP. So, $\mathcal{B}$ can just launch the UKS attack against the AP (i.e., AP thinks that he/she agrees upon a key with $\mathcal{B}$, but in fact he/she negotiates a key with STA, while STA correctly thinks that he/she negotiates a key with AP), that is, $\mathcal{B}$ just can forward the key agreement request message for him/her to STA. But in this request, AP's signature includes SPI which includes the MAC address of the $\mathcal{B}$, so STA will not accept this request forwarded from $\mathcal{B}$. Therefore the key-agreement protocol in WAI of implementation plan can resist the UKS attack.

From the previous analysis, we can see that the essential reasons that WAI in the implementation plan can resist the UKS attack are that: (1) the implementation plan requires that the key agreement request be sent from AP; (2) AP's signature includes SPI which includes the destination entity's address.

2. The key-agreement protocol in the WAI of the implementation plan can resist the KCI attack. KCI attacks for the protocol π have two manners. The first one is that AP's private key is compromised and the attacker can impersonate STA to AP. The second one is that STA's private key is compromised and the attacker can impersonate AP to STA. In the following, we will discuss these two cases respectively.

If AP's private key is compromised, the attacker can decrypt $ENC(PK_{AP}, r_2)$ to get r_2. In order to get r_1, he/she just has two possible methods: (1) attacks the encryption algorithm *ENC*; and (2) impersonates other entity to establish another session with STA, and sends $ENC(PK_{STA'}, r_1)$ to STA, then the attacker exposes this session and gets r_1 through some computations. But neither of these two methods is feasible. For the first method, we know that if the encryption algorithm *ENC* is CCA2 secure, the attacker cannot get r_1 from the attack of this algorithm directly. As for the second method, the implementation plan requires the key agreement request be sent by AP, and the attacker cannot forge AP's signature, so the attacker cannot impersonate other entity to establish another session with STA. Therefore the attacker cannot get r_1. Then he/she still cannot get the host key k and session key k_d.

If STA's private key is compromised, the attacker can decrypt $ENC(PK_{STA'}, r_1)$ to get r_1. In order to get session key r_2, he/she just has two possible methods: (1) attacks the encryption algorithm *ENC* directly to get r_2; and (2) impersonates another mobile guest STA' to establish a new session with AP and sends it $ENC(PK_{AP}, r_2)$ in the key agreement acknowledgement. From the previous analysis we get that the first method is infeasible. As for the second method, because r_2 and the host key k are just the ephemeral values, we assume that they are not the session states of AP. Therefore, the session states of the new session in AP are just the session key k_d^*, the message authentication key k_a^* and the message integration key. The attacker cannot get any information about r_2 from these session states because these three keys are the hash values of the host key k^*. Therefore the attacker cannot get r_2 either. (If the session key is not the hash value of k^*, the attacker can get k^*, futher can get r_2.) So the attacker still cannot get the host key k and the session key k_d.

As a whole, the essential reasons that the key-agreement protocol can resist KCI attack are that: (1) the implementation plan requires that the key agreement request be sent by AP; and (2) the session key in the implementation plan is derived through the hash function.

(3) The WAI module in the implementation plan realizes the mutual explicit identity authentication between STA and AP, which can withstand faulty charge. For AP, π is an explicit key authentication protocol. So AP can authenticate the identity of STA at the end of WAI. At the same time, STA can authenticate the identity of AP in the certificate authentication. Therefore WAI in the implementation plan realizes the mutual explicit identity authentication between AP and STA. Therefore it can withstand faulty charge.

FUTURE TRENDS

In the future, possible research "hot" points in formal analysis of key-agreement protocol include: (1) decrease in the basic assumptions of the protocol, such as the "perfect" cryptography assumptions, free encryption assumptions; such that the theory research is closer to the practice; (2) extension of the protocol analysis scope; (3) enhancement of the analysis capability of "protocol composition," which is the "hot" and difficult point; (4) integration of the characters of different analysis methods, such as the comparison and combination of CSP model, string space model, model check method, and linear logic methods; (5) the research in automatic generation and check of security protocol; (6) the research in the case that the party number is indefinitely increased; (7) solution to "state exploration" problem in the model check methods; and (8) the research in new areas, such as the DoS attack.

CONCLUSION

In this chapter we focused on the provably secure formal methods for the key-agreement protocols, especially the CK model and universally composable security model. First, these two models are introduced; then we gave a study of these two models. An analysis of CK model presented its security analysis, advantages, and disadvantages, and a bridge between this formal method and the informal method (heuristic method) is established; an extension of UC security model gives a universally composable anonymous hash certification model. Next, with the four-way handshake protocol in 802.11i and the Chinese WLAN security standard WAPI, we give the application of these two models. At last, the future trend of formal analysis method of key-agreement protocol was presented.

REFERENCES

Aboba, B., & Simon, D. (1999). *PPP EAP TLS authentication protocol* (RFC 2716). Retrieved from http://www.ietf.org/rfc/rfc2716.txt

Bellare, M., & Rogaway, P. (1993). Random Oracle are practical: A paradigm for designing efficient protocols. In *Proceedings of the First ACM Conference on Computer and Communications Security.*

Bellare, M., & Rogaway, P. (1995). Provably secure session key distribution: The three party case. In *Proceedings of the 27th ACM Symposium on the Theory of Computing—STOC 1995* (pp. 57-66). ACM Press.

Bellare, M., Canetti, R., & Krawczyk, H. (1998). A modular approach to the design and analysis of authentication and key-exchange protocols. In *Proceedings of the 30th Symposium on the Theory of Computing, STOC 1998* (pp. 419-428).

Birgit, P., & Michael, W. (2001, May). A model for asynchronous reactive systems and its application to secure message transmission. In *Proceedings of the IEEE Symposium on Security and Privacy*, Oakland, CA (pp. 184-200).

Blake-Wilson, S., Johnson, D., & Menezes, A. (1997). Key agreement protocols and their security analysis. In *Proceedings of the sixth IMA international Conference on Cryptography and Coding.*

Borisov, N., Goldberg, I., & Wagner, D. (2001). Intercepting mobile communications: The insecurity of 802.11. In *Proceedings of the 7th Annual International Conference on Mobile Computing and Networking,* Italy.

Brown, D. R. L. (2001). *The exact security of ECDSA* (IEEE 1363).

Burrows, M., Abadi, M., & Needham, R. M. (1990). A logic of authentication. *ACM Transactions on Computer Systems, 8*(1), 122-133.

Burton, S., & Kaliski, J. R. (2001). An unknown key-share attack on the MQV key agreement protocol. *ACM transactions on Information and System Security, 4*(3), 275-288.

Canetti, R. (2001). Universally composable security: A new paradigm for cryptographic protocols. In *Proceedings of the 42th IEEE Annual Symposium on Foundations of Computer Science* (pp. 136-145).

Canetti, R. (2004). Universally composable signature, certification, and authentication. In *Proceedings of 17th IEEE computer security foundations workshop (CSFW)* (pp. 219-245). IEEE Computer Society Press.

Canetti, R., & Krawczyk, H. (2001). Analysis of key exchange protocols and their use for building secure channels. In B. Pfitzmann (Ed.), *Advances in cryptology—EUROCRYPT 2001* (LNCS 2045, pp. 453-474) Berlin, Germany: Springer-Verlag.

Canetti, R., & Krawczyk, H. (2002). Universally composable notions of key exchange and secure channels. In *Proceedings of Eurocrypt 2002.*

Changhua, H., & Mitchell, C. J. (2004, October 1). Analysis of the 802.11i 4-way handshake. In *Proceedings of ACM Workshop on Wireless Security, WiSe'04,* Philadelphia, PA.

Choo, K. K. R, & Hitchcock, Y. (2005). Security requirement for key establishment proof models: Revisiting Bellare-Rogaway and Jeong-Katz-Lee protocols. In *Proceedings of the 10th Australasian Conference on Information Security and Privacy—ACISP.*

Diffie, W., & Hellman, M. (1976). New directions in cryptography. *IEEE Transactions on Information Theory, 22,* 644-654.

Diffie, W., Van Oorschot, P., & Wiener, M. (1992). Authentication and authenticated key exchanges. *Designs, Codes and Cryptography, 2,* 107-125.

Fan, Z., JianFeng, M., & Moon, S. (2007). A universally composable anonymous hash certification model. *Science in China (F serial) 50*(3), 440-445.

Güther, C. G. (1990). An identity-based key-exchange protocol. In *Advances in Cryptology-EUROCRYPT'89* (LNCS 434, pp. 29-37). Springer-Verlag.

IEEE 802.1X-2001. (2001). *IEEE standard for local and metropolitan area networks—Port-based network access control.*

IEEE P802.11i D3.0. (2002). *Specification for enhanced security.*

Krawczyk, H. (1996, February). SKEME: A versatile secure key exchange mechanism for Internet. In *Proceeding of the 1996 Internet Society Symposium on Network and Distributed System Security* (pp. 114-127).

Krawczyk, H. (2005). HMQV: A high-performance secure Diffie-Hellman protocol. In *Advances in Cryptology–CRYPTO 2005: 25th Annual International Cryptology Conference* (LNCS 3621, pp. 546-566). Springer-Verlag.

Law, L., Menezes, A., Qu, M., Solinas, J., & Vanstone, S. (1998). An efficient protocol for authenticated key agreement (Tech. Rep. CORR 98-05). Ontario, Canada: University of Waterloo, Department of Combinatorics & Optimization.

Li, X., Ma, J., & Moon, S. (2005). On the security of Canetti-Krawczyk model. (LNAI 3802, pp. 356-363). Springer-Verlag.

Martin, A., & Phillip, R. (2002). Reconciling two views of cryptography. *Journal of Cryptology, 15*(2), 103-127.

Meadows, C. (1996). Formal verification of cryptographic protocols: A survey. In *Proceedings of the Advances in Cryptology, Asiacrypt'96* (LNCS1163, pp. 135-150). Springer-Verlag.

Menezes, A., Van Oorschot, P., & Vanstone, S. (1996). Handbook of applied cryptography. In chapter 12. CRC Press.

Michael, B., & Dennis, H. (2004). How to break and repair a universally composable signature functionality. In *Information security conference—ISC 2004* (LNCS 3225, pp. 61-74).

Mitchell C. J., Ward M., & Wilson, P. (1998). Key control in key agreement protocols. *Electronics Letters, 34,* 980-981.

National Standard of the People's Republic of China. (2003). Information technology—Telecommunications and information exchange between systems—Local and metropolitan area networks—Specific requirements—Part 11: Wireless LAN medium access control (MAC) and physical layer (PHY) specifications (GB 15629.11-2003).

National Standard of the People's Republic of China. (2004). Guide for GB 15629.11-2003 Information technology—Telecommunications and information exchange between systems—Local and metropolitan area networks—Specific requirements—Part 11: Wireless LAN medium access control (MAC) and physical layer (PHY) specifications.and GB 15629.1102-2003 Information technology—Telecommunications and information exchange between systems—Local and metropolitan area networks—Specific requirements—Part 11: Wireless LAN medium access control (MAC) and physical layer (PHY) specifications: Higher-speed physical layer extension in the 2.4 GHz band.

Rigney, C., Willens, S., Rubens, A., & Simpson, W. (2000). *Remote authentication dial in user service (RADIUS)* (RFC 2865). Retrieved from http://www.ietf.org/rfc/rfc2865.txt

Shoup, V. (1999). *On formal models for secure key exchange.* Theory of Cryptography Library. Retrieved from http://citeseer.ist.psu.edu/cache/papers/cs2/769/http:zSzzSzeprint.iacr.orgzSz1999zSz012.pdf/shoup99formal.pdf

Tin, Y. S. T., Boyd, C., & Nieto, J. G. (2003). Provably secure key exchange: An engineering approach. In *Australasian Information Security Workshop 2003(AISW 2003)* (pp. 97-104).

Wenbo, M. (2004). *Modern cryptography: Theory and practice.* Prentice-Hall, PTR.

Yehuda, L. (2003). Composition of secure multi-party protocols—A comprehensive study (LNCS, 2815). Springer-Verlag.

KEY TERMS

Acknowledgment (ACK) property: Let $\mathcal{F}$ an ideal functionality and let π be an SK-secure KE protocol in the $\mathcal{F}$-hybrid model. An algorithm I is said to be an internal state simulator for π if for any environment machine Z and any adversary $\mathcal{A}$ we have $\mathrm{HYB}^{F}_{\pi\,\pi,A,Z} \approx \mathrm{HYB}^{F}_{\pi,A,Z,I}$

Protocol π is said to have the ACK property if there exists a good internal state simulator for π.

Composition Theorem: The key advantage of UC security is that we can create a complex protocol from already designed sub-protocols that securely achieves the given local tasks. This is very important since complex systems are usually divided into several sub-systems, each one performing a specific task securely. Canetti presented this feature as the composition theorem (Canetti, 2001). This theorem assures that we can generally construct a large size "UC-secure" cryptographic protocol by using sub-protocols which is proven as secure in UC-secure manner.

Explicit Key Authentication: Explicit key authentication is the property obtained when both (implicit) key authentication and key confirmation hold.

(Implicit) Key Authentication: (Implicit) key authentication is the property whereby one party is assured that no other party aside from a specifically identified second party (and possibly additional identified trusted parties) may gain access to a particular secret key.

Key-Agreement Protocol: A key-agreement protocol or mechanism is a key establishment technique in which a shared secret is derived by two (or more) parties as a function of information contributed by, or associated with, each of these, (ideally) such that no party can predetermine the resulting value.

Key Confirmation: Key confirmation is the property whereby one party is assured that a second (possibly unidentified) party actually has possession of a particular secret key.

Session-Key Security: *A KE protocol* π *is called Session-key secure (or SK-secure) if the following properties hold for any KE-adversary* M:

1. *Protocol* π *satisfies the property that if two uncorrupted parties complete matching sessions then they both output the same key; and*

2. *The probability that* $\mathcal{M}$ *distinguishes the session key from a random value is no more than 1/2 plus a negligible fraction* ε *in the security parameter.* ε *is called "advantage".*

Universally Composable (UC) Security: In UC framework, real protocol π securely realizes ideal functionality if $\mathcal{F}$, for $\forall$ $\mathcal{A}$ and $\forall$ $\mathcal{Z}$, it has the same "action" as $\mathcal{F}$,. Formally, a protocol π securely realizes an ideal functionality $\mathcal{F}$ is for any real-life adversary $\mathcal{A}$ there exists an ideal-process adversary $\mathcal{S}$ such that, for any environment $\mathcal{Z}$ and on any input, the probability that $\mathcal{Z}$ outputs "1" after interacting witha $\mathcal{A}$ and parties running π in the real-life model differs by at most a negligible fraction from the probability that $\mathcal{Z}$ outputs "1" after interacting with $\mathcal{S}$ and $\mathcal{F}$ in the ideal process.

Chapter XVI
Multimedia Encryption and Watermarking in Wireless Environment

Shiguo Lian
France Telecom R&D Beijing, China

ABSTRACT

In a wireless environment, multimedia transmission is often affected by the error rate; delaying; terminal's power or bandwidth; and so forth, which brings difficulties to multimedia content protection. In the past decade, wireless multimedia protection technologies have been attracting more and more researchers. Among them, wireless multimedia encryption and watermarking are two typical topics. Wireless multimedia encryption protects multimedia content's confidentiality in wireless networks, which emphasizes on improving the encryption efficiency and channel friendliness. Some means have been proposed, such as the format-independent encryption algorithms that are time efficient compared with traditional ciphers; the partial encryption algorithms that reduce the encrypted data volumes by leaving some information unchanged; the hardware-implemented algorithms that are more efficient than software based ones; the scalable encryption algorithms that are compliant with bandwidth changes; and the robust encryption algorithms that are compliant with error channels. Compared with wireless multimedia encryption, wireless multimedia watermarking is widely used in ownership protection, traitor tracing, content authentication, and so forth. To keep low cost, a mobile agent is used to partitioning some of the watermarking tasks. To counter transmission errors, some channel encoding methods are proposed to encode the watermark. To keep robust, some means are proposed to embed a watermark into media data of low bit rate. Based on both watermarking and encryption algorithms, some applications arise, such as secure multimedia sharing or secure multimedia distribution. In this chapter, the existing wireless multimedia encryption and watermarking algorithms are summarized according to the functionality and multimedia type; their performances are analyzed and compared; the related applications are presented; and some open issues are proposed.

INTRODUCTION

With the development of multimedia technology and network technology, multimedia data are used more and more widely in human's daily life, such as mp3 sharing, video conference, video telephone, video broadcasting, video-on-demand, p2p streaming, and so forth. For multimedia data may be in relation with privacy, profit, or copyright, multimedia content protection becomes necessary and urgent. It permits that only the authorized users could access and read the multimedia data, it can detect the modification of the multimedia data, it can prove the ownership of the multimedia data, it can even trace the illegal distribution of the multimedia data, and so forth.

During the past decades, some means have been proposed to protect multimedia data. Among them, multimedia encryption (Furht & Kirovski, 2006) and multimedia watermarking (Cox, Miller, & Bloom, 2002) are two typical ones. Multimedia encryption algorithms protect multimedia data's confidentiality by encoding or transforming multimedia data into unintelligible forms under the control of the key. Thus, only the authorized users who have the correct key can recover the multimedia data successfully. Till now, some multimedia encryption algorithms have been proposed, which focus on the security, time efficiency, and communication friendliness (Zeng, Zhuang, & Lan, 2004). Multimedia watermarking algorithms protect multimedia data's ownership by embedding ownership information into multimedia data under the control of the key. Thus, the authorized users can extract or detect the ownership information and authenticate it. Many watermarking algorithms (Barni & Bartolini, 2004) have been proposed during the last decade, which consider security, imperceptibility, robustness and capacity, and so forth.

Recently, mobile/wireless multimedia communication has become more and more popular, which benefits from the improvement of the capability of mobile terminals and the bandwidth of wireless channel. Compared with wired communication, wireless multimedia communication has some special properties (Salkintzis & Passas, 2005). Firstly, the bandwidth is still limited compared with wired channels. Secondly, there are many more transmission errors in wireless communication, such as channel error, loss, delay, jitter, and so forth, which are caused by path error, fading, noise or interference, and so forth. Thirdly, wireless or mobile terminals are often of limited memory. Fourthly, the terminals are often energy-constraint caused by the scale-limited battery. These properties push some requirements to multimedia encryption and watermarking algorithms.

To meet mobile/wireless multimedia content protection, some mobile digital rights management (DRM) systems (Kundur, Yu, & Lin, 2004) have been proposed, such as Nokia's Music Player, NEC VS-7810, Open Mobile Alliance (OMA), and so forth. In these systems, multimedia encryption and multimedia watermarking are two core technologies. Compared with wired environment, wireless multimedia encryption and watermarking should consider some extra requirements. For example, the algorithms should be lightweight in order to meet the constraint energy of the terminals. Additionally, the algorithms should be robust against transmission errors in some extent. Furthermore, the algorithms should be scalable to switch between wireless services and wired services.

During the past decade, some means have been proposed to make suitable wireless multimedia encryption and watermarking algorithms. These algorithms obtain the security, efficiency, and error robustness by considering the properties of wireless/mobile multimedia communication. In this chapter, they are classified into several types according to the functionalities, and their performances are analyzed and compared. Additionally, some open issues are presented.

The rest of the chapter is arranged as follows. In the next section, the requirements of wireless/mobile multimedia encryption and watermarking are presented respectively. The multimedia encryption algorithms are analyzed and compared in the third section, and the watermarking algorithms are analyzed and compared in the fourth section. In the fifth section, some research topics and applications based on the combination of watermarking and encryption are presented, followed by some open issues in the sixth section. Finally, in the last section, some conclusions are drawn.

GENERAL REQUIREMENTS OF MULTIMEDIA CONTENT PROTECTION

Requirements of Multimedia Encryption

Multimedia data are often of high redundancy, large volumes, real time operations, and the compressed data are of certain format. These properties require that wireless multimedia encryption algorithms should satisfy some requirements (Furht & Kirovski, 2006), such as content security, time efficiency, format compliance, and so forth. All of them are presented in detail as follows.

Security. In multimedia encryption, the security refers to content security. It is composed of two aspects, that is, cryptographic security and perceptual security. The former one refers to the security against such cryptographic attacks as brute-force attack, ciphertext-only attack, known-plaintext attack, and so forth. The latter one refers to the intelligibility of the encrypted multimedia content. Generally, for multimedia encryption, an encryption algorithm is regarded as secure if the cost for breaking it is no smaller than the one paid for the multimedia content's authorization. For example, in broadcasting, the news may be of no value after an hour. Thus, if the attacker can not break the encryption algorithm during an hour, then the encryption algorithm may be regarded as secure in this application. Thus, according to this case, encrypting only significant parts of multimedia data may be reasonable if the cryptographic security and perceptual security are both confirmed, which will decrease the encrypted data volumes.

Efficiency. The efficiency refers to both time efficiency and energy-consumption efficiency. Since real-time transmission or access is often required by multimedia-related applications, multimedia encryption algorithms should be time efficient so that they do not delay the transmission or access operations. Generally, two kinds of method can be adopted to improve time efficiency, that is, the first one is to reduce the encrypted data volumes, and the second one is to adopt fast encryption algorithms. Additionally, to adapt the energy-constraint devices, such as mobile terminals, handset, handheld, and so forth, the lightweight encryption algorithms are preferred to decrease the energy-consumption.

Compression ratio. Multimedia data are often compressed in order to reduce the storage space or transmission bandwidth. In this case, multimedia encryption algorithms should not change the compression ratio.

Format compliance. In multimedia data, the format information, such as file header, frame header, file tail, and so forth will be used by the decoder to realize synchronization. Encrypting multimedia data except the format information will keep the encrypted data stream format-compliant. Thus, the encrypted data can be previewed directly. Additionally, the format information can be used to resynchronize the transmission process in error environment.

Communication compliance. In wireless or mobile environment, transmission errors often happened, such as, channel error, loss, delay, or jitter. The good multimedia encryption algorithms should not cause error propagation. Thus, the error conditions will also be considered when designing a wireless/mobile multimedia encryption algorithm.

Direct operation. If the encrypted multimedia data can be operated directly, the decryption-operation-encryption triples can be avoided, and the efficiency can also be improved. A typical example is to support direct bit rate conversion, that is, the encrypted data stream can be cut off directly in order to adapt the channel bandwidth. This property brings convenience to the applications in wireless or mobile environment.

Requirements of Multimedia Watermarking

For multimedia watermarking algorithms, some performances are required, such as security, robustness, transparency, oblivious, vindicability, and efficiency (Cox et al., 2002). Here, only the ones related to wireless/mobile environment are emphasized.

Security. Similar to an encryption algorithm, the construction of a watermarking algorithm should consider the security against various attacks (Kutter, Volosphynovskiy, & Herrigel, 2000; Linnartz & Dijk, 1998; Petitcolas, Anderson, & Kuhn, 1999). According to the attacker's ability, the attacks can be classified into several types: attack under the condition of knowing nothing about the watermarking system, attack knowing some watermarked copies, attack knowing the embedding algorithm, and the attack knowing the watermark detector. Generally, some encryption operations are introduced to watermarking algorithms in order to keep secure.

Imperceptibility. Imperceptibility means that the watermarked media data have no difference with the original ones in perception. It is also named transparency or fidelity. This makes sure that the watermarked copy is still of high quality and suitable for practical applications.

Robustness. Multimedia data are often processed during transmission process, and some of the processing operations are acceptable. Thus, the watermark should still be detected after these operations. Generally, the robustness refers to the ability for the watermark to survive such operations including general signal processing operations (filtering, noising, A/D, D/A, re-sampling, recompression, etc.) and geometric attacks (rotation, scaling, shifting, transformation, etc.). For wireless/mobile multimedia, transmission errors should also be considered, such as loss, delay, jitter, and so forth.

Efficiency. Efficiency refers to both time efficiency and energy-consumption efficiency. The watermarking algorithm with high time efficiency is more suitable for real time applications, such as video-on-demand, broadcasting, per-per-view, and so forth. For some energy-limited devices, the lightweight watermarking algorithm is preferred, which costs less power and is more efficient in implementation.

Oblivious detection. Oblivious detection means that the detection process needs not the original copy. It is also named blind detection. On the contrary, non-blind detection means that the original copy is required by the detection process. In practical applications, especially in wireless/mobile environment, memory is limited, and thus blind or oblivious detection is preferred.

THE ENCRYPTION ALGORITHMS FOR WIRELESS MULTIMEDIA

Some encryption algorithms have been proposed with respect to image, audio, speech, or video in wireless environment. These algorithms adopt some means to meet wireless communication requirements. According to the functionality, the encryption algorithms are classified into four types: (1) format independent encryption, (2) format compliant encryption, (3) communication compliant encryption, and (4) direct-operation supported encryption. The first type supports the media data of arbitrary format, the second one combines the encryption operation with the compression process, the third one considers the transmission errors, and the fourth one supports some direct operations on the encrypted multimedia data. In the following content, they are introduced and analyzed in detail.

Format Independent Encryption

Format independent encryption algorithms regard multimedia data as binary data and encrypt multimedia data without considering of the file format. Traditional ciphers (Mollin, 2006), such as DES, IDEA, AES, RSA, and so forth, encrypt text or binary data directly without considering of the file format. These ciphers have been included in the protocols, IP security (IPsec) and secure socket layer (SSL), and the package CryptoAPI, and these protocols are also included in a multilayer security framework (Dutta, Das, Li, & Auley, 2004). The energy requirements of most of the encryption algorithms are analyzed in Potlapally, Raghunathan, and Jha (2003), some of which are suitable for wireless applications. However, for wireless multimedia, some means should be made to improve the efficiency. One solution is to implement the algorithms in hardware, and another one is to design lightweight algorithms.

Hardware implementation. To improve the encryption algorithms' efficiency, hardware implementation is a suitable solution. The security processing architectures are proposed in Raghunathan, Ravi, Hattangady, and Quisquater (2003), which include an embedded processor, a cryptographic hardware accelerator, and a programmable security protocol engine. For the core encryption algorithms, some experiments are done to show their suitability. For example, hardware implementation of triple data encryption standard (3DES) is proposed in Hamalainen, Hannikainen, Hamalainen, and Saarinen (2001). The experiments show that 3DES implementations with small area and reasonable throughput can be realized even though 3DES turns out to be quite large and resource-demanding. It is suitable for some applications in wireless LAN (WLAN). Compared with such block cipher as 3DES, stream ciphers have some good properties, such as immunity to error propagation, increased flexibility, and greater efficiency. The Linear Feedback Shift Register (LFSR)-based stream ciphers are implemented in hardware (Goodman & Chandrakasan, 1998), which are shown ideally suited to low power wireless communications as they can be constructed from very simple and power-efficient hardware. Additionally, some wireless suitable stream ciphers, for example, wired equivalent privacy (WEP), improved wired equivalent privacy (IWEP), and Ron's cipher #4 (RC4) are implemented in hardware and tested in WLAN (Tikkanen, Hannikainen, Hamalainen, & Saarinen, 2000). Among them, IWEP is more suitable for hardware and of lower cost than RC4 although it is of lower security than RC4. Generally, hardware implementation improves the computing efficiency, but it also brings some problems, for example, the high cost to upgrade the algorithms.

Lightweight encryption algorithms. Compared with hardware implementation, software implementation is cheaper and more flexible for upgrades. For wireless applications, some lightweight encryption algorithms have been proposed, such as WEP, IWEP, RC2, RC4, RC5, and so forth. Experiments are done in Ganz, Park, and Ganz (1998) to test the software efficiency of RC2, RC4, and RSA. It is shown that software implementation of these ciphers can meet the requirements of such wireless applications as multimedia e-mail, multimedia notes, telephone-quality audio, video conferencing or MPEG video interaction, and so forth. The disadvantage is that their performance is limited by the computer system configuration. Besides the experiments, some design guidelines (Ganz, Park, & Ganz, 1999) are proposed for real-time software encryption, which considers the WLAN throughput, quality of service (QoS) requirements, encryption throughput determined by computer configuration, and additional processing overhead incurred by other protocol layers.

Generally, the algorithms with higher security are often of higher computing complexity. In traditional applications, an encryption algorithm is evaluated in a one-or-nothing manner, for example, secure or insecure (Ong, Nahrstedt, & Yuan, 2003). In pervasive environments, it is insufficient, because the limited computing resources may limit the security requirement. Thus, a quality of protection (QoP) framework (Ong et al.) is proposed, which evaluates an encryption algorithm in an adaptive manner. That is, the security level can be tuned in order to meet some other performances suitable for wireless/mobile applications. For example, the QoP metadata may be <content type, interval of security, encryption algorithm, encryption key length, encryption block size>. By tuning these parameters in the metadata, the suitable performances can be obtained. This framework has the following properties: (1) it can tune the quality of protection, (2) it gets a balance between security and performance requirement, and (3) it is flexible and upgradable to support latest cryptographic standards. However, before using this scheme, some problems should be solved, for example, how and where to store or transmit the metadata.

Format Compliant Encryption

For multimedia data, partial encryption (Furht & Kirovski, 2006) can be used to reduce the en-

Figure 1. An example of partial encryption method

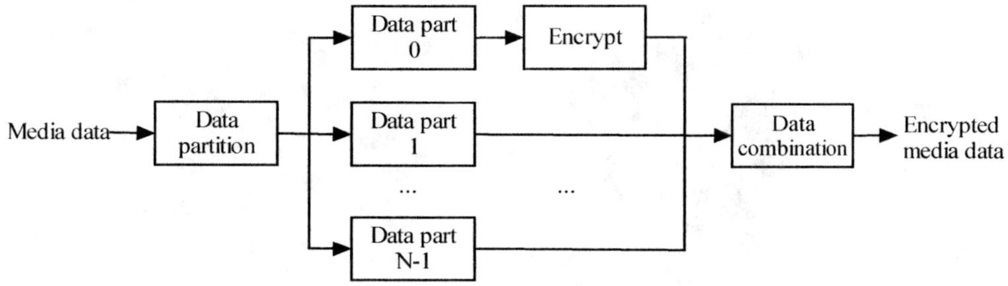

crypted data volumes, which keeps the file format unchanged. Additionally, the left format information can be used to synchronize the transmission process, especially in wireless/mobile environment where transmission errors often happen. The core of partial encryption is encrypting only the significant parameters in multimedia data while leaving other ones unchanged. Figure 1 gives an example for partial encryption, in which, media data are partitioned into N data parts, only the first data part is encrypted, while other parts are left unencrypted. The data part may be a block or region of the image, a frame of the video sequence, a bit-plane of the image pixels, a parameter of the compression codec, a segment of the compressed data stream, and so forth. The encrypted data part (Data part 0) and the other data parts are then combined together to generate the encrypted media data. The significance of the encrypted data part determines the security of the encryption scheme.

For multimedia data are often compressed before stored or transmitted, partial encryption often combines with compression codecs (Liu & Eskicioglu, 2003). That is, for different multimedia encoding codec, different partial encryption algorithm will be designed. During the past decade, some partial encryption algorithms have been proposed, which are classified and analyzed as follows according to the type of multimedia data and the codecs.

Partial audio encryption. Based on audio or speech codecs, some partial encryption algorithms have been proposed. For example, an algorithm based on G.729 (Servetti & Martin, 2002a, 2002b) is proposed to encrypt telephone-bandwidth speech. This algorithm partitions the code stream into two classes, for example, the most perceptually relevant one, and the other one. Among them, the former one is encrypted while the other one is left. It is reported that encrypting about 45% of the bitstream achieves content protection equivalent to full encryption. In another method (Sridharan, Dawson, & Goldburg, 1991), speech data are encrypted by encrypting only the parameters of Fast Fourier Transformation during speech encoding, and the correct parameters are used to recover the encrypted data in decryption. For MP3 (Gang, Akansu, Ramkumar, & Xie, 2001; Servetti, Testa, Carlos, & Martin, 2003) music, only the sensitive parameters of MP3 stream are encrypted, such as the bit allocation information, which saves much time or energy cost.

Partial image encryption. Some means are proposed to encrypt images partially or selectively. For raw images, only some of the most significant bit-planes are encrypted for secure transmission of image data in mobile environments (Podesser, Schmidt, & Uhl, 2002). Another image encryption algorithm (Scopigno & Belfiore, 2004) is proposed, which encrypts only the edge information in the image decomposition that produces three separate components: (1) edge location, (2) gray-tone or color inside the edges, and (3) residuum "smooth" image. For JPEG images, some significant bit-planes of discrete cosine transform (DCT)coefficients in JBIG are encrypted (Pfarrhofer & Uhl, 2005), and only DCT blocks are permuted and DCT coefficients' signs are encrypted in JPEG encoding

Figure 2. Experimental result of the image encryption algorithm

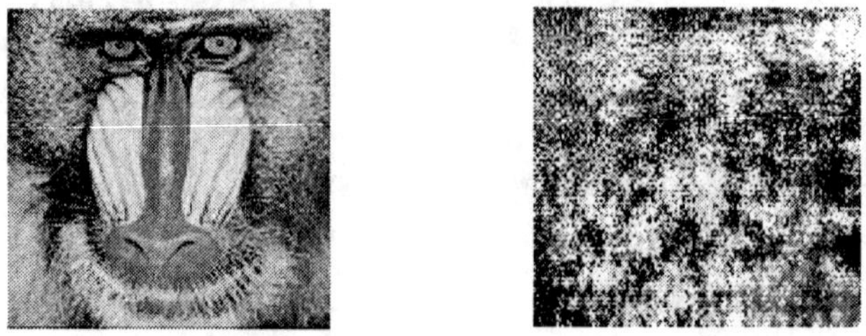

(a) Original image *(b) Encrypted image*

(Lian, Sun, & Wang, 2004a). These algorithms obtain high perceptual security and encryption efficiency. In JPEG2000 image encryption, only the significant streams in the encoded data stream are encrypted (Ando, Watanabe, & Kiya, 2001, 2002; Lian, Sun, & Zhang, 2004b; Norcen & Uhl, 2003; Pommer & Uhl, 2003), which is selected according to the scalability in space or frequency domain. These algorithms often keep secure in perception. Figure 2 gives the encryption result of the algorithm proposed in Lian et al., 2004b). As can be seen, the encrypted image is unintelligible. Additionally, in these algorithms, no more than 20% of the data stream is encrypted, which obtains high efficiency.

Partial video encryption. Compared with images or audios, videos are often of higher redundancy, which are compressed in order to save the transmission bandwidth. Among the video codecs, MPEG1/2, MPEG4, and H.264/AVC are more popular. Combined with them, some video encryption algorithms have been proposed, which saves time cost by encrypting the compressed video data selectively or partially.

In MPEG1/2 codec, the signs of DCT coefficients are encrypted with the video encryption algorithm (VEA) (Shi & Bhargava, 1998a), the signs of direct current coefficients (DCs) and motion vectors are encrypted with a secret key (Shi & Bhargava, 1998b), the base layer is encrypted while the enhancement layer is left unencrypted (Tosun & Feng, 2001a), the DCT coefficients are permuted (Lian, Wang, & Sun, 2004c; Tang, 1996), or the variable length coding (VLC) tables are modified by rearranging, random bit-flipping, or random bit-insertion (Wu & Kuo, 2000, 2001).

In MPEG4 codec, the Minimal Cost Encryption Scheme (Kim, Shin, & Shin, 2005) is proposed to encrypt only the first 8 bytes in the macroblocks (MBs) of a video object plane (VOP). It

Figure 3. Video encryption based on AVC codec

(a) Original *(b) Encrypted (Ahn et al., 2004)* *(c) Encrypted (Lian et al., 2005a)*

is implemented and proved suitable for wireless terminals. A format-compliant configurable encryption framework (Wen, Severa, Zeng, Luttrell, & Weiyin, 2002) is proposed for MPEG4 video encryption, which can be reconfigured for a given application scenario including wireless multimedia communication.

In H.264/AVC codec, the intra-prediction mode of each block is permuted with the control of the key (Ahn, Shim, Jeon, & Choi, 2004), which makes the video data degraded greatly. Some other algorithms (Lian, Liu, & Ren, 2005a; Lian, Liu, Ren, & Wang, 2006a) encrypt the DCT coefficients and motion vectors with sign encryption. For these algorithm encrypt both the texture information and motion information, they often obtain high security in human perception. Figure 3 shows the results of the algorithm proposed in Ahn et al. (2004) and the one proposed in Lian et al. (2005a). As can be seen, the video encrypted by the former algorithm is still intelligible, while the video encrypted by the latter algorithm is unintelligible. Thus, for high security, the latter encryption algorithm is preferred.

Communication Compliant Encryption

Multimedia data are often encrypted before being transmitted. In the encrypted data stream, transmission errors are often spread out due to encryption algorithms' ciphertext-sensitivity (Mollin, 2006). In wireless/mobile applications, some means should be taken to reduce the error propagation.

Constructing the encryption algorithms based on error correction code may be a solution. For example, the encryption algorithm based on forward error correction (FEC) code is proposed in Tosun & Feng, 2001b), which permutes the information-bits and complements a subset of the bits. The encryption algorithm can preserve the error robustness of the encrypted multimedia data, that is, the encrypted data stream can realize error correction itself. Additionally, the encryption algorithm is implemented very efficiently because of the simple encryption operations. Thus, it has some desirable properties suitable for wireless multimedia transmission. However, the disadvantage is also clear that it is not secure against known-plaintext attacks.

Another solution is to change the block length in data encryption. Generally, the block length is in close relation with the error propagation property. Taking stream cipher and block cipher for examples, the former one is of low error propagation, while the latter one is often of high error propagation. Generally, the bigger the block length is, the higher the error propagation is. Due to this case, a robust encryption scheme for secure image transmission over wireless channels is proposed in Nanjunda,

Figure 4. Robust video encryption based on segment

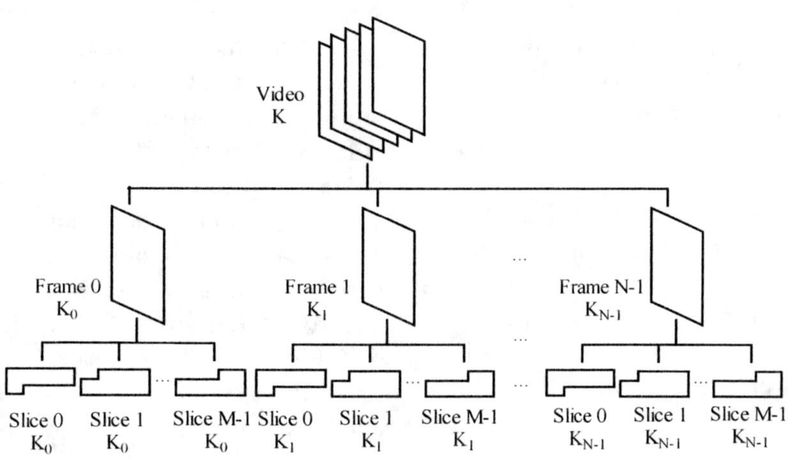

Figure 5. Scalable encryption scheme for MPEG2 video

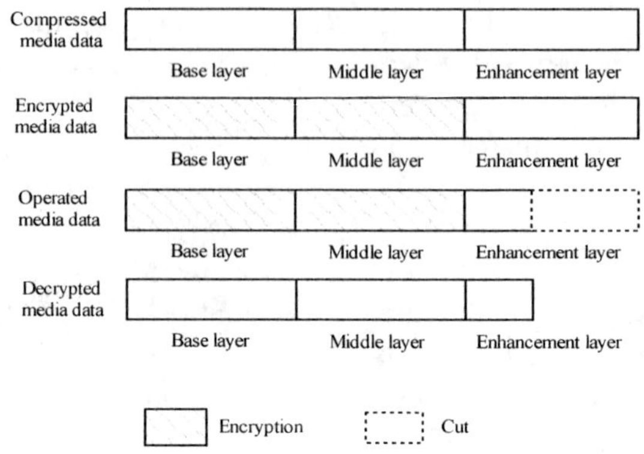

Haleem, and Chandramouli (2005), which varies the block length according to the channel's error properties. This method obtains a trade-off between the security and error robustness. However, some problems should be solved before hand, for example, how to transmit the parameters of varying the block length, and how to determine the channel's error properties in advance.

Additionally, segment-based encryption algorithms are proposed to reduce the effect cause by transmission errors. By partitioning the plaintext into segments and encrypting each segment independently, the transmission errors can be limited in a segment. The only difficulty is to synchronize the segments. An example proposed in Lian, Liu, Ren, and Wang (2005b) is shown in Figure 4. It encrypts advanced video coding (AVC) videos according to the following steps: (1) partition the video data into N frames (each frame acts as a segment), (2) partition each frame into M macroblocks (each macroblock acts as a subsegment), and (3) encrypt each frame with different keys (K_0, K_1,..., K_{N-1}), and encrypt all the macroblocks in a frame with the same key. Thus, if a macroblock is lost, the other macroblocks can still be recovered correctly. If a frame is lost, the frame index can be used to synchronize the key, and recover other frames correctly. Thus, if the synchronization problem is solved, the segment based encryption will be a good solution in wireless/mobile applications.

Direct Operation Supported Encryption

To operate the encrypted multimedia data directly without decryption is challenging while cost efficient. Especially in wireless/mobile environment, no decryption and re-encryption operations are required, which saves much cost. Some solutions have been proposed to realize direct transcoding or bit rate conversion.

A secure transcoding scheme is proposed in Chang, Han, Li, and Smith (2004). In this scheme, the multimedia data are decomposed into multiple streams at the source, each stream is encrypted independently, and each stream is annotated with cleartext metadata. In transcoding, lower priority streams are dropped directly based on the cleartext metadata. The receiver can decrypt the remaining streams and recombine them into the transcoded output stream.

As progressive and scalable encoding becomes more and more popular, such as JPEG2000, MPEG4 FGS, SVC, and so forth, scalable encryption is focused, which supports direct bit rate conversion. The scalable encryption algorithm encrypts the progressive or scalable data streams, for example, base layer, middle layer, or enhance layer, one by one from the significant ones to the least significant ones. Thus, the bit rate can be changed by cutting the insignificant streams directly. For example, Tosun and Feng (2000) proposed the algorithm shown in

Figure 5, which encrypts only the base layer and middle layer in the three layers (base layer, middle layer, and enhancement layer) of an MPEG2 video stream. In this algorithm, the enhancement layer is left unencrypted, which can be cut off directly. Wee and Apostolopoulos (2001, 2003) and Zhu, Yuan, Wang, and Li (2005) proposed the algorithms for secure scalable streaming enabling transcoding without decryption. Generally, the stream is partitioned into segments according to the cipher's code length. To change the bit-rate, some segments at the end of the stream are cut off directly.

THE WATERMARKING ALGORITHMS FOR WIRELESS MULTIMEDIA

Watermarking algorithms (Barni & Bartolini, 2004; Cox et al., 2002) are generally composed of two parts, that is, watermark embedding and watermark extraction/detection. Generally, watermarking algorithms should be robust to some operations, such as recompression, A/D or D/A conversion, noise, filtering, and so forth and can survive such attacks as geometric attack, collusion attack, copy attack, and so forth. Similar to encryption algorithms, some watermarking algorithms may be of high security and robustness, but they are also of high time or energy cost. On the contrary, the watermarking algorithms with lost cost are often of low security or robustness. This contradiction becomes a problem in wireless/mobile environment when the limited energy or computing capability is provided. Experiments have been done to analyze the energy consumption, complexity and security level of multimedia watermarking on mobile handheld devices (Kejariwal, Nicolau, Dutt, & Gupta, 2005). And some conclusions are drawn: (1) the security level often contradicts with energy consumption, (2) watermark extraction/detection may be of higher cost than watermark embedding, and (3) image resolution affects the energy consumption. To conquer these problems, some proposals are presented, for example, introduce the tunable parameter to obtain trade-offs between security level, energy consumption, and other performances, or move some computationally expensive tasks to mobile proxies.

Mobile Agent Based Task Partitioning

Mobile agents use the proxies as agents that can connect to a range of heterogeneous mobile terminals. Using mobile agents to reduce the load of the server or terminals has been widely studied (Burnside et al., 2002; Rao, Chang, Chen, & Chen, 2001). If the mobile agent can implement watermark embedding or extraction/detection, then the terminals' computing load will be greatly reduced.

Figure 6. Watermarking tasks partitioning based on mobile agents

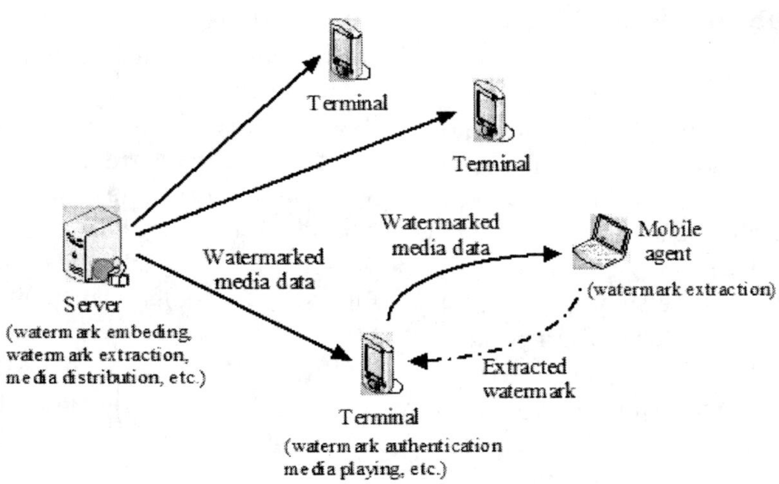

Figure 7. Architectures of some lightweight watermarking algorithms

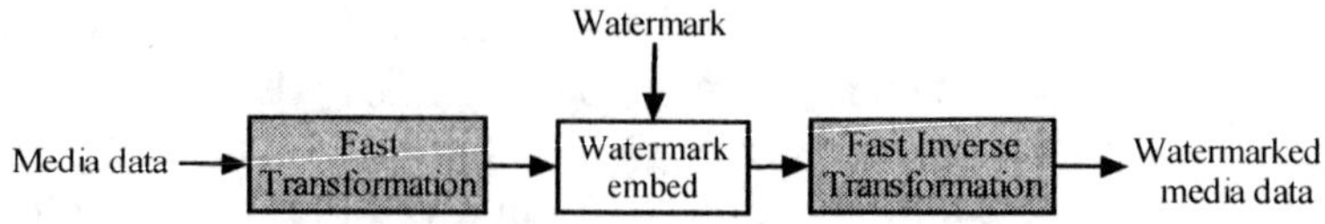

(a) Fast transformation based watermarking embedding

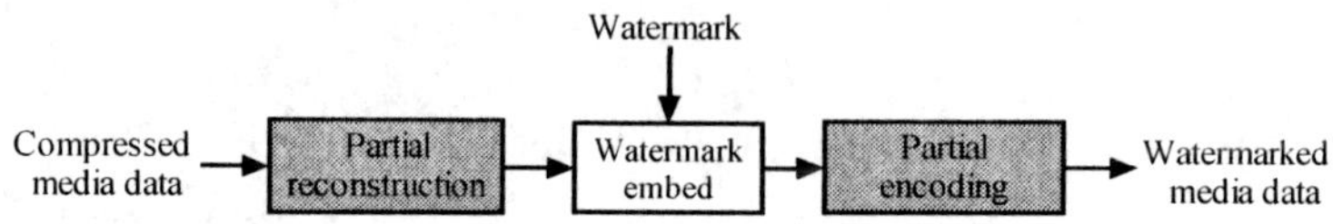

(b) Watermarking embedding in compressed media data

The scheme proposed in Liu and Jiang (2005), as shown in Figure 6, uses mobile agent to replace terminals to realize watermark detection, which decreases the server and network's load during detecting watermarks. In another scheme (Kejariwal, Gupta, Nicolau, Dutt, & Gupta, 2004), the watermark embedding and detection tasks are both partitioned and moved to mobile proxies completely or partially. For example, to keep secure, only some tasks not sensitive to the security are moved out, such as image transformation, bit decomposition, plane alignment, and so forth. The partitioning schemes make watermarking applications more practical in mobile environment.

Lightweight Watermarking Algorithms

Using mobile agents to implement some watermarking related tasks can reduce the load of the server or terminals in some extent. However, frequent interaction between mobile agent and terminals are still costly. To reduce the cost of the server or terminals, improving the efficiency of watermarking embedding, or extraction/detection algorithms is a key problem. Considering that the watermark is often embedded into the transformation domain, some lightweight algorithms are proposed to implement transformation domain watermarking. Two typical ones are shown in Figure 7. The first one, as shown in Figure 7a, uses fast transformations to reduce the cost of converting media data into frequency domain. The second one, as shown in Figure 7b, embeds the watermark into the compressed media data according to the following steps: (1) reconstruct the coefficients partially from the compressed data stream, (2) embed the watermark into the selected coefficients, and (3) re-encode the watermarked coefficients. In the following content, some lightweight watermarking algorithms are introduced and analyzed.

A scalable watermarking algorithm is proposed to mark the audio data encoded with Advanced Audio Zip (AAZ) (Li, Sun, & Lian, 2005). In this algorithm, the watermark is embedded into the quantized modified discrete cosine transform (MDCT) coefficients in the core layer adaptively, and detected by computing the correlation between the spreading sequence and the bitstream. A speech watermarking scheme is proposed in Arora and Emmanuel (2003), which is designed based on the adaptive modulation of spread spectrum sequences and is robust against some removal or impairment attacks. The experiments in global system for mobile communications (GSM) cellular communications show that the algorithm is suitable for mobile applications.

For images, an efficient steganography scheme (Pal, Saxena, & Muttoo, 2004) is proposed for resources constrained wireless networks. In this scheme, the coefficients in Hadamard transform-domain are manipulated to contain some hidden information. The Discrete Hadamard Transform can be implemented using fast algorithms, which makes the scheme computationally efficient and practical in mobile communications.

For videos, a spread spectrum watermarking algorithm (Petrescu, Mitrea, & Preteux, 2005) is proposed to protect low rate videos. In this algorithm, the DCT or wavelet coefficients of transformed video data are watermarked with spread spectrum sequences. Experiments are done for the videos varying from 64kbit/s to 256 kbit/s, and suitable transparency or robustness is obtained. Furthermore, a more efficient algorithm (Checcacci, Barni, Bartolini, & Basagni, 2000) is proposed to mark MPEG4 videos. In this algorithm, only the Luma macroblocks are watermarked by adjusting the coefficients' value in each coefficient pair. It is proved efficient in implementation and robust to transmission errors. Additionally, a more robust video watermarking algorithm (Alattar, Lin, & Celik, 2003) is proposed for low bit rate MPEG4 videos. In this algorithm, the watermark is composed of both the synchronization template and the watermark content combined with the template, and the watermark is embedded into the alternative current (AC) coefficients of the luminance plane of the VOPs. The template can survive geometric attacks, such as transcoding, cropping, scaling, rotation, noise, and so forth. Experiments on various videos are done, which show good performances for the video rate ranging from 128kbit/s to 768kbit/s.

Communication Compliant Algorithms

In wireless/mobile communication, transmission errors often happen, which may reduce the watermark detection rate. Generally, several means may be adopted to improve the watermarking algorithm's robustness against transmission errors. The first one, as shown in Figure 8a, is applying

Figure 8. Architectures of some robust watermarking algorithms

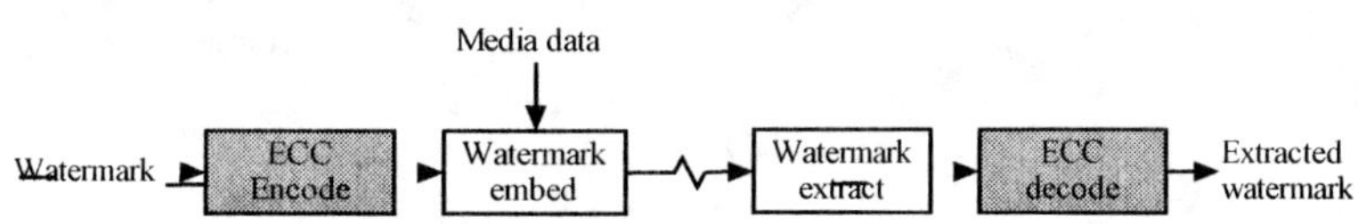

(a) ECC based watermarking scheme

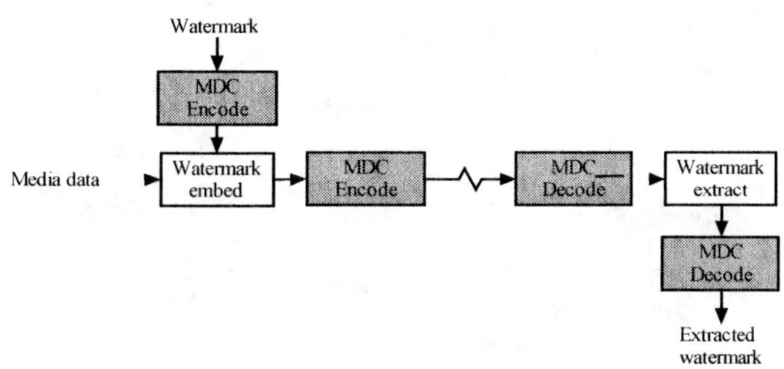

(b) MDC based watermarking scheme

error-correcting codes (ECC) to encode the watermark before embedding it into the multimedia data. For example, the watermark can be repeated for several times (Kundur, 2001), such codes as convolutional code, block code, or turbo code are used to encode the watermark (Ambroze et al., 2001), or the combination of watermark repetition and error-correcting code is used (Desset, Macq, & Vandendorpe, 2002). This kind of method improves the robustness by increasing the redundancy in the watermark. The second method, as shown in Figure 8b, is using multiple description code (MDC) to transmit the watermark or the watermarked multimedia data. For example, the watermark is encoded with MDC before being embedded (Hsia, Chang, & Liao, 2004), the watermarked media data are transmitted based on MDC (Chu, Hsin, Huang, Huang, & Pan, 2005; Pan, Hsin, Huang, & Huang, 2004), or both the watermark and the watermarked media data are encoded with MDC (Ashourian & Ho, 2003). This kind of method adopts the redundancy of multimedia data and is more suitable for the scenario of high error rate. Another method (Song, Kim, Lee, & Kim, 2002) partitions multimedia data into segments each of which fits for the packet in wireless transmission, and then embeds a watermark into each packet. Thus, it is robust to wireless packet error conditions including not only channel error but also delay and jitter.

COMBINATION OF MULTIMEDIA ENCRYPTION AND MULTIMEDIA WATERMARKING

Multimedia encryption and watermarking realize different functionalities, for example, confidentiality protection and ownership protection, they can be combined together to provide stronger security. This is also required by some applications, such as secure multimedia sharing, secure multimedia distribution, or exchange between watermarking and encryption.

Secure Multimedia Sharing

Multimedia sharing is more and more popular with the development of network technology, especially when such a network as p2p is developed. Generally, in these applications, the ownership information is embedded into the multimedia data with watermarking technology, and then the watermarked multimedia data are encrypted and distributed. The ownership information can be extracted later to prove the ownership right, and the encryption process prevents unauthorized users from accessing the real content of the multimedia data. A typical example is the music sharing system, named Music2Share (Kalker, Epema, Hartel, Lagendijk, & Steen, 2004), as shown in Figure 9. In this system, the watermark representing ownership information is embedded into music files, and the

Figure 9. Architecture of a multimedia sharing system

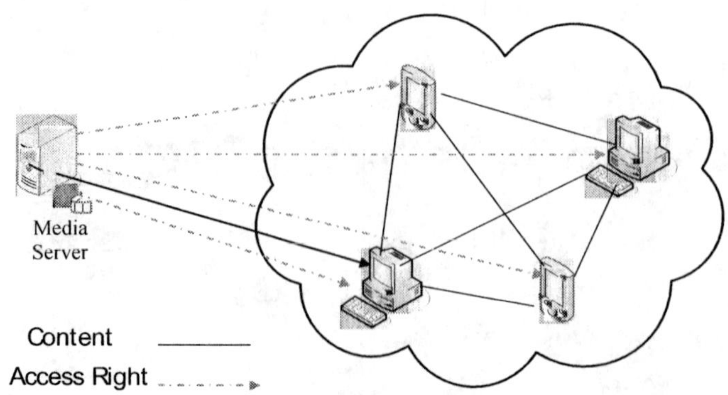

watermarked files are encrypted then distributed over p2p networks. The customer can access the encrypted music files, while must apply for the right from the server before he can decrypt the files. The watermark extracted from the music file can prove the legality of the music.

Secure Multimedia Distribution

In secure multimedia distribution, multimedia data are transmitted from the server to customers in a secure way. In this case, the confidentiality can be protected, and the illegal distributor who redistributes his/her copy to other customers can be traced. Generally, both encryption and watermarking technology are used. Till now, three kinds of schemes have been proposed, which embed watermarks at the server side, in the router or at the client side, respectively. In the first kind of scheme, the customer information is embedded into multimedia data at the server side before multimedia encryption. This scheme is more suitable for unicast than for multicast or broadcast because it is difficult for the server to assign different copies to different customers simultaneously. In the second kind of scheme, the customer information is embedded by the routers in lower level (Brown, Perkins, & Crowcroft, 1999), which distributes the server's loading to the routers. This scheme reduces the server's loading, but also changes the network protocols. In the third kind of scheme, the customer information is embedded at the customer side (Bloom, 2003). This scheme is time efficient, but the security is a problem because of the isolation between decryption and watermarking. Some means (Anderson & Manifavas, 1997; Kundur & Karthik, 2004; Lian, Liu, Ren, & Wang, 2006b) have been proposed to improve the security, which combine decryption with watermark embedding. These combined methods improve the system's security at the same time of keeping low cost.

Commutative Watermarking and Encryption

Generally, watermarking operation and encryption operation are separate. That is, the encrypted multimedia data should be decrypted before being watermarked. In some applications, if the operation triple decryption-watermarking-encryption can be avoided, the operation cost will be reduced greatly. In this case, the encrypted multimedia data can be watermarked directly without decryption, and the watermark can be extracted directly from the encrypted or decrypted multimedia data. This kind of watermarking-encryption pair is named commutative watermarking and encryption (CWE). A practical scheme is proposed in Lian, Liu, Ren, and Wang (2006c), which is based on partial encryption. In this scheme, multimedia data are partitioned into two parts, that is, the perception significant part and the robust part, among which, the perception significant part is encrypted, while the robust part is watermarked. Thus, the encryption and watermarking are independent of each other, and they support the commutative operations.

OPEN ISSUES

Contradiction Between Format Independence and Format Compliance

To keep low cost, partial encryption scheme is used to encrypt multimedia data, which keeps format compliant. Thus, for different multimedia data or different codec, the encryption algorithms are often different. If various multimedia data are included in an application, then various encryption algorithms should be used, and some extra information is required to tell which encryption algorithm has been used. Compared with format compliant encryption, format independent encryption regards multimedia data as binary data and is easy to support various data. Thus, for the applications with versatile data, format independent encryption is more suitable. For example, in such DRM systems as internet streaming media alliance (ISMA), advanced access content system (AACS), or open mobile alliance (OMA) (Kundur et al., 2004), the algorithms, advanced encryption standard (AES) and data encryption standard (DES), are recommended to encrypt multimedia data not considering the file format. Thus, for practical applications, the trade-off between computational cost and convenience is to be made, which determines which kind of algorithm should be used.

Standardization of Watermarking Algorithms

Compared with encryption algorithms that have been standardized to some extent, watermarking algorithms are still in study. For the diversity of multimedia content, the difficulty in multimedia understanding and the variety of applications, it is difficult to standardize multimedia watermarking algorithms. Generally, they have different performances in security, efficiency, robustness, capacity, and so forth. Using which watermarking algorithm depends on the performances required by the applications. Defining suitable watermarking algorithms will provide more convenience to wireless/mobile applications.

Fingerprint Algorithms Against Collusion Attacks

In secure multimedia distribution, collusion attack (Zhao, Wang, & Liu, 2005) threatens the system. That is, different customers combine their copies together through averaging, substitution, and so forth, which produces a copy without any customer information. To counter this attack, some fingerprint encoding methods (Boneh & James, 1998; Wu, Trappe, Wang, & Liu, 2004) have been proposed. These methods generate different fingerprint codes for different customers, and the colluded copy can still tell one or more of the colluders. However, there is still a trade-off between the watermark capacity and the supported customers, and some new attacks are still not predicted, such as the linear combination collusion attack (LCCA) attack (Wu, 2005). Thus, better fingerprint encoding methods with good efficiency are expected.

Key Management in Mobile Applications

Multimedia encryption and watermarking can both be controlled by the keys; key management needs to be investigated. For example, whether the encryption key should be independent of the watermarking key, and how to assign different decryption keys to different customers in multimedia distribution? Additionally, for multicast or p2p networks, key generation and distribution (Cherukuri, 2004; Eskicioglu, 2002) are important topics not only in fixed networks but also in mobile environments.

CONCLUSION

In this chapter, mobile/wireless multimedia encryption and watermarking algorithms are introduced and analyzed, including the general requirements, various multimedia encryption algorithms, some watermarking algorithms, the combination between encryption and watermarking, and some open issues. Among them, the multimedia encryption algorithms are classified and analyzed according to the functionalities, and the watermarking algorithms with low cost are emphasized. The combination between encryption and watermarking brings up some new research topics, for example, fingerprint or commutative watermarking and encryption. And some open issues are also presented, including the contradiction between format compliance and format independence, the standardization of watermarking algorithms, the fingerprint algorithms resisting collusion attacks, and the key management in mobile applications.

REFERENCES

Ahn, J., Shim, H., Jeon, B., & Choi, I. (2004). Digital video scrambling method using intra prediction mode. In *Pacific Rim Conference on Multimedia, PCM2004* (LNCS 3333, 386-393). Springer.

Alattar, A., Lin, E., & Celik, M. (2003). Digital watermarking of low bit-rate advanced simple profile MPEG-4 compressed video. *IEEE Transactions on Circuits and Systems for Video Technology, 13,* 787-800.

Ambroze, A., Wade, G., Serdean, C., Tomlinson, M., Stander, J., & Borda, M. (2001). Turbo code protection of video watermark channel. *IEE Proceedings of Vision and Image Signal Processing, 148,* 54-58.

Anderson, R., & Manifavas, C. (1997). Chameleon—A new kind of stream cipher. In *Fast Software Encryption* (LNCS, vol. 1267, pp. 107-113). Springer-Verlag.

Ando, K., Watanabe, O., & Kiya, H. (2001). Partial-scrambling of still images based on JPEG2000. In *Proceedings of the International Conference on Information, Communications, and Signal Processing,* Singapore.

Ando, K., Watanabe, O., & Kiya, H. (2002). Partial-scrambling of images encoded by JPEG2000. *IEICE Transactions, J85-D-11*(2), 282-290.

Arora, S., & Emmanuel, S. (2003). Real-time adaptive speech watermarking scheme for mobile applications. In *Proceedings of the International Conference on Information, Communications & Signalprocessing (ICICS)—IEEE Pacific-rim Conference on Multimedia (PCM)* (pp. 850-853).

Ashourian, M., & Ho, Y. (2003). Multiple description coding for image data hiding jointly in the spatial and DCT domains. In *ICICS 2003* (LNCS 2836, 179-190).

Barni, M., & Bartolini, F. (2004). *Watermark systems engineering.* Marcel Dekker.

Bloom, J. (2003). Security and rights management in digital cinema. *Proceedings of IEEE International Conference on Acoustic, Speech and Signal Processing, 4,* 712-715.

Boneh, D., & James, S. (1998). Collusion-secure fingerprinting for digital data. *IEEE Transactions on Information Theory, 44*(5), 1897-1905.

Brown, I., Perkins, C., & Crowcroft, J. (1999). Watercasting: Distributed watermarking for multicast media. In *Proceedings of the First International Workshop on Networked Group Communication* (LNCS 1736, pp. 286-300). Springer-Verlag.

Burnside, M., Clarke, D., Mills, T., Maywah, A., Devadas, S., & Rivest, R. (2002). Proxy-based security protocols in networked mobile devices. In *Proceedings of the 2002 ACM symposium on Applied Computing* (pp. 265-272).

Chang, Y., Han, R., Li, C., & Smith, J. R. (2004). Secure transcoding of Internet content. In *Proceedings of International Workshop on Intelligent Multimedia Computing and Networking (IMMCN)* (pp. 940-943).

Checcacci, N., Barni, M., Bartolini, F., & Basagni, S. (2000). Robust video watermarking for wireless multimedia communications. In *Proceedings of the 2000 IEEE Conference on Wireless Communications and Networking* (pp. 1530-1535).

Cherukuri, S. (2004). *An adaptive scheme to manage mobility for secure multicasting in wireless local area networks.* Unpublished masters thesis, Arizona State University, Tempe.

Chu, S., Hsin, Y., Huang, H., Huang, K., & Pan, J. (2005). Multiple description watermarking for lossy network. *IEEE Computer Society, 4,* 3990-3993.

Cox, I., Miller, M., & Bloom, J. (2002). *Digital watermarking.* San Francisco: Morgan Kaufmann.

Desset, C., Macq, B., & Vandendorpe, L. (2002). Block error-correcting codes for systems with a very high BER: Theoretical analysis and application to the protection of watermarks. *Signal Processing: Image Communication, 17,* 409-421.

Dutta, A., Das, S., Li, P., & Auley, A. (2004). Secured mobile multimedia communication for wireless Internet. In *Proceedings of 2004 IEEE International Conference on Networking, Sensing & Control* (pp. 181-186).

Eskicioglu, A. (2002). Multimedia security in group communications: Recent progress in wired and wireless networks. In *Proceedings of the IASTED*

International Conference on Communications and Computer Network s, Cambridge, MA (pp. 125-133).

Furht, B., & Kirovski, D. (Eds.). (2006). *Multimedia encryption and authentication techniques and applications.* Boca Raton, FL: Auerbach Publications.

Gang, L., Akansu, A., Ramkumar, M., & Xie, X. (2001). Online music protection and MP3 compression. In *Proceedings of the International Symposium on Intelligent Multimedia, Video and Speech Processing* (pp. 13-16).

Ganz, A., Park, S., & Ganz, Z. (1998). Inline network encryption for multimedia wireless LANs. In *Proceedings of the IEEE Military Communications Conference.*

Ganz, A., Park, S., & Ganz, Z. (1999). Experimental measurements and design guidelines for real-time software encryption in multimedia wireless LANs. *Cluster Computing, 2*(1), 35-43.

Goodman, J., & Chandrakasan, A. (1998). Low power scalable encryption for wireless systems. *Wireless Networks, 4,* 55-70.

Hamalainen, P., Hannikainen, M., Hamalainen, T., & Saarinen, J. (2001). Configurable hardware implementation of triple DES encryption algorithm for wireless local area network. In *Proceedings of the 2001 IEEE International Conference on Acoustics, Speech and Signal Processing* (pp. 1221-1224).

Hsia, Y., Chang, C., & Liao, J. (2004). Multiple-description coding for robust image watermarking. In *Proceedings of the 2004 International Conference on Image Processing* (pp. 2163-2166).

Kalker, T., Epema, D., Hartel, P., Lagendijk, R., & Steen, M. (2004). Music2Share—Copyright-compliant music sharing in P2P systems. *Proceedings of the IEEE, 92*(6), 961-970.

Kejariwal, A., Gupta, S., Nicolau, A., Dutt, N., & Gupta, R. (2004). Proxy-based task partitioning of watermarking algorithms for reducing energy consumption in mobile devices. In *Proceedings of the 2004 Design Automation Conference* (pp. 556-561).

Kejariwal, A., Nicolau, S., Dutt, A., & Gupta, N. (2005). Energy analysis of multimedia watermarking on mobile handheld devices. In *Proceedings of the International Conference on Embedded Systems for Real-Time Multimedia (ESTImedia 2005)* (pp. 33-38).

Kim, G., Shin, D., & Shin, D. (2005). Intellectual property management on MPEG-4 video for handheld device and mobile video streaming service. *IEEE Transactions on Consumer Electronics, 51*(1), 139-143.

Kundur, D. (2001). Watermarking with diversity: insights and implications. *IEEE Transactions on Multimedia, 8,* 46-52.

Kundur, D., & Karthik, K. (2004). Video fingerprinting and encryption principles for digital rights management. *Proceedings of the IEEE, 92*(6), 918-932.

Kundur, D., Yu, H., & Lin, C. (2004). Security and digital rights management for mobile content. In T. Wu & S. Dixit (Eds.), *Content delivery in the mobile Internet.* John Wiley & Sons.

Kutter, M., Volosphynovskiy, S., & Herrigel, A. (2000). The watermarking copy attack. In *Security and Watermarking of Multimedia Contents II* (SPIE 3971, pp. 371-380).

Li, Z., Sun, Q., & Lian, Y. (2005). An adaptive scalable watermark scheme for high-quality audio archiving and streaming applications. In *Proceedings of the IEEE International Conference on Multimedia and EXPO.*

Lian, S., Liu, Z., & Ren, Z. (2005a). Selective video encryption based on advanced video coding. In *Proceedings of 2005 Pacific-Rim Conference on Multimedia (PCM2005), Part II* (LNCS 3768, pp. 281-290).

Lian, S., Liu, Z., Ren, Z., & Wang, H. (2006b). Secure distribution scheme for compressed video stream. In *Proceedings of the 2006 IEEE International Conference on Image Processing (ICIP2006).*

Lian, S., Liu, Z., Ren, Z., & Wang, H. (2006c). Commutative watermarking and encryption for media data. *International Journal of Optical Engineering, 45*(8), 0805101-0805103.

Lian, S., Liu, Z., Ren, Z., & Wang, Z. (2005b). Selective video encryption based on advanced video coding. In *Proceedings of Pacific-Rim Conference on Multimedia (PCM2005)* (pp. 281-290).

Lian, S., Liu, Z., Ren, Z., & Wang, H. (2006a). Secure advanced video coding based on selective encryption algorithms. *IEEE Transactions on Consumer Electronics, 52*(2), 621-629.

Lian, S., Sun, J., & Wang, Z. (2004a). A novel image encryption scheme based-on JPEG encoding. In *Proceedings of International Conference on Information Visualization (IV 2004)* (pp. 217-220).

Lian, S., Sun, J., Zhang, D., & Wang, Z. (2004b). A selective image encryption scheme based on JPEG2000 codec. In *Proceedings of 2004 Pacific-Rim Conference on Multimedia (PCM2004)* (LNCS 3332, pp. 65-72). Springer.

Lian, S., Wang, Z., & Sun, J. (2004c). A fast video encryption scheme suitable for network applications. In *Proceedings of International Conference on Communications, Circuits and Systems, 1,* 566-570.

Linnartz, J., & Dijk, M. (1998, April 15-17). *Analysis of the sensitivity attack against electronic watermarks in images.* Paper presented at the Workshop on Information Hiding, Portland, OR.

Liu, Q., & Jiang, X. (2005). Applications of mobile agent and digital watermarking technologies in mobile communication network. In *Proceedings of the 2005 International Conference on Wireless Communications, Networking and Mobile Computing* (pp. 1168-1170).

Liu, X., & Eskicioglu, A. (2003). Selective encryption of multimedia content in distribution networks: Challenges and new directions. In *Proceedings of the IASTED International Conference on Communications, Internet and Information Technology (CIIT 2003).* Scottsdale, AZ: ACTA Press.

Mollin, R. (2006). *An introduction to cryptography.* CRC Press.

Nanjunda, C., Haleem, M., & Chandramouli, R. (2005). Robust encryption for secure image transmission over wireless channels. In *Proceedings of the IEEE International Conference on Communications (ICC)* (pp. 1287-1291).

Norcen, R., & Uhl, A. (2003). Selective encryption of the JPEG2000 bitstream. In *IFIP International Federation for Information Processing* (LNCS 2828, 194-204).

Ong, C., Nahrstedt, K., & Yuan, W. (2003). Quality of protection for mobile multimedia applications. In *Proceedings of the IEEE International Conference on Multimedia and Expo (ICME2003),* Baltimore, MD.

Pal, S., Saxena, P., & Muttoo, S. (2004). Image steganography for wireless networks using the hadamard transform. In *Proceedings of the 2004 International Conference on Signal Processing and Communications* (pp. 131-135).

Pan, J., Hsin, Y., Huang, H., & Huang, K. (2004). Robust image watermarking based on multiple description vector quantization. *Electronics Letters, 40*(22), 1409-1410.

Petitcolas, F., Anderson, R., & Kuhn, M. (1999). Information hiding—A survey. *Proceedings of IEEE, 87*(7), 1062-1078.

Petrescu, M., Mitrea, M., & Preteux, F. (2005). Low rate video protection: The opportunity of spread spectrum watermarking. *WSEAS Transactions on Communications, 7*(4), 478-485.

Pfarrhofer, R., & Uhl, A. (2005). Selective image encryption using JBIG. In *Proceedings of the IFIP TC-6 TC-11 international conference on communications and multimedia security (CMS 2005)* (pp. 98-107).

Podesser, M., Schmidt, H., & Uhl, A. (2002). Selective bitplane encryption for secure transmission of image data in mobile environments. In *CD-ROM Proceedings of the 5th IEEE Nordic Signal Processing Symposium (NORSIG 2002).*

Pommer, A., & Uhl, A. (2003). Selective encryption of wavelet-packet encoded image data: Efficiency and security. In *Proceedings of the Communications and Multimedia Security 2003* (pp. 194-204).

Potlapally, N., Raghunathan, A., & Jha, N. (2003). Analyzing the energy consumption of security protocols. In *Proceedings of the 2003 International Symposium on Low Power Electronics and Design*, Seoul, Korea (pp. 30-35).

Raghunathan, A., Ravi, S., Hattangady, S., & Quisquater, J. (2003). Securing mobile appliances: New challenges for the system designer. In *Proceedings of the 2003 Europe Conference and Exibition in Design, Automation and Test* (pp. 176-181).

Rao, H., Chang, D., Chen, Y., & Chen, M. (2001). iMobile: A proxy-based platform for mobile services. In *Proceedings of the Wireless Mobile Internet* (pp. 3-10).

Salkintzis, A., & Passas, N. (2005). *Emerging wireless multimedia: Services and technologies.* John Wiley & Sons.

Scopigno, R., & Belfiore, S. (2004). Image decomposition for selective encryption and flexible network services. In *Proceedings of the IEEE Globecom 2004,* Dallas, TX.

Servetti, A., & Martin, J. (2002a). Perception-based selective encryption of G. 729 speech. *Proceedings of IEEE ICASSP, 1,* 621-624.

Servetti, A., & Martin, J. (2002b). Perception-based selective encryption of compressed speech. *IEEE Transactions on Speech and Audio Processing, 10*(8), 637-643.

Servetti, A., Testa, C., Carlos, J., & Martin, D. (2003). *Frequency-selective partial encryption of compressed audio.* Paper presented at the International Conference on Audio, Speech and Signal Processing, Hong Kong.

Shi, C., & Bhargava, B. (1998a). A fast MPEG video encryption algorithm. In *Proceedings of the 6th ACM International Multimedia Conference,* Bristol, UK (pp. 81-88).

Shi, J., & Bhargava, B. (1998b). An efficient MPEG video encryption algorithm. In *Proceedings of the 6th ACM International Multimedia Conference,* Bristol, UK (pp. 381-386).

Song, G., Kim, S., Lee, W., & Kim, J. (2002). Meta-fragile watermarking for wireless networks. In *Proceedings of the International Conference of Communications, Circuits, and Systems.*

Sridharan, S., Dawson, E., & Goldburg, B. (1991). Fast Fourier transform based speech encryption system. *IEE Proceedings of Communications, Speech and Vision, 138*(3), 215-223.

Tang, L. (1996). Methods for encrypting and decrypting MPEG video data efficiently. In *Proceedings of the Fourth ACM International Multimedia Conference (ACM Multimedia'96),* Boston, MA (pp. 219-230).

Tikkanen, K., Hannikainen, M., Hamalainen, T., & Saarinen, J. (2000). Hardware implementation of the improved WEP and RC4 encryption algorithms for wireless terminals. In *Proceedings of European Signal Processing Conference* (pp. 2289-2292).

Tosun, A., & Feng, W. (2000). Efficient multi-layer coding and encryption of MPEG video streams. *IEEE International Conference on Multimedia and Expo, 1,* 119-122.

Tosun, A., & Feng, W. (2001a). Lightweight security mechanisms for wireless video transmission. In *Proceedings of International Conference on Information Technology: Coding and Computing,* Las Vegas, NV (pp. 157-161).

Tosun, A., & Feng, W. (2001b). On error preserving encryption algorithms for wireless video transmission. In *Proceedings of the ACM International Multimedia Conference and Exhibition.* Ottawa, Ontario, Canada (pp. 302-308). Elsevier Engineering Information Inc.

Wee, S., & Apostolopoulos, J. (2001). Secure scalable video streaming for wireless networks. In *Proceedings of the IEEE International Conference on Acoustics, Speech, and Signal Processing, 4,* 2049-2052.

Wee, S., & Apostolopoulos, J. (2003). Secure scalable streaming and secure transcoding with JPEG-2000. *IEEE International Conference on Image Processing, 1,* 205-208.

Wen, J., Severa, M., Zeng, W., Luttrell, M. H., & Weiyin J. (2002). A format-compliant configurable encryption framework for access control of video. *IEEE Transactions on Circuits and Systems for Video Technology, 12*(6), 545-557.

Wu, C., & Kuo, C. (2000). Fast encryption methods for audiovisual data confidentiality. *Proceedings of SPIE, 4209,* 284-295.

Wu, C., & Kuo, C. (2001). Efficient multimedia encryption via entropy codec design. *Proceedings of SPIE, 4314,* 128-138.

Wu, M., Trappe, W., Wang, Z., & Liu, K. (2004). Collusion-resistant fingerprinting for multimedia. *IEEE Signal Processing Magazine, 21*(2), 15-27.

Wu, Y. (2005). Linear combination collusion attack and its application on an anti-collusion fingerprinting. In *Proceedings of the IEEE International Conference on Audio, Speech and Signal Processing (ICASSP'05)* (pp. 13-16).

Zeng, W., Zhuang, X., & Lan, J. (2004). Network friendly media security rationales, solutions, and open issues. In *Proceedings of the International Conference on Image Processing (ICIP 2004)* (pp. 565-568).

Zhao, H., Wang, Z., & Liu, K. (2005). Forensic analysis of nonlinear collusion attacks for multimedia fingerprinting. *IEEE Transactions on Image Processing, 14*(5), 646-661.

Zhu, B., Yuan, C., Wang, Y., & Li, S. (2005). Scalable protection for MPEG-4 fine granularity scalability. *IEEE Transactions on Multimedia, 7*(2), 222-233.

KEY TERMS

Commutative Watermarking and Encryption: Commutative watermarking and encryption is the watermarking-encryption pair that supports the exchange between the encryption algorithm and the watermarking algorithm. Thus, the media data can either be watermarked followed by encryption or be encrypted followed by watermarking.

Digital Watermarking: Digital watermarking is the technology to embed information into the original data by modifying parts of the data. The produced data are still usable, from which the information can be detected or extracted.

Format Compliant Encryption: Format compliant encryption is the multimedia encryption method that keeps the format information unchanged. In this method, the encrypted media data can be decoded or browsed by a general decoder or player.

Joint Fingerprint Embedding and Decryption: Joint fingerprint embedding and decryption is the technology to implement fingerprint embedding and data decryption at the same time. The input is the encrypted media copy, while the output is the decrypted media copy with a unique fingerprint, for example, the customer ID.

Partial Encryption: Partial encryption is the encryption method that encrypts only parts of the original data while leaving the other parts unchanged. In this method, traditional ciphers can be used to encrypt the selected parts.

Robust Watermarking: Robust watermarking is the watermarking algorithm that can survive not only such general operations such as compression, adding noise, filtering, A/D or D/A conversion, and so forth, but also such geometric attacks such as rotation, scaling translation, shearing, and so forth. It is often used in ownership protection.

Scalable Encryption: Scalable encryption is the multimedia encryption method that keeps the scalability of the progressive or scalable media data. The scalable media data can be produced by such codecs as JPEG2000, MPEG4, scalable video coding (SVC), and so on.

Chapter XVII
System-on-Chip Design of the Whirlpool Hash Function

Paris Kitsos
Hellenic Open University (HOU), Patras, Greece

ABSTRACT

In this chapter, a system-on-chip design of the newest powerful standard in the hash families, named Whirlpool, is presented. With more details an architecture and two very large-scale integration (VLSI) implementations are presented. The first implementation is suitable for high speed applications while the second one is suitable for applications with constrained silicon area resources. The architecture permits a wide variety of implementation tradeoffs. Different implementations have been introduced and each specific application can choose the appropriate speed-area, trade-off implementation. The implementations are examined and compared in the security level and in the performance by using hardware terms. Whirlpool with RIPEMD, SHA-1, and SHA-2 hash functions are adopted by the International Organization for Standardization (ISO/IEC, 2003) 10118-3 standard. The Whirlpool implementations allow fast execution and effective substitution of any previous hash families' implementations in any cryptography application.

INTRODUCTION

Nowadays many financial and other electronic transactions are grown exponentially and they play an important role in our life. All these transactions have integrated data authentication processes. In addition many applications like the public key infrastructure (PKI) (Adams & Farrell, 1999; National Institute of Standards and Technology [NIST, 2005=http://csrc.nist.gov/publications/nistpubs/800-77/sp800-77pdf]) and many mobile communications include authentication services. All the aforementioned applications have integrated an authentication module including a hash function embedded in the system's implementation.

A hash function is a function that maps an input of arbitrary length into a fixed number of output bits, the hash value.

One of the most widely used hash function is RIPEMD (Dobbertin, Bosselaers, & Preneel, 1996). These are two different RIPEMD versions the RIPEMD-128 and the RIPEMD-160, with similar design philosophy but different word length of the produced message digest (128- and 160-bit, respectively). In August 2002, NIST announced the updated Federal Information Processing Standard (FIPS 180-2), which has introduced another three new hash functions referred to as SHA-2 (256, 384, 512). In addition, the new European schemes for signatures, integrity, and encryption (NESSIE) (2004), was responsible to introduce a hash function with high security level. In February 2003, it was announced that the hash function included in the NESSIE portfolio is Whirlpool (Barreto & Rijmen, 2003). Finally, the most known hash function is the secure hash algorithm-1 (SHA-1) (NIST, 1995=http://itl.nist.gov/fipspub/fip180-1.htm). However, some security problems have been raised as it has already (see Wang, Yin, & Yu, 2005) shown. This collision of SHA-1 can be found with complexity less than 2^{69} hash operations. This is the first attack on the full 80-step SHA-1 with complexity less than the 2^{80} theoretical bound. A collision in SHA-1 would cast doubt over the future viability of any system that relies on SHA-1. The result will cause a significant confusion and it will create reengineering of many systems, and incompatibility between new systems and old. In addition, the National Security Agency (NSA) did not disclose the SHA-2 design criteria and also its design philosophy is similar to the design of SHA-1 function. So, the attack against SHA-1 probably will have affected to the SHA-2 function. Also, this issue stands for RIPEMD hash families. On the other hand, the internal structure of Whirlpool is different from the structure of all the aforementioned hash functions. So, Whirlpool function does not suffer for that kind of problems and makes it a very good choice for electronics applications.

All the afore-mentioned hash functions are adopted by the International Organization for Standardization (ISO, 2003) 10118-3 standard.

In this chapter, an architecture and two VLSI implementations of the new hash function, Whirlpool, are proposed. The first implementation is suitable for high speed applications while the second one is suitable for applications with constrained silicon area resources.

The architecture and the implementations presented here were the first in scientific literature (Kitsos & Koufopavlou, 2004). Until then, two hardware architectures have been also presented. The first one (McLoone & McCanny, 2002) is a high speed hardware architecture and the second one (Pramstaller, Rechberger, & Rijmen, 2006) is a compact field-programmable gate array (FPGA) architecture and implementation of Whirlpool. Both architectures are efficient for specific applications; analytical comparisons with the proposed implementations will be given in the rest of this chapter. In addition, comparisons with other hash families' implementations (Ahmad & Shoba Das, 2005; Deepakumara, Heys, & Venkatesam, 2001; Dominikus, 2002; Grembowski et al., 2002; McLoone, McIvor, & Savage, 2005; Sklavos & Koufopavlou, 2003, 2005; Yiakoumis, Papadonikolakis, Michail, Kakarountas, & Goutis, 2005); are provided. From the comparison results it is proven that the proposed implementation performs better and composes an effective substitution of any previous hash families' such as MD5, RIPEMD-160, SHA-1, SHA-2, and so forth, in all the cases.

The organization of the chapter is the following: In the second section, fundamental for hash functions families, is presented. So, the (ISO/IEC) 10118-3 standard first is briefly described and secondly the Whirlpool hash function specifications are defined. In the third section, the proposed architecture and VLSI implementations are presented. Implementation results and discussion (comparison with other works) are reported in the fourth section. Finally, the fifth section concludes this chapter.

FUNDAMENTALS FOR HASH FUNCTIONS

In this section a brief description of the ISO/IEC 10118-3 standard is presented. This standard specifies dedicated hash functions. The hash functions are based on the iterative use of a round-function. Seven distinct round functions are specified, giving rise to distinct dedicated hash-functions. Six of them are briefly described and at last, Whirlpool is described in details.

Figure 1. The SHA-1 round function

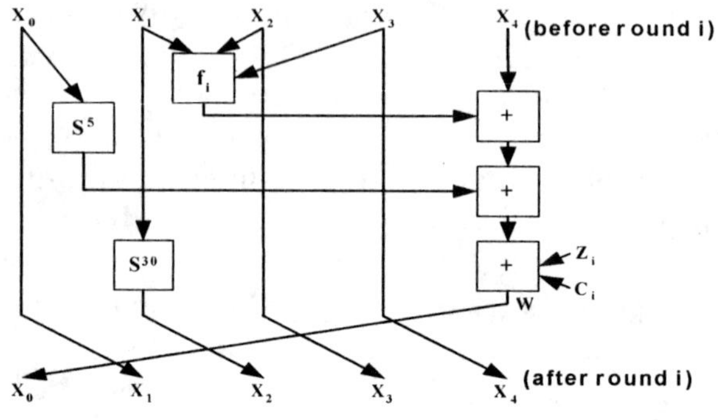

Dedicated Hash Functions

In each SHA-1 round, a hash operation is performed that takes as inputs five 32-bit variables, and two extra 32-bit words. The first one is the message schedule, Z_i, which is provided by the padding unit, and the other word is a constant, C_i, predefined by the standard. Figure 1 shows the diagram of the SHA-1 round function.

The sequence of functions $f_0, f_1, .., f_{79}$ is used in this round-function, where each function f_i, $0 \leq i \leq 79$ takes three words X_1, X_2 and X_3 as input and produces a single word as output. The operations S^5 and S^{30} means circular left shift by 5-bit and 30-bit positions respectively.

The algorithm for generation of message digest is identical for SHA-256 and SHA-512 and only the constants (C_i) and functions, e_i and d_i that they have been used differs (the functions e_i used by SHA-256 and functions d_i used by SHA-512), and hence, SHA-256 and SHA-512 are discussed simultaneously. The diagram of the SHA-256 and SHA-512 round function is depicted in Figure 2.

When a message of any length $<2^{64}$ bits, for SHA-256, or $<2^{128}$ bits, for SHA-512 is input, the hash functions SHA-256 and SHA-512 compute the message digest. The message digest generated by SHA-256 and SHA-512 are 256 and 512 bits long, respectively. The procedure consists of two stages, namely, preprocessing and hash computation. In

Figure 2. The SHA-256 and SHA-512 round function

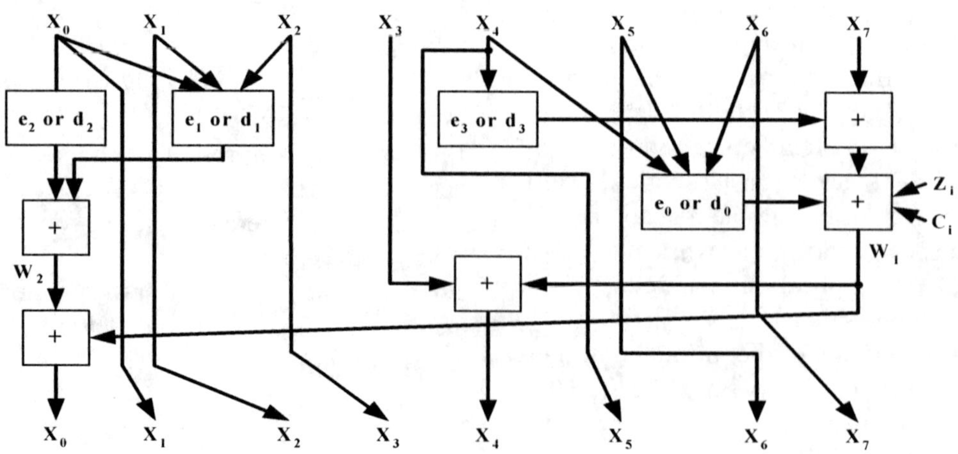

the preprocessing stage, the message is padded, parsed into m-bit blocks and initialization values are been set in order to hash computation. A message scheduler divides the m-bit block into 16 words and prepares a message schedule by passing one word at a time. A series of hash values are generated iteratively from functions, constants, and word operations and the final hash value is the message digest. SHA-256 requires 64 transformation steps (round-functions) while SHA-512 requires 80 round function transformations.

SHA-384 uses exactly the same round function as SHA-512 and requires 80 round function transformations. Only the initialization values are different. The 384-bit message digest is obtained by truncating the SHA-512-based hash output to its left-most 384-bit.

RIPEMD-160 and RIPEMD-128 replaces the previous published version of RIPEMD and overcomes the security problems that they have raised (see Dobbertin, 1997). The main design principle of both hash functions is to maximize the confidence gained by RIPEMD, but with as few changes as possible to the original structure. The produced message digest, ranges in length from 128- to 160-bit, depending on the selected hash function each time. These hash functions enable the determination of a message's integrity. Any change to the message will, with a very high probability, result in a different produced message digest.

The round-function of RIPEMD-160 is described in terms of operations on 32-bit words. A sequence of functions g_0, g_1, .., g_{79} is used in this round-function, where each function g_i, $0 \leq i \leq 79$, takes three words X_1, X_2 and X_3 as input and produces a single word as output. Two sequences of constant words C_0, C_1, .., C_{79} and C'_0, C'_1, .., C'_{79} are used in this round-function. Besides, two sequences of 80 shift-values are used in this round-function, where each shift-value is between 5 and 15. The diagram of the RIPEMD-160 round function is illustrated in Figure 3.

As Figure 4 shows, the round-function of RIPEMD-128 is described in terms of operations on 32-bit words. A sequence of functions g_0, g_1, .., g_{63} is used in this round-function, where each function g_i, $0 \leq i \leq 63$, takes three words X_1, X_2 and X_3 as input and produces a single word as output. Two sequences of constant words C_0, C_1, .., C_{63} and C'_0, C'_1, .., C'_{63} are used in this round-function. Two sequences of 64 shift-values are also used in this round-function, where each shift-value is between 5 and 15.

Whirlpool Hash Function Specifications

Whirlpool is a one-way, collision resistant 512-bit hash function operating on messages less than 2^{256} bits in length. It consists of the iterated application

Figure 3. The RIPEMD-160 round function

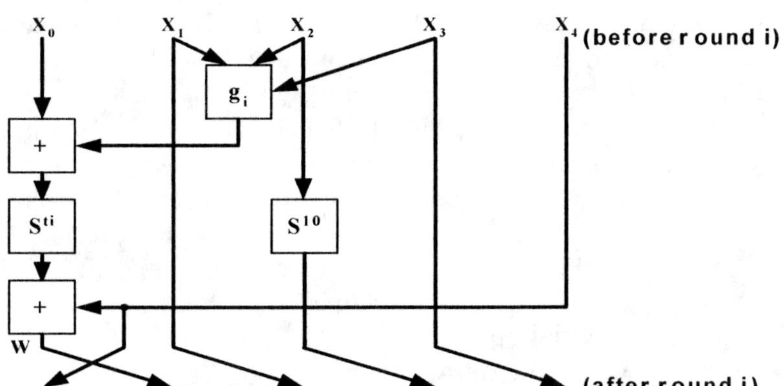

Figure 4. The RIPEMD-128 round function

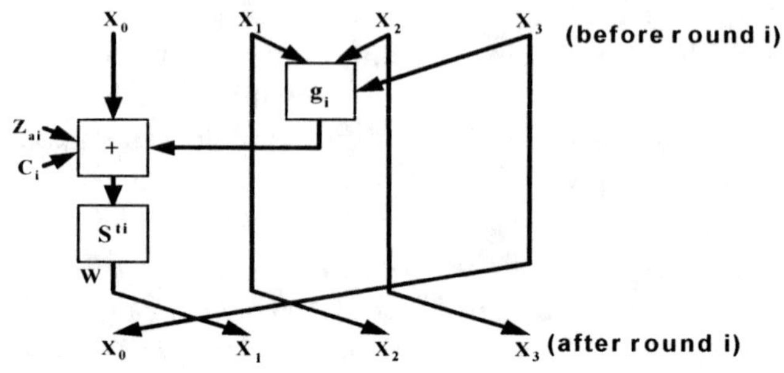

of a compression function, based on an underlying dedicated 512-bit block cipher that uses a 512-bit key. The Whirlpool is a Merkle hash function (Menezes, Van Oorschot, & Vastone, 1997) based on a 512-bit block cipher, W, using a chained 512-bit key state, both derived from the input data. The round function, of the W, is operating in the Miyaguchi-Preneel mode (Menezes et al.) as shown in Figure 5.

As Figure 5 shows, a 512-bit data block, m_i, with a 512-bit key, h_{i-1}, is used for the operation of W block cipher. The output of the block cipher with the original input data block and also with the input key are all together XORed in order to produce the hash value, h_i. This hash value is used as a key in the next input data block.

In the rest of this chapter, the round function of the block cipher, W, is defined. The block diagram of the W block cipher basic round is depicted in Figure 6. The round function, $\rho[k]$, is based on combined operations from three algebraic functions. These functions are the non-linear layer γ, the cyclical permutation π, and the linear diffusion layer θ. So, the round function is the composite mapping $\rho[k]$, parameterized by the key matrix k, and given by the following equation.

$$\rho[k] \equiv \sigma[k] \circ \theta \circ \pi \circ \gamma \qquad (1)$$

Symbol "$\circ$" denotes the sequential operation of each algebraic function where the right function is executed first.

The key addition $\sigma[k]$, consists of the bitwise addition (XOR) of a key matrix k such as:

$$\sigma[k](a) = b \Leftrightarrow b_{ij} = a_{ij} \oplus k_{ij}, 0 \le i, j \le 7 \qquad (2)$$

This mapping is also used to introduce round constants in the key schedule. The input data (hash state) is internally viewed as a 8x8 matrix over $GF(2^8)$. Therefore, 512-bit data string must be mapped to and from this matrix format. This can be done by function μ such as:

$$\mu(a) = b \Leftrightarrow b_{ij} = a_{8i+j}, 0 \le i, j \le 7 \qquad (3)$$

The first transformation of the hash state is through the non-linear layer γ, which consists of the parallel application of a non-linear substitution *S-Box* to all bytes of the argument individually. After, the hash state is passed through the permutation π that cyclical shifts each column of its argument independently, so that column j is shifted downwards by j positions. The final transformation is the linear diffusion layer θ, which the hash state is multiplied with a generator matrix. The effect of θ is the mix of the bytes in each state row.

So, the dedicated 512-bit block cipher $W[K]$, parameterized by the 512-bit cipher key K, is defined as:

$$W[K] = \left(O_1^{r=R} \rho[K^r]\right) \circ \sigma[K^0] \qquad (4)$$

Figure 5. Whirlpool hash function

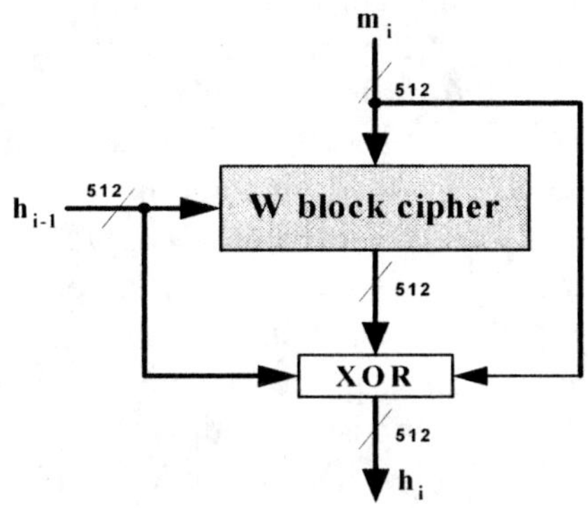

where, the round keys $K^0, \ldots, K^R$ are derived from K by the key schedule. The default number of rounds is R=10. The key schedule expands the 512-bit cipher key K onto a sequence of round keys $K^0, \ldots, K^R$ as:

$$K^0 = K$$
$$K^r = \rho[c^r](K^{r-1}), r > 0 \quad (5)$$

The round constant for the r-th round, r>0, is a matrix c^r defined by substitution box (S-Box) as:

$$c_{oj}^r \equiv S[8(r-1)+j], 0 \le j \le 7,$$
$$c_{ij}^r \equiv 0, \qquad 1 \le i \le 7, 0 \le j \le 7 \quad (6)$$

So, the Whirlpool iterates the Miyaguachi-Preneel hashing scheme over the t padded blocks m_i, $1 \le i \le t$, using the dedicated 512-bit block cipher W:

$$n_i = \mu(m_i),$$
$$H_0 = \mu(IV),$$
$$H_i = W[H_{i-1}](n_i) \oplus H_{i-1} \oplus n_i, 1 \le i \le t \quad (7)$$

where, *IV* (the Initialization Vector) is a string of 512 0-bits.

As Equations 4 and 5 show the internal block cipher W, comprises of a data randomizing part and a key schedule part. These parts consist of the same round function.

Before being subjected to the hashing operation, a message *M* of bit length $L < 2^{256}$ is padded with a 1-bit, then as few 0-bits as necessary to obtain a bit string whose length is an odd multiple of 256,

Figure 6. Block diagram of the W basic round with algebraic functions transformations

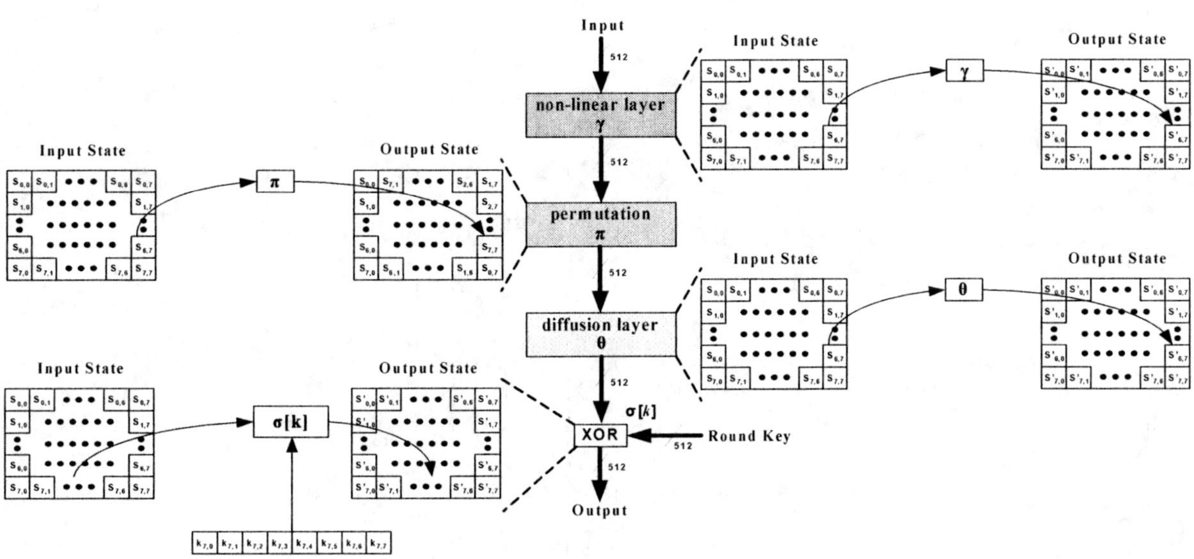

Figure 7. Whirlpool hash function architecture

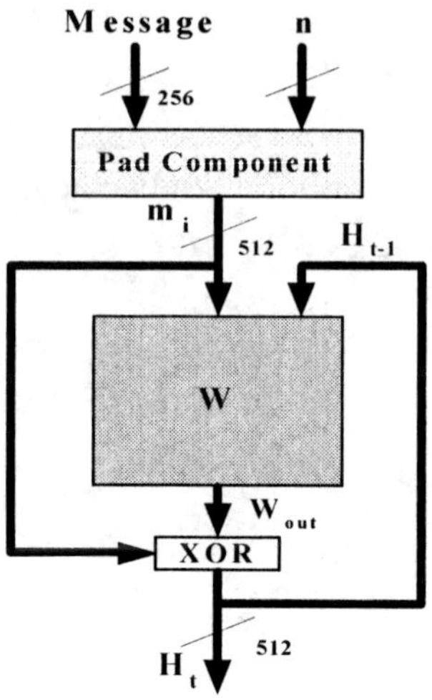

and finally with the 256-bit right-justified binary representation of L, resulting in the padded message m, partitioned in t blocks $m_1, m_2, \ldots, m_t$.

WHIRLPOOL ARCHITECTURES AND VLSI IMPLEMENTATIONS

In this paragraph the proposed architecture and implementations are explained in detail of the hash function Whirlpool. A general diagram of the architecture that performs the Whirlpool hash function is shown in Figure 7. The *Pad Component* pads the input data and converts them to n-bit padded message. In the proposed architecture an interface with 256-bit input for *Message* is considered. The input n, specifies the total length of the message. The padded message is partitioned into a sequence of t 512-bit blocks $m_1, m_2, \ldots, m_t$. This sequence is then used in order to generate a new sequence of 512-bit string, $H_1, H_2, \ldots, H_t$ in the following way. m_i is processed with H_{i-1} as key, and the resulting string is XORed with m_i in order to produce the H_i. H_0 is a string of 512 0-bits and H_t is the hash value.

The block cipher W, is mainly consists of the round function ρ. The implementation of the round function ρ is illustrated in Figure 8.

The non-linear layer γ, is composed of 64 substitution tables (S-Boxes). The internal structure of the S-Box is shown in Figure 8. It consists of five 4-bit mini boxes E, E^{-1}, and R. These mini boxes can be implemented either by using look-ip-tables (LUTs) or Boolean expressions. Next, the cyclical permutation π, is implemented by using combinational shifters. These shifters are cyclically shift (in downwards) each matrix column by a fixed number (equal to j), in one clock cycle. The linear diffusion layer θ, is a matrix multiplication between the hash state and a generator matrix. In Barreto and Rijmen (2003) an efficient method is provided in order to implement the matrix multiplication. However, in this chapter an alternative way is proposed which

Equation 8.

$$b_{i0} = a_{i0} \oplus a_{i1} \oplus a_{i3} \oplus a_{i5} \oplus a_{i7} \oplus X[a_{i2}] \oplus X^2[a_{i3} \oplus a_{i6}] \oplus X^3[a_{i1} \oplus a_{i4}]$$
$$b_{i1} = a_{i0} \oplus a_{i1} \oplus a_{i2} \oplus a_{i4} \oplus a_{i6} \oplus X[a_{i3}] \oplus X^2[a_{i4} \oplus a_{i7}] \oplus X^3[a_{i2} \oplus a_{i5}]$$
$$b_{i2} = a_{i1} \oplus a_{i2} \oplus a_{i3} \oplus a_{i5} \oplus a_{i7} \oplus X[a_{i4}] \oplus X^2[a_{i5} \oplus a_{i0}] \oplus X^3[a_{i3} \oplus a_{i6}]$$
$$b_{i3} = a_{i0} \oplus a_{i2} \oplus a_{i3} \oplus a_{i4} \oplus a_{i6} \oplus X[a_{i5}] \oplus X^2[a_{i6} \oplus a_{i1}] \oplus X^3[a_{i4} \oplus a_{i7}]$$
$$b_{i4} = a_{i1} \oplus a_{i3} \oplus a_{i4} \oplus a_{i5} \oplus a_{i7} \oplus X[a_{i6}] \oplus X^2[a_{i7} \oplus a_{i2}] \oplus X^3[a_{i5} \oplus a_{i0}]$$
$$b_{i5} = a_{i0} \oplus a_{i2} \oplus a_{i4} \oplus a_{i5} \oplus a_{i6} \oplus X[a_{i7}] \oplus X^2[a_{i0} \oplus a_{i3}] \oplus X^3[a_{i6} \oplus a_{i1}]$$
$$b_{i6} = a_{i1} \oplus a_{i3} \oplus a_{i5} \oplus a_{i6} \oplus a_{i7} \oplus X[a_{i0}] \oplus X^2[a_{i1} \oplus a_{i4}] \oplus X^3[a_{i7} \oplus a_{i2}]$$
$$b_{i7} = a_{i0} \oplus a_{i2} \oplus a_{i4} \oplus a_{i6} \oplus a_{i7} \oplus X[a_{i1}] \oplus X^2[a_{i2} \oplus a_{i5}] \oplus X^3[a_{i0} \oplus a_{i3}]$$

Figure 8. Implementation of the round function ρ

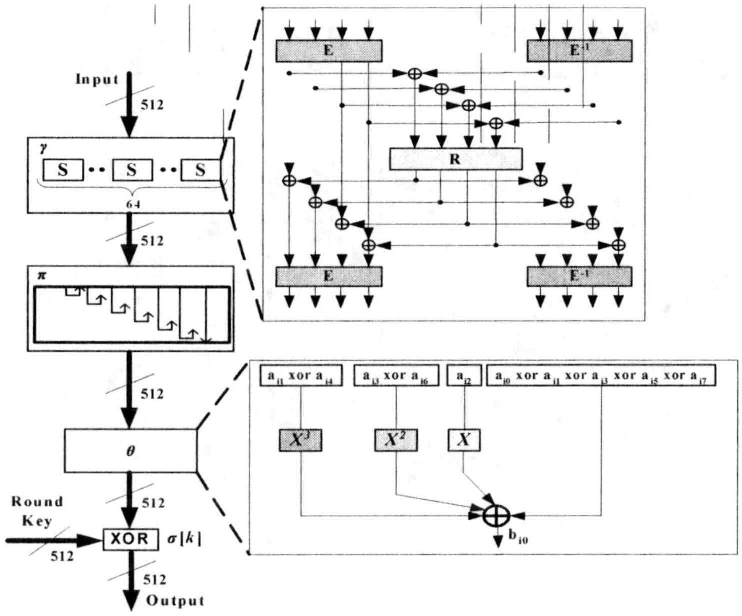

is suitable for hardware implementation. The transformation expressions of the diffusion layer are given next. (See Equation 8.)

Bytes $b_{i0}, b_{i1}, b_{i2},\ldots, b_{i7}$ represent the eight bytes of the i row of the output of the layer θ hash state. Table X implements the multiplication by the polynomial $g(x)=x$ modulo $(x^8+x^4+x^3+x^2+1)$ in $GF(2^8)$. Table X^2 is defined as $X^2 \equiv X \circ X$ and X^3 as $X^3 \equiv X \circ X \circ X$. In Figure 8, the implementation of the output byte b_{i0} is depicted in details. The other bytes are implemented in a similar way. The key addition ($\sigma[k]$) consists of eight 2-input XOR gates for any byte of the hash state. Every bit of the round key is XORed with the appropriate bit of the hash state.

The first implementation is depicted in Figure 9. This implementation has two similar parallel data paths, the *data randomizing* and the *key schedule*. The implementation details of the non-linear layer γ, the cyclical permutation π, and the linear diffusion layer θ are shown in Figure 8. The input block m_i is set to the *Input data* simultaneously with the *initial vector* (*IV*) to the *Key*. In the key schedule data path, the output data of the θ layer is bitwise XORed with the c^r constant. A round key is produced, on the fly, in one clock cycle. Each produced round key is used in the next clock cycle (through the multiplexer) for the production of the next round key. In the data randomizing data path, the hash state of the θ layer is bitwise XORed with the appropriate round key. After, the intermediate feedback data are used as input to the next round (through the multiplexer). After 10 execution rounds the *Output Register* latches the *temp* value. This is bitwise XORed with the H_{i-1} value in order to compute the W_{out}.

In a clock cycle, one execution round is executed and, simultaneously, the appropriate round key is calculated. The system needs 10 clock cycles per block. If another block m_{i+1} is required to be transformed, the previous process is repeated (by using as cipher key the H_i value). So, for t blocks the execution time is $10*t$ clock cycles.

The second implementation of the W block cipher architecture is shown in Figure 10. This implementation is suitable for applications with constrained silicon area resources. The appropriate key schedule part is integrated with the data

Figure 9. The implementation of the W block cipher suitable for high speed applications

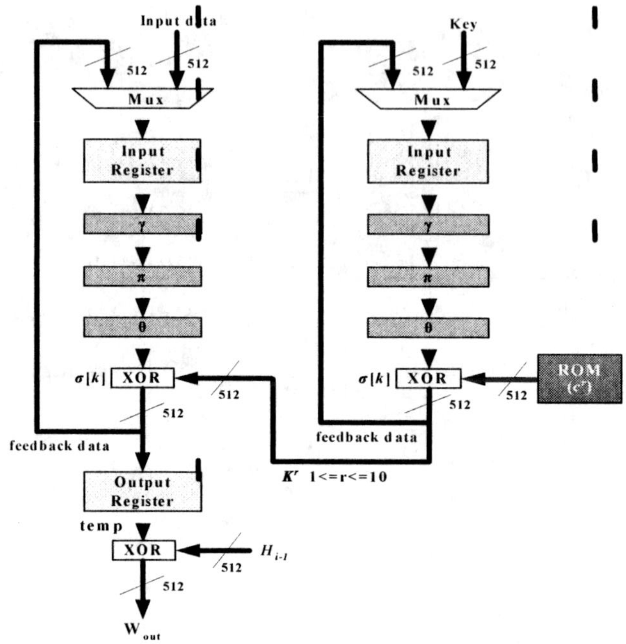

randomizing part in order to reduce the required hardware resources. The execution of the W block cipher on this implementation is performed in two phases. In the first phase, the round keys are produced and stored in the *RAM*. In the second phase, the hash value is computed. The algorithm specifies 10 rounds for the hash state. The *Input data* is the *initialization vector* (*IV*), in order to produce the round keys (first phase). The *Input Register* is used for buffering the algorithm *Input data*. The output data of the θ layer is bitwise XORed with the c^r constant. Each execution round lasts one clock cycle. After the first execution round, the first round key is stored in the *RAM*. It is used as input in the second execution round, through the multiplexer (feedback data), for the production of the second round key. This process is repeated 10 times (10 execution rounds) and lasts 10 clock cycles. The c^r constants are predefined and stored in the *ROM*. The multiplexer selects during the first phase the c^r constants, and during the second phase the round keys. The computation of the hash value is taking place during the second phase. In this phase, the Input data is the m_i block. The output data of the θ layer is bitwise XORed with the appropriate round key, which is stored in the *RAM*. After 10 execution rounds the *Output Register* latches the *temp* result. This result is bitwise XORed with the H_{i-1} value (in this case is equal to the IV) in order to compute the W_{out}. The W_{out} is XORed with the m_i (see figure 7), so the final, hash value H_i, is computed.

If another block m_{i+1} is required to be transformed, the previous process is repeated (by using as cipher key the H_i value). So, for t blocks the execution time is $20*t$ clock cycles. This has a result the total throughput of this implementation is half than the first implementation; however it needs almost half silicon area.

IMPLEMENTATION RESULTS AND DISCUSSION

The VIRTEX FPGA device used in order to evaluate the performance of the proposed implementations. Especially the XC4VLX100 device is used; this device belongs to a new family manufactured in 1.2 volts, 90nm triple-oxide technology and offers twice the performance, twice the density, and less than one-half the power consumption of

Figure 10. The implementation of the W block cipher suitable for applications with constrained silicon area resources

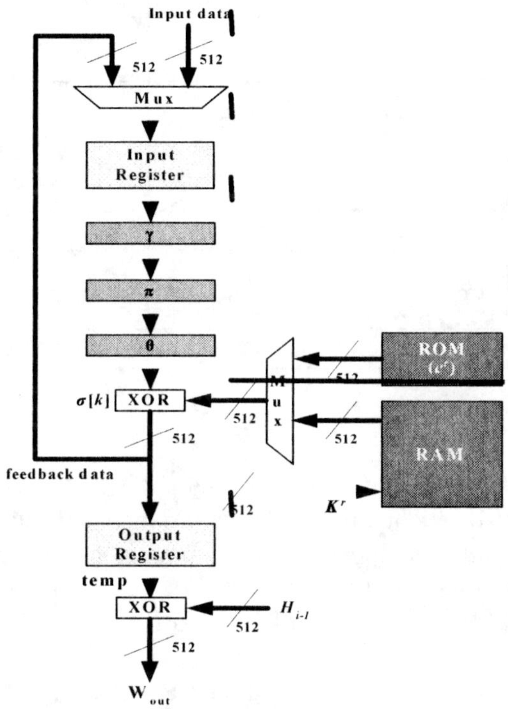

previous-generation devices. The basic building block of these devices is the DSP48 slice (see Xilinx, 2006). The purpose of this module is to deliver off-the-shelf programmable devices with the best mix of logic, memory, I/O, processors, clock management, and digital signal processing. In Figure 11 the DSP48 slice architecture is depicted. The Virtex-4 DSP slices are organized as vertical DSP columns. Within the DSP column, two vertical DSP slices are combined with extra logic and routing to form a DSP tile. The DSP tile is four CLBs tall. Each DSP48 slice has a two-input multiplier followed by multiplexers and a three-input adder/subtractor. The multiplier accepts two 18-bit, two's complement operands producing a 36-bit, two's complement result. The result is a sign extended to 48 bits that can optionally be fed to the adder/subtractor. The adder/subtractor accepts three 48-bit, two's complement operands, and produces a 48-bit two's complement result. Higher level DSP functions are supported by cascading individual DSP48 slices in a DSP48 column. One input (cascade B input bus) and the DSP48 slice output (cascade P output bus) provide the cascade capability.

The XC4VLX100 device used in this chapter contains 96 DSP48 slices.

Each one of the proposed implementations was captured by using VHSIC hardware description language (VHDL), with structural description logic. Both implementations were simulated to operating correctly by using the test vectors which are provided by the NESSIE submission package (NESSIE, 2004), and the ISO/IEC 10118-3 standard (ISO, 2003). Parts of the proposed implementations were designed by using two alternative techniques.

Figure 11. The DSP48 slice architecture

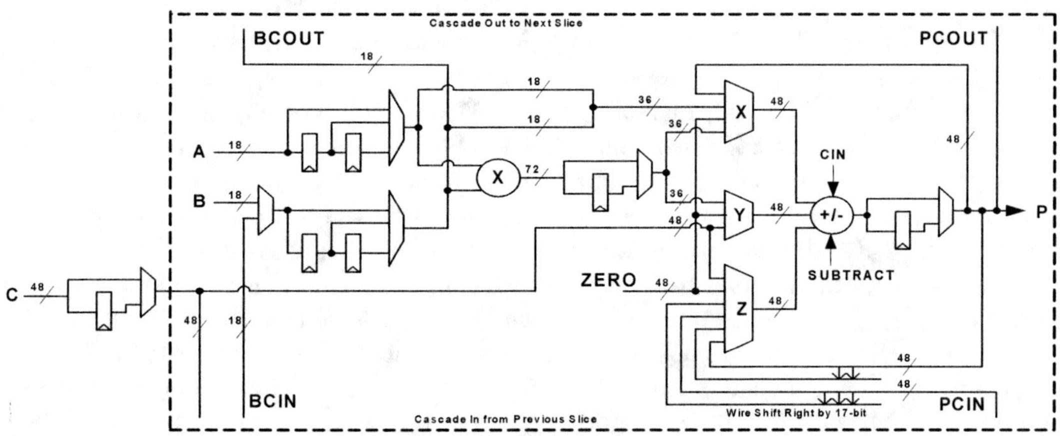

Table 1. Performance analysis measurements

Implementation	FPGA Device	Slices / BRAM	Frequency (MHz)	Throughput (Mbps)	Throughput / Slices
In McLoone et al. (2005) Unroll x 2	XC4VLX100	13210 / 0	47.8	4896	0.37
In McLoone et al. (2005) Iterative	XC4VLX100	4956 / 68	144	4790	0.96
In Pramstaller et al. (2006)	XC2VP40	1456 / 0	131	382	0.26
Author 1st_impl_BB	XC4VLX100	5021 / 0	236	12083	2.40
Author 1st_impl_LB	XC4VLX100	4848 / 0	337	17254	3.56
Author 2nd_impl _BB	XC4VLX100	3451 / 0	275	7040	2.04
Author 2nd_impl _LB	XC4VLX100	3376 / 0	313	8013	2.37

The 4-bit mini boxes (E, E^{-1}, and R) were designed by using LUTs and Boolean expressions. The usage of FPGA-LUTs does not increase the algorithm execution latency. Besides, the LUTs are implemented by using function generators. So, for the implementation of the Whirlpool hash function four alternative solutions are proposed.

Two performance metrics are considered: the area utilized and the throughput achieved by the implementations. The measurements of the performance analysis are shown in Table 1. And also, comparisons with other Whirlpool hash hardware implementations (McLoone et al., 2005; Pramstaller et al., 2006) are given. We symbolized as Boolean expressions based (BB) the mini boxes implementations by using Boolean expressions, and as LUT based (LB) the mini boxes implementations by using FPGA-LUTs.

Both implementations (1st and 2nd) were realized by the same FPGA device. The algorithm constants (c^r) are stored in a ROM which is implemented by using LUT. The 2nd implementation uses a 10x512-bit RAM in order to store the necessary round keys. This RAM is mapped to the 5K bits distributed RAM, and furthermore, none of the proposed implementations use block RAM (BRAM).

The 1st implementation requires 10 clock cycles for each block. So, the BB implementation throughput is 12 Gbps at 236 MHz clock frequency, and the LB implementation throughput is 17.2 Gbps at 337 MHz. The 2nd implementation was designed in order to support applications with area restrict requirements. It demands 20 clock cycles for each data block and requires less hardware resources. The BB implementation throughput is 7 Gbps at 275 MHz clock frequency and the LB implementation throughput is 8 Gb/s at 313 MHz.

In McLoone et al. (2005) two Whirlpool hash hardware implementations are presented. In the first one, two rounds of the block cipher W are unrolled and during one clock cycle two rounds are performed. This method reduces the overall latency of the design, but it will also result in a reduction in frequency. In order to compute the final hash output needs to be iterated five times. This implementation achieves a throughput equal to 4896 Mbps at 47.8 MHz. The second one is iterative implementations with algorithmic latency equal to 10 clock cycles. The major difference with previous and also with author implementations is that use BRAM in order to implement the S-boxes. The throughput of this implementation is 4790 Mbps at 144 MHz. An 68 BRAM is also used.

In Pramstaller et al. (2006) a very compact Whirlpool hash hardware implementation is discussed. This design has different philosophy than the implementations in this chapter and uses an innovative state representation that makes it possible to reduce the required hardware resources remarkably. The complete implementation into XC2VP40 VIRTEX FPGA requires 1456 CLB-slices and no BRAMs. It achieves a throughput equal to 382 Mbps at a clock frequency equal to 131 MHz.

As Table 1 shows that the author's proposed hardware implementations of the Whirlpool hash function clearly outperforms all the others implementations. The proposed implementations are faster by a factor range from 1.5 to 45 times. Especially comparing with implementations in McLoone et al. (2005) some important results can be extracted. Firstly, the two implementations in McLoone et al., use the same FPGA device with the proposed implementations reported in this chapter. So, any comparisons are absolutely fair and accurate. Secondly, by using FPGA-LUTs much better results are achieved in both time performance and area requirements. Finally, about the ratio throughput per slice, that measures the hardware resource cost associated with the implementation resulting throughput and it is proven that the proposed implementations in this chapter philosophy matches better than the implementations in McLoone et al., to the FPGA characteristics (due to the high throughput per slice ratio). The design in Pramstaller et al. (2006) achieves a throughput equal to 382 Mbps at 131 MHz slower by a factor range from 18 to 45 compared with the implementations in this chapter. Although, as I have already mentioned, this design has different philosophy and requires only a small amount of hardware resources.

Besides, comparisons with some other hash families' implementations (Ahmad & Shoba Das, 2005; Deepakumara et al., 2001; Dominikus, 2002; Grembowski et al. 2002; McLoone & McCanny, 2002; Sklavos & Koufopavlou, 2003, 2005; Yiakoumis et al., 2005) (the faster implementations of other hash families' are collected) are given in Table 2 in order to have a fair and detailed comparison with the proposed implementations.

From Table 2, it is obvious that the Whirlpool implementation performs much better in terms of throughput, comparing to all the previous hash families published implementations (Ahmad & Shoba Das, 2005; Deepakumara et al., 2001; Dominikus, 2002; Grembowski et al., 2002; McLoone & Mc-

Table 2. Comparisons with other hash families' implementations

Implementation	FPGA Device	Slices	Frequency (MHz)	Throughput (Mbps)
MD5 (Dominikus, 2002)	XV300E	1004	42.9	146
MD5 (Deepakumara et al., 2001)	XV1000FG680	4763	71.4	354
SHA-1 (Yiakoumis et al., 2005)	Virtex-II	854	162	1036.8
SHA-2 (512) (Sklavos & Koufopavlou, 2003)	XCV200	2237	75	480
SHA-2 (512) (Grembowski et al., 2002)	XCV1000	3441	55.5	670
SHA-2 (512) (McLoone & McCranny, 2002)	XC4VLX100	2734 + 2 BRAM	~	854
SHA-2 (512) (Ahmad & Shoba Das, 2005)	STRATIX EP1S10F484C5	4229 LEs	47.9	1226
RIPEMD-128 (Sklavos & Koufopavlou, 2005)	2V250FG456	1814	78	2300
RIPEMD-160 (Sklavos & Koufopavlou, 2005)	2V250FG456	2014	73	2100
Author 1st_impl_BB	XC4VLX100	5021	236	12083
Author 1st_impl_LB	XC4VLX100	4848	337	17254
Author 2nd_impl_BB	XC4VLX100	3451	275	7040
Author 2nd_impl_LB	XC4VLX100	3376	313	8013

Cranny, 2002; Sklavos & Koufopavlou, 2003, 2005; Yiakoumis et al., 2005) . The implementation in McLoone and McCranny (2002) uses the same FPGA device as the proposed implementations in this chapter. It also, requires more hardware resources compared with the other hash families' implementations. This is a logical result of the algorithm philosophy and not an implementation trade-off. Finally the Whirlpool has the smaller algorithm execution latency. It needs only 10 clock cycles in order to transform each block compared with the 64 clock cycles of the MD5, and SHA-2 (256), and 80 clock cycles of the RIPEMD-160, SHA-1, and SHA-2 (384, 512). This is an important advantage of the hardware implementation.

CONCLUSION

The Whirlpool hash function is the most recent hash function to be standardized. It was selected to be included in the NESSIE portfolio of cryptographic primitives. An efficient architecture and VLSI implementations for this hash function are presented in this chapter. Two architectures for *W* block cipher are introduced. The first one is appropriate for high speed applications since the round keys are produced on the fly while the second one is appropriate for area restricted devices. Parts of the proposed implementations were designed by using two alternative techniques. The 4-bit mini boxes (E, E^{-1}, and R) were designed by using LUTs and Boolean expressions. So, four implementations have been introduced and each specific application can choose the appropriate speed-area, trade-off implementation. The achieved throughput for the proposed implementations ranges from 7 Gbps to 17.2 Gbps. These hardware architectures and implementations are significantly faster than any other previous reported implementations of the algorithm and they are also up to 16.5 times faster than hardware implementations of other hash functions.

REFERENCES

Adams, C., & Farrell, S. (1999, March). *Internet X.509 PKI—Certificate management protocols (RFC 2510).* Retrieved March 1999, from, http://www.ietf.org/rfc/rfc2510.txt

Ahmad, I., & Shoba Das, A. (2005). Hardware implementation analysis of SHA-256 and SHA-512 algorithms on FPGAs. *Computers and Electrical Engineering, 31*, 345-360.

Barreto, P. S. L. M., & Rijmen, V. (2003, May). *The whirlpool hashing function* (Rev. ed.). Paper presented at the NESSIE.

Deepakumara, J., Heys, H. M., & Venkatesam, R. (2001). FPGA Implementation of MD5 hash algorithm. In *IEEE Canadian Conference on Electrical and Computer Engineering* (CCECE 2001) (Vol. 2, pp. 919-924).

Dobbertin, H. (1997). RIPEMD with two-round compress function is not collision free. *Journal of Cryptology, 10*, 51-69.

Dobbertin, H., Bosselaers, A., & Preneel, B. (1996). RIPEMD-160, a strengthened version of RIPEMD. In *Fast Software Encryption (FSE'96)* (LNCS 1039, pp. 71-82). Springer-Verlag.

Dominikus, S. (2002). A hardware implementation of MD4-Family algorithms. In *IEEE International Conference on Electronics Circuits and Systems (ICECS 2002)* (pp. 15-18).

Grembowski, T., Lien, R., Gaj, K., Nguyen, N., Bellows, P., Flidr, J., et al. (2002). Comparative analysis of the hardware implementations of hash functions SHA-1 and SHA-512. In *Fifth International Conference on Information Security* (LNCS 2433, pp. 75-89). Springer-Verlag.

International Organization for Standardization (ISO). (2004). *ISO/IEC 10118-3: Information technology—Security techniques—Hash functions—Part 3: Dedicated hash-functions.* Retrieved 2003, from http://www.iso.org/iso/en/CatalogueDetailPage.CatalogueDetail?CSNUMBER=39876.

Kitsos, P., & Koufopavlou, O. (2004). Efficient architecture and hardware implementation of the whirlpool hash function. *IEEE Transactions on Consumer Electronics, 50*(1), 208-213.

McLoone, M., & McCanny, J. V. (2002). Efficient single-chip implementation of SHA-384 & SHA-512. In *IEEE International Conference on Field-Programmable Technology (FPT)* (pp. 311-314).

McLoone, M., McIvor, C., & Savage, A. (2005). High-speed hardware architectures of the whirlpool hash function. In *IEEE International Conference on Field-Programmable Technology (FPT)* (pp. 13-18).

Menezes, A. J., Van Oorschot, P. C., & Vastone, S. A. (1997). *Handbook of applied cryptography.* CRC Press.

National Institute of Standards and Technology (NIST). (1995, April 17). *SHA-1 standard, secure hash standard (FIPS PUB 180-1).* Retrieved April 17, 1995, from http://www.itl.nist.gov/fipspubs/fip180-1.htm

National Institute of Standards and Technology (NIST). (2002, August 1). *SHA-2 standard, secure hash standard (FIPS PUB 180-2).* Retrieved August 1, 2002, from http://csrc.nist.gov/publications/fips/fips180-2/fips180-2.pdf

National Institute of Standards and Technology (NIST). (2005, December). *SP800-77, Guide to IPSec VPN's.* Retrieved December 2005, from http://csrc.nist.gov/publications/nistpubs/800-77/sp800-77.pdf

New European scheme for signatures, integrity, and encryption (NESSIE). (2004). Retrieved March 2004, from https://www.cosic.esat.kuleuven.ac.be/nessie

Pramstaller, N., Rechberger, C., & Rijmen, V. (2006). A compact FPGA implementation of the hash function whirlpool. In *14th ACM/SIGDA International Symposium on Field-Programmable Gate Arrays - FPGA* (pp. 159-166). ACM Press.

Sklavos, N., & Koufopavlou, O. (2003). On the hardware implementation of the SHA-2 (256, 384, 512) hash functions. In *IEEE International Symposium on Circuits and Systems (ISCAS 2003)* (Vol. V, pp. 153-156).

Sklavos, N., & Koufopavlou, O. (2005). On the hardware implementation of RIPEMD processor: Networking high speed hashing, up to 2 Gbps. *Computers and Electrical Engineering, 31,* 361-379.

Wang, X., Yin, Y. L., & Yu, H. (2005). Finding collisions in the full SHA-1. In *Advances in cryptology, 25th Annual International Cryptology Conference* (LNCS 3621, Santa Barbara, CA pp. 17-36).

Xilinx Incorporated. (2006). *Silicon solutions—Virtex series FPGAs.* Retrieved October 10, 2006, from http://www.xilinx.com/products/

Yiakoumis, I., Papadonikolakis, M., Michail, H., Kakarountas, A. P., & Goutis, C. E. (2005). Efficient small-sized implementation of the keyed-hash message authentication code. In *IEEE 2005 International Conference on "Computer as a tool"* (EUROCON) (pp. 1875-1878).

KEY TERMS

Cryptography: In modern times, cryptography has become a branch of information theory, as the mathematical study of information and especially its transmission from place to place. Cryptography is central to the techniques used in computer and network security for such things as access control and information confidentiality.

DSP48 Slice: DSP48 slice is the basic building block of XILINX VIRTEX-4 FPGAs.

Field-Programmable Gate Array (FPGA) Device: FPGA device is a semiconductor device used to process digital information, similar to a microprocessor. It uses gate array technology that can be reprogrammed after it is manufactured, rather than having its programming fixed during the manufacturing—a programmable logic device.

Hardware Implementation: Hardware implementation is the building of the blocks of digital chip (either ASIC or FPGA) design and it relates them to the hardware description languages that are used in their creation.

Hash Function: Hash function is a function that maps an input of arbitrary length into a fixed number of output bits, the hash value.

New European Schemes for Signatures, Integrity, and Encryption (NESSIE): NESSIE was a European project that was responsible to introduce new cryptographic primitives with high security levels.

Whirlpool Hash Function: Whirlpool hash function is the most recent hash function to be standardized. It was selected to be included in the NESSIE project of cryptographic primitives.

Section II
Security in 3G/B3G/4G

Chapter XVIII
Security in 4G

Artur Hecker
Ecole Nationale Supérieure des Télécommunications (ENST), France

Mohamad Badra
National Center for Scientific Research, France

ABSTRACT

The fourth generation (4G) of mobile networks will be a technology-opportunistic and user-centric system combining the economic and technological advantages of different transmission technologies to provide a context-aware and adaptive service access anywhere and at any time. Security turns out to be one of the major problems that arise at different interfaces when trying to realize such a heterogeneous system by integrating the existing wireless and mobile systems. Indeed, current wireless systems use very different and difficult to combine proprietary security mechanisms, typically relying on the associated user and infrastructure management means. It is generally impossible to apply a security policy to a system consisting of different heterogeneous subsystems. In this chapter, we first briefly present the security of candidate 4G access systems, such as 2/3G, wireless LAN (WLAN), WiMax, and so forth. In the next step, we discuss the arising security issues of the system interconnection. We namely define a logical access problem in heterogeneous systems and show that both the technology-bound, low-layer and the overlaid high-layer access architectures exhibit clear shortcomings. We present and discuss several proposed approaches aimed at achieving an adaptive, scalable, rapid, easy-to-manage, and secure 4G service access independently of the used operator and infrastructure. We then define general requirements on candidate systems to support such 4G security.

GENERATIONS OF PUBLIC LAND MOBILE NETWORKS

From 1G to 2G

The first generation of public land mobile networks (PLMN) is characterized by the fact that both control channels and traffic channels are analog. Voice (commonly at 3 kHz) and data (if any) are frequency-modulated on a carrier. Today, these networks are usually summarized under the common name first generation (1G) although there are different analog network standards like Nordic mobile telephony (NMT), American mobile phone system AMPS), and total access communication system (TACS).

NMT was the first commercially operated PLMN (1981). NMT uses two different frequency bands about 450 and about 900 MHz (NMT 450 and NMT 900). NMT900 was introduced in 1986 as a result of the fact that the number of channels in NMT 450 was insufficient. NMT 900 has been implemented in Europe, the Middle East, and Asia.

AMPS was specified by the U.S. consortium TIA/EIA/ANSI. The first AMPS network became operational in 1984. In 1988, an extension providing additional frequency bands was added (E-AMPS). AMPS networks are found in the Americas, Australia, and in Asia.

TACS is a modification of AMPS aiming at the British market, where the standard was operational in 1985. TACS also received a wider frequency band in 1988, E-TACS. Since that time, TACS has spread to many countries around the world.

In 1982, at the time of the commercialization of the first 1G networks, the Groupe Spécial Mobile was formed at CEPT (*Conférence européenne des Administrations des Postes et des Tèlécommunications*, the creator and standard-body predecessor of today's *European Telecommunications Standard Institute, ETSI*), with the task of developing a Europe-wide standard for cellular communication. In other words, the scope here was to provide the same service (voice) by a new, universal system.

In 1987 the CEPT working group decided to build a digital, narrowband time division multiple access (TDMA) system. In 1990, ETSI published Phase I of the GSM system specifications. Three frequency bands have been defined for global system for mobile communications (GSM) usage: 900MHz, 1800 MHz, and 1900 MHz. The corresponding standards are similar, aiming at

Table 1. Ten years cycles in the mobile networks (from a European view)

Year	Milestone	Cycles		
1981	*Commercial deployment of NMT:* ***1G start***	1G to 2G: 10 years		
1982	Creation of Groupe Spécial Mobile at CEPT			
1984	Commercial deployment of AMPS networks in the US			
1986	Big number of users leads to NMT extensions			
1988	Big number of users leads to AMPS extensions			
1989	European Union RACE Project "invents" UMTS		3G conception: 10 years	
1992	World Administrative Radio Conference (today: WRC) allocates 230 MHz to Future Public Land Mobile Telecommunication System (FPLMTS).			
1992	*Commercial deployment of GSM:* ***2G start***			2G to 3G: 10 years
1994	Second wave of UMTS research projects			
1995	RACE vision of UMTS			
1996	Creation of UMTS task force			
1996	*Digital overcomes analog*			
1997	Establishment of the UMTS Forum			
1999	UMTS decision			
2000	WRC designates IMT-2000 extension bands			
2002	*Commercial deployment of UMTS:* ***3G start***			

wide-range (GSM 900) and dense-area (GSM 1800/1900) deployments respectively. The first commercial GSM services were launched in the middle of 1991, thus marking the start of the second generation (2G) era.

GSM was the first completely digital PLMN. It is thus naturally a revolutionary approach, as compared to its analog predecessors. GSM defines a series of improvements and innovations compared to previous cellular networks; aiming for an efficient use of the available spectrum; secure transmissions; an improvement in voice quality; a reduction in the cost of handsets (using very large-scale integration [VLSI]); infrastructure and management; an ability to support new services; and a full compatibility with *Integrated Services Digital Network (ISDN)* and with other data transmission networks. Another basic characteristic of the system is called *international roaming*, that is, the possibility for the mobile user to access GSM service even when he/she finds himself/herself physically outside the coverage area for which he/she is subscribed, registering as a "visitor." Provided that the necessary business contracts exist, the roaming is completely automatic. In addition to roaming, GSM offers new user services, including data transmission, fax service, and short message service (SMS).

Thus, in Europe one completely new standard has replaced different existing ones. Almost the contrary happened in the U.S.: the quasi unique AMPS has been replaced by a variety of (at least partially) incompatible, (partially) digital systems: N-AMPS, D-AMPS (IS-54, IS-136), PCS (IS-95), GSM 1900, Omnipoint, and PACS.

The variety of incompatible networks and the increasing popularity of data services have motivated and much influenced the work on the third generation (3G) of mobiles. In 1992, at the same time as the commercial deployment of first 2G networks started, the International Telecommunications Union (ITU) allocated frequency ranges for the next generation of PLMN (then called FPLMTS) thus providing an international common base for the 3G. Finally, in 2002 the first commercial 3G networks were commercially deployed in Japan.

Table 1 summarizes the history of the PLMN development from the European point of view as presented in Pereira (2000). In particular, it illustrates the repeating approximate 10-year cycles both in the conception phases and in the generation lifetimes.

The Third Generation of PLMN

The 3G of mobiles was expected to be the future global standard for the integrated voice and data communications. 3G was designed in the last decade of the 20th century with the goal to provide enhanced wide-range voice and data services. But it turns out that it changes little in the actual user experience.

Technically, 3G design mainly aimed at the improvement of the radio link performance in the 2G scope. Although the developed standard features drastically improved data rates as compared to 2G, from the point of view of the data services the practically offered data rates can be still considered scarce. This can be observed in a direct comparison to the development of the wired technologies providing home Internet access. From 1994 until 2004, the phone-line Internet access technologies have evolved from V.34 modems (28.8 kbps) over V.90 (56 kbps) to cable (1-2 Mbps shared) and ADSL (originally 500 kbps, 2004 up to 10 Mbps). This means an almost 350-fold increase in 10 years. In the same period, the data rate of the wireless cellular access has not been able to keep up the pace. From the original GSM CSD service introduced in 1994 and providing 9.6 kbps, the cellular systems evolved over General Packet Radio Service (GPRS) (about 64 kbps in practice) to EDGE/cdma2000 RTT-1X (typically about 100-130 kbps). The 3G (e.g., UMTS) provides about 300 kbps in practice. This corresponds to a 30-50 fold increase in the same decade. Moreover, the provided data rates highly depend on the network operator's overall capacity, the number of users in the cell and the distance to the base station.

However, the relatively limited data rate is not the only problem of the 3G data service. Because of the vast, national-scope infrastructure, and many intermediate nodes, the user experiences high network latency (e.g., from the point of view of IP, the whole 3G infrastructure is one link). GRPS and EDGE often exhibit network *round trip times* (RTT) of 600 ms and more. UMTS links are expected to be better, but they still have RTT of about 200-250 ms. The wideband code division multiple access (W-CDMA)/high speed downlink packet access (HSDPA) service, defined in the fifth release, is expected to have more than 100 ms RTT, that is, almost an Internet-level RTT. Such high network latencies are inappropriate for certain application classes: interactive applications imply latency constraints that typically lie under the 300-400 ms overall RTT proposed by 3G (the end system is not necessarily within the UMTS backbone and thus the typical Internet RTT of about 100-200 ms has to be added to the 3G latency). For example, voice over IP (VoIP) and similar applications (videoconferencing, etc.) require an RTT to be under 250 ms; in some existing popular interactive online games (e.g., id Software's Quake, etc.), the maximal acceptable RTT to the game server is required to be under 100 ms in order to provide a fair chance of winning and a good game experience.

Furthermore, by its design 3G targets telecommunications providers. Like 2G, 3G uses a license model to prevent random medium access by non-authorized parties. Since the licenses are expensive (Van Damme, 2002), in reality this implies a major telecom operator with a mammoth infrastructure behind every 3G RAN. To fulfill the requirements of such an authority, the 3G RANs are designed to be reliable and manageable and to support different qualities of service. This justifies the high cost of the 3G equipment. At the same time, this limits the competition on the market to few license holders who not only have invested a lot in the infrastructure but also have paid a high price for the license. The operators have to amortize this fixed cost and the current variable maintenance and management cost over the user services provided by the infrastructure. Thus, 3G RAN access is likely to remain costly. It is unclear if attractive unlimited flat-line pricing models (like in xDSL) are applicable to such infrastructures. Current per

Figure 1. 3G: Current and planned UMTS launches

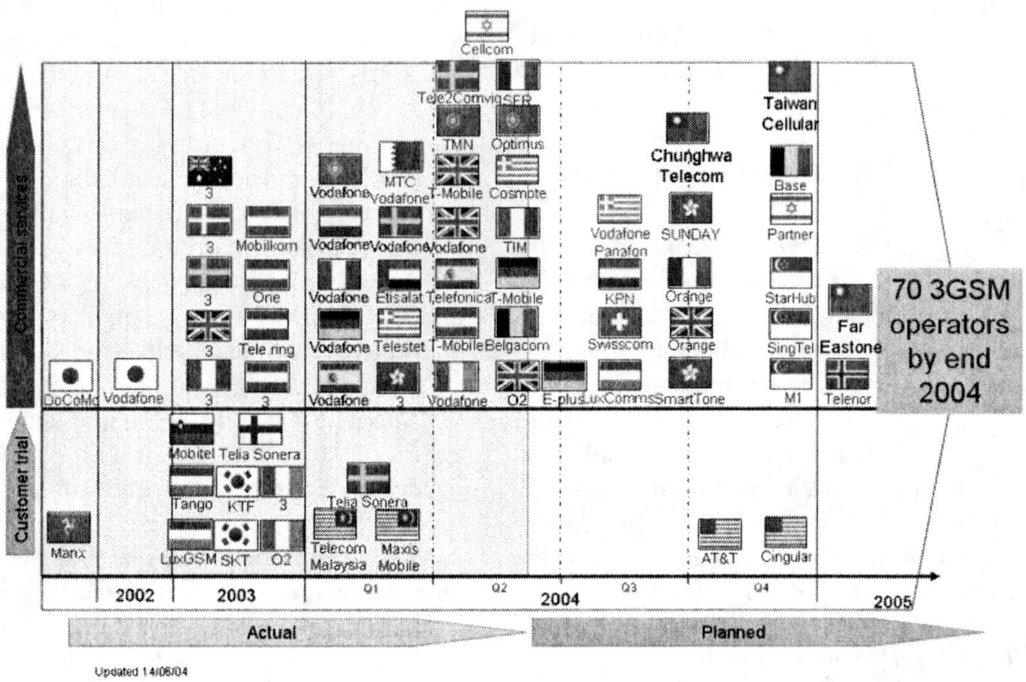

byte (or even per minute!) pricing seems hardly suitable for the always-on paradigm.

A consequent national-scope investment is needed for 3G advantages to materialize (both for users and for providers). This is however difficult to afford, especially in developing countries where big investments are particularly risky. In a focused coverage, 3G comes at a very high cost per bit compared to other, more data-centric technologies like local or metropolitan area networks. That is one of the reasons why the 3G systems had a difficult start. They are primarily being deployed in Japan, South Korea, Taiwan, Hong Kong, Indonesia, a few countries of South America, Australia, New Zealand, western Europe, and North America (CDMA Development Group, n.d.; GSM Association, n.d.). Figure 1 (GSM Association, n.d.) summarizes the actual and planned commercial launches of the 3G system from the 2004 European point of view (W-CDMA/UMTS). It shows that the developed countries prevail.

Although the slow 2G-3G transition process started in 2003-2004, so far the 3G systems do not seem suitable to provide a broadband data access service deployment. In the developed world, these are often considered technologically inadequate (users perceive it as a better 2G). For the developing world, the technology needs major investments. Thus, a new, more flexible technology is necessary, allowing new usage scenarios and business models.

The Anticipated 3G to 4G Transition

In regards to 3G, the observed 10-year cycles seems to continue. The first research concepts aiming at 3G appeared about 1989. The spectrum was reserved by ITU-R's World Radiocommunication Conference (ITU-R Radiocommunication Conference, 1992), that is, at the same time as the first 2G networks were deployed. The active technological development of 3G started with the creation of the UMTS task force in 1996 and culminated in the UMTS decision in 1999. The largest parts of the standards were accomplished by then.

Consequently, the first projects naming fourth generation (4G) started in 1999 and the first dedicated thoughts about beyond 3G (B3G) and 4G systems appeared in the international research press about 2000-2001 (Bria et al., 2001; Evans & Baughan, 2000; Pereira, 2000; Raivio, 2001; Varshney & Jain, 2001), that is, just before the first commercial 3G networks were deployed in Japan. In 2000, the WRC allocated 3G extension bands, which were to be used in the B3G scope. All this corresponds to the 10-year cycles illustrated in Table 1.

Continuing along this line, the concrete shapes of 4G should be clarified by the end of 2007 and the active 4G vision refinement should start about 2007-2008. This should be finished roughly by 2010, with several detail issues being addressed in the following years. The first commercial systems could then be operational by 2012. However, this presumes that no additional delays occur.

Possible Delays

At least in Europe and in the U.S., the 3G deployment seems to be delayed. Indeed, by the end of 2004, not all western European countries started the 3G deployment. Also, the deployment process is starting quite slowly, often being limited to some few centers. The critics of 3G claim that the reasons for this could be in the developed technology itself. Indeed, one could argue that 3G (in Europe: UMTS) is too complicated and too costly to become successful. One could also criticize the fact that the original goal of creating one common global standard has not been achieved since different concurrent versions of 3G are being standardized and deployed, in some extreme cases within the same country (e.g., Japan has deployed both cdma2000 and W-CDMA). However, the deployment of the alternative technologies (like e.g., 802.11 hotspots or WiMax) also lags behind the expectations that have predicted a WiFi-boom and hotspot number explosions by 2005, which so far have failed to become true. There is no doubt about the popularity of WiFi. However it is not booming, it is being carefully developed. The real reasons thus could be either of a social (e.g., a simple current disinterest in mobile data) or of an economic nature (too costly in deployment, too risky for operators; too costly, too complicated for users, etc.).

We tend to think that economic barriers prevail. Indeed, businesses have so far often expressed their need for mobile communications development. This has been much discussed in different business scopes: home- and telework, instant data access for mobile sales personnel, fleet management, reduction of infrastructural costs, globalization, and so forth. With the further development of the Internet and the associated technologies, private users are also likely to be interested in services such as mobile e-commerce, online gaming, private communications (e.g., voice or instant messaging), various personal and business data exchanges, and so forth.

The telecommunication crisis initiated by the complete flop of the exaggerated initial Internet business activities (often referred to as the *bursting of the Internet bubble*) could have been one of the key economic factors responsible for the observed 3G deployment delays. Indeed, the investments in the IT and telecommunication sectors have since radically switched from headlong promiscuity to skeptical cautiousness. From the European point of view, the starting crisis was amplified by the UMTS license auctions in 2000-2001 raising cumulatively over 100 billion USD in the Western European countries (Van Damme, 2002).

The paid spectrum prices washed away much of the liquidity of the Western-European telecommunications operators. Yet, this liquidity was necessary for the deployment of the network (infrastructure updates and add-ons). Since the UMTS cannot substantially improve the GSM voice service as such, the only added value of the UMTS is in the improved data services. Hence, compared to classic GSM offerings, the paid auction price for the UMTS licenses must be amortized over time over the new services, which UMTS is just about to propose. However, this could render these new services particular expensive.

4G: A TECHNOLOGY-OPPORTUNISTIC, USER-CENTRIC SYSTEM

4G Expectations

With the ongoing globalization, world-wide communications become an essential service. The 3G, meant to provide a global communications standard, has mostly failed to do so. Instead, it now uses different standards in different countries. Moreover, 3G remains a closed "big company" telecommunications forum. That results in the situation where users still need costlier multi-band, multi-technology handsets, yet they cannot access the 3G services using other devices over newer radio access networks (RANs). To provide users with a world-wide service we need open flexible standards, also suitable for the Internet and data communications deployment in the developing countries.

At the other end, personal communications are being rapidly developed using short range radios. These need to be considered for the next generation communications because their rapid development is a fact (Raychaudhuri, 2002). The existing personal area networks (PAN) and LAN technologies are often used for device-to-device data transfers but can easily do more than that. Wireless headphones, handsets, and PDAs can already build personal networks capable of data and voice transport. In the home area or in vehicles (e.g., personal cars), this can be extended to LAN-like communications. The aim here is to give users access to their data independently of the device currently in use. So, handsets can be asked to dial numbers stored in the home PC and to direct the voice flow to the wireless headphones. Wireless sensors are already available, for example, for outdoor weather condition measurements. Wireless sensors are used more and more in cars. They are also expected to be further developed for home users (intelligent home). This underlines the increasing part of the machine-to-machine (M2M) and network-to-network (N2N) communications in the future communications landscape.

The obviously challenging scenario is to provide users with a bidirectional communications possibility to their personal Intranets independently of their location (anywhere), thus combining the two topics discussed previously. These WAN/MAN/LAN/PAN spanning communication sessions have to be secure, reliable, and economically reasonable. Also, communications become ubiquitous. The used technology needs to be able to reply to this challenge, providing the best available connection anytime, any place. Existing standards do not allow for this usage.

However, it is not a matter of contention between these existing standards. They are more and more understood as complementary. Indeed, the WLANs can easily provide a true LAN experience in limited areas at a low cost while 3Gs RANs are designed to provide true mobility, quality of services and vast coverage. The idea to try to integrate both technologies is thus straightforward.

Taking into account the previously observed cycles and the current delays, we could try to compile a prognosis on the B3G and 4G development in the next decade. The current situation and our forecast are illustrated in Table 2.

The convergence between the different infrastructures will start because of the economic and technological limits of the used technologies. Big telcos will try to reduce their service cost by integrating alternative transmission technologies as radio access networks (RAN) into their 3G infrastructure (e.g. UMA-like). However, this integration will still be much more complicated and costly than a new deployment possible for a small wireless internet service provider (WISP). At the same time, the small WISPs will encounter increasing management problems with the growing user basis and the user traffic. It will hardly be reasonable to add a 3G infrastructure upon the existing one as the control plane. Given the lack of standardized methods, the alternative infrastructures are thus likely to be managed in a proprietary way, requiring specific access methods. This will produce the demand for standardization.

Because of the true need for mobile broadband data access and the scarce spectrum of 2G, the 3G will be eventually deployed in the business centers of the developed countries despite the currently observed delays. In Europe, this process could be further promoted by governmental policy in some countries planning to partly reimburse some license fees. However, the delays and the high license fee (Van Damme, 2002) have already motivated the development of and the investments in the alternative transmission technologies, for example, IEEE 802.11 and IEEE 802.16.

Table 2. Possible 3G development in the next years

Year	Milestone	Cycles
2003	*European 3G start*	3G to 4G: 10 years
Until 2005	Different 4G visions and early 4G research projects	
2006	3G deployment in all business areas in the developed world	
2006	Broad deployment of alternative technologies (from WiFi to WiMax, etc.)	
2007	Further deployment, different UMTS updates (HSDPA, HSUPA) and integration of alternative technologies in the UMTS infrastructure	
2009	Convergence of different 4G views implied by the economic and technological factors	
2010	The high popularity of data services shows 3G transport limits and WiMax/WiFi management limits (security, mobility, usability, etc.)	
2011	Deployment of first B3G (3.5G) systems	
2011	Establishment of a 4G forum	
2012	Mature technical drafts of 4G systems integrating different technologies	
2014	*First commercial 4G services*	

This development, if commercially successful, will lead to a situation with several parallel infrastructures installed in the European centers by 2008-2009. While the 3G infrastructures will be homogeneous, they are likely to remain more expensive. The alternative offerings will be cheaper but are not likely to provide neither the same service quality nor the same coverage. Because of the required spectrum licenses, the same national-scope operators will own the 3G systems. The alternative technologies are license-free and thus enable a free network deployment. These can be owned by both global big telcos and small local WISPs.

Users will buy newer products equipped with further wireless technologies. Deploying these products at home, users will be interested in accessing the combined service offers. Different devices will be capable of several access methods (e.g., a wireless ADSL router). Users will be incited to open their hotspots for the usage by the others. For instance, a major French telecom provider proposes a reimbursement plan for its ADSL users if they provide WLAN access to its cellular customers over such devices. At the same time, alternative technology operators are forming roaming organizations and user communities, aiming for the same results (see WeROAM, Fon communities, etc.)

Meanwhile, the research will push towards unified and concrete B3G and 4G views. To protect the investments, the deployed alternative infrastructures are likely to be given the necessary attention in this development process. The result will likely be a system providing for a convergence between the different technologies.

While the new 4G architecture is being conceived and is maturing technologically, 3.5G systems are likely to appear on the market by 2010 at the latest, filling the gap between LAN-experience and manageable. These updates of the radio link and of the backbone infrastructure could provide the basis for the later expected 4G much in the same manner as GPRS/EDGE (2.5G) have required and accomplished the necessary infrastructural changes for the transition process from 2G to 3G. The commercial and technological convergence and the available B3G systems will provide the drivers for the establishment of an industry group (e.g., 4G forum) that will be given the task of 4G system standard development. Based on the situation and the previously accomplished research, it could produce mature system drafts by 2012 and the first commercial 4G deployments could start about 2014.

Our 4G Vision

Our vision is motivated by the previous work and the ongoing development of the global telecommunications networks, in particular of the Internet. It respects the fact of the proliferation of the Internet technology in all telecommunications branches and is similar to the All-IP approach when used for data transport.

Learning from 2G and 3G experiences, 4G envisages an architecture that allows the maximum possible infrastructure reuse. The idea is to minimize a risky engagement with a particular technology and to guarantee the long-term flexibility for the involved authorities. We believe that the versatility here can provide an enhanced flexibility both technologically and from the business point of view. This ultimately market-driven solution should be capable of providing any service in any manner, restricted solely by user's demand and not by any technological factors.

From Service-Centric to Data-Centric Approaches, from Technology-Centric to User-Centric Approaches

The classic telecommunications industry approach dominated by the national-scope telecom operators with the well-managed infrastructures currently cannot provide a cost-effective focused access to Internet services. This is particularly true for the developing countries where neither new installations nor massive updates of the existing infrastructure can be afforded.

In its initial collaborative work, the telecommunications industry was much influenced by the dominating demand for the voice telecommunications. The 1G and 2G systems were originally designed to provide one single service: the mobile voice telephony. Their system design was *service-oriented.* As a result, the conceived core infrastructure is circuit-oriented and the wireless link's capacity is tailored to the voice-implied bandwidth requirements. Due to these properties, 2G currently provides a reliable voice service; it is however quite difficult to reuse this infrastructure for other purposes. However, deploying a new infrastructure for every service is not scalable and financially impossible. Especially with the modern digital technologies, it is much more efficient to reuse the same infrastructure for different services.

3G development is an example of a *network-oriented design* process (sometimes also called operator-oriented design). It is a step ahead from the service-oriented design of the 2G system since it explicitly provides for infrastructure reuse for various services. Principally aiming at operators and networks, such design tries to respond to operator's management requirements. It thus specifies parts of the network core, producing homogeneous technologies comprising everything the operator has requested. According to this design paradigm, the 3G technologies deliver voice and data within the same infrastructure. In presence of an existing voice-oriented 2G infrastructure this renders the only added service—the mobile broadband data—quite expensive in itself. The operators have to amortize the network deployment and the license cost over the new service. Thus, from the user's point of view, this new service is often perceived as too expensive.

To be able to provide cost-effective data services at any chosen place in the world we need more user-oriented and data-centric approaches than what 2G and 3G paradigms deliver. At least in the mid-term, the hope here lies in a more opportunistic approach from the technological point of view. Indeed, the user typically does not care about who provides a particular service and how. The user cares about the availability of services, their performance (throughput, latency, etc.), the quality of service (QoS) (i.e. the performance and the variation of the performance factors), the ease of use, and service prices. Accordingly, the *user-oriented design* tries to respond to these user wishes assuring the possibility to freely choose an available service. Choice, as the driving factor for the competition, plays a crucial role in this scope since it results in better and cheaper technology.

From the system's point of view, the resulting overall architecture delivers very different services through completely heterogeneous access networks (ANs). User-oriented design has to cope with the question how to manage the system and how to provide user services with an expected quality. The management is important because a good management reduces the operational costs. The provision of the expected quality is the main factor for the user satisfaction.

Such architecture could help to achieve more infrastructural and architectural flexibility providing a free technology choice for the local operators and thus, in the final run, reducing the costs and offering more choices for the users. By featuring more flexibility, this step to further diversification gives new opportunities and could help, for example, to reduce the cost or to mitigate some aspects of the digital divide problem.

At the same time, this task is not technologically simple. As could be seen from the previous examples, the service-oriented design approach is a straightforward technological way to conceive a network dedicated to the needs of one single service. Provision of more services within the same infrastructure makes it more difficult to assure that every service individually is provided in a satisfactory way. We can generally allege that the QoS in the multi-services network is more difficult to maintain because very different requirements have to be fulfilled by the same infrastructure. Yet, owing to this common homogeneous infrastructure, with the network-oriented design it is still relatively easy to conceive systems enabling a comprehensive network management. The necessary dynamic infrastructure-to-service adaptation (e.g., for QoS) can then be achieved using the integrated management functions.

The step to the user-oriented design potentially implies a broad diversification of data transport technologies providing different services. Thus, the resulting systems inherit the problems of the dynamic per-service QoS provision. Additionally, we run into difficulties trying to consolidate all these different technologies and make them do what the operator wants. This applies to the network management in general. In particular, it concerns the mentioned QoS provision problematic and also raises diverse security considerations, both of the operators (infrastructure control and protection, resource usage control, accounting and billing) and of users (data confidentiality, location privacy, flawless billing).

Hence, the user-oriented design opens new possibilities but potentially results in a heterogeneous environment. To be deployed and maintained by the operators, this environment needs to be understandable, manageable, flexible, and secure. To be used, it needs to be user-friendly, reliable, and fair. In particular, users should be able to use different services over different infrastructures in the same, familiar manner.

Thus, we need to develop more flexible infrastructures and more sophisticated mechanisms for infrastructure access incorporating but hiding the whole technological complexity. These mechanisms should provide adaptability to both users and contents. Here we concentrate on heterogeneous network access mechanisms and the necessary corresponding network management functions in the scope of the future integrated environments.

Multi-Provider Network Environment

For 4G, the accent lies on users and the requested services (Pereira, 2000). For the flexibility and cost reasons, the 4G architecture has to be able to integrate different technologies to provide *services* to users. Services are divers offerings, commercial or free, ranging from a basic connectivity (e.g., to the Internet) to more sophisticated services such as voice calls or instant messaging (IM). To provide more complex services, some providers can use services proposed by other providers.

We see 4G as a potentially open, heterogeneous, user-oriented architecture, consisting of different service and ANs. These networks are operated by different authorities. We call such authorities *service providers*[1] if access to services is possible over their respective infrastructures or *networks*. The global 4G architecture is shown in Figure 0-1. It is composed of a panoply of service provider networks (SPNs) connected by an IP-based core network for any global data exchanges. SPNs principally support different wireless ANs. AN technology can range from personal to wide area networks.

Each provider may, but is not required to, have its own *users* and propose multiple services over different ANs. Users are defined as logical system identities subject to the service contract between two legal bodies, one representing the provider and the other representing the served user. This definition implies that every user corresponds to a *service contract* with exactly one provider.[2] Note that this contract requirement does not imply any price models or restrictions. Since every user corresponds to one legal body, we use these terms interchangeably in the rest of the document unless explicitly distinguished.

The service contract provides the trust relationship and the set of authorizations. From the user's point of view, the provider from the corresponding service contract is called *home provider*. If a user uses a provider only for user identification, authorization, and billing services, we call this provider a *virtual operator*[3] *(Zhang, Li, Weinstein, & Tu, 2002)*. Virtual operators (VOs) can but do not need to have their own infrastructures. Typical VOs are, for example, 2G or 3G providers (because of their existent user database), miscellaneous resellers but also credit card issuers, banks, public remote authentication services, and so forth.[4]

Providers may (but are not required to) serve users for whom they are not home providers. Providers may propose access to services in their own and in other infrastructures (e.g., in the Internet or in user's home network). The necessary network interconnection can be based upon private infrastructure interconnections of several providers or it can be based on a public backbone like the Internet. This and other definitions, for example, service level agreements, price agreements, mutual agreements on user authorization in visited networks, and so forth are subjects of so-called *roaming agreements* signed between the legal bodies representing the providers. Using these roaming agreements, providers can verify identities and profiles of visiting users whom we call *visitors*.

Figure 2. Global system architecture

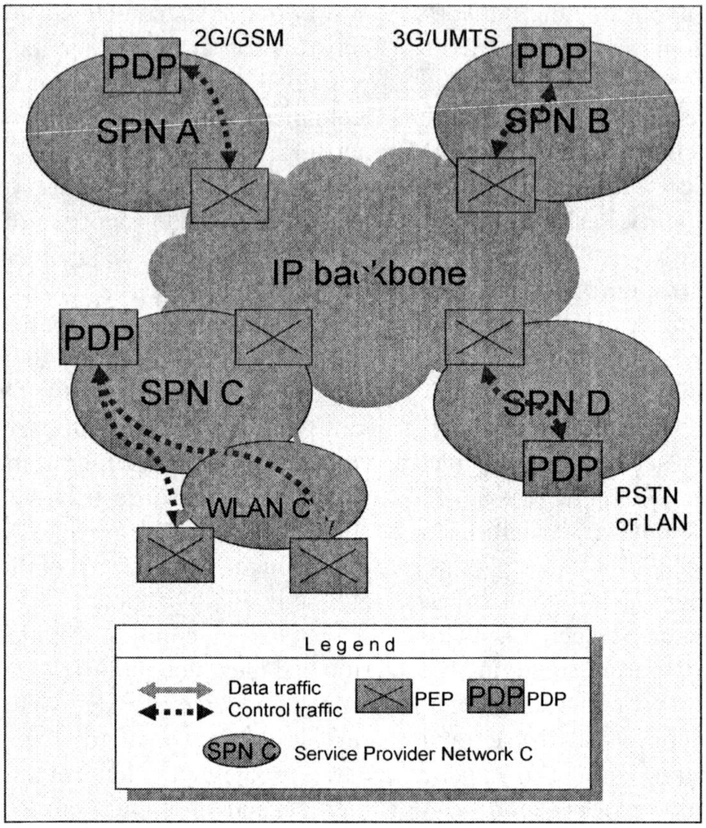

The users who do not have any verifiable service contracts may be treated as *guests*. Guests are users with special authorizations (profiles), locally and freely defined by any operator. These are thus local users and will not be treated differently in the following.

SPN Organization and Management

Management tasks in the SPN are carried out by the SPN owner, that is, the provider. The actions are based on the management policies that reflect provider and user requirements. For this purpose, providers deploy policy decision points (PDP), that is, logical entities capable of taking completely automated or assisted decisions based on the observed network situation and the defined policy. Policy enforcement points (PEP) are installed in the control equipment to enforce made decisions. In particular, PEPs are installed in the edge equipment.

SPNs are supposed to be trusted, non-public networks with appropriate protection measures. User traffic is to be strictly separated from the management traffic. The internal communications are IP-based. Inter-SPN management traffic can be protected by IP security (IPsec) (Kent & Atkinson, 1998), or by using dedicated protected links (L2 virtual private network [VPN] services, trusted sub-infrastructures, etc.)

Internal SPN architectures are deliberately left open. The protocols and mechanisms regarding PEP, PDP, measurements, and so forth do not need to be defined at the system level, because this complexity can be hidden within the SPN entity. Our main concern is to define architectures that do not impose any specific solutions. In a heterogeneous 4G system with its different providers (in terms of size, available resources, locality, services, capital, etc.), this is an additional degree of freedom. Different approaches are principally suitable for management purposes such as proprietary console or Web-based management, SNMP (Case, Fedor, Schoffstall, & Davin, 1990), COPS (Durham et al., 2000), GMPLS, and so forth.

Possible Approaches to 4G

On a high abstraction level, three approaches to 4G are theoretically possible 4G (Varshney & Jain, 2001).

Multimode Devices

Multimode devices (which already exist on the market, e.g., GSM/WiFi phones, PDAs with 802.11 WLAN, Bluetooth and GSM access modules, smartphones with Bluetooth capabilities, etc.) easily expand the effective coverage area managing the cooperation issues by the installed software. This concept pushes the 4G connection management complexity to the terminals, that is, it does not require any additional complexity in the wireless networks. However, the terminal equipment has to integrate operational logics including not only every technology-specific treatment but also the translation of quite different technological parameters to be able to make decisions. It is not clear if this can be done in an economically reasonable fashion for multiple, very different technologies, in particular taking into account the vertical (in the sense of the ISO/OSI model) complexity of QoS, security, and mobility management.

Overlay Networks

Another possibility is the installation of an overlay network of 4G access points situated above the actually available wireless networks. Note that in this approach the devices will still need to have several network interfaces to be able to access the entire infrastructure. The distinction lies in the additional complexity, which is completely shifted to the overlay. The requirements on the underlying technology are minimal. The overlay has to define the necessary signaling and transport functions. Besides the physical access to the used technology, the wireless device has to implement the overlay access module that will implement 4G signaling, 4G management, 4G security, 4G transport, and so forth functions. An example for such architecture would be the well-known All-IP approach discussed in the following sections.

Common Access Protocol

The third possibility is to unify the access protocols of the wireless networks, thus enabling users to access the 4G network by some standard means. This possibility implies separation of the transport and the control planes. Further, it is necessary to identify technology-specific functions that are part of the control plane. These functions have to be externalized and reflected by an abstraction layer/abstraction application program interface (API) that could then implement this common access protocol.

Note that this list is exhaustive (meaning that there are no other possible approaches to an integrated 4G system in the sense of the previous section). However, the mentioned alternative approaches are not necessarily mutually exclusive. It is imaginable to have some combinations of these general high-level approaches in a final solution. In the following, we present some of the proposed 4G architectures classifying these according to the previous scheme.

Related Work

Related Work on 4G Architectures

In Raivio (2001) the author discusses the currently most popular approach to 4G. This approach is based on a common Internet core for different networks, unifying everything over IP and the related Internet Engineering Task Force (IETF) technologies. With respect to this so-called All-IP (sometimes Full-IP) approach, the author briefly discusses the possibilities and the deficiencies in the concerned IETF protocols including the authentication, authorization, and accounting framework (AAA), Mobile IP, IPv6, IPsec, and SIP. The author points out that this approach is straightforward but also problematic in terms of QoS, security, and mobility management.

The presented All-IP idea is the current state of the art approach in the high-level 4G research. In the classification given in the previous section, All-IP represents an overlay network approach. The IP network is used as an overlay that integrates different technologies. IP technologies are used for both control and transport planes. IP base stations are used as access points in that 4G vision.

In Otsu, Okajima, Umeda, and Yamao (2001) the authors research a possible core network design for 4G systems. Describing the current situation of the telecommunications and the predominance of IP-based applications, they give an outlook on estimated traffic in the future generation of wireless systems. Then they discuss possible wireless transmission characteristics in terms of transmission bit rate, spectrum, area coverage, and hierarchical service area and define such network requirements as seamless connections, reduction in the number of control messages, short delay at handover, reduction of cost per bit, service integration based on IP, and movable network support. The network architecture is then defined as a core network (CN) connecting different ANs like a future, yet-to-be-defined 4G-RAN, and already existing WLAN, 3G, and PSTN to the Internet. CN and 4G-RAN are completely IP-based. The terminals have IP-addresses assigned. The CN is directly connected to 4G-RANs and the Internet and uses gateways to connect to the public switched telephone network (PSTN) and 3G. Mobility management is done by using the hierarchical Mobile IPv6 approach. Additionally, the article discusses some issues in the 4G-RAN configuration. In other words, this proposal is an instantiation of the All-IP approach.

Another All-IP proposal is discussed in Yumiba, Imai, and Yabusaki (2001). The recognized requirements here are huge (IP-) multimedia traffic handling, advanced mobility management (MM), diversified radio access support, seamless service, and application service support. The authors then discuss possible solutions for MM and seamless services and name Mobile IP, Cellular IP, and similar techniques. However, they recognize the deficiencies of such systems since they are hardly suited to provide a mobility management of the same quality as is the case in 3G. The authors claim that the networks beyond IMT2000[5] should be much more location-registration oriented and should identify the location registration management as a study topic. For instance, hierarchical or concatenated location registration techniques have to be studied. Then they discuss handover issues distinguishing local handovers and overall network handovers and identify this feature as a further study object.

Trying to provide an infrastructure-independent access to services and applications for highly mobile users, Kellerer, Vögel, and Steinberg (2002) present a solution based on a communication gateway. Originally driven by an automobile environment, the basic idea is to install an intermediate element between the actual user equipment and the serving networks. From the network point of view, such a communication gateway thus resides within the end-system. Including caching and switching units, the gateway provides a general middleware interface to the applications. Thus, this approach pushes the intelligence towards the end-systems, trying to map user requests at their origin to available networks and services. In our classification, this proposal represents the multimode device approach.

Becchetti, Priscoli, Inzerillli, Mähönen, and Muñoz (2001) take a slightly different approach. Mainly dealing with QoS support over different wireless infrastructures, they define a new intermediate layer between the IP and the second layers. This wireless application layer (WAL) then provides a QoS-generic interface for IP featuring uniform guaranteed link reliability and traffic control. The position of WAL in the ISO/OSI model implies a hop-by-hop QoS agreement logic. The details on the modular architecture of WAL, its class and association based QoS provision, Snoop TCP method to avoid congestions in the TCP layer can be found in the paper. In our classification this proposal is an overlay proposal, since WAL instances have to be integrated in the terminals and in the access points. IP is used as a general transport in the All-IP manner, but the technological heterogeneity is hidden within the WAL, which acts as a convergence sub-layer. WAL instances rely on SNMP to build the necessary decision bases and so forth.

Related Work on 4G Security

The user verification and network access in heterogeneous environments represents one of the major 4G problems. This is discussed later in detail. One of the problems is the access protocol but there are only some open questions concerning the back-end trust architectures and multi-domain, multi-party AAA.

An interesting related work seems to be Zhang et al. (2002). Introducing the concept of a so-called *virtual operator*, the authors describe how an authentication service reachable over the Internet could authenticate its users in a foreign hot spot environment using AAA. As potential virtual operators the authors see ISPs, content providers, cellular operators, or pre-paid card issuers. To reduce the number of necessary trust relationships between potentially numerous hot spot operators and diverse virtual operators, the authors propose a commonly trusted broker entity.

IETF currently works on the protocol for carrying authentication for network access (Forsber, Ohba, Pati, Tschofenig, & Yegin, 2003) in its PANA working group. PANA specifies an architecture very similar to the IEEE 802.1X architecture used in this work for LAN/WLAN access. PANA is link layer agnostic transporting authentication information between the PANA client and PANA authentication agent at higher layers. Since it is principally capable of identifying users, PANA could thus be used as a common access protocol to heterogeneous networks. However, since PANA has to access a higher level element, the L2 mostly remains unprotected. Also, after the (unprotected) L2 establishment, the local PANA client needs to discover its network's pendant, the PANA authentication agent (PAA). This involves discovery broadcasts and round trips. PANA here nicely illustrates the problems inherent to higher layer network access: questionable security, holes in the access controllers, broadcasting in the access phase, and high network access latency.

Besides, PANA does not optimally support mobility: Without additional mechanisms, the authentication has to be completely restarted at the next visited PAA (even within the same network). Such mechanisms could be a L3 (i.e., in the 4G scope typically IP) context transfer protocol that would allow arbitrary context transfers between different PAAs. IETF will shortly publish its context transfer protocol (CTP) specification (Nakhjiri, Perkins, & Koodli, 2004) as an experimental standard. However, the payload formats for CTP have to be specified too.

The work on the public access wireless networks (PAWNs) can be interesting in the 4G scope since it has to practically resolve several problems very similar to the anticipated 4G problems. PAWNs are typically implemented with IEEE 802.11 technology. Since the integrated 802.11 mechanisms are insufficient for almost all typical PAWN areas (per user quality of service, system-wide mobility, security, user network access, etc.), the solutions proposed for PAWNs are typically completely decoupled from the underlying technology. Hence, the practical experiences gained in such installations are of tremendous importance for the 4G research.

An approach for WLAN hot spots providing a secure wireless Internet access in public places is Microsoft's CHOICE (Bahl, Balachandran, & Venkatachary, 2001). The authors build a network that globally authenticates users and then securely connects them to the Internet via a serving 802.11 WLAN. A reasonable argumentation against IPsec for this purpose can be found in the publication. Introducing a new software module (PANS) instead of IPsec, the architecture promises authorization, access control, privacy, security, last hop quality of services, and accounting. However, this software (responsible for packet marking on mobile hosts) has to be installed on all mobile terminals, effectively modifying protocol stacks. The WLAN itself is open but does not allow any connections to any other networks, except for HTTPS connections to the global authenticator (global MS Passport service) and HTTP to the local Web server where, for example, the software module can be downloaded. Network's PANS authorizer module obtains key information from the global authenticator after successful user authentication. The authorizer can also install all required policies.

It then reroutes the traffic to a PANS verifier. The latter actively processes every packet checking the mark/tag added by the PANS module running on the mobile and providing, for example, per user access control and accounting.

Mobility support for public WLANs is presented in Friday et al. (2001). Using a similar packet tagging approach as in CHOICE, the authors describe their GUIDE/GUIDE II systems. Originally meant for a metropolitan scale access using modified client protocol stacks, GUIDE offers ordinary citizens secure and accountable Internet access over the deployed 802.11 WLAN-infrastructure. GUIDE II adds handover management using Mobile IPv6. IPv6 datagrams are tagged by clients using the modified MobileIPv6 stack. Programmable access routers ensure that only packets containing valid access tokens get to the trusted core network. Over an access router, users authenticate at an AAA authentication server. The latter distributes session keys to the access router group and the mobile terminal. User payload encryption is optionally possible between the router and the user equipment.

4G SECURITY REQUIREMENTS

4G security measures have to provide protection for 4G users and 4G providers.

There is no particular and evident reason why 4G security could be easier to achieve than 3G or 2G security. On the contrary, there are several reasons why it could be indeed more difficult, some of which are discussed in the *4G Vulnerabilities* section. One of the obvious reasons is the heterogeneity of the 4G system. Other reasons are provider inequality and the envisioned connection ubiquity.

Main security considerations in our 4G vision refer to the open system interfaces. One of the security targets is thus provider-provider interface. However, the most important and 4G-characteristic target is the user-network interface (including the user-service interface[6]). In the following sections we discuss these topics, specifically dealing with the user-network interface.

Important, but not necessarily new, security provisions must be considered in the internal provider network organization. The latter point is not discussed in the following.

4G Vulnerabilities

Vulnerabilities of Wireless Networks

Wireless networks are generally more vulnerable than their wired equivalents. Wireless security is a difficult problem that has to take into account the vulnerable medium per se (unclear network perimeter, shared medium, naturally broadcast, invisible/virtual network access), performance (security overhead, group communications), limited handset capabilities (human-machine interface, CPU, and memory), battery constraints (sleep management, on/off behavior), and different user services (roaming, mobility, localization). These problems have been discussed in this work per wireless technology in the (Hecker, 2005) per wireless technology.

Heterogeneousness adds a new dimension to this discussion. It multiplies the number of available mechanisms and, from the point of view of attacker, caters to more opportunities to attack the overall system (weakest link). New attack scenarios are conceivable: an attacker could use a weakness within one access network and the systemic interdependencies to gain access to another access network. Terminals can be attacked over several available interfaces at the same time. The services have to be provided over several interfaces, thus resulting in tighter performance constraints and complexity. A typical example is a handover between two different technologies (called *vertical handover*), but the same has to be considered for sleep management (*paging*) and generally for signaling.

Vulnerabilities of Service Provider Networks

A 4G system encompassing different technologies has to support complex management mechanisms (control systems, signaling, etc.), which considerably add to the system complexity and thus represent a major vulnerability per se. This is especially true for a multi-provider and thus multi-authority environment where a mutual preliminary user-network trust does not necessarily exist and must be established by some means (typically involving management subsystems and signaling before the user identity can be verified).

The serving network protection is one of the critical points to ensure service continuity and investment in new infrastructures. From the secure mobility discussions (such as Mobile IP security), we know that visited networks are often overexposed to resource consumption and denial of service. In our 4G vision, an SPN has to be protected from the users on the user-network interface and from the outer world on its backbone interface(s), including protection from other providers.

User Vulnerabilities

As a wireless user is vulnerable to unauthorized data access, traps/impostors, and desinformation, the user must be protected from abuse by third parties and from the part of the serving SPNs.

Given a rising part of the M2M communications and the wish for infrastructureless communications, the user device is also vulnerable to attacks by other devices involved in the provision of the consumed services (impostors, data modifications, data sniffing, man-in-the-middle) and by devices consuming services provided by the user device (denial of service, abuse).

Connected to multiple interfaces over several providers the device is naturally multi-homed. It is potentially exposed to all attacks over the established connections, including malicious code intrusion (viruses, spyware, and worms).

User vulnerability includes headset vulnerability. A typical 4G headset featuring several active interfaces is naturally exposed to different kinds of attacks, such as attacks on device drivers of the communication interfaces, attacks against the transport and signaling communication stacks, and attacks against all services potentially provided or assisted by the headset itself (e.g., file sharing, localization, auto-update). An important and often forgotten point is device theft. Today, mobile devices are trendy and, having a rich and versatile feature set, can be quite expensive. They have become an important accessory and manufacturers are doing their best to render them more portable and more powerful at the same time. It is obvious that these devices have become an interesting target for thieves. Thus, physical device security is an important but insufficient subject. Mobile handsets can store important personal user data (address books, access codes, professional data, personal medical information). Remote device deactivation, blocking, and erasure seem important future security features.

A 4G user needs a particular protection to ensure his/her anonymity and an offer-consistent and verifiable billing. Without any protection, in an international multi-provider 4G environment, a user can be an easy target for both price fraud (charging wrong prices, charging incorrect usage) and user tracking.

Heterogeneous Security

Current wireless technologies have different security considerations and provide corresponding security definitions in the standards. The latter are naturally dedicated to the respective link layer and thus concentrate on the implementation within the network interface cards, adapters, and so forth. In 4G, different link layer technologies are likely to coexist for the reasons explained in the previous sections. Also, the focus changes: in the personal communications the security focus should be on users, not on network devices.

The problem with the characteristic 4G security is twofold. On the one hand, there are very basic open questions that have to be answered by the ongoing research by weighing practical constraints against the required security level. What is security in 4G if we do not know what 4G looks like, what services it is supposed to provide, and in which environments it is going to operate? The system architecture is crucial for the security considerations. Additionally, we need trust and threat models. What are the capabilities of potential attackers? Which ANs will be used and how? Trust models should correspond to the probable usage scenarios. For instance, if users are not "owned" by providers (Pereira, 2000), how can trust be established and to whom? With all that, a consistent security policy has to be defined along with the security architecture, identifying technology-independent subjects, objects, relationships, authorizations, threats, and protective measures. This is however difficult and defines a problem known as *heterogeneous security*).

On the other hand, there are practical problems concerning the technical applicability of solutions. The security solutions proposed by the wireless technologies are limited to the identified needs. They are thus different from technology to technology reflecting its expected usage. Very often, they fail to fulfill the security requirements, typically because of conceptual or implementational flaws. But even if their implementation is correct, their scope is naturally wrong: as access security, they aim to provide link security, but ultimately providers need service access security and users need personal data security.

How can the defined security policy for the entire system be applied and enforced to all system entities given that the available solutions are different, potentially flawed, and limited to system parts? For instance, if the security policy identifies link encryption as a necessary confidentiality implementation, how can this be universally activated and with which keys and properties? How can we guarantee an adequate, comparable strength of the different encryption mechanisms? What to do with the technologies that do not provide link encryption? The security policy must consider these cases and provide answers to such questions.

4G Security Layer

The aforementioned practical problems with the 4G security can be avoided if the technology-dependent security measures are not used. Instead, all security measures could be applied in the overlaid technology. However, it is often insecure or at least inefficient to enforce security in the overlay. For example, 2G/3G network providers rely on L2 security measures for network access control, frame integrity and link encryption. While the link encryption is not important for the provider, the access control is primordial for infrastructure protection and revenue guarantees. Moreover, the L2 security measures are often implemented in the network interface hardware. Their design includes power consumption and computational resource considerations. A higher level solution would be implemented in the device control logics, that is, typically software. Given the constraints with the 4G terminals (wireless security processing gap), it would be wise to use the hardwired security solutions in the network adapter. Furthermore, in the OSI logic, multiple links could lie between the user and the used L3 device (router), but only one link is possible between the user and any used L2 device. Thus, the L2 security measures are guaranteed to be implemented in the first network entity (the access device), that is, next to the user, at the very edge of the network. That brings the security as close to the user as possible and thus guarantees physical infrastructure protection. Moreover, it potentially scales better since the access devices are designed to support a fixed number of connections, including the connection properties to be enforced. Another point is that higher level security solutions cannot achieve the same user privacy. For instance, user location privacy is in danger since lower layer addresses (such as world-wide unique MAC addresses) cannot be hidden by higher layer security measures.[7]

For reasons stated previously, we think that L2 security is indispensable in 4G. This is by the way also the most characteristic point of 4G: whatever the 4G vision, everybody seems to agree that 4G will be technology-opportunistic, incorporating different wireless ANs in one system. The network access security is thus one of the major challenges, typical and characteristic for 4G.

NETWORK ACCESS SECURITY

A particular security problem is bound to the user network access. The 4G user has a terminal with multiple network interfaces. The security measures for each interface have been designed according to an initial security analysis during the technology standardization phase. Since the technologies are meant for different purposes, the risks and the defined security functions are likely to be different.

The security mechanisms are definitely different. Thus, every interface has different requirements on credentials in terms of identities, expiration policies, initial trust representation, and so forth. These requirements have to be fulfilled since otherwise the interface could be unusable or the access by the means of this interface impossible. If the user definition in the system is consistent, then the 4G user cannot be expected to use multiple identities: in 4G, every network provider needs to be able to identify any given user correctly, in particular in the different ANs, which the user might be using simultaneously. That is important for the authorizations defined in the security policy. It is equally an important requirement for a consistent billing. Network access can thus be divided into various sub-problems that are treated in more details in following.

Network Selection

In the outlined 4G vision, a free service choice is an important design criterion. To provide that choice, users must be able to collect information on the ANs of all available providers. Most importantly, this is required for the decision of which network the user should connect to. For instance, it cannot be generally assumed that every network is accessible for every user (e.g., because the user's home provider does not have any roaming agreement with the provider of the detected network).

Network selection is a problem since some preliminary network access is necessary prior to authentication, which however should be limited so as not to contradict the security policy. Network selection thus represents a security-usability compromise.

In a dynamic multi-provider multi-technology 4G environment, active exchanges (through signaling, like network discovery) are necessary since the existence of system-wide coherent network identifiers do cannot be relied upon. These identifiers have very different meanings in different technologies. For instance, if a 2G provider wants to deploy a supplementary data service over an 802.11 WLANs, what should be used as a network identifier? There is no regulation on service set identification (SSID) naming in the 802.11 WLANs. Besides, in a dynamic 4G environment with the very different proposed services, over different technologies and with different prices, it is difficult to believe that a network identifier alone is a sufficient base for a reasonable network selection decision.

In a user-centric environment, the network selection decision should be made based on physically available networks and channel qualities, user identity and user service authorizations within the encountered networks, and on offered service prices. Especially price display for a given user appears as one of the critical issues in a multi-provider environment characterized by continuous roaming between several different (big/small, national/local, etc.) providers. Indeed, even in 2G with a typical limitation to a handful of providers per location (2-8), users traveling to foreign countries have been known to feel badly informed about pricing of out- and incoming calls. In 4G with multiple-interface terminals and possibly new business models, several providers can be used at the same time, possibly offering similar services at prices depending on dynamic factors such as current network usage (per-session price determination).

The involvement in such rather complex *pre-authenticated* (Hecker & Labiod, 2004) user-network signaling represents major risks for both network operators (infrastructure intelligence, unpaid resource consumption, denial of service) and users (localization, tracking). Additionally, optimizations are necessary to that recurrent process, which in 4G can be repeated in-session, since it can have an important impact on mobility performance (vertical handover).

User-Network Authentication

A user-network authentication is necessary from network provider's point of view to be able to enforce a reliable access control to its resources and to authorize requested service sessions in its infrastructure or at least a transport (connectivity service) over its infrastructure. It is also required by the user's home provider for authorization and billing.

From the user's point of view, network authentication permits to verify the received network identity information, guarantees access to the correct environment, and thus permits to establish trust to the serving provider. It helps to eliminate impostors and to protect against man-in-the-middle attacks.

After the service information collection, some networks can be eliminated by policy or user wish (e.g., a pre-configuration of the type "never use provider X" or rules like "always choose the cheapest available service", etc.) Now, the user can actually access the required services over available networks. A reliable user-network authentication is required at this moment at latest.

The L2 user-network authentication is a problem in 4G since the logical and technological requirements are very different from technology to technology. We illustrate this on an arbitrary example, comparing UMTS and standard 802.11 security.

UMTS uses an external module (USIM) that hides the actual authentication method from the used device and the visited network. The authenticated logical entities are the USIM and the visited network, represented by the authentication center (AuC). USIM is supposed to grant network access to the device (i.e., also to the user). The USIM is capable of key derivation after a successful authentication.

IEEE 802.11 defines a handshake procedure based on credentials existing between the network (the access point) and the user. The whole procedure (i.e., the authentication method, the exchanges, the cryptographic functions and the success conditions) is hardwired in the network interfaces. The only authenticated entity is the network interface of the user device (i.e., the access point is not authenticated). The authentication does not derive any key material. Moreover, the procedures are almost useless because of several concept errors.

As can be seen, the provided services are very different in terms of capabilities and the achieved security level. However, the purpose of this example is not to blame WLAN security. Today, other security models and methods are available for WLANs (notably the 802.11i introducing a different security model). Nevertheless, this situation exemplifies the normality of a heterogeneous 4G: the security models, the trust presumed relationships, the technical possibilities and the vulnerabilities are very different from technology to technology. The resolution of this problem must not lead per se to security problems. Thus, if the L2 authentication is to be used in the 4G scope, every technology has to fulfill a minimal common requirement set. Otherwise, higher level security has to be used and the associated higher level access controllers have to be collocated with the L2 access devices. If that cannot be guaranteed, this technology should be considered unsuitable for 4G.

From today's perspective, the requirements on the L2 authentication are cryptographic strength, mutuality, and dynamic key material negotiation for the subsequent session protection. The key material negotiation should provide *perfect forward secrecy* (PFS), that is, a successful attack on the produced key material should not give any clues on the long-term secret such as the used credentials. User location privacy should be supported, that is, if possible, any user-specific identifiers should be unreadable for a third party.

Note that we do not formulate any requirements on the authentication logic (how many parties involved and how), used protocols, implementation, method placement, or on the used trust representation. However, authentication methods are generally hard to conceive and represent one of the most vulnerable parts of modern cryptosystems. Due to the flaws typically found in the authentication methods during their lifetime, and given the number of different authentication methods in 4G, we additionally require that the authentication method be easily updateable.

Whatever the actual mechanisms is, it has to correspond to the performance requirements in terms of possible vertical and horizontal mobility. Fast re-authentication (less RTT) and particularly pre-authentication (over the same or a different interface) seem useful in the 4G context.

Data Encryption and Integrity Functions

Different wireless technologies use very different link encryption and data integrity techniques based on different mechanisms. Typically, shared-key mechanisms are used for both link encryption and data integrity. The actually used key is usually derived from the key material established by the authentication function.

Very often proprietary solutions are implemented both for encryption and data integrity. The needs of the used encryption and integrity mechanisms in terms of key properties (format, length, known weak keys, etc.) and the optionally used initialization vectors are very different. The provided security levels are also quite different. Thus, the situation of these functions is similar to the user-network authentication. If some minimum requirements cannot be fulfilled, these have to be replaced (e.g., in the overlay) or the technology could not be used.

Simultaneously, both functions are in use during the whole session. Thus, their power and resource consumption is particularly critical. For that reason, we think that both encryption and integrity functions should be implemented in the associated network adapter (hardwired or in form of hardwired cryptographic bricks connected by the soft-wired firmware definitions). Both functions must use the key material derived during the last authentication session and support rapid re-keying, both periodical and on-demand. Ideally, both functions should be cryptographically strong. However, if flaws are detected, the rapid re-keying can help mitigate the problem by changing the encryption keys very often.

Provider-Provider Security

In 2G/3G providers usually sign preliminary bilateral contracts known as *roaming agreements*. Such agreements build the basis for mutual user authentication, authorization, service and charging. Every provider thus sets up a special subsystem serving AAA requests from other providers, acting as peers. Such requests are as such subject to prudent access control, authorization, and extensive logging.

Principally, that mechanism can also be used in 4G. However, the differences between the provider size and financial weight must be accounted for. In a multi-provider environment, it cannot be reasonably assumed that all providers will still trust each other. Another point is that bilateral agreements are not a scalable approach for a big number of providers ($O(n^2)$).

From the WiFi network experience, we know that instead providers use additional trusted entities as their official roaming contract partner. Such trusted entities are either special brokers or provider associations acting as separate legal bodies. This approach permits to reestablish trust and to minimize the number of bilateral contracts.

To ensure correct billing and charging providers often rely on external billing services and involve third party clearing houses (e.g., financial audit institutions certifying the correctness of the bills and the processes).

Other Security Problems

The remaining security issues mainly concern the SPN, its integrity, and its internal interfaces. Network engineering techniques such as flow and traffic separation, filtering, and continuous monitoring are classically used to achieve a good security level.

This is not an easy problem to solve. However, its exact resolution highly depends on the actually responsible provider: both the security needs and the technical capabilities will change depending on the provider size. That is why, at this moment, we prefer to hide the complexity within the SPN body.

APPROACHES TO 4G SECURITY

In the previous section we defined several requirements on 4G security. 4G systems are still in an early concept phase and specific realizations do not yet exist. However, different approaches are possible to achieve the technology-spanning security mechanisms, often required.

Virtualization

Virtualization is an important means of integrating flexibility in the system design. Virtualization specifies *what* is to be done but not *how* it should be done. In other words, different behaviors accessible and corresponding to the same specified interface can be used (sometimes interchangeably) during the system runtime. The instantiation can happen by pre-configuration, through soft- and firmware updates or even dynamically, at request.

Examples for virtualization include the GSM and UMTS security (GSM 11.11, n.d.; 3rd Generation Partnership Project [3GPP] TS 33.102, n.d.) but also, for example, EAP in IEEE 802.1X (2001) and 802.11i (IEEE Draft, n.d.). The approaches to the virtualization are very different.

2G/3G security relies upon smart cards that represent the network counterparts for authentication and per packet encryption. The actual algorithms and methods are hidden and implemented within the closed card; they are always run against the home provider, who also acts as smart card issuer. In that manner, every home provider has a free choice of standard or better mechanisms to fulfill his/her particular security requirements. The whole transport and signaling infrastructure, including the visited network provider, is independent of that implementation. It is thus feasible to enforce different authentication and per packet security on a per user basis.

The usage of 802.1X/EAP protocol in 802.11i for access control provides a generic authentication function: by specifying how to control, transport, and evaluate user authentication frames instead of specifying how to authenticate users, this standard is now open to authentication method choices. The deployed infrastructure is freed from any authentication logic; only the central authentication server has to implement the actual mechanism, such as, for example, EAP-TLS specifying TLS (Dierks & Allen, 1999) transport over EAP. That enables an authentication method choice on a per-session, per-user basis.

Virtualization is thus a strong design principle for open systems. It thus seems very interesting for 4G security. Nevertheless, even with virtualization, requirements on all mechanisms need to be thoroughly specified.

Adaptation

Adaptation refers to dynamic changes within the implementations of the communicating parties. This could concern the user terminal security measures and the provided network and service environment.

Adaptation could rely on different profiling mechanisms, including machine learning. However, in the telecommunications context, it could also involve more pragmatic signaling-based adaptation. Given the current network situation, the used access network technology and an extensive user-network signaling capable of expressing user needs, the network could actually create a (virtual) SPN corresponding exactly to users' expectations in the chosen access network technology. Different virtualization techniques can be used in that scope, from the previous security virtualization examples to infrastructure virtualization techniques such as VLAN (IEEE 802.1q) or MPLS (Rosen, Viswanathan, & Callon, 2001). This approach could ultimately provide users with their own virtual environments, inaccessible by others, and, at the same time, render the actual physical SPN and its management subsystems inaccessible by the users. Hence, this approach could come handy to solve parts of the problem with the heterogeneity of the ANs.

On the terminal, adaptation can help bridge the differences in the available implementations of the access network security. By inspecting the available (active) interfaces, the really activated measures, and taking into account the available information on recently discovered low level vulnerabilities (user input, secure home network signaling channel, etc.), the terminal implementations could preprocess the sent data, so it still fulfills the overall security policy in spite the insufficiencies.

Standardization

Since the 4G does not yet exist, standardization could be used at these early stages to build a good base for future 4G security. Basically, such standardization efforts should apply to new definitions and adapt the existing technologies, so these could be used in the future 4G landscape.

In 4G, standardization is one of the central discussions. Not everything can be standard, since otherwise we migrate back from the technology-opportunistic vision to a monolithic one-technology-vision. On the other hand, without any standards, hardly any communication is possible. The compromise between what we standardize in the 4G scope and what we leave to the respective technology is the most critical design decision.

The standardization should respect the three introduced interfaces, differentiating user-network, provider-provider, and internal SPN interfaces (mainly management plane).

Virtualization plays an important role for 4G standardization. We can learn from the former experiences that specifying what and how separately is more flexible. To provide adaptation, we need at least a common signaling standard. This represents a seemingly viable alternative approach to the current pure overlay solutions such as All-IP. We could standardize a common 4G signaling protocol, including virtual definitions for network access and data protection phases, and then use the access technologies *as is*, without any additional changes, as a pure data transport.

CONCLUSION

The 4G reflections started about 2000-2001 are not yet mature enough to present a sound overview of the 4G security. At the current state, there is no common 4G vision and what will eventually be called 4G is an open question.

Independent of that, we believe that the technology-opportunistic system as the one presented in this chapter will eventually be built. That is the reason why the new security problems related to the high system heterogeneity and the new usage scenarios and presented in this chapter seem to be of major importance for the understanding of the vulnerabilities and design of future telecom systems.

More specifically, in this chapter, we present the development process from 1G to 4G discussing telecommunications landscape changes and time scales. We then introduce the current state of the 4G discussion and present our vision of 4G as a technology-opportunistic, user-centric mobile services system built of multi-interface terminals and heterogeneous ANs, bound by a decent management subsystem. Given that 4G shape, conform to the main trend in the current 4G research, we introduce main system interfaces, its links and entities to discuss its vulnerabilities.

We then introduce 4G security requirements, justifying the special character of and insisting on the network access phase. Finally, we propose several high level approaches to 4G security, including virtualization, adaptation and standardization.

REFERENCES

3rd Generation Partnership Project (3GGP) TS 33.102 Release 99. (n.d.). *3GPP: Technical specification group (TSG), 3G security: Security architecture.* Sophia Antipolis Cedex, France: Author.

Al-Muhtadi, I., Mickunas, D., & Campbell, R. (2002, April). A lightweight reconfigurable security mechanism for 3G/4G mobile devices. *IEEE Wireless Communications, 9*(2), 60-65.

Bahl, P., Balachandran, A., & Venkatachary, S. (2001, June). Secure wireless Internet access in public places. In *Proceedings of the IEEE International Conference on Communications (IEEE ICC 2001),* Finland.

Becchetti, L., Priscoli, F. D., Inzerillli, T., Mähönen, P., & Muñoz, L. (2001, August). Enhancing IP service provision over heterogeneous wireless networks: A path towards 4G. *IEEE Communications Magazine, 39*(8), 74-81.

Blake, S., Black, D., Carlson, M., Davies, E., Wang, Z., & Weiss, W. (1998, June). *An architecture for differentiated services* (RFC 2475). Retrieved from http://www.ietf.org/rfc/rfc2475.txt

Braden, R., Clark, D., & Shenker, S. (1994, June). *Integrated services in the Internet architecture: An overview* (RFC 1633). Retrieved from http://tools.ietf.org/html/rfc1633

Bria, A., Gessler, F., Queseth, O., Stridh, R., Unbehaun, M., & Wu, J. (2001, December). 4th-generation wireless infrastructures: Scenarios and research challenges. *IEEE Personal Communications, 8*(6), 25-31.

Case, J. D., Fedor, M., Schoffstall, M. L., & Davin, J. (1990, May). *Simple network management protocol (SNMP)* (RFC 1157). Retrieved from http://www.ietf.org/rfc/rfc1157.txt

Code Division Multiple Access (CDMA) Development Group. (n.d.). *Technology: 3G—cdma2000.* Retrieved from http://www.cdg.org/technology/3g.asp

Dell'Uomo, L., & Scarrone, E. (2001, September). The mobility management and authentication/authorization mechanisms in mobile networks beyond 3G. *IEEE Personal, Indoor and Mobile Radio Communications, 1,* C44-C48.

Dierks, T., & Allen, C. (1999, June). *The TLS protocol version 1.0* (RFC 2246). Retrieved from http://www.ietf.org/rfc/rfc2246.txt

Durham, D. (Ed.), Boyle, J., Cohen, R., Herzog, S., Rajan, R. & Sastry, A. (2000, January). *The COPS (common open policy service) protocol* (RFC 2748). Retrieved from http://www.rfc-editor.org/rfc/rfc2748.txt

Emmerich, W. (2000, June). *Engineering distributed objects.* John Wiley & Sons.

Evans, B. G., & Baughan, K. (2000, December). 4G visions. *IEEE Electronics & Communications Engineering Journal, 12*(6), 293-303.

Forsber, D., Ohba, Y., Pati, B., Tschofenig, H., & Yegin, A. (2003, March). *Protocol for carrying authentication for network access.* IETF PANA Working Group Draft, work in progress. Internet Engineering Task Force.

Friday, A., Wu, M., Schmid, S., Finney, J., Cheverst, K., & Davies, N. (2001, July). A wireless public access infrastructure for supporting mobile context-aware IPv6 applications. In *Proceedings of the ACM 1st Workshop on Wireless Mobile Internet,* Rome, Italy (pp. 11-18).

Ginzboorg, P. (2000, November). Seven comments on charging and billing. *Communications of the ACM, 43*(11), 89-92.

Global System for Mobile Communications (GSM) 11.11 (n.d.). *Digital cellular telecommunication system (Phase 2+), specification of the subscriber identity module—Mobile equipment (SIM-ME) interface.* Author.

Global System for Mobile Communications (GSM) Association. (n.d.). *3GSMplatform.* Retrieved from http://www.gsmworld.com/technology/3g/index.shtml

Gupta, V., & Gupta, S. (2002, March). KSSL: Experiments in wireless Internet security. In *Proceedings of the Wireless Communications and Networking Conference* (pp. 860-864).

Hecker, A. (2005, March 16). *On logical network access control and the associated user and network management in future heterogeneous 4G wireless systems.* Computer Science and Networing Department, Ecole Nationale Supérieure des Télécommunications (ENST), Paris, France.

Hecker, A., & Labiod, H. (2004). Pre-authenticated signaling in wireless LANs using 802.1X access control. In *Proceedings of the IEEE GLOBECOM 2004,* Dallas, TX.

IEEE Draft 802.11e. (2003, February). Draft supplement to standard for telecommunications and information exchange between systems—LAN/MAN specific requirements—Part 11: Wireless medium access control (MAC) and physical layer (PHY) specifications: Medium access control (MAC) enhancements for quality of service (QoS). Author.

IEEE Draft 802.11i. (n.d.). *Draft supplement to IEEE Std 802.11. Part 11: Specifications for enhanced security.* Author.

IEEE Standard 802.11F. (2003, July). *Trial-use recommended practice for multi-vendor access point interoperability via an inter-access point protocol across distribution systems supporting IEEE 802.11 operation.* Author.

IEEE Standard 802.1X. (2001, June). *Port-based network access control.* Author.

International Telecommunication Union-Radio Communication Sector (ITU-R) World Radiocommunication Conference, Retrieved from http://www.itu.int/ITU-R/index.asp?category=conferences&link=wrc&lang=en

Kellerer, W., Vögel, H.-J., & Steinberg, K.-E. (2002, March). A communication gateway for infrastructure-independent 4G wireless access. *IEEE Communications Magazine, 40*(3), 126-131.

Kent, S., & Atkinson, R. (1998, November). *Security architecture for the Internet protocol* (RFC 2401). Retrieved from http://www.ietf.org/rfc/rfc2401.txt

Misra, A., Das, S., Dutta, A., McAuley, A., & Das, S. K. (2002, March). IDMP-based fast handoffs and paging in IP-based 4G mobile networks. *IEEE Communications Magazine, 40*(3), 138-145.

Nakhjiri, M., Perkins, C., & Koodli, R. (2004, August). Context transfer protocol. In J. Loughney (Ed.), *Approved IETF draft, work in progress.* Internet Engineering Task Force.

Otsu, T., Okajima, I., Umeda, N., & Yamao, Y. (2001, October). Network architecture for mobile communications systems beyond IMT-2000. *IEEE Personal Communications Magazine, 8*(5), 31-37.

Peirce, M. (2000, October). *Multi-party electronic payments for mobile communications.* Unpublished PhD thesis, Department of Computer Science, University of Dublin, Trinity College.

Pereira, J. M. (2000, September). Fourth generation: Now, it is personal! In *Proceedings of the IEEE* International Symposium on Personal, Indoor and Mobile Radio Communications *(PIMRC)* London *(*Vol. 2, 1009-1016).

Raivio, Y. (2001, March). 4G—Hype or reality (Conference Publication No. 477). In *IEE 3G Mobile Communication Technologies* (pp. 346-350).

Raychaudhuri, D. (2002, September). 4G network architectures: WLAN hot-spots, infostations and beyond... In *IEEE PIMRC 2002 Keynote Talk,* Lisbon, Portugal.

Rosen, E., Viswanathan, A., & Callon, R. (2001, January). *Multiprotocol label switching architecture* (RFC 3031). Retrieved from http://tools.ietf.org/html/rfc3031

Schulzrinne, H., & Wedlund, E. (2000, July). Application-layer mobility using SIP. *ACM Mobile Computing and Communications Review, 4*(3), 47-57.

Tsao, S.-L., & Lin, C.-C. (2002, September). Design and evaluation of UMTS-WLAN interworking strategies. In *Proceedings of the IEEE 56th Vehicular Technology Conference (VTC),* Vancouver, Canada.

Van Damme, E. (2002, May 4-5). The European UMTS-auctions. *European Economic Review, 46,* 846-858.

Varshney, U., & Jain, R. (2001, June). Issues in emerging 4G wireless networks. *IEEE Computer, 34*(6), 94-96.

Yahalom, R., Klein, B., & Beth, Th. (1993, May). Trust relationships in secure systems—A distributed authentication perspective. In *Proceedings of the IEEE ComSoc Symposium on Research in Security and Privacy,* Oakland, CA (pp. 150-164).

Yumiba, H., Imai, K., & Yabusaki, M. (2001, October). IP-based IMT network platform. *IEEE Personal Communications Magazine, 8*(5), 18-23.

Zhang, T., & Agrawal, P., & Chen, J.-C. (2001, October). IP-based base stations and soft handoff in all-IP wireless networks. *IEEE Personal Communications Magazine, 8*(5), 24-30.

Zhang, J., Li, J., Weinstein, S., & Tu, N. (2002, July). Virtual operator based AAA in wireless LAN hot spots with ad-hoc networking support. *ACM Mobile Computing and Communications Review, 6*(3), 10-21.

ENDNOTES

1. Since users are the main focus of our work, we prefer this term to the synonymic *operator,* which refers to the infrastructure.
2. This is not limiting since any legal body can have multiple user assignments.
3. This is used consistently to the original definition given in Zhang et al. (2002). However, since in this special case no infrastructure exists, the actually "operated" entity is the user. This term is thus also consistent with our strictly user-oriented view.
4. That underlines the fact that our model mainly requires the service contract as a means for a reliable user identification. Indeed, without any pre-established trust, no reliable billing is possible.
5. International Mobile Telecommunications 2000, ITU's common name for different 3G variants.
6. Note that generally these two problems are not equivalent. However, in our 4G vision we suppose that SPNs are organized as integrated transport and services networks run by the same authority. In that view, the difference between the two is of a very technical nature; it is merely limited to and by the internal SPN organization.
7. Although the lower layer address and the user identity are two completely different identifiers, one initial passive network observation in the proximity of a victim allows an establishment of a direct relationship.

Chapter XIX
Security Architectures for B3G Mobile Networks

Christoforos Ntantogian
University of Athens, Greece

Christos Xenakis
University of Piraeus, Greece

ABSTRACT

The integration of heterogeneous mobile/wireless networks using an IP-based core network materializes the beyond third generation (B3G) mobile networks. Along with a variety of new perspectives, the new network model raises new security concerns, mainly, because of the complexity of the deployed architecture and the heterogeneity of the employed technologies. In this chapter, we examine and analyze the security architectures and the related security protocols, which are employed in B3G networks focusing on their functionality and the supported security services. The objectives of these protocols are to protect the involved parties and the data exchanged among them. To achieve these, they employ mechanisms that provide mutual authentication as well as ensure the confidentiality and integrity of the data transferred over the wireless interface and specific parts of the core network. Finally, based on the analysis of the security mechanisms, we present a comparison of them that aims at highlighting the deployment advantages of each one and classifies the latter in terms of: (1) security, (2) mobility, and (3) reliability.

INTRODUCTION

The evolution and successful deployment of wireless LANs (WLANs) worldwide has yielded a demand to integrate them with third generation (3G) mobile networks. The key goal of this integration is to develop heterogeneous mobile data networks, named as beyond 3G (B3G) networks, capable of supporting ubiquitous computing. Currently, the network architecture (3rd Generation Partnership Project [3GPP] TS 23.234, 2006) that integrates 3G and WLAN specifies two different access scenarios: (1) the *WLAN Direct IP Access* and (2) the *WLAN 3GPP IP Access*. The first scenario provides to a user an IP connection to the public Internet or to an intranet via the WLAN access network

(WLAN-AN), while the second allows a user to connect to packet switch (PS) based services (such as wireless application protocol [WAP], mobile multimedia services [MMS], location-based services [LBS] etc.) or to the public Internet, through the 3G public land mobile network (PLMN).

Along with a variety of new perspectives, the new network model (3G-WLAN) raises new security concerns, mainly, because of the complexity of the deployed architecture and the heterogeneity of the employed technologies. In addition, new security vulnerabilities are emerging, which might be exploited by adversaries to perform malicious actions that result in fraud attacks, inappropriate resource management, and loss of revenue. Thus, the proper design and a comprehensive evaluation of the security mechanisms used in the 3G-WLAN network architecture is of vital importance for the effective integration of the different technologies in a secure manner.

In this chapter we examine and analyze the security architectures and the related security protocols, which are employed in B3G, focusing on their functionality and the supported security services for both WLAN Direct IP Access and 3GPP IP Access scenarios. Each access scenario (i.e., WLAN Direct Access and WLAN 3GPP IP Access) in B3G networks incorporates a specific security architecture, which aims at protecting the involved parties (i.e., the mobile users, the WLAN, and the 3G network) and the data exchanged among them. We elaborate on the various security protocols of the B3G security architectures that provide mutual authentication (i.e., user and network authentication) as well as confidentiality and integrity services to the data transferred over the air interface of the deployed WLANs and specific parts of the core network. Finally, based on the analysis of the two access scenarios and the security architecture that each one employs, we present a comparison of them. This comparison aims at highlighting the deployment advantages of each scenario and classifying them in terms of: (1) security, (2) mobility, and (3) reliability.

The rest of this chapter is organized as follows. The next section outlines the B3G network architectures and presents the WLAN Direct IP Access and the 3GPP IP Access scenarios. The third section elaborates on the B3G security architectures analyzing the related security protocols for each scenario. The fourth section compares the security architectures and consequently, the two access scenarios. Finally, the fifth section contains the conclusions.

BACKGROUND

The B3G Network Architecture

As shown in Figure 1, the B3G network architecture includes three individual networks: (I) the WLAN-AN, (II) the visited 3G PLMN, and (III) the home 3G PLMN. Note that Figure 1 illustrates the architecture for a general case where the WLAN is not directly connected to the user's home 3G PLMN. The WLAN-AN includes the wireless access points (APs), the network access servers (NAS), the authentication, authorization, accounting (AAA) proxy (Laat, Gross, Gommans, Vollbrecht, & Spence, 2000), and the WLAN-access gateway (WLAN-AG). The wireless APs provide connectivity to mobile users and act like AAA clients, which communicate with an AAA proxy via the Diameter (Calhoun, Loughney, Guttman, Zorn, & Arkko, 2003) or the Radius (Rigney, Rubens, Simpson, & Willens, 1997) protocol to convey user subscription and authentication information. The AAA proxy relays AAA information between the WLAN and the home 3G PLMN. The NAS allows only legitimate users to have access to the public Internet, and finally, the WLAN-AG is a gateway to 3G PLMN networks. It is assumed that WLAN is based on the IEEE 802.11 standard (IEEE std 802.11, 1999).

On the other hand, the visited 3G PLMN includes an AAA proxy that forwards AAA information to the AAA server (located in the home 3G PLMN), and a wireless access gateway (WAG), which is a data gateway that routes users' data to the home 3G PLMN. On the other hand, the home 3G PLMN includes the AAA server, the packed data gateway (PDG) and the core network elements of the universal mobile telecommunications system (UMTS), such as the home subscriber service (HSS) or the home location register (HLR), the Gateway GPRS support node (GGSN) and the Serving GPRS support node (SGSN). The AAA server retrieves authentication information from the HSS/HLR and validates authentication credentials provided by users. The PDG routes user data traffic between a user and an external packet data network, which is selected based on the 3G PS-services requested by the user. The latter identifies these services by means of a WLAN-access point name (W-APN), which represents a reference point to the external IP network that supports the PS services to be accessed by the user.

As mentioned previously, the integrated architecture of B3G networks specifies two different network access scenarios: (1) the WLAN direct IP access and (2) the WLAN 3GPP IP Access. The first scenario provides to a user connection to the public Internet or to an intranet via the WLAN-AN. In this scenario both the user and the network are authenticated to each other using the extensible authentication protocol method for GSM subscriber identity modules (EAP-SIM) (Haverinen & Saloway, 2006) or the Extensible Authentication Protocol-Authentication and Key Agreement (EAP-AKA) (Arkko & Haverinen, 2006) protocol. Moreover, in this scenario, the confidentiality and integrity of users data transferred over the air interface is ensured by the 802.11i security framework (IEEE std 802.11i, 2004). On the other hand, the WLAN 3GPP IP Access scenario allows a WLAN user to connect to the PS services (like WAP, MMS, LBS, etc.) or to the public Internet through the 3G PLMN. In this scenario, the user is authenticated to the 3G PLMN using the EAP-SIM or alternatively the EAP-AKA protocol encapsulated within IKEv2 (Kaufman, 2005) messages. The execution of IKEv2 is also used for the establishment of an

Figure 1. The B3G network architecture

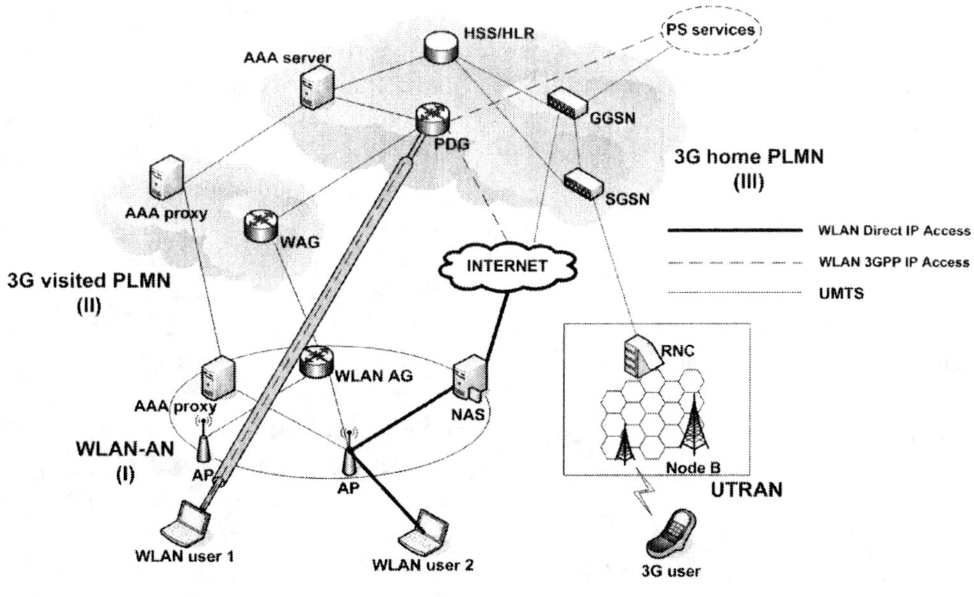

Table 1. 3G-WLAN interworking security mechanisms

Security	*WLAN Direct IP Access*	*3GPP IP Access*
Authentication	EAP-SIM or EAP-AKA	IKEv2 with EAP-SIM or EAP-AKA
Data protection	CCMP or TKIP protocol	IPsec based VPN tunnel using the ESP protocol

CCMP = Counter-Mode/CBC-Mac Protocol TKIP = Temporal Key Integrity Protocol

IP security (Ipsec)-based virtual private network (VPN) (Kent & Atkinson, 1998a) tunnel between the user and the PDG that provides confidentiality and integrity services to the data exchanged between them (see Figure 1). Table 1 summarizes the security protocols employed in each access scenario.

SECURITY ARCHITECTURES FOR B3G NETWORKS

Each network access scenario (i.e., WLAN direct access and WLAN 3GPP IP access) in B3G networks incorporates a specific security architecture, which aims at protecting the involved parties (i.e., the mobile users, the WLAN, and the 3G network) and the data exchanged among them. These architectures (3GPP TS 23.234, 2006) consist of various security protocols that provide mutual authentication (i.e., user and network authentication) as well as confidentiality and integrity services to the data sent over the air interface of the deployed WLANs and specific parts of the core network. In the following, the security architectures and the involved security protocols, which are employed in B3G networks, are presented and analyzed focusing on their functionality and the supported security services.

WLAN Direct IP Access Scenario

In the WLAN Direct IP Access scenario, both the user and the network are authenticated to each other using EAP-SIM or EAP-AKA, which are based on the 802.1X port access control (IEEE std 802.1X, 2001). After a successful authentication, the user obtains an IP address from the WLAN-AN and then, he/she gets access to the public Internet or an intranet, depending on the requested service. In this scenario, the confidentiality and integrity of user's data conveyed over the air interface of WLAN (IEEE std 802.11, 1999) are ensured by 802.11i (IEEE std 802.11i, 2004), which is analyzed next.

Authentication in WLAN Direct IP Access

The specific security protocol (i.e., EAP-AKA or EAP-SIM) that will be used for mutual authentication between the user and the network depends on the user's subscription. If the user possesses a UMTS subscribers identity module (USIM) card (3GPP TS 22.100, 2001), then, the EAP-AKA protocol is employed. Otherwise, EAP-SIM is used in cases that the user has a SIM-card (European Telecommunications Standards Institute [ETSI] TS 100 922, 1999) of global system for mobile communications (GSM)/general packet radio service (GPRS) (3GPP TS 0.3.6, 2002). When the AAA server receives the user's identity, it fetches from the HSS/HLR the user's profile in order to determine the employed authentication protocol that will be employed (i.e., EAP-SIM or EAP-AKA). In the following, we analyze the functionality of these two protocols focusing on the security services that each one provides.

EAP-SIM. EAP-SIM (Haverinen & Saloway, 2006) provides mutual authentication in a network environment that integrates 3G and WLANs using the credentials included in a SIM-card of a GSM/GPRS subscription. It involves a user, an AAA client (which is actually a wireless AP), and an AAA server that obtains authentication information (i.e., authentication triplets) from the HSS/HLR of the network where the user is subscribed (see Figure 2). EAP-SIM incorporates two basic enhancements that eliminate known security weaknesses of the authentication and key agreement procedure of GSM/GPRS (Haverinen & Saloway, 2006). First, the keys used in EAP-SIM are enhanced to have 128-bits security, in contrast to the 64-bit security of the original GSM/GPRS keys. Second, EAP-SIM supports mutual authentication, in contrast to the GSM/GPRS authentication, which performs only user to network authentication.

For the generation of stronger keys, the EAP-SIM protocol combines n (n=2 or n=3) individual random challenge (RANDs) that result in the derivation of n session keys, Kc. These keys are combined with a random number (NONCE pay-

Figure 2. The EAP-SIM authentication and session key agreement procedure

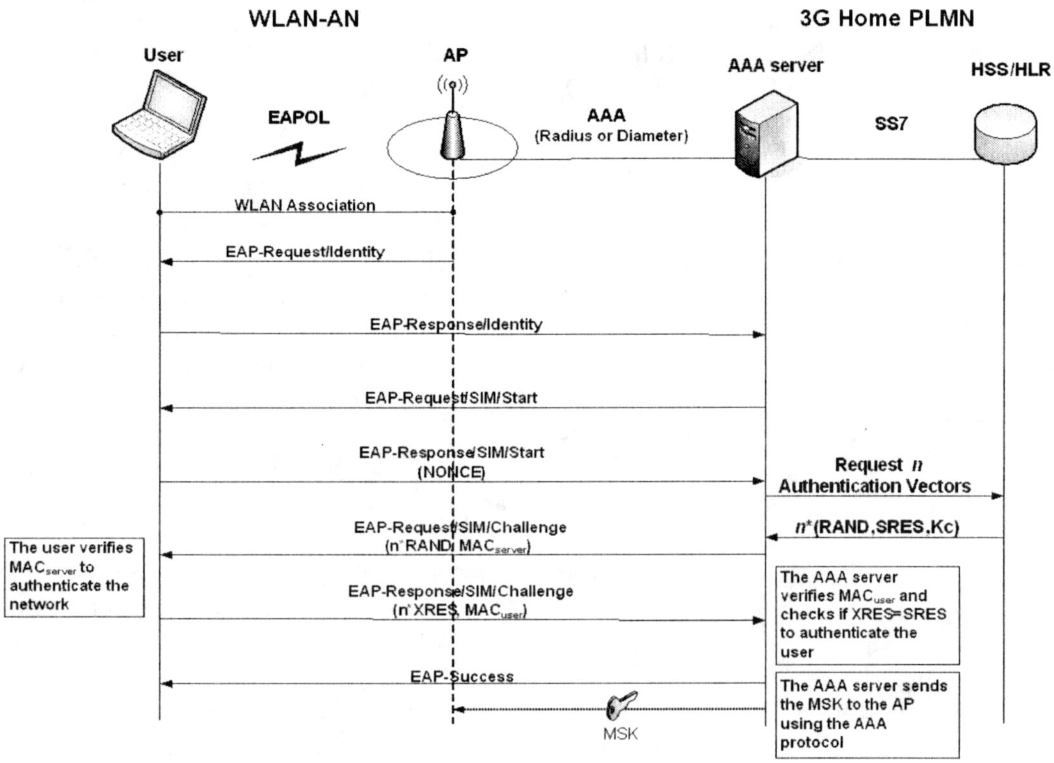

load), the user identity and other context-related information in order to generate the master key (*MK*) of the EAP-SIM protocol, as shown in the following formula:

*MK= SHA1(Identity| n*Kc| NONCE| Version List| Selected Version)*[1], (1)

where SHA1 is a hash function (Eastlake & Jones, 2001). In the sequel, the produced key *MK* is fed into a pseudo random function (prf) that generates other keys used in EAP-SIM. From these keys the most important are: (1) the master session key (*MSK*), which is used in 802.11i to generate the encryption keys, as described later on, and (2) the *K_auth* key, which is used in EAP-SIM for the generation of keyed message authentication codes (MACs) for authentication purposes.

Figure 2 shows the message exchange of EAP-SIM between the user and the AAA server, which is analyzed next. Note that the user communicates with the wireless AP via the EAP over LAN (EAPOL) protocol (IEEE std 802.1X, 2004).

- First, the user associates with the wireless AP and the latter sends an EAP-Request/Identity message to the user asking for his/her identity.
- The user responds with a message (EAP-Response/Identity) that includes his/her identity in the format of network access identifier (NAI) (Aboba & Beadles, 1999). This identity can be either the International Mobile Subscriber Identity (IMSI), or a temporary identity (i.e., pseudonym).
- Knowing the user's identity, the AAA server issues an EAP-Request/SIM/Start message, which actually starts the authentication procedure.
- The user sends back an EAP-Response/SIM/Start message that includes a nonce parameter

(NONCE), which is the user's challenge to the network.

- Upon receiving this message, the AAA server communicates with HSS/HLR and obtains n (n=2 or n=3) authentication triplets (RAND, SRES, Kc) for the specific user (the holder of the SIM-card). The generation of the GSM authentication triplets is based on a permanent, pre-shared (between the user and the network) secret key, *Ki,* which is assigned to the user when the latter is subscribed to the GSM/GPRS network.
- Then, the AAA server sends to the user an EAP-Request/SIM/Challenge message, which contains the n RANDs and the MAC_{server} of the message payload, which is calculated using the *K_auth* key as follows:

$$MAC_{server}=HMAC\_SHA1_{K\_auth}(EAP\text{-}Request/SIM/Challenge(n*RAND)|\ NONCE)^{2}, \quad (2)$$

where *NONCE* is the nonce sent by the user to the AAA server, and HMAC-SHA1 (Krawczyk, Bellare, & Canetti, 1997) is the MAC algorithm that generates the keyed hash value. Before the calculation of the MAC_{server} value, the AAA server must first generate the *MK* key (see Eq. 1), and, subsequently, the *K_auth* and *MSK* keys.

- Upon receiving the EAP-Request/SIM/Challenge, the user executes the GSM/GPRS authentication algorithms n times (one for each received RAND), in order to produce the n Kc keys and the n expected response (XRES) values. In the sequel, using the produced n *Kc* keys he/she generates the *MK* (see Eq. 1), and, consequently, the *K_auth* and the *MSK* keys, similarly, to the AAA server.
- Next, the user verifies the MAC_{server} using the *K_auth* key, and if this check is successful, then, the network is authenticated to the user, and the latter conveys to the AAA server the generated n XRES values within a EAP-Response/SIM/Challenge message. This message also includes the MAC_{user} value generated as follows:

$$MAC_{user}=HMAC\text{-}SHA1_{K\_auth}(EAP\text{-}Response/SIM/Challenge(n*XRES)\ |\ n*XRES)^{3}, \quad (3)$$

- Upon receiving this message, the AAA server examines whether the produced MAC_{user} is valid and if the n XRES values are equal to the n SRES values received from HSS/HLR for authentication. If these checks are successful, the AAA server sends an EAP-Success message to the user indicating the successful completion of the authentication procedure. In addition, the AAA server sends to the wireless AP the session key *MSK* within an AAA message (e.g., Radius or Diameter).

At this point, both the user and the network are mutually authenticated, and the user and the wireless AP share the key *MSK*, which is used for encryption purposes in the employed 802.11i security framework (see the *Data protection-802.11i standard* section).

EAP-AKA. EAP-AKA (Arkko & Haverinen, 2006) is an alternative to the EAP-SIM authentication protocol that uses a USIM-card and the UMTS AKA procedure. It involves the same network components with EAP-SIM (i.e., a user, an AAA client and an AAA server) and uses the same protocols for communication between them (i.e., EAPOL, Radius, Diameter, etc.). In the following, the EAP-AKA message exchange is analyzed:

- Likewise EAP-SIM, in the first two messages in the EAP-AKA negotiation (see Figure 3) the wireless AP requests for the user's identity (EAP request/identity message), and the latter replies by sending an EAP response/identity message, which contains his/her permanent IMSI or a temporary identity in an NAI format.
- After obtaining the user's identity, the AAA-server checks whether it possesses a 3G authentication vector, stored from a previous authentication with the specific user. If not, the AAA server sends the users IMSI to the HSS/HLR. The latter generates n 3G authentication vectors for the specific user

Figure 3. The EAP-AKA authentication procedure and session key agreement

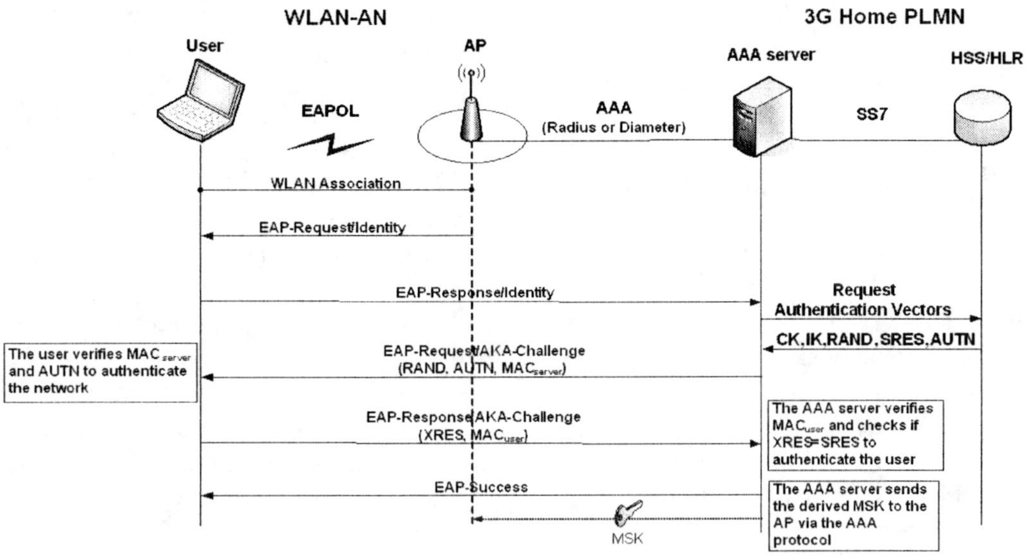

by using the UMTS permanent secret key, *K*, which is assigned to the user when he/she is subscribed to the network, and sends it to the AAA-server. Note that an authentication vector includes a RAND, the authentication token (AUTN), the XRES, the encryption key (*CK*), and the integrity key (*IK*) (Xenakis & Merakos, 2004).

- In the sequel, the AAA server selects one out of *n* obtained authentication vectors to proceed with the EAP-AKA authentication procedure and stores the remaining *n*-1 for future use. From the selected authentication vector, the AAA server uses the keys *CK* and *IK* and the identity of the user to compute the *MK* of EAP-AKA as shown in the following formula:

$$MK = SHA1(Identity|IK|CK), \quad (4)$$

MK is used as a keying material to generate the *MSK* and the *K_auth* key. The AAA server uses the *K_auth* key to calculate a keyed MAC_{server} (see Eq. 5), which verifies the integrity of the next EAP-AKA message (EAP-Request/AKA-Challenge).

$$MAC_{server} = HMAC\text{-}SHA1_{K\_auth}(EAP\text{-}Request/AKA/Challenge(RAND, AUTN)), \quad (5)$$

- The AAA server sends this message (EAP-Request/AKA-Challenge) to the user that contains the RAND, AUTN, and MAC_{server} payload. After receiving this information message, the user executes the UMTS-AKA algorithms and verifies the AUTN payload (Xenakis & Merakos, 2004). In the sequel, he/she generates the *IK* and *CK* keys and uses these two keys, as shown in Equation 4, to calculate the key *MK*. Subsequently, he/she uses *MK* to calculate the key *MSK* and the key *K_auth*, in order to verify the received MAC_{server} value.
- If these verifications (i.e., AUTN, MAC_{server}) are successful, the user computes the user's response to the challenge, noted as XRES payload, and sends an EAP-Response/AKA-Challenge message to the AAA server that includes the XRES and a new MAC_{user} value, which covers the whole EAP message and it is calculated using the *K_auth* key as follows:

MAC_{user}=HMAC-SHA1$_{K_auth}$(EAP-Response/AKA/Challenge(n*XRES)), (6)

- Upon receiving the EAP-Response/AKA-Challenge message the AAA server verifies the received MAC_{user} value and checks if the received user's response to the challenge (XRES) matches with the response (i.e., SRES) received from the HLR/HSS.
- If all these checks are successful, the AAA server sends an EAP-Success message along with the key *MSK* to the wireless AP. The latter stores the key and forwards the EAP-Success message to the user.

Finalizing the EAP-AKA protocol, both the user and the network have been authenticated to each other, and the user and the wireless AP share the key *MSK*, which is used in the security framework of 802.11i for generating the session encryption keys, as described in the next section.

Data Protection-802.11i Standard

As mentioned previously, 802.11i is employed to provide confidentiality and integrity services to users' data conveyed over the radio interface of the deployed WLAN in the WLAN Direct IP Access scenario. The 802.11i standard was developed to enhance the security services provided in WLANs. Its design was motivated by the fact that the wired equivalent privacy (WEP) protocol, due to its security flaws, could not fulfil the security requirements of WLANs (Borisov, Goldberg, & Wagner, 2001). The design goal of 802.11i is twofold: (1) to provide session key management by specifying a four-way handshake and group key handshake procedures, and (2) to enhance the confidentiality and integrity services provided to users' data by incorporating two security protocols (1) the counter-mode/CBC-MAC protocol (CCMP), which employs the advanced encryption standard (AES), and (2) the temporal key integrity protocol (TKIP), which uses the same encryption (RC4) with the WEP protocol. In the following, we analyze the four-way and group key handshake procedures of 802.11i and we present the functional details of the CCMP protocol. Since the TKIP protocol is considered to be a short term solution and it is merely a software enhancement of WEP, we do not elaborate further on it.

Four-way and group key handshakes. After a successful completion of the authentication procedure of EAP-SIM or EAP-AKA, the user and the AP perform the four-way and group key handshakes of 802.11i (IEEE std 802.11i, 2004) in order to generate the session keys. In the four-way handshake, both the user and the AP derive the pairwise transient key (*PTK*) from the *MSK* key that was generated in EAP-SIM or EAP-AKA to protect the four-way handshake messages and the unicast messages. In addition, the AP delivers to the user a group temporal key (*GTK*), which is used to protect broadcast/multicast messages. The *GTK* key is generated from the group master key (*GMK*), which is stored and maintained in the AP. The group key handshake is executed whenever the AP wants to deliver a new *GTK* key to the connected users. Note that all the messages exchanged during the four-way and the group key handshakes comply with the EAPOL-Key message format (IEEE std 802.1X, 2004).

As its name implies, the 802.11i four-way handshake consists of a total of four EAPOL-Key messages, which are analyzed next. Each of these messages includes key information (key_info payload), such as key identity, key replay counter, and so forth.

- At the beginning of the four-way handshake, the AP sends an EAPOL-Key message to the user that includes the A_{nonce}, which is a random number used as input for the generation of the *PTK* key, as described later on.
- Upon receiving the first EAPOL-Key message, the user generates a new random number called S_{nonce}. Then, he/she calculates the 384-bits *PTK* key using the first 265 bits of the *MSK* key (*MSK* was generated during the authentication procedure of EAP-SIM or EAP-AKA as described in the *Authentication in the WLAN Direct AP Access* section), the

Figure 4. The CCMP protocol key hierarchy

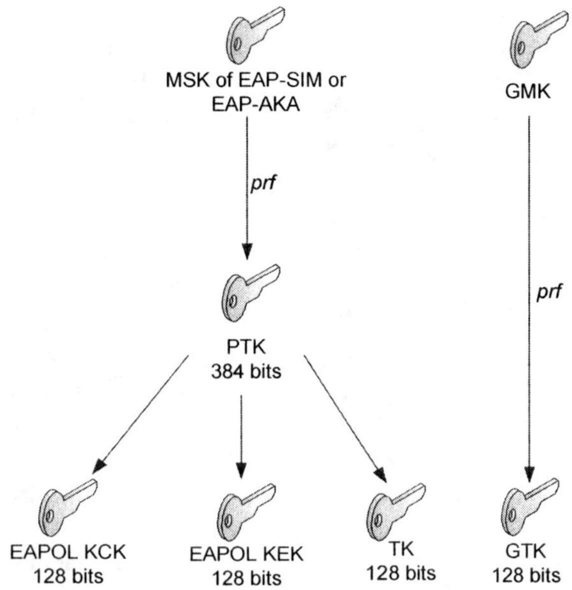

user's address, the AP's address, the S_{nonce} value, and the A_{nonce} value, as follows:

$$PTK=prf(MSK, \text{"Pairwise key expansion"}, Min(AP\ address, user's\ address) \mid Max(AP\ address, user's\ address) \mid Min(A_{nonce}, S_{nonce}) \mid Max(A_{nonce}, S_{nonce})), \quad (7)$$

where *prf* is a pseudo random function, "*Pairwise key expansion*" is a set of characters, and, finally, the Min and Max functions provide the minimum and maximum value, respectively, between two inputs. In the sequel, the generated *PTK* key is partitioned to derive three other keys: (1) a 128-bits key confirmation key (*KCK*) that provides integrity services to EAPOL-Key messages, (2) a 128-bits key encryption key (*KEK*) used to encrypt the *GTK* key as described next, and, (3) a 128-bits temporal key (*TK*) used for user's data encryption (see Figure 4).

- After the calculation of these keys, the user forwards to the AP the second EAPOL-Key message (step 2-Figure 5) that includes the S_{nonce}, the user's Robust Security Network Information Element (RSN IE) payload, which denotes the set of authentication and cipher algorithms that the user supports, and a message integrity code (MIC), which is a cryptographic digest used to provide integrity services to the messages of the four-way handshake and it is computed as follows:

$$MIC= HASH_{KCK}\ (EAPOL\text{-}Key\ message), \quad (8)$$

where $HASH_{KCK}$ denotes a hash function (i.e., HMAC-MD5 or HMAC-SHA-128) that uses the *KCK* key to generate the cryptographic hash value over the second EAPOL-Key message.

- Upon receiving this message, the AP calculates the key *PTK* and the related keys (i.e., *KCK, KEK,* and *TK* keys), (the same with the user), and, then, verifies the integrity of the message (producing the MIC value). Next, it generates the 128-bits *GTK* key from the *GMK* key as follows:

$$GTK=prf(GMK, \text{"Group key expansion"} \mid AP\ address \mid G_{nonce}), \quad (9)$$

where G_{nonce} is a random number generated from the AP to derive the *GTK* key

- In the sequel, the AP replies to the user by sending the third EAPOL-Key message (step 3), which includes the A_{nonce} value (the same with the first EAPOL-Key message), an MIC over the third EAPOL-Key message, the AP's RSN IE, and the *GTK* key, which is used to protect the broadcast/multicast messages and it is conveyed encrypted using the *KEK* key, as follows:

$$Encrypted\ GTK = ENC_{KEK}\ (GTK), \qquad (10)$$

where ENC_{KEK} denotes the encryption algorithm (i.e., AES or RC4), which uses the *KEK* key to encrypt the *GTK* key.

- By receiving this message, the user checks whether the MIC is valid and compares his/her RSN IE with the AP's RSN IE ensuring that they support the same cryptographic algorithms. If all these checks are correct, the user decrypts the *GTK* key using the *KEK* key and sends to the AP the last message of the four-way handshake (step 4), which includes an MIC payload over the fourth EAPOL-Key message, to acknowledge to the AP that he/she has installed the *PTK* key and the related keys (i.e., *KEK*, *KCK*, and *TK* keys), as well as the *GTK* key.
- Once the AP receives the fourth EAPOL-Key message, it verifies the MIC as previously. If this final check is successful, the four-way handshake is completed successfully, and both the user and the AP share: (1) the *TK* key to encrypt/decrypt unicast messages, and (2) the *GTK* key to encrypt/decrypt broadcast/multicast messages.

In case that the AP wants to provide a new *GTK* key to the connected users, it executes the group key handshake, as shown in Figure 5.

Figure 5. The four-way and group key handshakes of 802.11i

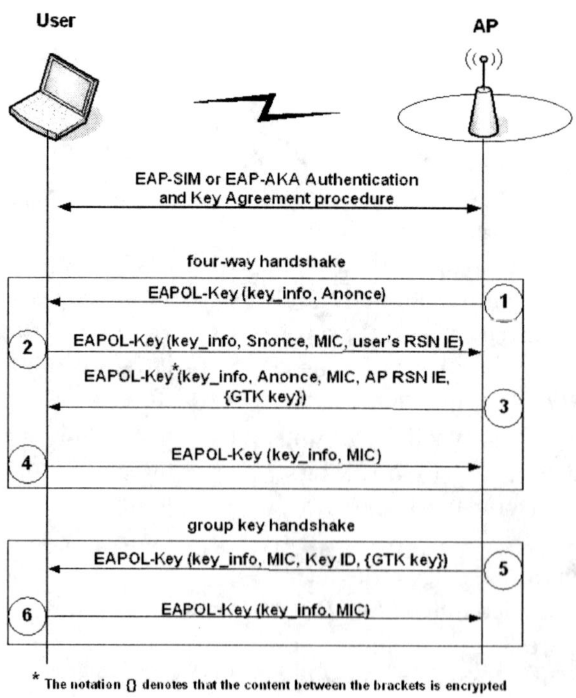

Figure 6. The CCMP protocol

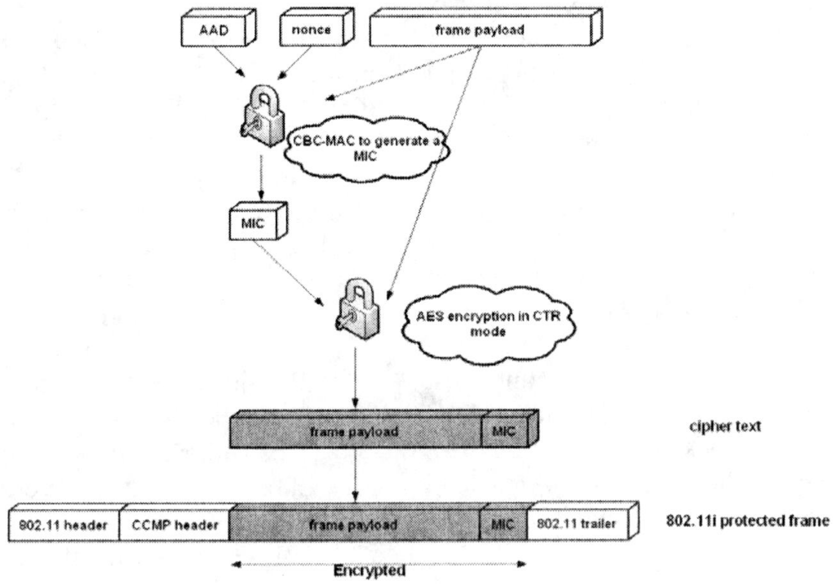

- The AP first generates a fresh *GTK* key from the *GMK* key and sends an EAPOL-Key message that includes an MIC value and the new *GTK* key to the users. Note that MIC is computed over the body of this EAPOL-Key message using the *KCK* key, and the *GTK* key is conveyed encrypted using the *KEK* key. Recall that both the user and the AP share the *KEK* and *KCK* keys, which were generated in the four-way handshake.
- Upon receiving the previous message, the user employs the *KCK* key to verify whether the MIC is valid and then, he/she decrypts the *GTK* key using the *KEK* key. Finally, he/she replies to the AP with an EAPOL-Key message, which includes an MIC that acknowledges to the AP that he/she has installed the *GTK* key.
- Once the AP receives this message, it verifies the MIC. If this final verification is successful, then, the group key handshake is completed successfully and the user can encrypt broadcast/multicast messages using the new *GTK* key.

CCMP Protocol. 802.11i incorporates the CCMP protocol to provide confidentiality and integrity services to users' data conveyed over the radio interface of WLANs. The CCMP protocol combines the AES encryption algorithm in CounTeR mode (CTR-AES) to provide data confidentiality and the Cipher Block Chaining Message Authentication Code (CBC-MAC) protocol to compute an MIC over the transmitted user's data that provides message integrity (Whiting, Housley, & Ferguson, 2003).

The operation of the CCMP protocol can be divided into three distinct phases. In phase 1, the CCMP protocol constructs an additional authentication data (AAD) value from constant fields of the 802.11 frame header (IEEE std 802.11, 1999). In addition, it creates a nonce value from the priority field of the 802.11 frame header and from the packet number (PN) parameter, which is a 48-bit counter incremented for each 802.11i protected frame. In phase 2, the CCMP protocol computes an MIC value over the 802.11 frame header, the AAD, the nonce, and the 802.11 frame payload using the CBC-MAC algorithm and the *TK* key (or the *GTK* key for broadcast/mulitcast communication). Recall that the *TK* key is part of the *PTK* key that is generated in the four-way handshake. In the sequel, CCMP forms the cipher text of the 802.11 frame payload and the produced MIC, using the CTR-AES encryption algorithm and the *TK* key (or the *GTK* key). Finally, in phase 3, the CCMP protocol constructs the 802.11i frame from the concatenation of: (1) the 802.11 header, (2) the CCMP header, which is created from the PN parameter and the identity of the encryption key, (3) the cipher text, and (4) the 802.11 trailer, which is the frame check sequence (FCS) (see Figure 6). The receiver of the 802.11i frame must verify that the PN parameter is fresh and the MIC value is valid. If these checks are successful, then, the receiver decrypts the 802.11i frame payload using the *TK* key (or the *GTK* key).

WLAN 3GPP IP Access

In contrast to the WLAN Direct IP Access scenario, in which a user gets access to the public Internet, directly, through the WLAN-AN, the WLAN 3GPP IP Access scenario provides to the WLAN user access to the PS services or the Internet through the 3G PLMN. Before getting access to them, the user must perform the six (6) discrete steps, presented in Figure 7 and described as follows:

1. **Initial authentication.** The user and the network are authenticated to each other using either the EAP-SIM or EAP-AKA protocol. This authentication step enables the user to obtain a local IP address, called transport IP address, which is used for access to the WLAN environment and the PDG. Note that this initial authentication can be omitted, if the PDG trusts the WLAN network and its users.

Figure 7. 3GPP IP access authentication procedure

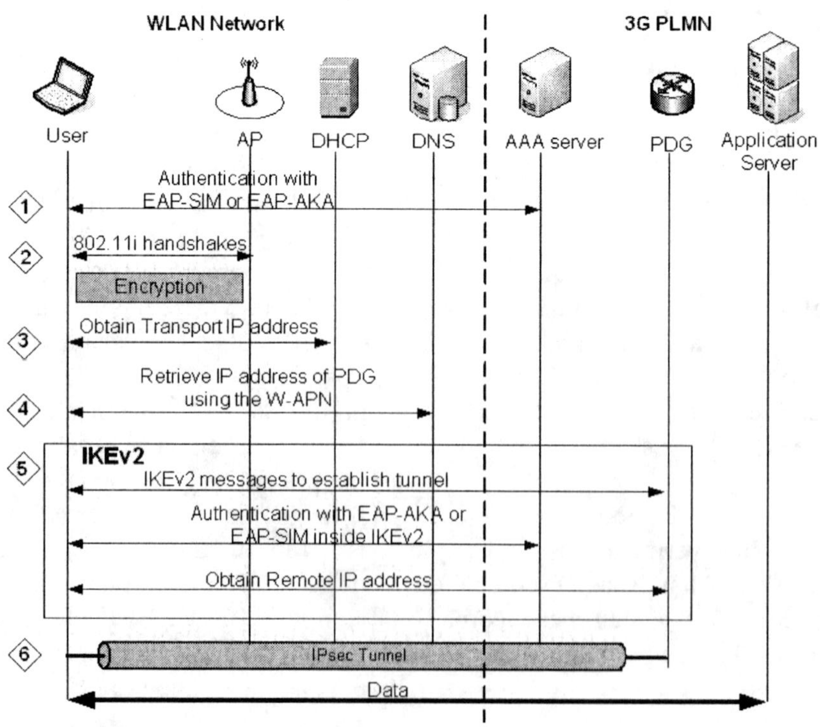

Figure 8. 3GPP IP access authentication protocol stack

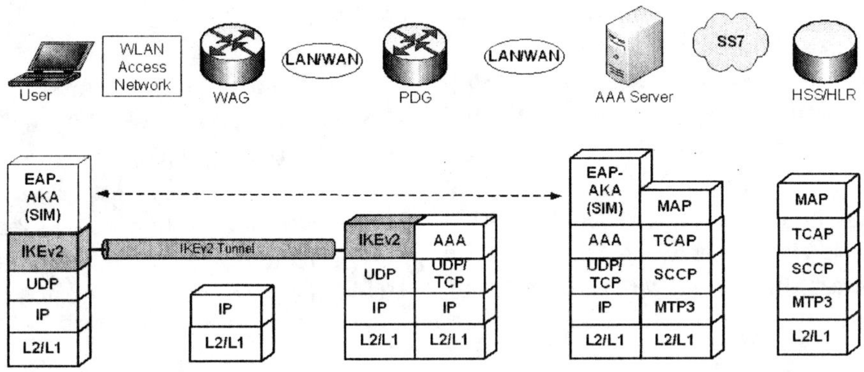

2. After the EAP-SIM or EAP-AKA execution, the four-way handshake and optionally the group key handshake follow to provide the 802.11i session keys. Then, the communication between the user and the wireless AP is encrypted using the CCMP or alternatively the TKIP protocol.
3. After the completion of the initial authentication step and the 802.11i handshakes, the user communicates with the Dynamic Host Configuration Protocol (DHCP) server to obtain the transport IP address. This local address is used by the user to execute the IKEv2 in step 4.
4. The user retrieves the IP address of the PDG using the W-APN identity and the domain name system (DNS) protocol. Thus, both the user and the PDG participate in a second authentication step that combines IKEv2 and EAP-SIM or EAP-AKA.
5. **Second authentication.** The user and the PDG execute the IKEv2 negotiation protocol, which encapsulates either EAP-SIM or EAP-AKA for authentication of the negotiating peers. After authentication completion, the user obtains a global IP address, called remote IP address, which is used for access to the PS services and the public Internet via the 3G PLMN. In addition, the execution of IKEv2 results in the establishment of a pair of IPsec security associations (SAs) between the user and the PDG, which are used for the deployment of an IPsec-based VPN.
6. The deployed IPsec based VPN protects user's data exchanged between the user and the PDG (in both directions) ensuring data origin authentication, data confidentiality and message integrity.

Figure 8 presents the protocol stack used in the 3GPP IP Access scenario for each entity that participates in the authentication procedure. The main authentication protocol is EAP-SIM or EAP-AKA, which is executed between the user and the AAA server. The user encapsulates EAP-SIM or EAP-AKA messages within IKEv2 and conveys them to the PDG. The latter acting as an AAA client transfers the EAP-SIM or EAP-AKA messages to the AAA server using an AAA protocol. Note that the AAA protocol can be either RADIUS, which runs over the user datagram protocol (UDP) or Diameter, which runs typically over the TCP protocol. The AAA server also includes the mobile application part (MAP) protocol stack to be able to communicate with the HSS/HLR and obtain authentication triplets and authorization information.

From the previous steps that a user has to perform to get access to the PS services or the public Internet in the WLAN 3GPP IP Access scenario, the initial authentication using either EAP-SIM or EAP-AKA (step 1) and the 802.11i handshakes (step 2) are the same with these of the WLAN Direct IP Access scenario, which has been analyzed in the *Authentication in the WLAN Direct IP Access* and *Data protection-802.11i standard* sections. Moreover, the acquisition of a local IP address (step 3) and the retrieval of the PDG address (step 4) do not present any significant interest from a security point of view. Thus, in the following sections we analyze the second authentication step (step 5), which includes a combined execution of IKEv2 with EAP-SIM or EAP-AKA, and the deployment of a bidirectional VPN that protects data exchanged.

Authentication in WLAN 3GPP IP Access

IKEv2 (Kaufman, 2005) is a simplified redesign of IKE (Harkins & Carrel, 1998) that allows two peers to authenticate each other (i.e., mutual authentication) and derive keys for secure communication with IPsec. The exchanged messages within IKEv2 are protected ensuring confidentiality and integrity, while the peers are authenticated using certificates, pre-shared keys, or the EAP protocol. In the con-

Figure 9. The execution of IKEv2 based on EAP-SIM or EAP-AKA

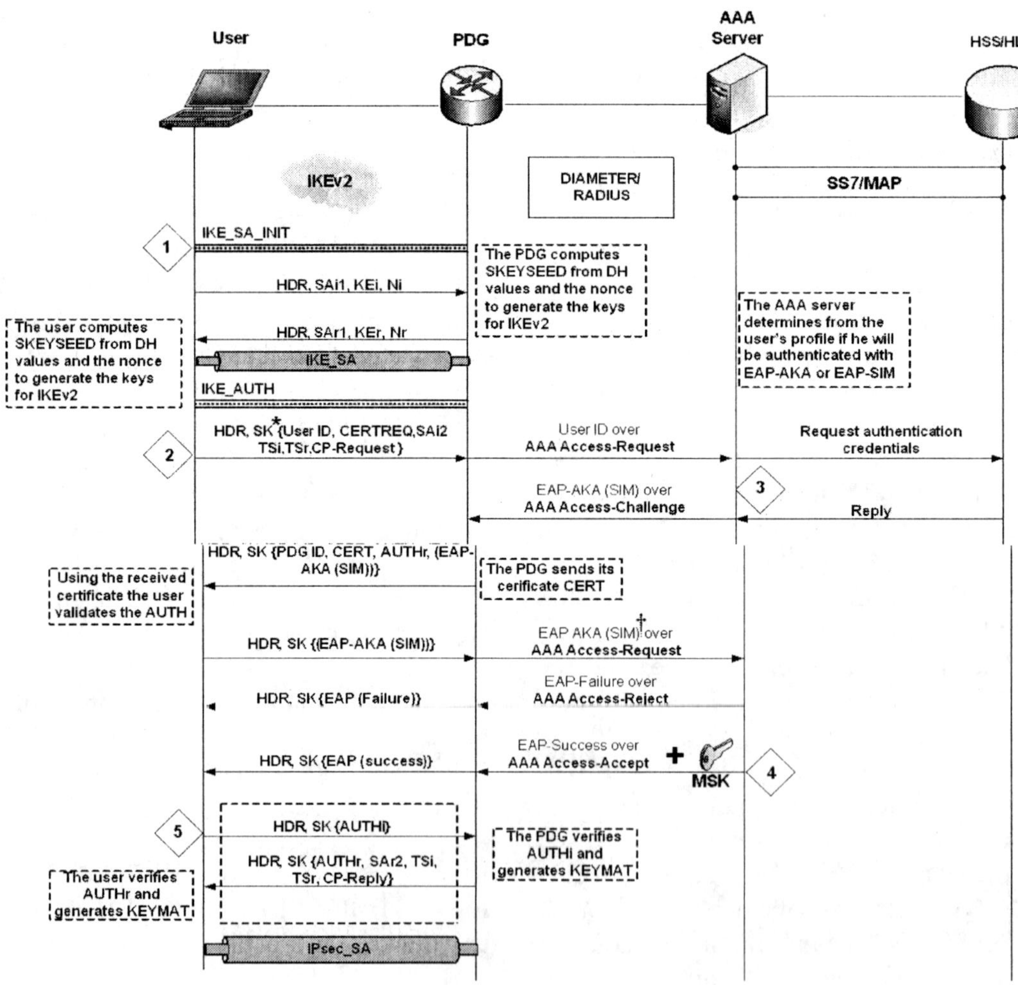

* The notation SK{. .} indicates that the content between brackets is encrypted and integrity protected

† The EAP-AKA (SIM) payload indicates that this payload can be either an EAP -AKA or an EAP-SIM message

text of WLAN 3GPP IP Access scenario, the user and the PDG execute IKEv2. The authentication of the user is based on EAP-SIM or EAP-AKA, while the authentication of the PDG is based on certificates.

The IKEv2 protocol is executed in two sequential phases (i.e., phase 1 and phase 2). In phase 1, the user and the PDG establish two distinct SAs: (1) a bidirectional IKE_SA that protects the messages of phase 2, and (2) an one-way IPsec_SA that protects user's data. During phase 2, the user and the PDG using the established IKE_SA can securely negotiate a second IPsec_SA that is employed for the establishment of a bidirectional IPsec based VPN tunnel between them.

The IKEv2 phase 1 negotiation between the user and the PDG is executed in two sub-phases: (1) the IKE_SA_INIT, and (2) the IKE_AUTH exchange, as shown in Figure 9. The IKE_SA_INIT exchange (noted as step 1 in Figure 9) consists of a single request and reply messages, which negotiate cryptographic algorithms, exchange nonces, and do a Diffie-Hellman exchange. In the context of this sub-phase, four cryptographic algorithms are negotiated: (1) an encryption algorithm, (2) an integrity protection algorithm, (3) a Diffie-Hellman group, and (4) a prf. The latter prf is employed for the construction of keying material for all of the cryptographic algorithms used. After the execution of the IKE_SA_INIT, an IKE_SA is established that protects the IKE_AUTH exchange. The second sub-phase (i.e., IKE_AUTH) authenticates the previous messages; exchanges identities and certificates; encapsulates EAP-SIM or alternatively EAP-AKA messages; and establishes an IPsec_SA (step 2-5 in Figure 9). All the messages of IKEv2 include a header payload (HDR), which contains a security parameter index (SPI), a version number, and security-related flags. The SPI is a value chosen by the user and the PDG to identify a unique SA. In the following, the IKEv2 negotiation is analyzed:

- At the beginning of the IKEv2 negotiation (step 1 in Figure 9), the user sends to the PDG the SAi1, which denotes the set of cryptographic algorithms for the IKE_SA that he/she supports, the KEi that is the Diffie-Hellman value, and an Ni value that represents the nonce. The nonce (i.e., a random number at least 128 bits) is used as input to the cryptographic functions employed by IKEv2 to ensure liveliness of the keying material and protect against replay attacks.
- The PDG answers with a message that contains its choice from the set of cryptographic algorithms for the IKE SA (SAr1), its value to complete the Diffie-Hellman exchange (KEr) and its nonce (Nr). At this point, both the user and the PDG can calculate the SKEYSEED value as follows:

$$SKEYSEED = prf((Ni \mid Nr), g \wedge ir)^{4}, \qquad (11)$$

where prf is the pseudo random function negotiated in the previous messages, and g^ir is the shared secret key that derives from the Diffie-Hellman exchange. The SKEYSEED value is used to calculate various secret keys. The most important are: the *SK_d* used for providing the keying material for the IPsec SA; *SK_ei* and *SK_ai* used for encrypting and providing integrity services, respectively, to the IKEv2 messages from the user to the PDG (IKE_SA); and, finally, *SK_er* and *SK_ar* that provide security services in the opposite direction (IKE_SA).

Finalizing the IKE_SA_INIT exchange, the IKE_AUTH exchange can start. It is worth noting that from this point all the payloads of the following IKEv2 messages, excluding the message header (HDR payload), are encrypted and integrity protected using the IKE_SA (see step 2 in Figure 9).

- The IKE_AUTH exchange of messages starts when the user sends to the PDG a message that includes his/her identity (IDi), which could be in an NAI format, the CERTREQ payload (optionally), which is a list of the certificate authorities (CA) whose public keys the user trusts, and the traffic selectors (TSi and TSr), which allow the peers to identify the packet flows that require processing by IPsec. In addition, in the same message the user must include the Configuration Payload Request (CP-Request), which is used to obtain a remote IP address from the PDG and get access to the 3G-PLMN.

- After receiving this information, the PDG forwards to the AAA server the user identity (IDi) including a parameter, which indicates that the authentication is being performed for VPN (tunnel) establishment. This will facilitate the AAA server to distinguish between authentications for WLAN access and authentications for VPN setup.
- Upon receiving the IDi, the AAA server fetches the user's profile and authentication credentials (GSM triplets if authentication is based on EAP-SIM, or 3G authentication vectors if authentication is based on EAP-AKA) from HSS/HLR (if these are not available in the AAA server in advance).
- Based on the user's profile, the AAA server initiates an EAP-AKA (if the user possesses a USIM card) or an EAP-SIM authentication (if the user possesses a GSM/GPRS SIM card) by sending to the PDG the first message of the related procedure (i.e., EAP-SIM or EAP-AKA) included in a AAA protocol (i.e., Radius or Diameter) (step 3 in Figure 9). Note that since there is no functional difference between the EAP-SIM and the EAP-AKA authentication when these protocols are encapsulated in IKEv2, we present them in a generic way. Thus, we introduce the EAP-AKA (SIM) payload notation (see Figure 9) to indicate that this payload can be an EAP-SIM or an EAP-AKA message.
- Upon receiving the first EAP-AKA (SIM) message, the PDG encapsulate it within an IKEv2 message and forwards the encapsulated message to the user. Except for the EAP-AKA (SIM) payload, this message also includes the PDG's identity, which identifies the provided 3G services (W-APN) (see the *Background* section), the PDG's certificate (CERT), and the AUTHr field. The latter contains signed data used by the user to authenticate the PDG. Similarly to the previous messages, the payload of this IKEv2 message, except for the message header, is encrypted using the IKE_SA.
- Upon receiving the EAP-AKA (SIM) payload, the user verifies the AUTHr field by using the public key of the PDG included in the certificate field (CERT), and answers by sending an EAP-AKA (SIM) response message encapsulated again within an IKEv2 message. From this point, the IKEv2 messages contain only EAP-AKA (SIM) payloads, which are encrypted and integrity protected as described previously.
- The EAP-SIM or EAP-AKA exchange continues, normally, until an EAP-SUCCESS message (or an EAP-FAILURE in case of a failure) is sent from the AAA server to the PDG, which ends the EAP-AKA or the EAP-SIM dialogue. Together with the EAP-SUCCESS message, the key *MSK* is sent from the AAA server to the PDG via the AAA protocol, as shown in Figure 9 (step 4).
- After finishing the EAP-AKA or EAP-SIM dialogue, the last step (step 5) of IKEv2 re-authenticates the peers, in order to establish an IPsec_SA. This authentication step is necessary in order defeat man-in-the-middle attacks, which might take place because the authentication protocol (e.g., EAP-SIM or EAP-AKA) runs inside the secure protocol (e.g., IKEv2). This combination creates a security hole since the initiator and the responder have no way to verify that their peer in the authentication procedure is the entity at the other end of the outer protocol (Asokan, Niemi, & Nyberg, 2002). Thus, in order to prevent possible attacks against IKEv2 (i.e., man-in-the-middle attacks), both the user and the PDG have to calculate the AUTHi and the AUTHr payloads, respectively, using the *MSK* key that was generated from the EAP-SIM or EAP-AKA protocol. Then, both the user and the PDG send each other the AUTHi and AUTHr payloads to achieve a security binding between the inner protocol (EAP-SIM or EAP-AKA) and the outer protocol (IKEv2).

Note that the PDG together with the AUTHr payload sends also its traffic selector payloads (TSi and TSr), the SAr2 payload, which contains the chosen cryptographic suit for the IPsec_SA and the assigned user's remote IP address in the Configuration Payload Reply (CP-REPLY) payload.

After the establishment of the IPsec_SA the keying material (*KEYMAT*) for this SA is calculated as follows:

$$KEYMAT = prf(SK\_d, Ni \mid Nr), \qquad (12)$$

where Ni and Nr are the nonces from the IKE_SA_INIT exchange, and *SK_d* is the key that is calculated from the SKEYSEED value (see eq. 11). The *KEYMAT* is used to extract the keys that the IPsec protocol uses for security purposes. Note that the deployed IPsec_SA protects the one-way communication between the user and the PDG. For bidirectional secure communication, one more SA needs to be established between them (the user and the PDG) by executing the IKEv2 phase 2 over the established IKE_SA.

Data Protection

After the completion of the authentication procedure and the execution of IKEv2 between the PDG and the user, a pair of IPsec_SAs has been established between these two nodes. This pair deploys a bidirectional VPN between them that allows for secure data exchange over the underlying network path. At the same time, the user has been subscribed to the 3G PLMN network for charging and billing purposes using either the EAP-AKA or EAP-SIM protocol.

The deployed VPN runs on top of the wireless link and extends from the user's computer to the PDG, which is located in the user's home 3G PLMN (see Figure 1 and 10). It is based on IPsec (Kent & Atkinson, 1998a), which is a developing standard for providing security at the network layer. IPsec provides two choices of security service through two distinct security protocols: the Authentication Header (AH) protocol (Kent & Atkinson, 1998c), and the encapsulating security payload (ESP) protocol (Kent & Atkinson, 1998b). The AH protocol provides support for connectionless integrity, data origin authentication, and protection against replays, but it does not support confidentiality. The ESP protocol supports confidentiality, connectionless integrity, anti-replay protection, and optional data origin authentication. Both AH and ESP support two modes of operation: transport and tunnel. The transport mode of operation provides end-to-end protection between the communicating end points by encrypting the IP packet payload. The tunnel mode encrypts the entire IP packet (both IP header and payload) and encapsulates the encrypted original IP packet in the payload of a new IP packet.

Figure 10. 3GPP IP access data plane

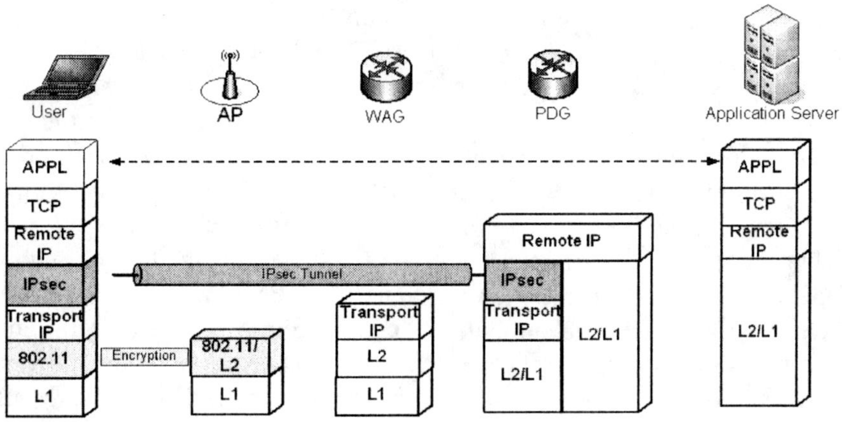

In the deployed VPN of the WLAN 3GPP IP Access scenario, IPsec employs the ESP protocol and is configured to operate in the tunnel mode. Thus, VPN provides confidentiality, integrity, data origin authentication, and anti-reply protection services protecting the payload and the header of the exchanged IP packets. From the two IP addresses (i.e., transport and remote IP address) of each authenticated user, the remote IP address serves as the inner IP address, which is protected by IPsec, and the transport IP address serves as the IP address of the new packets, which encapsulate the original IP packets and carry them between the user and the PDG (see Figure 10). Thus, an adversary can not disclose, fabricate unnoticed, or perform traffic analysis to the data exchanged between the user and the PDG. Finally, IPsec can use different cryptographic algorithms (i.e., DES, 3DES, AES, etc.) depending on the level of security required by the two peers and the data that they exchange.

COMPARISON OF THE SCENARIOS

Based on the presentation of the two access scenarios (i.e., WLAN Direct IP Access and 3GPP IP Access) that integrate B3G networks and the analysis of the security measures that each one employs, this section provides a brief comparison of them. The comparison aims at highlighting the deployment advantages of each scenario and classifies them in terms of: (1) security, (2) mobility, and (3) reliability.

Regarding the provided security services, both scenarios support mutual authentication. In the WLAN Direct IP Access scenario, the authentication procedure employs either EAP-SIM or EAP-AKA, depending on the user's subscription. However, both protocols present the same security weaknesses, which can be exploited by adversaries to perform several attacks such as identity spoofing, denial of service (DoS) attacks, replay attacks, and so forth (Arkko & Haverinen, 2006; Haverinen & Saloway, 2006). On the other hand, the authentication procedure of the 3GPP IP Access scenario is more secured, since it combines the aforementioned protocols (i.e., EAP-SIM and EAP-AKA) with IKEv2. Specifically, the PDG is authenticated using its certificate, and the user is authenticated using EAP-SIM or EAP-AKA. It is worth noting that since the EAP-SIM and EAP-AKA messages are encapsulated in protected IKEv2 messages, the identified security weaknesses associated with them are eliminated.

Regarding confidentiality and data integrity services, both scenarios protect sensitive data conveyed over the air interface. More specifically, in the WLAN Direct IP Access scenario, high level security services are provided only in cases that the CCMP security protocol is applied, since it incorporates the strong AES encryption algorithm. A downside of applying CCMP is that it requires hardware changes to the wireless APs, which might be replaced. In the WLAN 3GPP IP Access scenario, data encryption is applied at the layer 2 (using WEP, TKIP, or CCMP) and layer 3 (using IPsec), simultaneously (see Figure 10). This duplicate encryption provides advanced security services to the data conveyed over the WLAN radio interface, but at the same time it may cause bandwidth consumption, longer delays, and energy consumption issues at the level of mobile devices.

Another deployment feature, which can be used for comparing the two scenarios, has to do with mobility. The WLAN Direct IP Access scenario may support user mobility by employing one of the mobility protocols, proposed for seamless mobility in wireless networks (Saha, Mukherjee, Misra, & Chakraborty, 2004). On the other hand, in the WLAN 3GPP IP Access scenario, the established VPN between a user and the PDG adds an extra layer of complexity to the associated mobility management protocols of this scenario. This complexity arises from the fact that as the mobile user moves from one access network to another and his/her IP address changes, the mobility protocols must incorporate mechanisms that maintain, dynamically, the established VPN, enabling the notion of mobile VPN. An attempt to address this problem can be found in Dutta et al., 2004) that designs and implements a secure universal mobility architecture, which incorporates standard mobility management protocols, such as mobile IP for achieving mobile VPN deployment.

Finally, the deployed IPsec-based VPNs between the users and the PDG in the 3GPP IP Access scenario may raise reliability issues. Reliability is perceived as the ability to use VPN services at all times, and it is highly related to the network connectivity and the capacity of the underlying technology to provide VPN services. In the 3GPP IP Access scenario, all data traffic passes through the VPN tunnels that are extend from the users to the PDG. The number of the deployed VPNs can grow significantly, due to the fact that each user can establish multiple VPNs at the same time to access different services. Thus, the PDG must be able to support a large number of simultaneous VPNs in order to provide reliable security services.

CONCLUSION

This chapter has analyzed the security architectures employed in the interworking model that integrates 3G and WLANs, materializing B3G networks. The integrated architecture of B3G networks specifies two different network access scenarios: (1) the WLAN Direct IP Access, and (2) the WLAN 3GPP IP Access. The first scenario provides to a user connection to the public Internet or to an intranet via the WLAN-AN. In this scenario both the user and the network are authenticated to each other using EAP-SIM or EAP-AKA, depending on the user's subscription. Moreover, the confidentiality and integrity of the user's data transferred over the air interface are ensured by the 802.11i security framework. On the other hand, the WLAN 3GPP IP Access scenario allows a user to connect to the PS services (like WAP, MMS, LBS, etc.) or to the public Internet through the 3G PLMN. In this scenario, the user is authenticated to the 3G PLMN using EAP-SIM or alternatively EAP-AKA encapsulated within IKEv2, while the network is authenticated to the user using its certificate. In addition, the execution of IKEv2 is used for the establishment of an IPsec-based VPN between the user and the network that provides extra confidentiality and integrity services to the data exchanged between them.

ACKNOWLEDGMENT

Work supported by the project CASCADAS (IST-027807) funded by the FET Program of the European Commission.

REFERENCES

3rd Generation Partnership Project (3GPP) TS 22.100. (v3.7.0). (2001). *UMTS Phase 1 Release '99.* Sophia Antipolis Cedex, France: Author.

3rd Generation Partnership Project (3GPP) TS 0.3.6. (V7.9.0). (2002). *GPRS service description, Stage 2.* Sophia Antipolis Cedex, France: Author.

3rd Generation Partnership Project (3GPP) TS 23.234 (v7.3.0). (2006). *3GPP system to WLAN interworking. System description. Release 7.* Sophia Antipolis Cedex, France: Author.

3rd Generation Partnership Project (3GPP) TS 33.234 (v7.2.0). (2006). *3G security and WLAN interworking security. System description. Release 7.* Sophia Antipolis Cedex, France: Author.

Aboba, B., & Beadles, M. (1999). *The network access identifier* (RFC 2486). Retrieved from http://tools.ietf.org/html/rfc2486

Aboba, B., Blunk, L., Vollbrecht, J., Carlson, J., & Levkowetz, H. (2004). *The extensible authentication protocol* (RFC 3748). Retrieved from http://www.ietf.org/rfc/rfc3748.txt

Arkko, J., & Haverinen, H. (2006). *EAP-AKA authentication* (RFC 4187). Retrieved from http://www.rfc-editor.org/rfc/rfc4187.txt

Asokan, N., Niemi, V., & Nyberg, K. (2002). *Man-in-the-middle in tunneled authentication protocols.* Cryptology ePrint Archive, Report 2002/163. Retrieved from http://eprint.iacr.org/2002/163

Borisov, N., Goldberg, I., & Wagner, D. (2001, July). *Intercepting mobile communications: The insecurity of 802.11.* Paper presented at the 7th ACM/IEEE International Conference on Mobile Computing and Networking (MOBICOM), Rome, Italy.

Calhoun, P., Loughney, J., Guttman, E., Zorn, G., & Arkko, J. (2003). *Diameter base protocol* (RFC 3588). Retrieved from http://www.rfc-editor.org/rfc/rfc3588.txt

Dutta, A., Zhang, T., Madhani, S., Taniuchi, K., Fujimoto, K., Katsube, Y., et al. (2004, October). Secure universal mobility for wireless Internet. In *Proceedings of the 2nd ACM international workshop on Wireless mobile applications and services on WLAN hotspots (WMASH),* Philadelphia, PA.

Eastlake, D., & Jones, P. (2001). *US secure hash algorithm 1 (SHA1)* (RFC 3174). Retrieved from http://www.ietf.org/rfc/rfc3174.txt

Eronen, P. (2006). *IKEv2 mobility and multihoming protocol (MOBIKE)* (RFC 4555). Retrieved from http://www.ietf.org/rfc/rfc4555.txt

European Telecommunications Standards Institute (ETSI) TS 100 922 (v7.1.1). (1999). *Subscriber identity modules (SIM) functional characteristics.*

Harkins, D., & Carrel, D. (1998). *The Internet key exchange (IKE)* (RFC 2409). Retrieved from http://faqs.org/rfcs/rfc2409.html

Haverinen, H., & Saloway, J. (2006). *EAP-SIM authentication* (RFC 4186). Retrieved from http://www.ietf.org/rfc/rfc4186.txt

IEEE std 802.11 (1999). *Wireless LAN medium access control (MAC) and physical layer (PHY) specifications.*

IEEE std 802.11i. (2004). *Wireless medium access control (MAC) and physical layer (PHY) specifications: Medium access control (MAC) security enhancements.*

IEEE std 802.1X. (2004). *Port based access control.*

Kaufman, C. (2005). *The Internet key exchange (IKEv2) protocol* (RFC 4306). Retrieved from http://www.rfc-editor.org/rfc/rfc4306.txt

Kent, S., & Atkinson, R. (1998a). *Security architecture for Internet protocol* (RFC 2401). Retrieved from http://www.faqs.org/rfcs/rfc2401.html

Kent, S., & Atkinson, R. (1998b). *IP encapsulating security payload (ESP)* (RFC 2406). Retrieved from http://www.faqs.org/rfcs/rfc2406.html

Kent, S., & Atkinson, R. (1998c). *IP authentication header* (RFC 2402). Retrieved from http://www.rfc-editor.org/rfc/rfc2402.txt

Kivinen, T., & Tschofenig, H. (2006). *Design of the Mobike protocol* (RFC 4621). Retrieved from http://www.ietf.org/rfc/rfc4621.txt

Krawczyk, H., Bellare, M., & Canetti, R. (1997). *HMAC: Keyed-hashing for message authentication* (RFC 2104). Retrieved from http://www.faqs.org/rfcs/rfc2104.html

Laat, C., Gross, G., Gommans, L., Vollbrecht, J., & Spence, D. (2000). *Generic AAA architecture* (RFC 2903). Retrieved from http://isc.faqs.org/rfcs/rfc2903.html

Rigney, C., Rubens, A., Simpson, W., & Willens, S. (1997). *Remote authentication dial in user services (RADIUS)* (RFC 2138). Retrieved from http://tools.ietf.org/html/rfc2138

Saha, D., Mukherjee, A., Misra, I. S., & Chakraborty, M. (2004). Mobility support in IP: A survey of related protocols. *IEEE Network, 18*(6), 34-40.

Whiting, D., Housley, R., & Ferguson, N. (2003). *Counter with CBC MAC (CCM)* (RFC 3610). Retrieved from http://www.ietf.org/rfc/rfc3610.txt

Xenakis, C., & Merakos, L. (2004). Security in third generation mobile networks. *Computer Communications, 27*(7), 638-650.

KEY TERMS

Authentication, Authorization, and Accounting (AAA): AAA is a security framework which provides authentication, authorization, and accounting services. The two most prominent AAA protocols are Radius and Diameter.

Beyond Third Generation (B3G): B3G is the integration of heterogeneous mobile networks through an IP-based common core network.

Counter-Mode/CBC-MAC Protocol (CCMP): CCMP is a security protocol defined in 802.11i, which employs the AES encryption to provide confidentiality and data integrity services.

Extensible Authentication Protocol (EAP): EAP is a security framework used to provide a plethora of authentications options, called EAP methods.

Extensible Authentication Protocol-Authentication and Key Agreement (EAP-AKA): EAP-AKA is an EAP method based on UMTS authentication of USIM cards.

Extensible Authentication Protocol method for GSM Subscriber Identity Modules (EAP-SIM): EAP-SIM is an EAP method based on GSM authentication of SIM cards.

802.11i: 802.11i is a security framework that incorporates the four-way handshake and group-key handshake for session key management and specifies the TKIP and CCMP security protocols to provide confidentiality and integrity services in 802.11 WLAN.

IKEv2: IKEv2 is a security association (SA) negotiation protocol used to establish an IPsec-based VPN tunnel between two entities.

IP security (IPsec): IPsec is a security protocol used to provide VPN services.

ENDNOTES

1 (| means string concatenetation and the notation n*Kc denotes the n Kc keys concatenated)

2 (The notation n*RAND denotes the n RAND values concatenated)

3 (The notation n*XRES denotes the n XRES values concatenated)

4 | means string concatenation

Chapter XX
Security in UMTS 3G Mobile Networks

Christos Xenakis
University of Piraeus, Greece

ABSTRACT

This chapter analyzes the security architecture designed for the protection of the universal mobile telecommunication system (UMTS). This architecture is built on the security principles of second generation (2G) systems with improvements and enhancements in certain points in order to provide advanced security services. The main objective of the third generation (3G) security architecture is to ensure that all information generated by or relating to a user, as well as the resources and services provided by the serving network and the home environment are adequately protected against misuse or misappropriation. Based on the carried analysis the critical points of the 3G security architecture, which might cause network and service vulnerability are identified. In addition, the current research on the UMTS security and the proposed enhancements that aim at improving the UMTS security architecture are briefly presented and analyzed.

INTRODUCTION

The universal mobile telecommunication system (UMTS) (3rd Generation Partnership Project [3GPP] TS 23.002, 2002) is a realization of third generation (3G) networks, which intend to establish a single integrated system that supports a wide spectrum of operating environments. Users have seamless access to a wide range of new telecommunication services, such as high data rate transmission for high-speed Internet/intranet applications, independently of their location. Thus, mobile networks comprise a natural extension of the wired Internet computing world, enabling access for mobile users to multimedia services that already exist for non-mobile users and fixed networking.

Along with the variety of new perspectives, UMTS also raises new concerns on security issues. Wireless access is inherently less secure and mobility implies higher security risks compared to those encountered in fixed networks. The advanced wireless and wired network infrastructure, which supports higher access rates, and the complex network topologies, which enable "anywhere-anytime" connectivity, may increase the number and the ferocity of potential attacks. Furthermore, the potential intruders are able to launch malicious attacks from mobile devices with enhanced processing capabilities, which are difficult to trace. To defeat the possible vulnerable points, UMTS has incorporated a specific security architecture named as 3G security architecture.

This chapter analyzes the security architecture designed for the protection of UMTS. This architecture is built on the security principles of second generation (2G) systems with improvements and enhancements in certain points in order to provide advanced security services. The main objective of the 3G security architecture is to ensure that all information generated by or relating to a user, as well as the resources and services provided by the serving network (SN) and the home environment (HE) are adequately protected against misuse or misappropriation. Based on the carried analysis the critical points of the 3G security architecture, which might cause network and service vulnerability are identified. In addition, the current research on the UMTS security and the proposed enhancements that aim at improving the UMTS security architecture are briefly presented and analyzed.

The rest of this chapter is organized as follows. The next section outlines the UMTS network architecture and the 3G security architecture. The third section elaborates on the network access security features, and the fourth section examines the network domain security. The fifth section presents the user domain security, the application domain security, the visibility of security operation and configurability, and the network-wide confidentiality option. The sixth section analyzes potential weaknesses concerning the 3G security architecture and the seventh section presents the current research on the UMTS security. Finally, the last section contains the conclusions.

Figure 1. UMTS network architecture

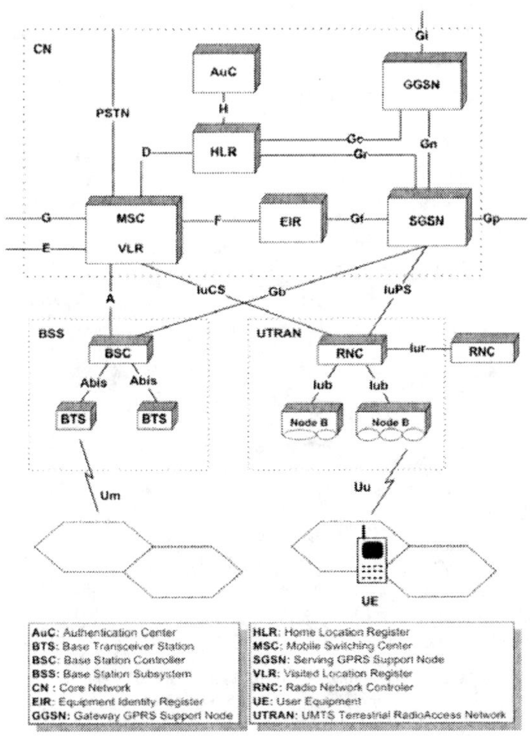

BACKGROUND

UMTS Network

UMTS has been standardized in several releases, starting from Release 1999 (R99), and moving forward to Release 4 (Rel-4), Release 5 (Rel-5), Release 6 (Rel-6), supporting compatibility with the evolved global system for mobile communications (GSM)/general packet radio service (GPRS) network. The UMTS network architecture includes the core network (CN), the radio access network, and the user equipment, as can be seen in Figure 1. This division provides the necessary flexibility by allowing the coexistence of different access techniques and different core network technologies, thus facilitating the migration form 2G to 3G networks. The fundamental difference between GSM/GPRS and UMTS R99 is that the latter supports higher bit rates (up to 2Mbps). This is achieved through a new wideband code division multiple access (WCDMA) radio interface for the land-based communications system, named UMTS terrestrial radio access network (UTRAN) (3GPP TS 25.401, 2002). UTRAN consists of two distinct elements, Node B, and the radio network controller (RNC). Node B converts the data flows between the lu-b and Uu interfaces and participates in radio resource management. The RNC owns and controls the radio resources of the Nodes B connected to it. The user equipment, which mainly comprises a mobile station (MS) with limited processing, memory, and power capabilities is connected to the UTRAN through the Uu radio interface (3GPP TS 23.002, 2002). The CN of the UMTS R99 uses the network elements of GSM/GPRS such as the home location register (HLR), the visitor location register (VLR), the authentication centre (AuC), the equipment identity register (EIR), the mobile service switching centre (MSC), the Serving GPRS support node (SGSN) and the Gateway GPRS support node (GGSN) (3GPP TS 23.002, 2002).

UMTS Security Architecture

3G security is built on the security principles of 2G systems, with improvements and enhancements in certain points in order to provide advanced security services. The elementary security features employed in 2G, such as subscriber authentication, radio interface encryption, and subscriber identity confidentiality are retained and enhanced where needed. The main objective of 3G security is to ensure that all information generated by or relating to a user, as well as the resources and services provided by the SN and the HE are adequately protected against misuse or misappropriation. The level of protection is better than that provided in

Figure 2. 3G-security Architecture

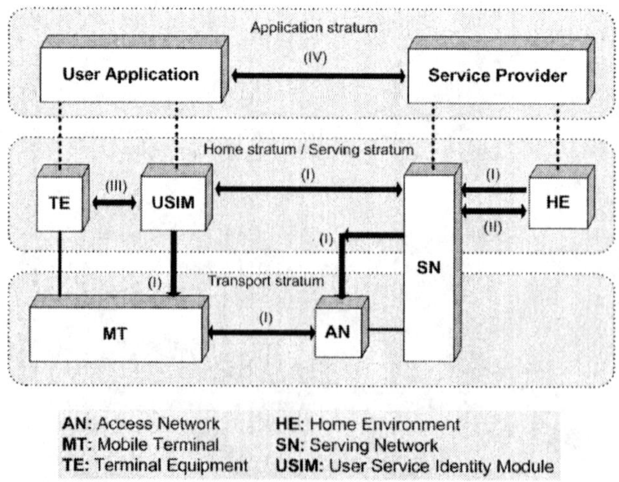

the contemporary fixed and mobile networks. The security features have been adequately standardized in order to ensure worldwide availability, interoperability, and roaming between different SNs. Furthermore, 3G security features and mechanisms can be extended and enhanced as required by new threats and services (Xenakis & Merakos, 2004b).

Figure 2 gives an overview of the 3G security architecture, illustrating five major security classes (3GPP TS 33.102, 2002):

- Network access security (I)
- Network domain security (II)
- User domain security (III)
- Application domain security (IV)
- Visibility and configurability of security (V)

NETWORK ACCESS SECURITY

Network access security is a key component in the 3G security architecture. This class deals with the set of security mechanisms that provide users with secure access to 3G services, as well as protect against attacks on the radio interface. Such mechanisms include: (1) user identity confidentiality, (2) authentication and key agreement, (3) data confidentiality, and (4) integrity protection of signaling messages. Network access security takes place independently in each service domain.

User Identity Confidentiality

User identity confidentiality allows the identification of a user on the radio access link by means of a temporary mobile subscriber identity (TMSI). This implies that confidentiality of the user identity is protected almost always against passive eavesdroppers. Initial registration is an exceptional case where a temporary identity cannot be used, since the network does not yet know the permanent identity of the user.

The allocated temporary identity is transferred to the user once the encryption is turned on. A TMSI in the circuit switched (CS) domain or P-TMSI in the packet switched (PS) domain has a local significance only in the location area or the routing area, in which the user is registered. The association between the permanent and temporary user identities is stored in the VLR or the SGSN (VLR/SGSN). If the mobile user arrives into a new area, then the association between the permanent and the temporary identity can be fetched from the old location or routing area. If the address of the old area is not known or the connection cannot be established, then, the permanent identity must be requested from the mobile user.

To avoid user traceability, which may lead to the compromise of user identity confidentiality as well as to user location tracking, the user should not be identified for a long period by means of the same temporary identity. Additionally, any signaling or user data that might reveal the user's identity are ciphered on the radio access link.

Authentication and Key Agreement

Authentication and key agreement mechanism achieves mutual authentication between the mobile user and the SN showing knowledge of a secret key (K), as well derives ciphering and integrity keys. The authentication method is composed of a challenge/response protocol (see Figure 3) and was chosen in such a way as to achieve maximum compatibility with the GSM/GPRS security architecture facilitating the migration from GSM/GPRS to UMTS. Furthermore, the user service identity module (USIM) (3GPP TS 22.100, 2001) and the HE keep track of counters SQN_{MS} and SQN_{HE}, respectively, to support the network authentication. The sequence number SQN_{HE} is an individual counter for each user, while the SQN_{MS} denotes the highest sequence number that the USIM has accepted. Whenever the SQN_{HE} is not in the correct range, the mobile station decides that a synchronization failure has occurred in the HE and consequently initiates a resynchronization to the HE.

Upon receipt of a request from the VLR/SGSN, the HE authentication center (HE/AuC) forwards an ordered array of authentication vectors (AV) to the VLR/SGSN. Each AV, which is used in the authentication and key agreement procedure

Figure 3. 3G authentication and key agreement

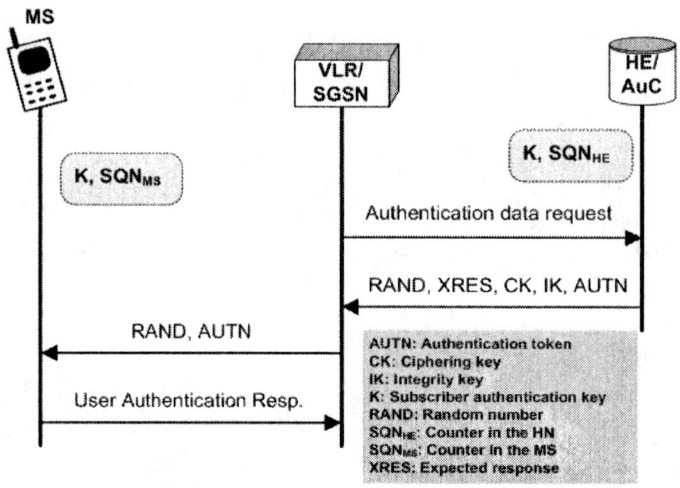

Figure 4. Generation of authentication vectors

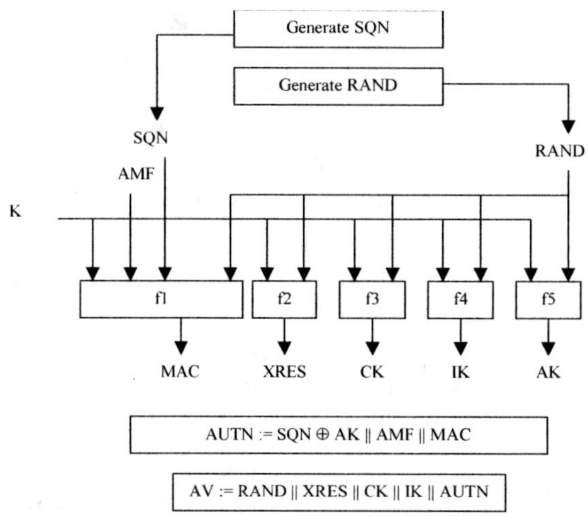

between the VLR/SGSN and the USIM consists of a random number (RAND), an expected response (XRES), a cipher key (CK), an integrity key (IK), and an authentication token (AUTN).

Figure 4 shows an AV generation by the HE/AuC. The HE/AuC starts with generating a fresh sequence number (SQN), which proves to the user that the generated AV has not been used before and an unpredictable challenge RAND. Then, using the secret key (K) it computes:

- The Message Authentication Code *(MAC) = f1k (SQN ||' RAND || AMF)*, where f1 is a message authentication function and the authentication and key management field (AMF) is used to fine tune the performance or bring a mew authentication key stored in the USIM into use.
- The expected response *XRES = f2k (RAND)* where f2 is a (possibly truncated) message authentication function.
- The cipher key *CK = f3k (RAND),*

- the integrity key *IK = f4k (RAND),*
- and the anonymity key *AK = f5k (RAND)* where f3, f4, and f5 are key generating functions.
- Finally, the HE/AuC assembles the authentication token $AUTN = SQN \oplus^2 AK \| AMF \| MAC$

It has to be noted that the authentication and key generation functions f1, f2, f3, f4, and f5, and the consequent AV computation follow the one-way property. This means that if the output is known there exists no efficient algorithm to deduce any input that would produce the output. Although the f1-f5 functions are based on the same basic algorithm, they differ from each other in a fundamental way in order to be impossible to deduce any information about the output of one function from the output of the others. Since they are used in the AuC and in the USIM, which are controlled by the home operator, the selection of the algorithms (f1-f5) is in principal operator specific. However, an example algorithm set has been proposed called MILENAGE (3GPP TS 35.205, 2001).

When the VLR/SGSN initiates an authentication and key agreement procedure, it selects the next AV from the ordered array and forwards the parameters RAND and AUTN to the user. The USIM using also the secret key (K) computes the AK,

AK = f5k (RAND),

and retrieves the SQN,

$SQN = (SQN \oplus AK) \oplus AK.$

Then, it computes $XMAC = f1k (SQN \| RAND \| AMF)$ and checks whether the received AUTN and the retrieved SQN values were indeed generated in AuC (3GPP TS 33.102, 2002). If so, the USIM computes the user response to the challange *RES=f2k (RAND)*, and triggers the mobile station (MS) to send back a user authentication response. Afterwards, the USIM computes the CK,

CK = f3k (RAND),

and the IK,

IK = f4k (RAND).

The VLR/SGSN compares the received RES with the XRES field of the AV. If they match, it considers that the authentication and key agreement exchange has been successfully completed. Finally, the USIM and the VLR/SGSN transfer the established encryption and integrity protection keys (CK and IK) to the mobile equipment and the RNC that perform ciphering and integrity functions.

Data Confidentiality

Once the user and the network have authenticated each other, they may begin secure communication. As described previously, a cipher key is shared between the core network and the terminal after a successful authentication event. User and signaling data sent over the radio interface are subject to ciphering using the function (f8). The encryption/decryption process takes place in the MS and the

Figure 5. Ciphering over the radio access link

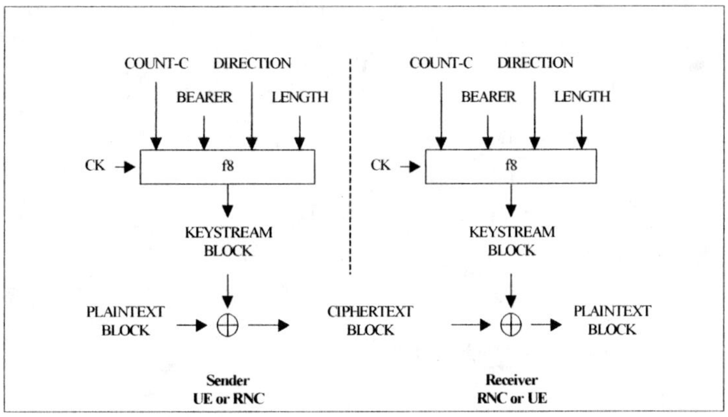

RNC on the network side. The f8 is a symmetric synchronous stream cipher algorithm that is used to encrypt frames of variable length. The main input to the f8 is a 128-bit secret cipher key CK. Additional inputs, which are used to ensure that two frames are encrypted using different keystreams are a 32-bit value COUNT, a 5-bit value BEARER, and a 1-bit value DIRECTION (see Figure 5). The output is a sequence of bits (the "keystream") of the same length as the frame. The frame is encrypted by XORing the data with the keystream. For UMTS R99, f8 is based on the Kasumi algorithm (3GPP TR 33.908, 2000).

Integrity Protection of Signaling Messages

The radio interface in 3G mobile systems has also been designed to support integrity protection on the signaling channels. This enables the receiving entity to be able to verify that the signaling data have not been modified in an unauthorized way since they were sent. Furthermore, it ensures that the origin of the received signaling data is indeed the one claimed. The integrity protection mechanism is not applied for the user plane due to performance reasons.

The function (f9) is used to authenticate the integrity and the origin of signaling data between the MS and the RNC in UMTS. It computes a 32-bit MAC (see Figure 6), which is appended to the frame and is checked by the receiver. The main inputs to the algorithm are a 128-bit secret IK and the variable-length frame content MESSAGE. Additional inputs, which are used to ensure that MACs for two frames with identical content are different, are a 32-bit value COUNT, a 32-bit value FRESH, and an 1-bit value DIRECTION. In the UMTS R99, the f9 is based on the Kasumi algorithm (3GPP TR 33.908, 2000).

NETWORK DOMAIN SECURITY

Network domain security (NDS) features ensure that signaling exchanges within the UMTS core as well as in the whole wireline network are protected. Various protocols and interfaces are used for the control plane signaling inside, and between core networks, such as the mobile application part (MAP) and the GPRS tunneling protocol (GTP) protocols, and the Iu (IuPS, IuCS) and Iur interfaces (3GPP TS 23.002, 2002). These will be protected by standard procedures based on the existing cryptographic techniques. Specifically, the IP-based protocols shall be protected at network level by means of IP security (IPsec) (Kent & Atkinson, 1998), while the realization of protection for the signaling system 7 (SS7)-based protocols and the Iu and Iur interfaces shall be accomplished at the application layer. In the following, the NDS context for IP-based (3GPP TS 33.210, 2002) and SS7-based (3GPP TS 33.200, 2002) protocols is presented. Moreover, the employment of traditional security technologies, originally designed for fixed networking, such as firewalls and static virtual private networks (VPNs) are examined. The application of these technologies safeguards the UMTS core network from external attacks and protects users' data when are conveyed over the public Internet.

Figure 6. Derivation of MAC on a signaling message

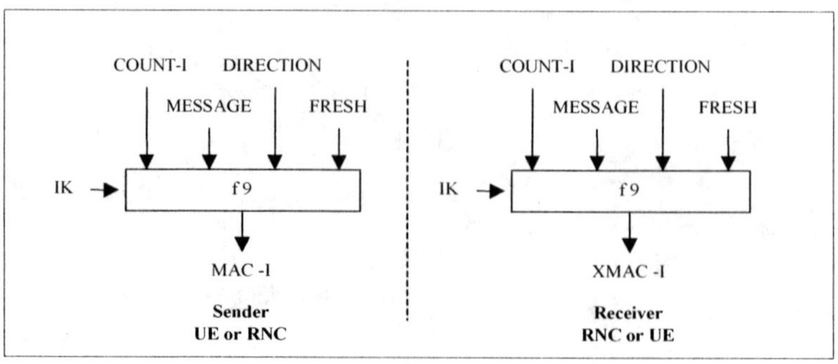

IP-Based Protocol

The UMTS network domain control plane is sectioned into security domains, which typically coincide with the operator borders. Security gateways (SEGs) are entities at the borders of the IP security domains used for securing native IP-based protocols. It is noted that NDS does not extend to the user plane, which means that packet flows over the Gi (3GPP TS 23.002, 2002) interface will not be protected by the SEGs. The key management functionality is logically separate from the SEG. Key administration centers (KACs) negotiate the IPsec security associations (SAs) by using the Internet key exchange (IKE) protocol (Harkins & Carrel, 1998) in a client mode, on behalf of the network entities (NEs) and the SEGs. The KACs also distribute SAs parameters to the NEs or the SEGs through standard interfaces. In Figure 7 the UMTS NDS architecture for IP-based protocols is depicted.

To secure the IP traffic between two NEs, either a hop-by-hop or an end-to-end scheme may be applied. The first requires that the originating NE establishes an IPsec tunnel to the appropriate SEG in the same security domain and forwards the data to it. The SEG terminates this tunnel and sends the data through another IPsec tunnel to the receiving network. The second tunnel is terminated by the SEG in the receiving domain, which in turn uses IPsec to pass the data to its final destination (path (a) in Figure 7). The end-to-end scheme implies that an IPsec SA is established between the two NEs (path (b) in Figure 7). This scheme can also be applied in case the two parties belong to the same security domain.

Node authentication can be accomplished using either pre-shared symmetric keys or public keys (Harkins & Carrel, 1998). Using pre-shared symmetric keys means that the KACs or the NEs do not have to perform public key operations as well as there is no need for establishing a public key infrastructure. The IPsec is configured either in transport mode or in tunnel mode (Kent & Atkinson, 1998). Whenever at least one end point is a gateway then the tunnel mode suits better. Finally, the IPsec protocol shall always be encapsulation security payload (ESP) (Kent & Atkinson, 1998), given that it can provide confidentiality and integrity protection as well.

SS7-Based Protocols

NDS for SS7-based protocols is mainly found at the application layer. Specifically, in case that the transport relies on SS7 or on a combination of SS7 and IP, then security shall be provided at the application layer. On the other hand, whenever the

Figure 7. NDS architecture for IP-based protocols

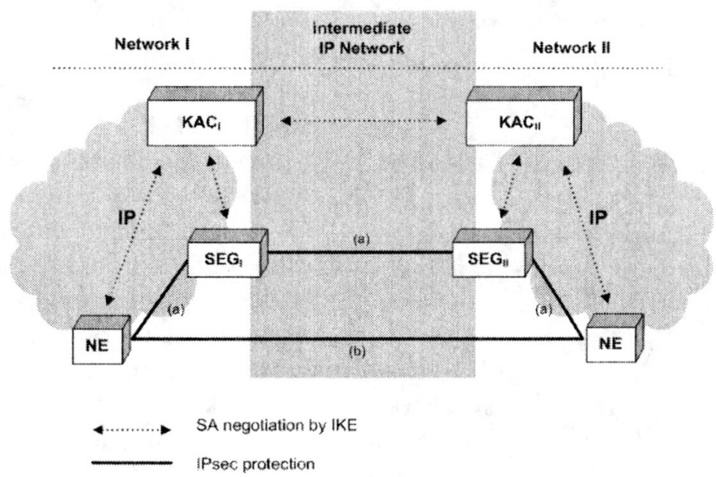

Figure 8. NDS architecture for SS7 and mixed SS7/IP-based protocols

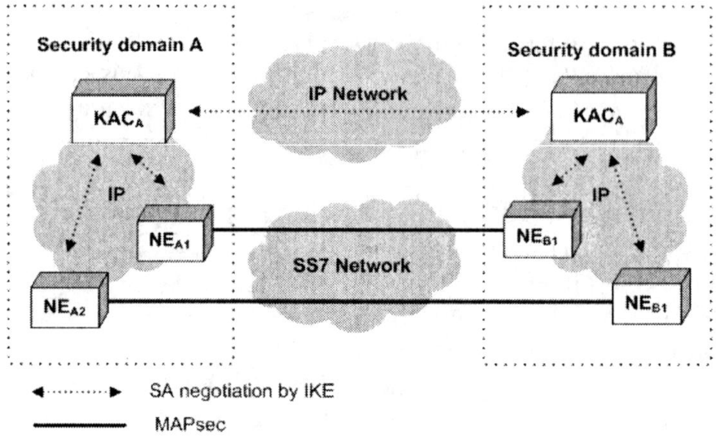

transport is based only on IP, then security may be provided either at the network layer exclusively using IPsec or in a combination of the application and network layer. For signaling protection at the application layer the necessary SAs will be network-wide and they are negotiated by KAC similarly to the IP-based architecture (see Figure 8). End-to-end protected signaling will be indistinguishable to unprotected signaling traffic to all parties, except for the sending and receiving sides.

It is worth noting that in Rel-4 the only protocol that is to be protected is the MAP. The complete set of enhancements and extensions that facilitate the MAP security is termed MAPsec (3GPP TS 33.200, 2002). The MAPsec covers the security management procedures, as well as the security of the transport protocol including data integrity, data origin authentication, anti-reply protection, and confidentiality. Finally, for IKE adaptation a specific Domain of Interpretation is required.

Traditional Network Security Features

Besides the security features that are included in the 3G security architecture, the mobile network operators can apply traditional security technologies used in terrestrial networking to safeguard the UMTS core network as well as the inter-network communications. User data in the UMTS backbone network are conveyed in clear-text exposing them to various external threats. Moreover, inter-network communications are based on the public Internet, which enables IP spoofing to any malicious third party who gets access to it. In order to defeat these vulnerable points, the mobile operators can use two complementary technologies: firewalls and VPNs (Gleeson, Lin, Heinanen, Armitage, & Malis, 2000).

Firewalls can be characterized as a technology providing a set of mechanisms to enforce a security policy on data from and to a corporate network. They are established at the borders of the core network allowing traffic originating from specific foreign IP addresses. Thus, firewalls protect the UMTS backbone from unauthorized penetration. Furthermore, application firewalls prevent direct access through the use of proxies for services, which analyze application commands, perform authentication, and keeps logs.

Since firewalls do not provide privacy and confidentiality, VPNs have to complement them to protect data in transit. VPN establishes a secure tunnel between two points, encapsulates and encrypts data, and authenticates and authorizes user access of the corporate resources on the network. Thus, they extend dedicated connections between remote branches or remote access to mobile users, over a shared infrastructure. Implementing a VPN makes security issues such as confidentiality, integrity, and authentication paramount. There is a two-fold benefit that arises from VPN deployment: the low cost and security.

The border gateway is an element that resides at the border of the UMTS core network and provides the appropriate level of security policy (e.g., firewall), as well as maintaining static pre-configured security tunnels (e.g., IPsec tunnels) granting VPN services to specific peers. It serves as a gateway between the PS domain and an external IP network that is used to provide connectivity with other PS domains located in other core networks.

USER AND APPLICATION DOMAIN SECURITY FEATURES

User Domain Security

User domain security (3GPP TS 33.102, 2002) ensures secure access to the MS. It is based on a physical device called UMTS integrated circuit card (UICC), which can be easily inserted and removed from terminal equipment, containing security applications such as the USIM (3GPP TS 22.100, 2001). The USIM represents and identifies a user and his/her association to an HE. It is responsible for performing subscriber and network authentication, as well as key agreement when 3G services are accessed. It may also contain a copy of the user's profile.

The USIM access is restricted to an authorized user or to a number of authorized users. To accomplish this feature, the user and the USIM must share a secret (e.g., a PIN). The user gets access to the USIM only if he/she proves knowledge of the secret. Furthermore, access to a terminal or to other user equipment can be restricted to an authorized USIM. To this end, the USIM and the terminal must also share a secret. If a USIM fails to prove its knowledge of the secret then access to the terminal is denied.

Application Domain Security

Application domain security (3GPP TS 33.102, 2002) deals with secure messaging between the MS and the SN or the SP over the network with a level of security chosen by the network operator or the application provider. A remote application should authenticate a user before allowing him/her to utilize the application services and it could also provide for application-level data confidentiality. Application-level security mechanisms are needed because the lower layers' functionality may not guarantee end-to-end security provision. The lack of end-to-end security could be envisioned when for instance the remote party is accessible through the Internet.

USIM application toolkit (3GPP TS 33.111, 2001) provides the capability for operators or third party providers to create applications that are resident on the USIM. To assure secure transactions between the MS and the SN or the service provider (SP), a number of basic security mechanisms such as entity authentication, message authentication, replay detection, sequence integrity, confidentiality assurance, and proof of receipt have been specified and integrated in the USIM Application Toolkit.

Figure 9a. WAP 1.2.1 architecture

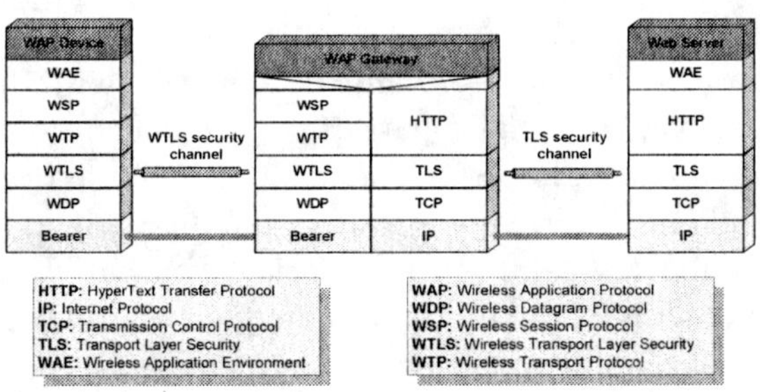

Figure 9b. WAP 2.0 architecture

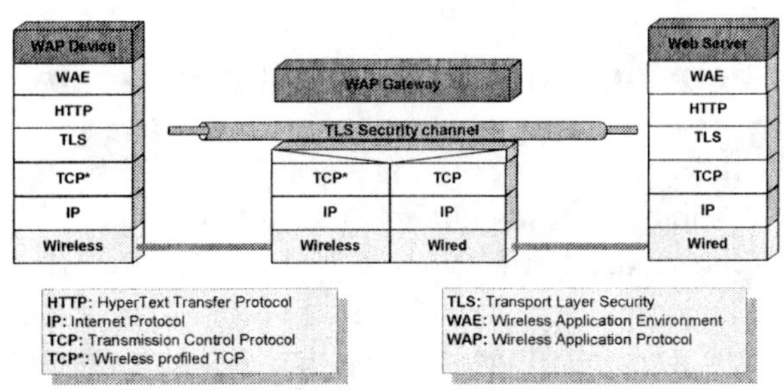

Wireless Application Protocol (WAP) is a suite of standards for delivery and presentation of Internet services on wireless terminals, taking into account the limited bandwidth of mobile networks as well as the limited processing capabilities of mobile devices. It separates the network in two domains (i.e., the wireless and the Internet domain) and introduces a WAP gateway that translates the protocols used in each domain. The WAP architecture has been standardized in two releases (ver. 1.2.1 and ver. 2.0) (Wireless Application Forum, n.d.).

In WAP 1.2.1 (see Figure 9a), security is applied by using the wireless transport layer security (WTLS) protocol (wireless application forum, n.d.) over the wireless domain and the transport layer security (TLS) protocol over the Internet domain. WTLS, which is based on TLS, provides peers authentication, data integrity, data privacy, and protection against denial-of-service in an optimized way for use over narrow-band communication channels. However, WAP 1.2.1 does not support end-to-end security, since the conveyed data are protected by two separate security channels (i.e., WTLS security channel and TLS security channel).

On the other hand, WAP 2.0 (see Figure 9b) introduces the Internet protocol stack into the WAP environment. It allows a range of different gateways, which enable conversion between the two protocol stacks anywhere from the top to the bottom of the stack. A TCP-level gateway allows for two versions of TCP, one for the wired and another for the wireless network domain. On the top of the TCP layer, TLS can establish a secure channel all the way from the MS to the remote server. Thus, the availability of a wireless profile for TLS enables end-to-end security allowing interoperability for secure transactions.

Security Visibility and Configurability

Although the security measures provided by the SN should be transparent to the end user, visibility of the security operations as well as the supported security features should be provided. This may include: (1) indication of access network encryption; (2) indication of network wide encryption; and (3) indication of the level of security (e.g., when a user moves from 3G to 2G).

Configurability enables the mobile user and the HE to configure whether a service provision should depend on the activation of certain security features. A service can only be used when all the relevant security features are in operation. The configurability features that are suggested include: (1) enabling/disabling user-USIM authentication for certain services; (2) accepting/rejecting incoming non-ciphered calls; (3) setting up or not setting up non-ciphered calls; and (4) accepting/rejecting the use of certain ciphering algorithms.

Network-Wide User Data Confidentiality

Network-wide confidentiality is an option that provides a protected mode of transmission of user data across the entire network. It protects data against eavesdropping on every link within the network and not only on the vulnerable radio links. Whenever network-wide confidentiality is applied, access link confidentiality on user data between the MS and the RNC is disabled to avoid replication. However, access link confidentiality for signaling information as well as user identity confidentiality are retained to facilitate the establishment of the encryption process. In Figure 10, the network-wide encryption deployment is depicted.

Network-wide confidentiality uses a synchronous stream cipher algorithm similar to that employed in the access link encryption. Initially, a data channel is established between the communicating peers indicating also the intention for network-wide encryption. VLRa and VLRb exchange cipher keys (Ka and Kb) for users a and b, respectively, using cross boundaries signaling protection, and then, pass them to the MSs over protected signaling channels. When each MS has received the other party's key, the end-to-end session key, Ks, is calculated as a function of Ka and Kb. Alternatively, VLRs can mutually agree on the Ks using an appropriate key agreement protocol. Both key management schemes satisfy the lawful interception requirement, since Ks can be generated by the VLRs.

Figure 10. Network-wide encryption deployment

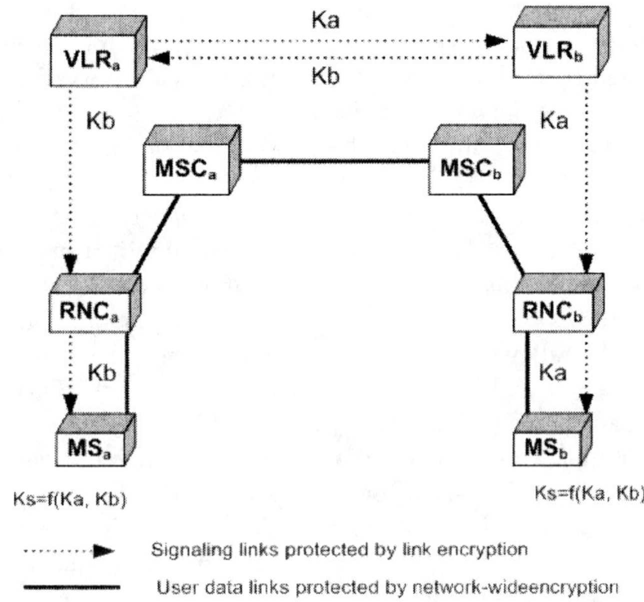

SECURITY WEAKNESSES

The analyzed 3G security architecture provides advanced security services and addresses many of the security concerns that have been listed in the context of next generation mobile networks. However, there are some critical points that need further elaboration and improvements. In the following, the identified security weaknesses of the 3G security architecture, which might cause network and service vulnerability, are briefly presented.

As mentioned previously, the mobile user identity and location is valuable information that requires protection. A possible weakness in the 3G security architecture is the backup procedure for TMSI reallocation (3GPP TS 24.008, 2002). Specifically, whenever the SN/VLR cannot associate the TMSI with the international mobile subscribers identity (IMSI) because of TMSI corruption or database failure, the VLR should request the user to identify himself by means of IMSI on the radio path. Furthermore, when the user roams and the new SN/VLRn cannot contact the previous (old) VLRo or cannot retrieve the user identity; the SN/VLRn should also request the user to identify himself by means of IMSI on the radio path (3GPP TS 33.102, 2002). This may lead an active attacker to pretend to be a new SN to which the user has to reveal his/her permanent identity. In both cases, the IMSI that represents the permanent user identity is conveyed in clear-text on the radio interface, violating user identity confidentiality.

Another critical point is that the users may be identified by means of the IMSI in signaling conversations in the wireline path. For example, the SN/VLR may use the IMSI to request the authentication data for a single user from his/her HE. Thus, user identity confidentiality and user location privacy rely on the security of the wireline signaling connections. NDS features protect signaling exchange in the wireline network architecture with IP and SS7 technologies, but these features are considered for the later versions of the UMTS standardization process, leaving the first one (R99) unprotected.

The authentication and key agreement procedure of UMTS presents two critical security flaws presented in Zhang and Fang (2005). The first one allows an adversary to redirect user traffic from one network to another. This can be achieved because the user (i.e., using the sequence numbers, SQN) can only verify whether an authentication vector was generated by the HE. On the other hand, he/she cannot determine if an authentication vector was requested by the SN, since the authentication vector could have been requested by any SN. Thus, the adversary owing a false base/mobile station device (i.e., a device that emulates a base station and a mobile station) can impersonate as a genuine base station and entices a legitimate user to camp on the radio channels of the false base station. The adversary can also impersonate as a legitimate mobile station and establishes connection with a genuine base station. This fact allows the adversary to relay messages in between a legitimate mobile station and a genuine base station realizing the redirection attack. This attack represents a real threat since the security levels provided by different networks are not always the same. In addition, it could cause billing problems as the service rates offered by different networks are not always the same, either.

The second security flaw that is related to the UMTS authentication (Zhang & Fang, 2005) allows an adversary to use the authentication vectors corrupted from one network to impersonate other networks. When a network is corrupted, an adversary could forge an authentication data request from the corrupted network to obtain authentication vectors for any user, independent of the actual location of the user. Then, the adversary could use the obtained authentication vectors to impersonate uncorrupted networks and to mount false base station attack against legitimate users. Therefore, the corruption of one network may jeopardize the entire system. For this reason, it is critical that security measures are in place in every network.

The application of firewalls in 3G systems presents some weaknesses since they were originally conceived to address security issues for fixed networks. Firewalls attempt to protect the cleartext transmitted data in the UMTS backbone from external attacks, but they are inadequate against attacks that originate from other mobile network malicious subscribers, as well as from network operator personnel or any other third party that gets access to the UMTS core network. Mobility may imply roaming between networks and operators possibly changing the source address, which because of the static configuration of firewalls may potentially lead to discontinuity of service connectivity for the mobile user. Moreover, the firewalls security value is limited because they allow direct connection to ports and cannot distinguish services.

Similarly to firewalls, the VPN technology fails to provide the necessary flexibility required by typical mobile users. Currently, VPNs for UMTS subscribers are established in a static manner between the border gateway of a UMTS network and a remote security gateway of a corporate private network. This fact allows the realization of VPNs only between a security gateway of a large organization and a mobile operator, when a considerable amount of traffic requires protection. Thus, this scheme can provide VPN services neither to individual mobile users that may require on demand VPN establishment, nor to enterprise users that may roam internationally. In addition, static VPNs have to be reconfigured every time the VPN topology or VPN parameters change.

On the other hand, if a mobile user uses the WAP architecture (ver. 1.2.1), data privacy is not guaranteed. Although encryption is used, the WAP gateway constitutes a security hole since inside the gateway data are transmitted un-encrypted. WTLS is only used between the mobile device and the gateway, while TLS can be used between the gateway and the Web server. From a security point of view, the gateway should be considered as an entity-in-the-middle. This means that data exchanged may be available to people with privileged access to the WAP gateway and thus, the privacy of the data depends on the gateway's internal security policy.

WAP 2.0 does address the "gap" in security caused by protocol translation at the WAP gateway of the previous version (ver. 1.2.1). However, the mobile phone would have to use an IP protocol stack at the expense of larger latency and bandwidth consumption. Although TLS can be used to secure the communication of any application, it must be integrated into the application and thus, to a large extent it is used for Web-based applications. Interaction with the end user is needed, for example, to check with whom a secure session has been established or to explicitly request the client to authenticate with the server. TLS is generally a resource consuming protocol for deployment in mobile devices with limited processing capabilities and low bandwidth/high latency wireless networks. Moreover, the operation overhead may be increased by complex key-exchange procedures in case the protected service contains cross-references to other services.

Finally, the network-wide encryption may also encounter problems when transcoding is used. Voice calls may need to be transcoded when they cross network borders, meaning that voice data may have to undergo change such as bit-rate change or some other transformation. It is not possible to apply such transformation on an encrypted signal, which implies that the signal has to be decrypted before transcoding. Furthermore, the network-wide confidentiality lacks flexibility and it is not applicable to all types of service in different mobile scenarios. Specifically, it is limited to protecting the communication between mobile subscribers.

CURRENT RESEARCH ON UMTS SECURITY

The weak points of the UMTS security architecture may lead to compromises of end users and network security of the UMTS system. These compromises may influence the system deployment and the users' trend to utilize UMTS for the provision of advanced multimedia services, which realizes the concept of mobile Internet. In the following, the current research on the UMTS security and the proposed enhancements that aim at improving the UMTS security architecture are briefly presented and analyzed.

Identity Confidentiality

To limit the exposure of the permanent identities (IMSI) of mobile users over the vulnerable radio interface, the additional usage of two complementary temporary identities for each mobile subscriber that is attached to the network has been proposed (Xenakis & Merakos, 2004b). One of these temporary identities will reside at the SN ($TMSI_{ALT}$), and the second one at the home network of the mobile user ($TMSI_{HE}$). When the VLR of the SN fail to page a mobile user using the current TMSI, it can try to page him/her using the alternative temporary identity ($TMSI_{ALT}$), which also resides in the VLR. In case of a VLR database failure or a corruption of the temporary identities (i.e., TMSI and $TMSI_{ALT}$) that resides in the VLR, the VLR requests the temporary identity (i.e., $TMSI_{HE}$) from the home network by which it can page the mobile user. This identity resides in the user's home network in order to avoid a possible corruption after a database (VLR) failure. In case that none of the TMSI is valid or all of them are corrupted, the user is not attached to the network.

Both the additional temporary identities (i.e., $TMSI_{ALT}$ and $TMSI_{HE}$) derive from the current TMSI. The latter consists of four octets and its generation procedure is chosen by the mobile operator. However, some general guidelines are applied in all implementations in order to avoid double allocation of TMSIs, after a restart of the allocating node (i.e., VLR or SGSN). For this reason, some part of the TMSI may be related to the time when it was allocated or contained a bit field, which is changed when the allocating node has recovered from the restart. After the generation of a TMSI, the allocating node applies two individual hash functions (i.e., $HASH_{ALT}$ and $HASH_{HE}$), which produce the corresponding $TMSI_{ALT}$ and $TMSI_{HE}$, respectively. Then, the allocating node forwards the three temporary identities to the involved mobile user and the $TMSI_{HE}$ to its home network. In cases that the home and the SN are the same, the $TMSI_{HE}$ can be stored in HLR, which is not affected by the reasons that corrupt the other two temporary identities. Finally, each time that the current TMSI is renewed, the two additional temporary identities change in order to eliminate the possibility of an adversary to link them to the permanent user's identity.

Authentication and Key Agreement

To address the security issues involved with the authentication and key agreement procedure Zhang and Fang (2005) have proposed an adaptive protocol for mobile authentication and key agreement, called AP-AKA. The proposed protocol can defeat the redirection attack and may drastically lower the impact of network corruption. An overview of AP-AKA is shown in Figure 11.

The AP-AKA protocol retains the framework of the legacy authentication and key agreement, but eliminates the synchronization required between the mobile station and its home network (i.e., SQN_{MS} and SQN_{HE}). In AP-AKA, each mobile station and its home network share an authentication key K and three cryptographic algorithms F, G, and H, where F and H are MACs and G is a key generation function. In practice, the authentication key is usually generated by the home network and programmed into the mobile station during service provisioning. Unlike the legacy authentication and key agreement, the home network in AP-AKA does not maintain a dynamic state, for example, the counter, for each individual subscriber. The mobile station can verify whether an AV was indeed requested by a SN and was not used before by the SN. The AP-AKA protocol specifies a sequence of six flows. Each flow defines a message type and format sent or received by an entity. Depending on the execution environment, entities have the flexibility of adaptively selecting flows for execution, and thus the AP-AKA is called an adaptive protocol.

User Data Security

Another weakness of the current UMTS security architecture that can be overcome is related to the lack of effective protection of user data in the fixed part of the UMTS network. To address this problem, two alternative security solutions, which are based on existing security technologies, can be used: (1) the application layer security, and (2) the establishment of mobile VPNs, dynamically, that satisfy users' needs.

Application layer security solutions integrate security into applications at the level of end users. The most prominent protocol that provides security at this layer for the Internet technology is the Secure Sockets Layer (SSL) protocol (Gupta & Gupta, 2001). SSL supports server authentication using certificates, data confidentiality, and message integrity. Since SSL is relatively "heavy" for implementations on mobile devices, which are characterized by limited processing capabilities, a lightweight version of SSL named "KiloByte" SSL (KSSL) has been proposed (Gupta & Gupta, 2001). This SSL implementation (KSSL) provides an advantage by enabling mobile devices (UMTS MS) to communicate directly and securely with a considerable number of Internet Web servers that support SSL.

Figure 11. Overview of AP-AKA

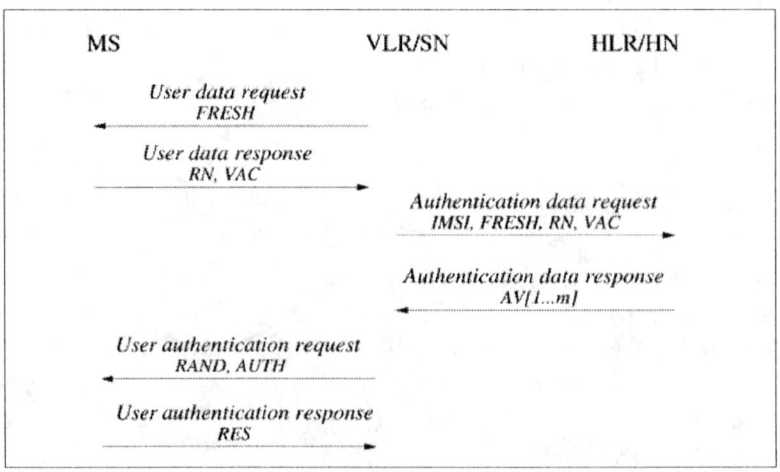

An alternative approach to the previous solutions that employ security at the application layer pertains to these that employ security at the network layer. The most prominent technique for providing security at the network layer is IPsec (Kent & Atkinson, 1998). As a network layer security mechanism, IPsec protects traffic on a per connection basis and thus, is independent from the applications that run above it. In addition, IPsec is used for implementation of VPNs (Gleeson et al., 2000). An IPsec-based VPN is used for the authentication and the authorization of user access to corporate resources, the establishment of secure tunnels between the communicating parties and the encapsulation and protection of the data transmitted by the network. On-demand VPNs that are tailored to specific security needs are especially useful for UMTS users, which require any-to-any connectivity in an ad hoc fashion. Regarding the deployment of VPNs over the UMTS infrastructure, three alternative security schemes have been proposed: (1) the end-to-end (Xenakis & Merakos, 2004a), (2) the network-wide (Xenakis & Merakos, 2006), and (3) the border-based (Xenakis, Loukas, & Merakos 2006). These schemes mainly differ in the position where the security functionality is placed within the UMTS network architecture (MS, RNC, and GGSN), and whether data in transit are ever in cleartext or available to be tapped by outsiders.

The end-to-end security scheme integrates the VPN functionality into the communicating peers, which negotiate and apply security. More specifically, an MS and a remote security gateway (SG) of a corporate private network establish a pair of IPsec SAs between them, which are extended over the entire multi-nature communication path, as shown in Figure 12. Thus, sensitive data are secured as they leave the originator site (MS or SG) and remain protected while they are conveyed over the radio interface, the GPRS backbone network, and the public Internet eliminating the possibilities of being intercepted or to be altered by anyone.

The deployed end-to-end VPN has no interrelation with the underlying network operation

Figure 12. The end-to-end security scheme

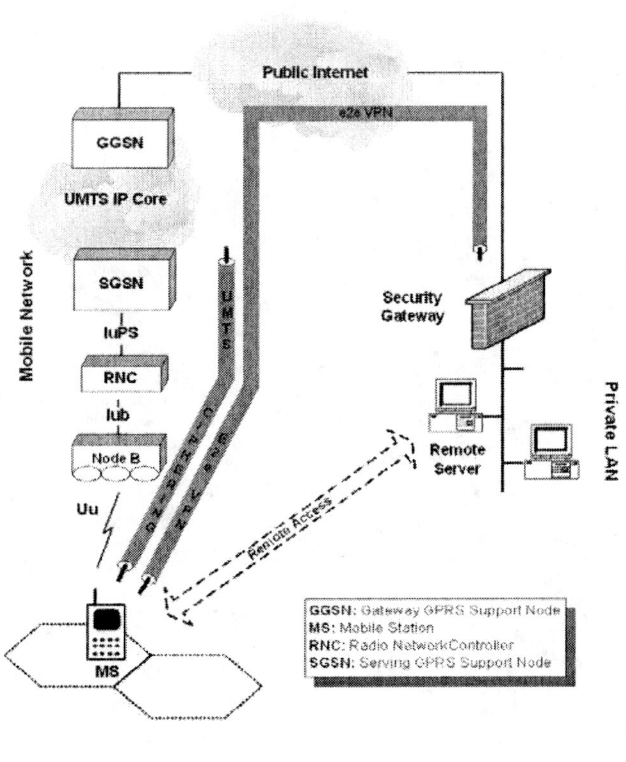

and the provided network connectivity. It operates above the network layer and thus, the security parameters, which are contained within the IPsec SA, are not affected by the MS movement. For this reason the MS may freely move within the UMTS coverage area maintaining network connectivity and VPN service provision. The UMTS mobility management procedures keep track of the user location and therefore, the incoming packets are routed to the MS. On the other hand, the end-to-end security scheme is not compatible with the legal interception option or any other application that requires access to the traversing data within the mobile network. The enforcement of network security policy, traditionally performed by border firewalls, is devolved to end hosts, which establish VPN overlays. Despite this, the border firewalls remain to perform packet filtering and counteract against denial of service attacks.

Contrary to the end-to-end security scheme, the network-wide (Xenakis & Merakos, 2006) and the border-based (Xenakis et al., 2006) schemes integrate the VPN functionality into the UMTS network infrastructure following a network-assisted security model. In both schemes a MS initiates a VPN that is negotiated and established by the network infrastructure thus minimizing the impact to end users and their devices. The network operators provide the security aggregation facilities, which are shared among the network subscribers, as a complementary service, granting-added value. They have solid network management expertise and more resources to effectively create, deploy, and manage VPN services originating from mobile subscribers.

For the deployment of both security schemes (i.e., network-wide and border-based) the MS must be enhanced with a security client (SecC) and the UMTS core network should incorporate a security server (SecS). The SecC is employed by the user to request for VPN services and express his preferences. It is a lightweight module that does not

Figure 13. The network-wide security scheme

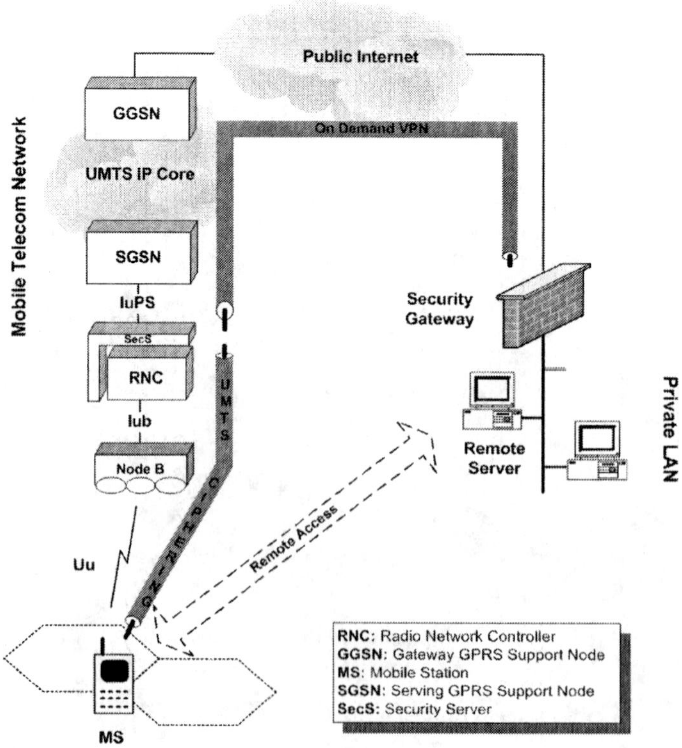

Figure 14. The border-based security scheme

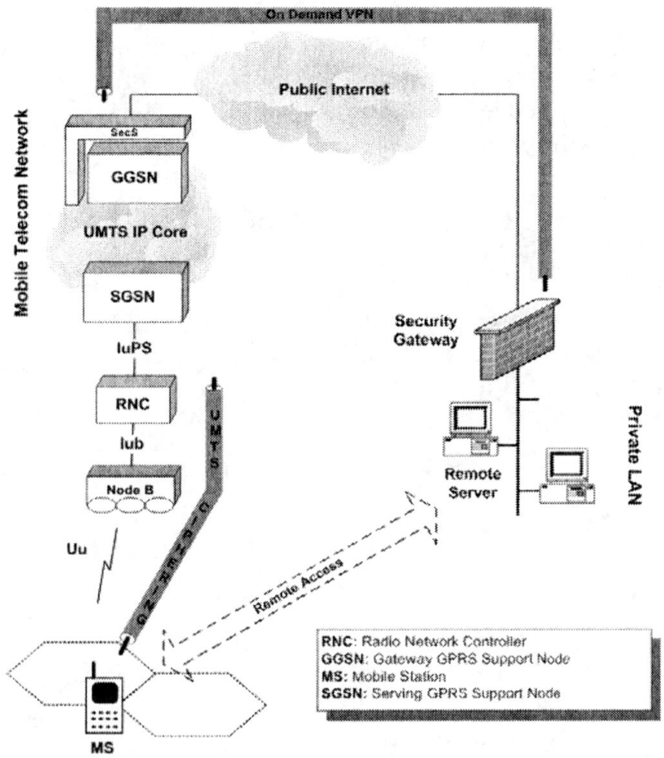

entail considerable processing and memory capabilities and thus, it can be easily integrated in any type of mobile device causing minor performance overhead. On the other side, the SecS establishes, controls, and manages VPNs between itself and remote SGs at corporate LANs on behalf of the mobile users. The SecS comprises an IPsec implementation modified to adapt to the client-initiated VPN scheme and the security service provision in a mobile UMTS environment. It can be readily integrated in the existing network infrastructure and thus, both schemes can be employed as add-on features of UMTS.

The network-wide scheme (see Figure 13) integrates the SecS into the RNC of the UMTS network infrastructure. This scheme provides maximal security services to the communicating peers by employing the existing UMTS ciphering over the radio interface and extending a VPN over the UMTS backbone and the public Internet. Thus, sensitive user data remains encrypted for the entire network route between the originator and the recipient. In order to achieve VPN continuity as a mobile user moves and roams, the standard UMTS mobility management procedures needs to be enhanced. The enhancements include the transfer of the related context (named as security context), which contains the details of the deployed security associations that pertain to the moving user, to the new visited access point. This transfer enables the reconstruction of the security associations of the moving user to the new visited access point, when the user connects to it, providing continuous VPN services from the end-user perspective. The network-wide scheme is compatible with legal interception; however, User Datagram Protocol (UDP) encapsulation is applied for Network Address Translation (NAT) traversal. Finally, the network security policy is enforced by the SGSN, which incorporates the SecS.

By placing the SecS in the GGSN, the border-based VPN deployment scheme is realized (see

Figure 14). This scheme protects data conveyance over the public Internet, which is a vulnerable network segment. The user mobility is transparent to the VPN operation, as long as the user remains under the same network operator coverage and is served by the same GGSN. However, whenever the mobile user roams to another GGSN, the existing security association cannot be used and a new VPN should be established. The border-based scheme is compatible with the legal interception option and NAT presence. Moreover, since the SecS resides at the GGSN, it also provides firewall services to the UMTS network applying network security policy.

CONCLUSION

The evolution of 3G networks signifies a shift towards open and easily accessible network architectures, which raise major security concerns. To address these concerns, a specific security architecture named as 3G security architecture has been designed. This chapter has presented an analysis of the 3G security architecture. This architecture comprises a set of mechanisms that attempt to ensure that all information generated by or relating to a user, as well as the resources and services provided by the serving network and the home environment, are adequately protected against misuse or misappropriation. In addition to these mechanisms, a set of traditional security technologies designed for fixed and wireless networks can also be applied to protect 3G networks. Based on the carried analysis, the critical points in the 3G-security architecture, which might cause network and service vulnerability, have been outlined. Finally, the current research activities on the UMTS security that aim at improving the UMTS security architecture have been briefly presented.

ACKNOWLEDGMENT

Work supported by the project CASCADAS (IST-027807) funded by the FET Program of the European Commission.

REFERENCES

3rd Generation Partnership Project (3GPP) TS 22.100 (v3.7.0). (2001). *UMTS phase 1 Release 99*. Sophia-Antipolis Cedex, France: Author. Retrieved from ftp://ftp.3gpp.org/specs/2006-12/R1999/22_series

3rd Generation Partnership Project (3GPP) TS 23.002 (v3.6.0). (2002). *Network architecture*. Sophia-Antipolis Cedex, France: Author. Retrieved from ftp://ftp.3gpp.org/specs/2006-12/R1999/23_series

3rd Generation Partnership Project (3GPP) TS 24.008 (v3.13.0). (2002). *Mobile radio interface signaling layer 3 specification; Core network protocols—Stage 3*. Sophia-Antipolis Cedex, France: Author. Retrieved from ftp://ftp.3gpp.org/specs/2006-12/R1999/24_series

3rd Generation Partnership Project (3GPP) TS 25.401 (v3.10.0). (2002). *UTRAN overall description*. Sophia-Antipolis Cedex, France: Author. Retrieved from ftp://ftp.3gpp.org/specs/2006-12/R1999/25_series

3rd Generation Partnership Project (3GPP) TS 31.111 (v3.7.0). (2001). *USIM application toolkit (USAT)*. Sophia-Antipolis Cedex, France: Author. Retrieved from ftp://ftp.3gpp.org/specs/2006-12/R1999/31_series

3rd Generation Partnership Project (3GPP) TS 33.102 (v3.12.0). (2002). *3G security, security architecture*. Sophia-Antipolis Cedex, France: Author. Retrieved from ftp://ftp.3gpp.org/specs/2006-12/R1999/33_series

3rd Generation Partnership Project (3GPP) TS 33.200 (v4.3.0). (2002). *3G security; Network domain security; MAP application layer security.* Sophia-Antipolis Cedex, France: Author. Retrieved from ftp://ftp.3gpp.org/specs/2006-12/Rel-4/33_series

3rd Generation Partnership Project (3GPP) TS 33.210 (v5.1.0). (2002). *3G security; Network domain security: IP network layer security.* Sophia-Antipolis Cedex, France: Author. Retrieved from ftp://ftp.3gpp.org/specs/2006-12/Rel-5/33_series

3rd Generation Partnership Project (3GPP) TR 33.908 (v3.0.0). (2000). *3G security; General report on the design, specification and evaluation of 3GPP standards confidentiality and integrity algorithms.* Sophia-Antipolis Cedex, France: Author. Retrieved from ftp://ftp.3gpp.org/specs/2006-12/R1999/33_series

3rd Generation Partnership Project (3GPP) TS 35.205 (v3.0.0). (2001). *3G security; Specification of the MILENAGE set: An example algorithm set for the 3GPP authentication and key generation functions f1, f1*, f2, f3, f4, f5, and f5*.* Sophia-Antipolis Cedex, France: Author. Retrieved from ftp://ftp.3gpp.org/specs/2006-12/Rel-4/35_series

Gleeson, B., Lin, A., Heinanen, J., Armitage, G., & Malis, A. (2000). *A framework for IP based virtual private networks* (RFC 2764). Retrieved from http://tools.ietf.org/html/rfc2764

Gupta, V., & Gupta, S. (2001). Securing the wireless Internet. *IEEE Communications Magazine, 39*(12), 68-74.

Harkins, D., & Carrel, D. (1998). *The Internet key exchange (IKE)* (RFC 2409). Retrieved from http://www.ietf.org/rfc/rfc2409.txt

Kent, S., & Atkinson, R. (1998). *Security architecture for the Internet Protocol* (RFC 2401). Retrieved from http://www.ietf.org/rfc/rfc2401.txt

Wireless Application Forum (WAP). (n.d.). *WAP specifications.* Retrieved from http://www.wapforum.org/what/technical.htm

Xenakis, C., Loukas, N., & Merakos, L. (2006, April). A secure mobile VPN scheme for UMTS. In *Proceedings of European Wireless 2006,* Athens, Greece.

Xenakis, C., & Merakos, L. (2004a). IPsec-based end-to-end VPN deployment over UMTS. *Computer Communications, 27*(17), 1693-1708.

Xenakis, C., & Merakos, L. (2004b). Security in third generation mobile networks. *Computer Communications, 27*(7), 638-650.

Xenakis, C., & Merakos, L. (2006). Alternative schemes for dynamic secure VPN deployment over UMTS. *Wireless Personal Communications, 36*(2), 163-194.

Zhang, M., & Fang, Y. (2005). Security analysis and enhancements of 3GPP authentication and key agreement protocol. *IEEE Transactions on Wireless Communications, 4*(2), 734-742.

KEY TERMS

International mobile subscriber identity (IMSI): IMSI is a unique number associated with all UMTS network mobile phone users.

Internet key exchange (IKE): IKE is a protocol used to set up a security association (SA) in the IPsec protocol suite.

IP security (IPsec): IPsec is a suite of protocols for securing IP communications by authenticating and/or encrypting each IP packet in a data stream.

Temporary mobile subscriber identity (TMSI): TMSI is a randomly allocated number that is given to the mobile the moment it is switched on and serves as a temporary identity between the mobile and the network.

Third generation (3G): 3G is a technology in the context of mobile phone standards. The services associated with 3G include wide-area wireless voice telephony and broadband wireless data, all in a mobile environment.

Universal mobile telecommunications system (UMTS): UMTS is one of the 3G mobile phone technologies.

Universal subscriber identity module (USIM): USIM is an application for UMTS mobile telephony running on a UICC smart card which is inserted in a 3G mobile phone and stores user subscriber information and authentication information.

Wideband code division multiple access (WCDMA): WCDMA is a wideband spread-spectrum mobile air interface that utilizes the direct sequence code division multiple access (CDMA) signaling method to achieve higher speeds and support more users compared to the implementation of time division multiplexing (TDMA) used by 2G GSM networks.

ENDNOTES

1 $\|$String concatenation.

2 $\oplus$Exclusive or

Chapter XXI
Access Security in UMTS and IMS

Yan Zhang
Simula Research Laboratory, Norway

Yifan Chen
University of Greenwich, UK

Rong Yu
South China University of Technology, China

Supeng Leng
University of Electronic Science and Technology of China, China

Huansheng Ning
Beihang University, China

Tao Jiang
Huazhong University of Science and Technology, China

INTRODUCTION

Motivated by the requirements for higher data rate, richer multimedia services, and broader radio range wireless mobile networks are currently in the stage evolving from the second-generation (2G), for example, global system for mobile communications (GSM), into the era of third-generation (3G) or beyond 3G or fourth-generation (4G). Universal mobile telecommunications system (UMTS) is the natural successor of the current popular GSM (http://www.3gpp.org) code division multiple access 2000 (CDMA2000) is the next generation version for the CDMA-95, which is predominantly deployed in North America and North Korea. Time division-sychrononous CDMA (TD-SCDMA) is in the framework of 3rd generation partnership project 2 (3GPP2) and is expected to be one of the principle wireless technologies employed in China in the future (http://www.3gpp.org; 3G TS 35.206). It is envisioned that each of three standards in the framework of international mobile telecommunications-2000 (IMT-2000) will play a significant role in the future due to the backward compatibility, investment, maintenance cost, and even politics. In all of the potential standards, access security is one of the primary demands as well as challenges to resolve the deficiency existing in the second generation wireless mobile networks such as GSM, in which only one-way authentication is performed for the core network part to verify the user equipment (UE) (3G TS 24.008). Such access security may lead to the "man-in-middle" problem, which is a type of attack that can take place when two clients are communicating remotely and exchange public keys in order to initialize secure communications. If both of the two public keys are intercepted in the route by someone, he/she can act as a conduit and send in the messages with his/her own faked public key. As a result, the secure communication is eavesdropped by a third party.

Multimedia service provisioning is one of the primary demands and motivations for the next generation wireless networks. To achieve this goal, the IP multimedia subsystem (IMS) is added as the core network in UMTS providing the multimedia service, for example, voice telephony, video conference, real-time streaming media, interactive game, voice over IP, picture, HTTP, and instant messaging (3G TS 33.203). The multimedia session management, initialization, and termination are specified and implemented in the session initiation protocol (SIP) (3G TS 29.228; Zhang & Fang, 2005). To ensure the secure communication in a multimedia session, an efficient access security mechanism shall be also provided.

In this chapter, we make an introduction to the access security in the next generation wireless mobile networks, including the mechanisms in the circuited-switched domain, packet-switched domain, and also the emerging IMS domain.

BACKGROUND OVERVIEW

Figure 1 shows the UMTS network architecture with most related components in security management (3G TS 29.002; 3G TS 33.102). User terminal (UE) utilizes the circuited-switched or packet-switched service through the radio interface between base station (BS) and itself. BS locates in the center of a cell which coveres a radio range. BS provides the wireless access point for UEs to the core network. Radio network controller (RNC) monitors and supervises the activities of several BS under its management. Radio access network (RAN) consists of the RNC and the associated BS under the RNC. Home location register (HLR) stores the permanent information for the subscrib-

Figure 1. UMTS network architecture

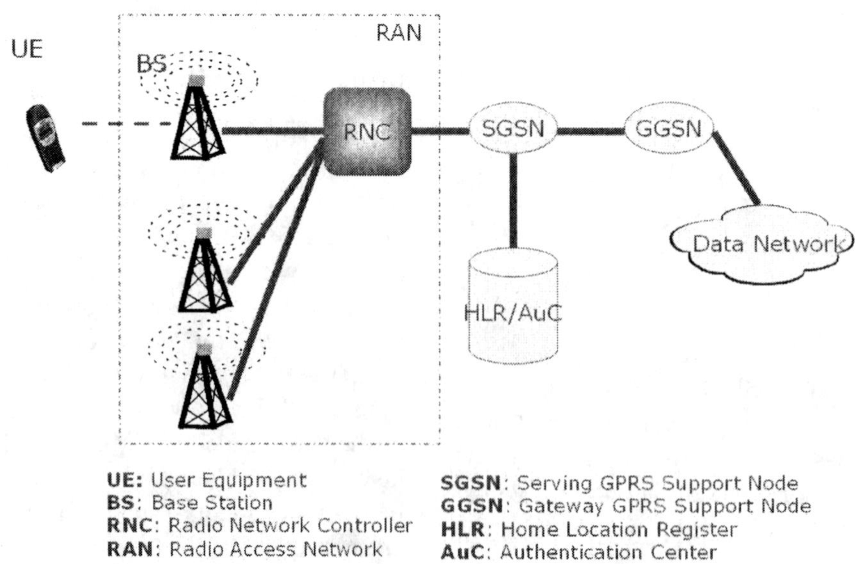

Figure 2. UMTS network authentication and key agreement (AKA)

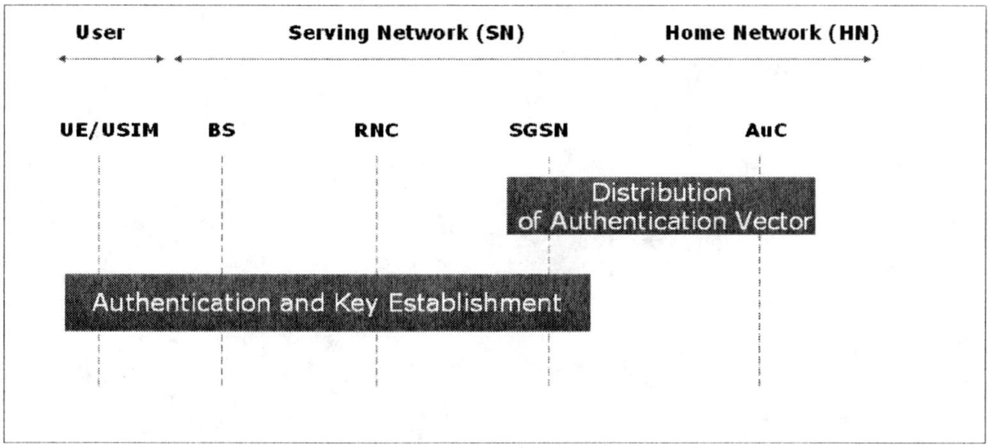

Figure 3. AKA phase 1: Distribution of authentication vector

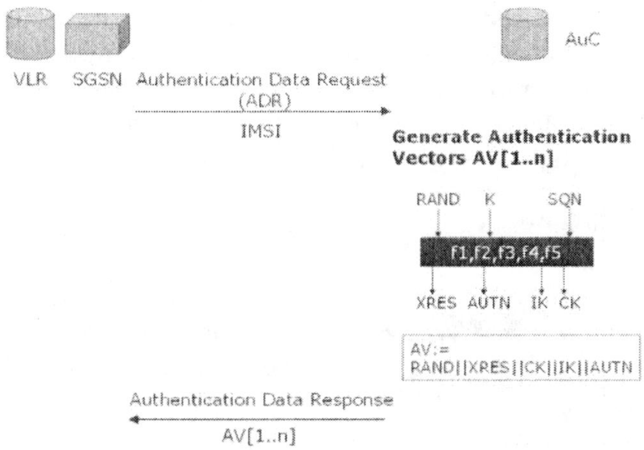

ers, for example, International Mobile Subscriber Identity (IMSI), subscribed service profile, and identity of current location area. Authentication center (AuC) is responsible to verify the validness of user's activity including call behavior and location management. Normally, HLR and AuC locates in the same database/server. Serving GPRS support node (SGSN) connects the core network and the radio access network and is responsible for location management and for delivering packets between UE and the core network. Gateway GPRS support node (GGSN) acts similar as a gateway between core network and the external IP networks such as Internet, telecommunication networks, and enterprise intranets.

ACCESS SECURITY IN UMTS

Figure 2 shows the most signicant feature in the framework of UMTS security management, that is, authentication and key agreement (AKA) (3G TS 24.008). Authentication refers to the mutual authentication mechanism that the subscriber is able to use to authenticate the network, and the network is also able to authenticate the user. Key agreement refers to the mechanism to generate the cipher key and integrity key. The events for triggering the AKA process include location update request, user registration, service request, attach request, detach request, and connection re-establishment request.

Figure 4. AKA phase 2: Authentication and key establishment

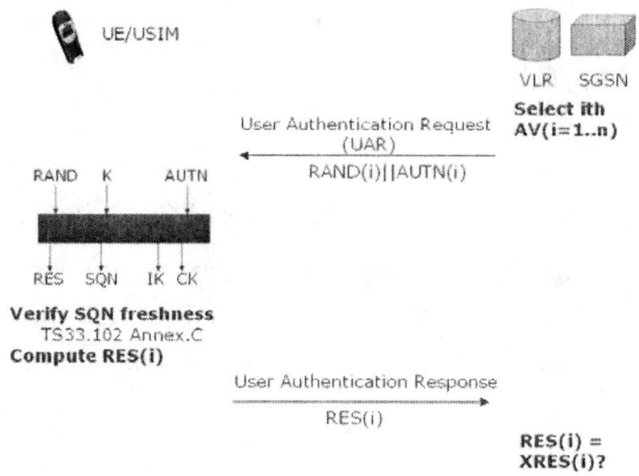

The authentication protocol is based on a permanent secret key **K** (128-bit) that is shared between the UE and HLR/AuC. The AKA mechanism can be divided into two phases: the distribution of authentication vector (DAV) from the HLR/AuC to the SGSN as shown in Figure 3, and the authentication and key establishment between the UE and the core network as illustrated in Figure 4.

Distribution of Authentication Vector

When a UE leaves an old SGSN (SGSNo) and moves into the coverage of a new SGSN (SGSNn), SGSNn has no corresponding record for the UE, which makes it necessary to authenticate the UE prior to the subsequent behavior. SGSNn will delivery the message authentication data request (ADR) to the HLR/AuC with the UE's unique IMSI. Based on the received IMSI, AuC can find the associated record in its database and hence the according master key **K** for this particular UE. Then, HLR/AuC generates the number of **n** AV instead of single one AV for the sake of saving signaling overhead. The AV structure is comprised of five components: (1) a random number **RAND**, (2) an expected authentication response **XRES**, (3) a cipher key **CK**, (4) an integrity key **IK**, and (5) a network authentication token **AUTN** (3G TS 23.060). In each generation, an AV is calculated by means of the authentication function f1-f5, where for instance the function f1 is employed to compute XRES, the function f2 is used to compute **RES**, and the function f3 is used to compute **CK** (3G TS 33.105; 3G TS 35.205; 3G TS 35.206). After successfully generating **n** AVs, AuC sends back the AV array to SGSNn via the message authentication data response, and SGSNn saves these **n** AVs for the particular UE. It is noteworthy that this phase executes not only upon UE entering a new SGSN area, but also when there are no AVs available upon an action arrival which requires authentication.

Authentication and Key Establishment

For each activity triggering authentication request such as call origination, paging, or location update the SGSN initiates the challenge user authentication request (UAR) message to the UE with the parameters **RAND** and **AUTN**, which is retrieved from the ith (i = 1,2,…n) AV in the first-in-first-out (FIFO) manner. Upon receiving the AV, the UE checks the validity of **AUTN**. For this goal, the UE retrieves SQN component from **AUTN** and calculates expected message authentication code for authentication (XMAC-A). The UE then compares X-MAC-A and message authentication code for authentication (MAC-A) component in **AUTN**, if they are equal to each other, then the network is verified. Otherwise, the UE rejects the UAR and hence the network. After the network is identified, UE checks the SQN freshness, that is, the SQN has never been used before. When the network succeeds, the UE then computes the authentication response **RES** from the received **RAND** value and sends it in a *user authentication response* message to the SGSN. If **RES** equals the expected response **XRES**, then the UE is successfully authenticated. Since there are **n** AVs generated and recorded in SGSNn during each operation of DAV while only one AV is used during an authentication event, the signaling between SGSN and **HLR/AuC** during DAV is not needed for every authentication event.

It is believed that, after the AKA procedure, all messages are claimed integrity protection, and the signaling data as well as user data are confidentiality protection. In the sense of integrity protection, the content of signaling messages should not be manipulated. With regard to confidentiality protection, the subscriber identification, location, user data, and signaling data should be encrypted.

ACCESS SECURITY IN IMS

There are three entities relevant to the IMS security architecture (see Figure 5). A proxy call session control function (P-CSCF) locates in the serving network of a UE and acts as the first access point in the serving network. P-CSCF is responsible for forwarding SIP messages of an UE to the home network. A serving call session control function (S-CSCF) locates in the home network to provide session control of multimedia services and acts as SIP registrar or SIP proxy server. The S-CSCF sends messages toward the home subscriber server (HSS) and the AuC to receive subscriber data and authentication information. An interrogating call session control function (I-CSCF) locates in the home network and acts as a SIP proxy toward the home network. I-CSCF is responsible for selecting an appropriate S-CSCF for the UE and forwarding SIP requests/responses toward the S-CSCF.

Figure 5. IMS network architecture

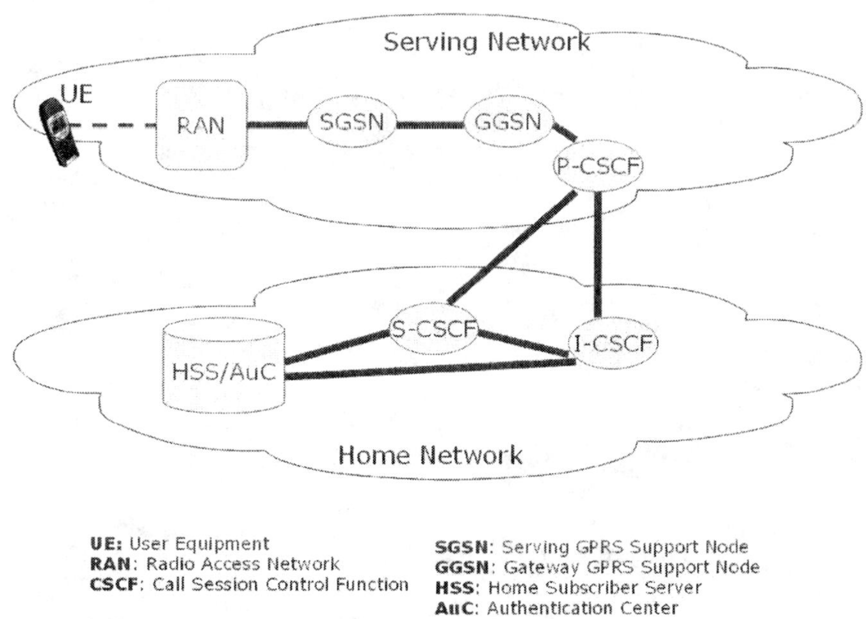

Different from the one-pass authentication procedure in AKA illustrated in Figure 2, the security in IMS is a two-pass authentication procedure, including general packet radio servcie (GPRS) authentication and IMS authentication (3G TS 29.229). Before utilizing the IMS service, a UE should first setup a data connection to know the IP address of P-CSCF and to carry the SIP signaling messages through the P-CSCF. The data connection establishment is comprised of two steps, that is, attach and packet data protocol (PDP) context activation. The first phase attach is used to establish mobility management context between the UE and SGSN. During this procedure, the UE should perform GPRS authentication and GPRS registration to verify its validness and retrieve the subscriber profile including subscribed services, quality of service (QoS) profile, IP address, and so on. Once the UE is attached, the second step PDP context activation is followed to activate a PDP address and build the association between the SGSN and GGSN. Only after attached and PDP context activation, an UE can access IMS services through registration process. The registration is necessary to inform the HSS the location, authenticate and download subscriber profile to S-CSCF. We will discuss the GPRS authentication and IMS authentication in the following two subsections. The discussion of either GPRS registration or PDP context activation is out of the scope and the readers are suggested to refer to the related technical specifications (3G TS 33.102).

GPRS AUTHENTICATION

GPRS authentication is performed in the framework of GPRS mobility management (GMM) (http://www.3gpp2.org; 3G TS 33.102). Figure 6 shows the messages sequence in GPRS authentication. In particular, the steps include:

1. The UE sends **GMM Attach Request (IMSI)** to the SGSN with the unique identity **IMSI**.
2. If the SGSN has at least one AV for the UE, then step 2 and 3 are skipped. Otherwise, the SGSN has to obtain AVs from the entity HSS/AuC. SGSN triggers the procedure DAV by sending a **MAP-SEND-AUTHENTICA-**

Figure 6. GPRS authentication

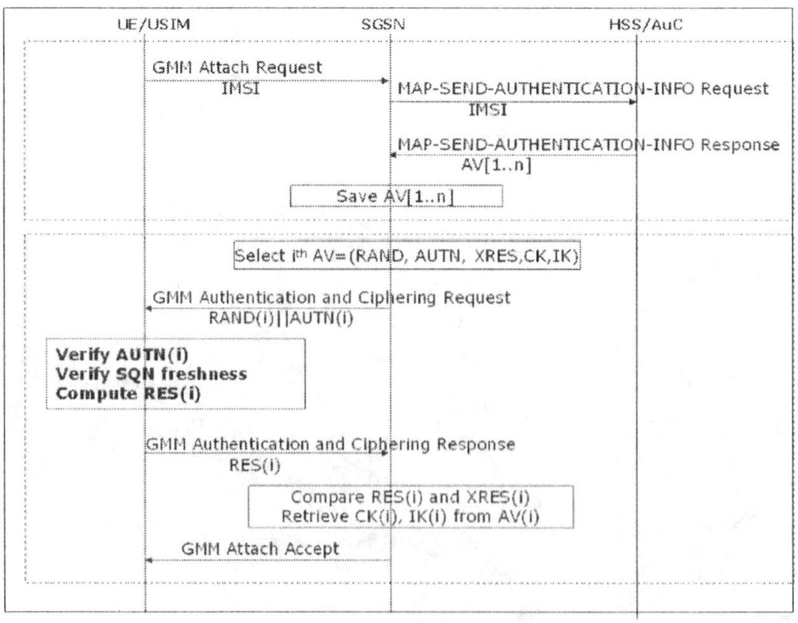

TION-INFO Request (IMSI) message to the HLR/AuC with the parameter **IMSI** uniquely identifying the UE.

3. Upon receiving the authentication request, the HSS/AuC searches the according record in the database on the basis of **IMSI**. Then, HSS/AuC generates an ordered array of **n** AVs for the specific UE. Each AV consists of the following components: a random number **RAND**, an expected response **XRES**, a cipher key **CK**, an integrity key **IK,** and an authentication token **AUTN**. The HSS/AuC then sends back the message **MAP-SEND-AUTHENTICATION-INFO Response** to SGSN with the AV array as parameters.
4. SGSN stores these **n** AVs for the particular UE and shall choose the next unused AV in the ordered AV array. Subsequently, the SGSN shall challenge the UE and sends message **GMM Authentication and Ciphering Request** with parameters **RAND** and **AUTN** populated from the selected AV.
5. The UE checks the validness of the received **AUTN**. In case it is acceptable, the UE shall calculate a response **RES** and send back to the SGSN through the message **GMM Authentication and Ciphering Response**. The SGSN retrieves the expected response **XRES** from the selected AV and compares **XRES** with the received response **RES**. If they match, the authentication and key agreement is successfully completed and the keys **CK** and **IK** are retrieved for the following signaling confidentiality and integrity protection.
6. The SGSN sends a **GMM Attach Accept** message to the UE to indicate the completion of the successful attach procedure.

IMS AUTHENTICATION

After the procedures of GPRS authentication, GPRS registration and PDP context activation, the UE has the IP address of the P-CSCF and is able to access the IMS services through the registration procedure using SIP and Cx commands as shown in Figure 7 (CWTS TSM 03.20; 3G TS 29.229). This procedure includes the IMS authentication and the IMS registration. In particular, the steps include:

1. To start registration, the UE sends a **SIP REGISTER (IMPI, IMPU)** message to the P-CSCF in the serving network. On the receipt, the P-CSCF forwards the registration request to the I-CSCF of the home network. I-CSCF then delivers the message to a chosen S-CSCF.
2. If the S-CSCF has at least one AV for the UE, then steps 2 and 3 are skipped. Otherwise, the S-CSCF has to obtain AVs from the entity HSS/AuC. S-CSCF triggers the procedure DAV by sending a **Cx-AV-Req(IMPI, n)** message to the HSS/AuC with the parameter **IMPI** uniquely identifying the UE and the number of **n** AVs wanted.
3. Upon receipt of a request from the S-CSCF, the HSS/AuC searches the database on the basis of the unique **IMPI**, obtains the subscriber profile, and generates an ordered array of **n** AVs for the specific UE. Each AV consists of the following components: a random number **RAND**, an expected response **XRES**, a cipher key **CK**, an integrity key **IK,** and an authentication token **AUTN**. Each AV is good for only one authentication and key agreement between the IMS subscriber and the S-CSCF. The HSS/AuC then sends back the message **Cx-AV-Req-Resp(IMPI, RAND1||AUTN1||XRES1||CK1||IK1,..., RANDn||AUTNn||XRESn||CKn||IKn)** to the S-CSCF with the array of AV as parameters.
4. The S-CSCF chooses the first unused AV in the array of AVs based on FIFO policy. From the selected AV, the items **RAND, AUTN, IK,** and **CK** are populated. The S-CSCF sends the message **SIP 4xx-Auth-Challenge (IMPI, RAND, AUTN, IK, CK)** to the I-CSCF, which then forwards the message to P-CSCF. Upon the receipt, the P-CSCF shall store the two keys **IK** and **CK** and remove the key information and finally forward the

Figure 7. IMS authentication

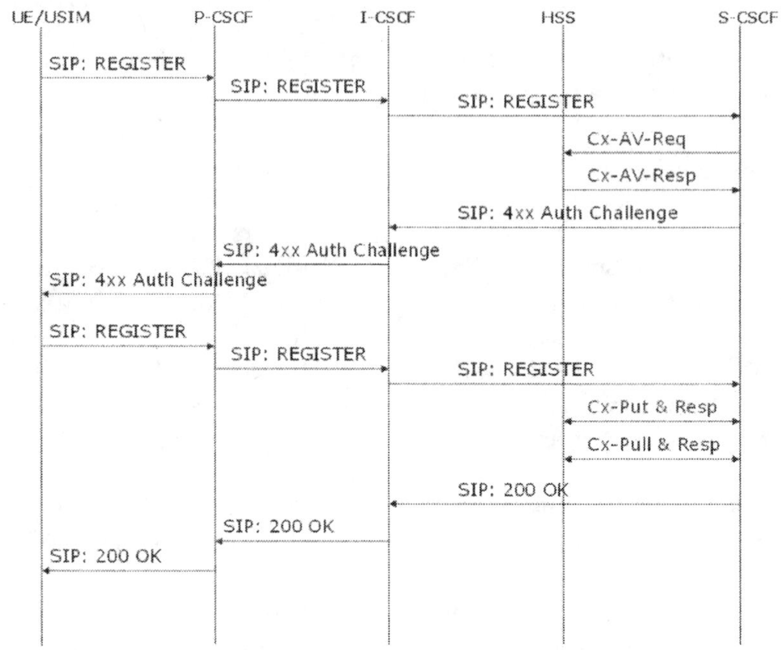

rest of the message **SIP 4xx-Auth-Challenge (IMPI, RAND, AUTN)** to the UE.

5. The UE verifies the freshness of the received **AUTN** and calculates a response **RES**. This result **RES** is sent back from the UE to the P-CSCF through the message **SIP REGISTER (IMPI, RES)**. After receiving the request, the P-CSCF forwards it to the I-CSCF, which further forwards the authentication response to the S-CSCF. The S-CSCF retrieves the expected response **XRES** and compares **XRES** and the received response **RES**. If they match, the authentication and key agreement is successfully completed. Next three steps perform registration.
6. The S-CSCF sends a **Cx-Put** message to the HSS/AuC with the UE identity. The HSS shall store the S-CSCF name, which is presently serving the UE, and then sends the **Cx-Put Response** for acknowledgement.
7. Next, the S-CSCF sends a **Cx-Pull** to the HSS/AuC with the UE identity in order to download the related information in the subscriber profile to the S-CSCF. HSS shall send a **Cx-Pull Response** to the S-CSCF with the indicated information.
8. The S-CSCF sends **SIP 200 OK** message to the UE through the I-CSCF and P-CSCF. After this step, a security associate (SA) is active for the protection of subsequent SIP messages between the UE and the P-CSCF.

FUTURE TRENDS

Security Management in Heterogeneous Network

The next generation wireless mobile networks are characterized as the co-existent of the variety of network architectures, protocols, and applications due to the diverse requirements for data rate, radio coverage, deployment cost, and multimedia service. The 3GPP is actively specifying the roaming mechanism in the integrated wireless LAN (WLAN)/UMTS networks. It should be noted that this scenario is only a specific heterogeneous network. The IEEE 802.16 standard is an emerging broadband wireless access system specified for wireless metropolitan area networks (WMAN) with the aim to bridge the last mile, replacing costly wireline and also providing high speed multimedia services in fast moving transportation. The recently amended 802.16e adds a mobility component for WMAN and defines both physical and MAC layers for combined fixed and mobile operations in licensed bands. It is envisaged that the future generation wireless networks is the flexible and seamless integration of the three technologies WLAN, WMAN, and wireless wide area network (WWAN), where WLAN (e.g., IEEE 802.11 Wi-Fi) serves as the hot-spot access area for short-range and very high speed; WMAN (e.g., IEEE 802.16 WiMAX) serves as the metropolitan-wide access network with high data rate and WWAN (e.g., UMTS) provides the national-wide network with relatively low data rate. The substantial technical challenge is to design and implement the security architectures and protocols across such heterogeneous networks taking into account the seamless mobility, scalability, and performance efficiency.

Security-Mobility Management Interaction and Security-Energy Tradeoff

The performance of security management has a close interaction with the framework of mobility management. Mobility management includes two components: location management and handoff management (http:www.3gpp2.org). There are two operations in the location management: updating the UE location and paging the UE. In UMTS, SGSN shall authenticate a UE when the SGSN receives an "Initial L3 message" sent from UE. This message is triggered by the actions, including location update request, connection management request, routing area update request, attach request, and paging response. It is clear that all these events are closely relevant to the user's mobility management architecture and mechanism. Liang and Wang (2005) constructed an analytical model to evaluate the impact of authentication on the security and QoS. The authors introduced the system model based on the widely used challenge/response mechanism. Then, a concept of security level is introduced to describe the different level of communication protection with regard to the nature of security, that is, information secrecy, data integrity, and resource availability. By taking traffic and mobility patterns into account, the technique establishes a quantitative connection between the security and QoS through the authentication and facilitates the evaluation of overall system performance under diverse security levels, mobility and traffic processes.

Generally, a UE is powered by battery and hence the mechanism in efficiently utilizing the limited energy is becoming very important. In case of more frequent authentication to increase the security, the UE will consume more energy. With fewer authentications incurring potential vulnerability, the UE is able to enlarge its lifetime before re-charging. As a consequence, there is a trade-off between the security and energy management. Potlapally, Ravi, Raghunathan, and Jha (2003) provided energy consumption empirical measurements for a variety of ciphers, hash functions, and signature algorithms. Based on the observations, the study presented some reasoning about the energy-security trade-offs in determining key length. However, no analytical models have been proposed to evaluate the energy-security trade-offs or make the intelligent decision on trade-off.

Higher Security Protocols

Although AKA has been standardized, the protocol has two significant weaknesses: (1) HLR/AuC does not verify whether the information sent from the visiting location register (VLR)/SGSN is valid or not. That is, AKA has assumed that the link between VLR/SGSN and HLR/AuC is adequately reliable; and (2) for the UMTS integrity protection mechanism, integrity key is transmitted without encryption and the user data are not protected. New strategies shall be designed to address these issues.

Harn and Hsin (2003) identified and discussed the inefficiency and complexity in keeping and managing the sequence number during the network authentication. Based on the combination of hash chaining and keyed-hash message authentication code techniques, an enhanced scheme is proposed to simplify the protocol implementation and simultaneously provide strong periodically mutual authentication.

Zhang and Fang (2005) showed that the 3GPP AKA protocol is vulnerable to a variant of the fake BS attack. The vulnerability allows an adversary to redirect user traffic from one network to another and to re-use corrupted AVs from one network to all other networks. To address such security problems in the current 3GPP AKA, the authors presented a new authentication and key agreement protocol AP-AKA which defeats redirection attack and drastically lowers the impact of network corruption.

Security Protocols Performance

Security architecture and protocol are normally evaluated to guarantee the security, confidentiality, and integrity requirement. Recently, a few studies have appeared to investigate the authentication signaling traffic performance due to the rapidly increasing number of subscribers and consequently potentially high authentication requests and heavy burden on the signaling networks. Lin and Chen (2003) argue the disadvantages in fetching the constant number of AV from HLR/AuC. Based on the observations of the mobility pattern, the authors proposed an adaptive scheme to generate an optimal number of AV array, which is able to significantly reduce the authentication signaling traffic and hence save the limited bandwidth utilization. Zhang and Fujise (2006) argue the long delay problem and proposed a mechanism to address the issue. In particular, when the two entities SGSN and HLR/AuC locate far away from each other, the response for an available AV may be potentially very long. The consequence of long delay includes call blocking and location update failure, and hence degraded QoS. To address this problem, the study proposed an enhanced scheme to fetch AV earlier before all AVs are used up. Comparing with the original 3GPP Technical Specification TS33.102 (2000), the proposed strategy is able to achieve very low probability in waiting for an available AV with negligible increased signaling overhead and low storage cost. The study in Al-Saraireh and Yousef (2006) also analyzes the transmission overhead during the procedure of AKA. It is proposed that security protocols performance should be evaluated from the security perspective and also from the signaling overhead point of view. New security protocols should consider to combat potential vulnerability as well as to introduce low additional signaling cost.

CONCLUSION

This chapter gives an overview on the security management in the next generation wireless networks. The AKA process is described and its extension in GPRS authentication and IMS authentication are further discussed in detail. The identified research challenges shall serve as the guidance for the further study to propose more efficient security protocols taking into account the network architecture heterogeneity, the energy-security trade-offs, the mobility-security interaction, and comprehensive performance evaluation.

REFERENCES

3rd Generation Partnership Project (3GPP) (1999). *Technical specification core network; Mobile application part (MAP) specification.* Technical Specification 3G TS 29.002 V3.7.0 (2000-12). Sophia Antipolis Cedex, France:_Author.

3rd Generation Partnership Project (3GPP). *Technical Specification Group Core Network; Mobile Radio Interface Layer 3 Specification; Core Network Protocols Stage 3 for Release 1999, 2000.* 3G TS 24.008 version 3.6.0 (2000-12). Sophia Antipolis Cedex, France:_Author.

3rd Generation Partnership Project (3GPP). *Technical Specification Group Services and Systems*

Aspects; 3G Security; Security Architecture, 2000, Technical Specification 3G TS 33.102 V3.7.0 (2000-12). Sophia Antipolis Cedex, France: Author.

3rd Generation Partnership Project (3GPP). *Technical Specification Group Services and Systems Aspects; General Packet Radio Service (GPRS); Service Description; Stage 2, 2000, Technical Specification 3G TS 23.060 version 3.6.0 (2001-01).* Sophia Antipolis Cedex, France: Author.

3G TS 33.105, 3G Security; Cryptographic Algorithm Requirements.

3G TS 35.205, 3G Security; *Specification of the MILENAGE Algorithm Set: An Example Algorithm Set for the 3GPP Authentication and Key Generation Functions f1, f1*, f2, f3, f4, f5, and f5*;* Document 1: General.

3G TS 35.206, *3G Security; Specification of the MILENAGE Algorithm Set: An Example Algorithm Set for the 3GPP Authentication and Key Generation Functions f1, f1*, f2, f3, f4, f5, and f5*;* Document 2: Algorithm Specification.

China Wireless Telecommunication Standard; 3G digital cellular telecommunications system; Security related network functions (Release 3); CWTS TSM 03.20 V3.0.0 (2002-08).

3rd Generation Partnership Project (3GPP). (2003). *Technical specification core network; Cx and Dx interfaces based on the diameter protocol; Protocol details*, Tech. Spec. 3G TS 29.229 V5.3.0 (2003-03).

3rd Generation Partnership Project (3GPP). (2003). *Technical specification core network; IP multimedia subsystem Cx and Dx interfaces; Signaling flows and message contents (Release 5),* Tech. Spec. 3G TS 29.228 V5.4.0 (2003-06).

3rd Generation Partnership Project (3GPP). (2003). *Technical specification group core network; Signaling flows for the IP multimedia call control based on SIP and SDP; Stage 3, version 5.5.0 (2003-06).* 3GPP TS 24.228.

3rd Generation Partnership Project (3GPP). (2003). Technical specification group services and systems aspects; 3G security; Access security for IP-based services, Tech. Spec. 3G TS 33.203 V5.5.0 (2003-03).

3rd Generation Partnership Project (3GPP). (2003). Technical specification group services and systems aspects; IP Multimedia subsystem stage 2, Tech. Spec. 3G TS 23.228 version 6.2.0 (2003-06).

Al-Saraireh, J., & Yousef, S. (2006). Authentication transmission overhead between entities in mobile networks. *International Journal of Computer Science and Network Security, 6*(3B), 150-154.

Harn, L., & Hsin, W. (2003). On the security of wireless networks access with enhancements. In *Proceedings of Web Information Systems Engineering (WiSE'03)* (pp. 88-95).

Liang, W., & Wang, W. (2005). A quantitative study of authentication and QoS in wireless IP networks. In *Proceedings of IEEE INFOCOM'05*, 2005.

Lin, Y., & Chen, Y. (2003). Reducing authentication signaling traffic in third-generation mobile network. *IEEE Transactions on Wireless Communication, 2*(3), 493-501.

Potlapally, N. R., Ravi, S., Raghunathan, A., & Jha, N. K. (2003). Analyzing the energy consumption of security protocols. In *Proceedings of the international symposium on Low power electronics and design* (pp. 30-35). ACM Press.

Rosenberg, J., Schulzrinne, H., Camarillo, G., Johnston, A., Peterson, J., Sparks, R., et al. (2002). *SIP: Session initiation protocol* (RFC 3261). Retrieved from http://www.ietf.org/rfc/rfc3261.txt

Zhang, M., & Fang, Y. (2005). Security analysis and enhancements of 3GPP authentication and key agreement protocol. *IEEE Transactions on Wireless Communication, 4*(2), 734-742.

Zhang, Y., & Fujise, M. (2006). An Improvement for Authentication Protocol in Third-Generation Wireless Networks. *IEEE Transactions on Wireless Communications. 5* (9), 2348-2352.

KEY TERMS

Access security: Access security is the mechanism that provides mobile users with secure access to wireless services and protects against attacks on the radio access interface.

General Packet Radio Service (GPRS): GPRS is regarded as 2.5 generation mobile system. It provides mobile data service to GSM users.

IP multimedia subsystem (IMS): IMS is the component to support multimedia services in 3G system.

Third generation (3G): 3G wireless communication systems is standardized to support multimedia services with high data rate.

Universal mobile telecommunications system (UMTS): UMTS is one of the third-generation wireless communication systems.

Chapter XXII
Security in 2.5G Mobile Systems

Christos Xenakis
University of Piraeus, Greece

ABSTRACT

The global system for mobile communications (GSM) is the most popular standard that implements second generation (2G) cellular systems. 2G systems combined with general packet radio services (GPRS) are often described as 2.5G, that is, a technology between the 2G and third generation (3G) of mobile systems. GPRS is a service that provides packet radio access for GSM users. This chapter presents the security architecture employed in 2.5G mobile systems focusing on GPRS. More specifically, the security measures applied to protect the mobile users, the radio access network, the fixed part of the network, and the related data of GPRS are presented and analyzed in detail. This analysis reveals the security weaknesses of the applied measures that may lead to the realization of security attacks by adversaries. These attacks threaten network operation and data transfer through it, compromising end users and network security. To defeat the identified risks, current research activities on the GPRS security propose a set of security improvements to the existing GPRS security architecture.

INTRODUCTION

The global system for mobile communications, (GSM) is the most popular standard that implements second generation (2G) cellular systems. 2G systems combined with general packet radio services (GPRS) (3GPP TS 03.6, 2002) are often described as 2.5G, that is, a technology between the 2G and third generation (3G) of mobile systems. GPRS is a service that provides packet radio access for GSM users. The GPRS network architecture, which constitutes a migration step toward 3G systems, consists of an overlay network onto the GSM network. In the wireless part, the GPRS technology reserves radio resources only when there is data to be sent, thus, ensuring the optimized utilization of radio resources. The fixed part of the network employs the IP technology and is connected to the public Internet. Taking advantage of these features, GPRS enables the provision of a variety of packet-oriented multimedia applications and services to mobile users, realizing the concept of the mobile Internet.

For the successful implementation of the new emerging applications and services over GPRS, security is considered as a vital factor. This is because of the fact that wireless access is inherently less secure and the radio transmission is by nature more susceptible to eavesdropping and fraud in use than wire-line transmission. In addition, users' mobility and the universal access to the network imply higher security risks compared to those encountered in fixed networks. In order to meet security objectives, GPRS uses a specific security architecture, which aims at protecting the network against unauthorized access and the privacy of users. This architecture is mainly based on the security measures applied in GSM, since the GPRS system is built on the GSM infrastructure.

Based on the aforementioned consideration, the majority of the existing literature on security in 2.5G systems refers to GSM (Mitchell, 2001; Pagliusi, 2002). However, GPRS differs from GSM in certain operational and service points, which require a different security analysis. This is because GPRS is based on IP, which is an open and wide deployed technology that presents many vulnerable points. Similarly to IP networks, intruders to the GPRS system may attempt to breach the confidentiality, integrity, or availability, or otherwise attempt to abuse the system in order to compromise services, defraud users, or any part of it. Thus, the GPRS system is more exposed to intruders compared to GSM.

This chapter presents the security architecture employed in 2.5G mobile systems focusing on GPRS. More specifically, the security measures applied to protect the mobile users, the radio access network, the fixed part of the network, and the related data of GPRS are presented and analyzed in details. This analysis reveals the security weaknesses of the applied measures that may lead to the realization of security attacks by adversaries. These attacks threaten network operation and data transfer through it, compromising end users and network security. To defeat the identified risks, current research activities on the GPRS security propose a set of security improvements to the existing GPRS security architecture. The rest of this chapter is organized as follows. The next section describes briefly the GPRS network architecture. The third section presents the security architecture applied to GPRS and the fourth section analyzes its security weaknesses. The fifth section elaborates on the current research activities on the GPRS security and the sixth section presents the conclusions.

GPRS NETWORK ARCHITECTURE

The network architecture of GPRS (3GPP TS 03.6, 2002) is presented in Figure 1. A GPRS user owns a mobile station (MS) that provides access to the wireless network. From the network side, the base station subsystem (BSS) is a network part that is responsible for the control of the radio path. BSS consists of two types of nodes: the base station controller (BSC) and the base transceiver station (BTS). BTS is responsible for the radio coverage of a given geographical area, while BSC maintains radio connections towards MSs and terrestrial connections towards the fixed part of the network (core network).

The GPRS core network (CN) uses the network elements of GSM such as the home location register (HLR), the visitor location register (VLR), the authentication centre (AuC) and the equipment identity register (EIR). HLR is a database used for the management of permanent data of mobile users. VLR is a database of the service area visited by an MS and contains all the related information required for the MS service handling. AuC maintains security information related to subscribers identity, while EIR maintains information related to mobile equipments' identity. Finally, the mobile service switching centre (MSC) is a network element responsible for circuit-switched services (e.g., voice call) (3GPP TS 03.6, 2002).

As presented previously, GPRS reuses the majority of the GSM network infrastructure. However, in order to build a packet-oriented mobile network some new network elements (nodes) are required, which handle packet-based traffic. The new class of nodes, called GPRS support nodes (GSN), is responsible for the delivery and routing of data packets between an MS and an external packet data network (PDN). More specifically, a serving

Figure 1. GPRS network architecture

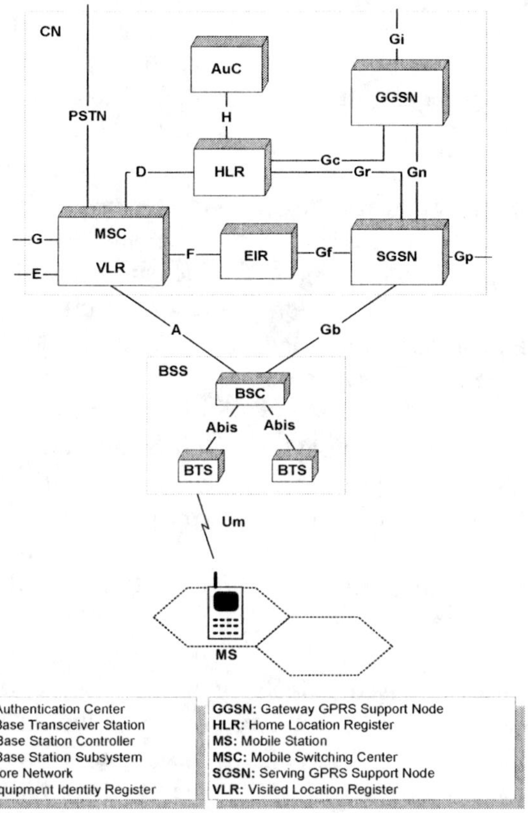

GSN (SGSN) is responsible for the delivery of data packets from, and to, an MS within its service area. Its tasks include packet routing and transfer, mobility management, logical link management, and authentication and charging functions. A gateway GSN (GGSN) acts as an interface between the GPRS backbone and an external PDN. It converts the GPRS packets coming from the SGSN into the appropriate packet data protocol (PDP) format (e.g., IP), and forwards them to the corresponding PDN. Similar is the functionality of GGSN in the opposite direction. The communication between GSNs (i.e., SGSN and GGSN) is based on IP tunnels through the use of the GPRS tunneling protocol (GTP) (3GPP TS 09.60, 2002).

GPRS SECURITY ARCHITECTURE

In order to meet security objectives, GPRS employs a set of security mechanisms that constitutes the GPRS security architecture. Most of these mechanisms have been originally designed for GSM, but they have been modified to adapt to the packet-oriented traffic nature and the GPRS network components. The GPRS security architecture, mainly, aims at two goals: (1) to protect the network against unauthorized access, and (2) to protect the privacy of users. It includes the following components (GSM 03.20, 1999):

- Subscriber identity module (SIM)
- Subscriber identity confidentiality
- Subscriber identity authentication

- User data and signaling confidentiality between the MS and the SGSN
- GPRS backbone security

Subscriber Identity Module (SIM)

The subscription of a mobile user to a network is personalized through the use of a smart card named SIM (ETSI TS 100 922, 1999). Each SIM card is unique and related to a user. It has a microcomputer with a processor, ROM, persistent EPROM memory, volatile RAM, and an I/O interface. Its software consists of an operating system, file system, and application programs (e.g., SIM application toolkit). The SIM card is responsible for the authentication of the user by prompting for a code (PIN), the identification of the user to a network through keys, and the protection of user data through cryptography. To achieve these functions it contains a set of security objects including:

- A (4-digit) PIN code, which is used to lock the card preventing misuse;
- A unique permanent identity of the mobile user, named international mobile subscriber identity (IMSI) (3GPP TS 03.03, 2003);
- A secret key, Ki, (128 bit) that is used for authentication; and
- An authentication algorithm (A3) and an algorithm that generates encryption keys (A8) (GSM 03.20, 1999).

Since the SIM card of a GSM/GPRS subscriber contains security critical information, it should be manufactured, provisioned, distributed, and managed in trusted environments.

Subscriber Identity Confidentiality

The subscriber identity confidentiality deals with the privacy of the IMSI and the location of a mobile user. It includes mechanisms for the protection of the permanent identity (IMSI) when it is transferred in signaling messages, as well as measures that preclude the possibility to derive it indirectly from listening to specific information, such as addresses, at the radio path.

The subscriber identity confidentiality is mainly achieved by using a temporary mobile subscriber identity (TMSI) (3GPP TS 03.03, 2003; GSM 03.20, 1999), which identifies the mobile user in both the wireless and wired network segments. The TMSI has a local significance and thus it must be accompanied by the routing area identity (RAI) in order to avoid confusions. The MS and the serving VLR and SGSN only know the relation between the active TMSI and the IMSI. The allocation of a new TMSI corresponds implicitly for the MS to the de-allocation of the previous one. When a new TMSI is allocated to the MS, it is transmitted to it in a ciphered mode. The MS stores the current TMSI and the associated RAI in a non-volatile memory, so that these data are not lost when the MS is switched off.

Further to the TMSI, a temporary logical link identity (TLLI) (3GPP TS 03.03, 2003) identifies also a GPRS user on the radio interface of a routing area. Since the TLLI has a local significance, when it is exchanged between the MS and the SGSN, it should be accompanied by the RAI. The TLLI is either derived from the TMSI allocated by the SGSN or built by the MS randomly and thus, provides identity confidentiality. The relationship between the TLLI and the IMSI is only known in the MS and in the SGSN.

Subscriber Identity Authentication

A mobile user that attempts to access the network must first prove his/her identity to it. User authentication (3GPP TS 03.6, 2002) protects against fraudulent use and ensures correct billing. GPRS uses the authentication procedure already defined in GSM with the same algorithms for authentication and generation of encryption key, and the same secret key, Ki, (see Figure 2). However, from the network side, the whole procedure is executed by the SGSN (instead of the BS) and employs a different random number (GPRS-RAND) and thus, it produces a different signed response (GPRS-SRES) and encryption key (GPRS-Kc) than the GSM voice counterpart.

To achieve authentication of a mobile user, the serving SGSN must possess security-related

Figure 2. GPRS authentication

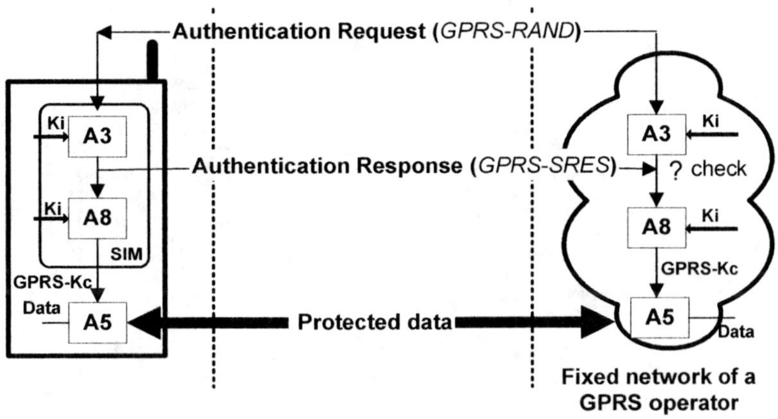

information for the specific user. This information is obtained by requesting the HLR/AuC of the home network that the mobile user is subscribed. It includes a set of authentication vectors, each of which includes a random challenge (GPRS-RAND), the related signed response (GPRS-SRES), and the encryption key (GPRS-Kc) for the specific subscriber. The authentication vectors are produced by the home HLR/AuC using the secret key Ki of the mobile subscriber.

During authentication the SGSN of the serving network sends the random challenge (GPRS-RAND) of a chosen authentication vector to the MS. The latter encrypts the GPRS-RAND by using the A3 hash algorithm, which is implemented in the SIM card, and the secret key, Ki. The first 32 bits of the A3 output are used as a signed response (GPRS-SRES) to the challenge (GPRS-RAND) and are sent back to the network. The SGSN checks if the MS has the correct key, Ki, and, then, the mobile subscriber is recognized as an authorized user. Otherwise, the serving network (SN) rejects the subscriber's access to the system. The remaining 64 bits of the A3 output together with the secret key, Ki, are used as input to the A8 algorithm that produces the GPRS encryption key (GPRS-Kc).

Data and Signalling Protection

User data and signaling protection over the GPRS radio access network is based on the GPRS ciphering algorithm (GPRS-A5) (3GPP TS 01.61, 2001), which is also referred to as GPRS encryption algorithm (GEA) and is similar to the GSM A5. Currently, there are three versions of this algorithm: GEA1, GEA2, and GEA3 (that is actually A5/3), which are not publicly known and thus, it is difficult to perform attacks on them. The MS device (not the SIM-card) performs GEA using the encryption key (GPRS-Kc), since it is a strong algorithm that requires relatively high processing capabilities. From the network side, the serving SGSN performs the ciphering/deciphering functionality protecting signaling and user data over the Um, Abis, and Gb interfaces.

During authentication the MS indicates which version(s) of the GEA supports and the network (SGSN) decides on a mutually acceptable version that will be used. If there is not a commonly accepted algorithm, the network (SGSN) may decide to release the connection. Both the MS and the SGSN must cooperate in order to initiate the ciphering over the radio access network. More specifically, the SGSN indicates whether ciphering should be used or not (which is also a possible option) in the *Authentication Request* message, and the MS starts ciphering after sending the *Authentication Response* message (see Figure 2).

GEA is a symmetric stream cipher algorithm (see Figure 3) that uses three input parameters (GPRS-Kc, INPUT, and DIRECTION) and produces an OUTPUT string, which varies between 5

Figure 3. GPRS ciphering

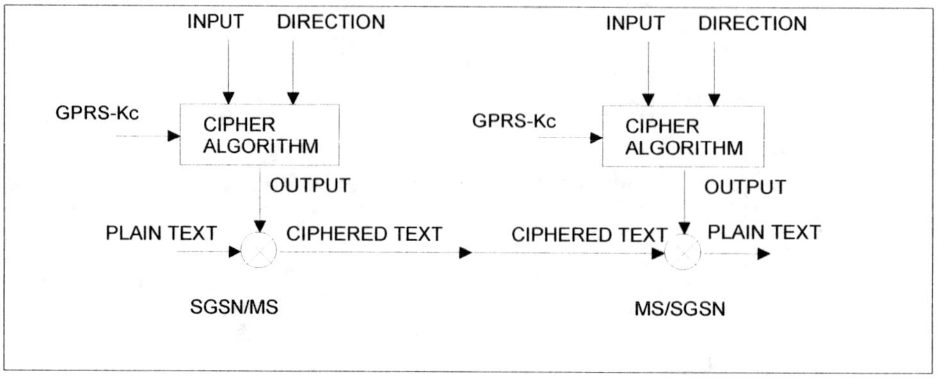

and 1,600 bytes. GPRS-Kc (64 bits) is the encryption key generated by the GPRS authentication procedure and is never transmitted over the radio interface. The input (INPUT) parameter (32 bits) is used as an additional input so that each frame is ciphered with a different output string. This parameter is calculated from the logical link control (LLC) frame number, a frame counter, and a value supplied by the SGSN called the input offset value (IOV). The IOV is set up during the negotiation of LLC and layer 3 parameters. Finally, the direction bit (DIRECTION) specifies whether the output string is used for upstream or downstream communication.

After the initiation of ciphering, the sender (MS or SGSN) processes (bit-wise XOR) the OUTPUT string with the payload (PLAIN TEXT) to produce the CIPHERED TEXT, which is sent over the radio interface. In the receiving entity (SGSN or MS), the original PLAIN TEXT is obtained by bit-wise XORed the OUTPUT string with the CIPHERED TEXT. When the MS changes SGSN, the encryption parameters (e.g., GPRS-Kc, INPUT) are transferred from the old SGSN to the new SGSN, through the (inter) routing area update procedure in order to guarantee service continuity.

GPRS Backbone Security

The GPRS backbone network includes the fixed network elements and their physical connections that convey user data and signaling information. signaling exchange in GPRS is mainly based on the signaling system 7 (SS7) technology (3GPP TS 09.02, 2004), which does not support any security measure for the GPRS deployment. Similarly, the GTP protocol that is employed for communication between GSNs does not support security. Thus, user data and signaling information in the GPRS backbone network are conveyed in cleartext exposing them to various security threats. In addition, inter-network communications (between different operators) are based on the public Internet, which enables IP spoofing to any malicious third party who gets access to it. In the sequel, the security measures applied to the GPRS backbone network are presented.

The responsibility for security protection of the GPRS backbone as well as inter-network communications belongs to mobile operators. They utilize private IP addressing and network address translation (NAT) (Srisuresh & Holdrege, 1999) to restrict unauthorized access to the GPRS backbone. They may also apply firewalls at the borders of the GPRS backbone network in order to protect it from unauthorized penetrations. Firewalls protect the network by enforcing security policies (e.g., user traffic addressed to a network element is discarded). Using security policies the GPRS operator may ensure that only traffic initiated from the MS and not from the Internet should pass through a firewall. This is done for two reasons: (1) to restrict traffic in order to protect the MS and the network elements from external attacks; and (2) to protect the MS from receiving unrequested traffic. Unrequested traffic may be unwanted for the mobile subscribers since they pay for the traffic received as well. The GPRS operator may also want to disallow some bandwidth-demanding protocols preventing a group of subscribers to consume so much bandwidth that other subscribers are noticeably affected. In addition, application-level firewalls prevent direct access through the use of proxies for services, which analyze application commands, perform authentication, and keep logs.

Since firewalls do not provide privacy and confidentiality, the virtual private network (VPN) technology (Gleeson, Lin, Heinanen, Armitage, & Malis, 2000) has to complement them to protect data in transit. A VPN is used for the authentication and the authorization of user access to corporate resources, the establishment of secure tunnels between the communicating parties, and the encapsulation and protection of the data transmitted by the network. In current GPRS implementations, pre-configured, static VPNs can be employed to protect data transfer between GPRS network elements (e.g., an SGSN and a GGSN that belong to the same backbone), between different GPRS backbone networks that belong to different mobile operators, or between a GPRS backbone and a remote corporate private network. The border gateway, which resides at the border of the GPRS backbone, is a network element that provides firewall capabilities and also maintains static, pre-configured VPNs to specific peers.

GPRS SECURITY WEAKNESSES

Although GPRS have been designed with security in mind, it presents some essential security weaknesses, which may lead to the realization of security attacks that threaten network operation and data transfer through it. In the following, the most prominent security weaknesses of the GPRS security architecture are briefly presented and analyzed.

Subscriber Identity Confidentiality

A serious weakness of the GPRS security architecture is related to the compromise of the confidentiality of subscriber identity. Specifically, whenever the serving network (VLR or SGSN) cannot associate the TMSI with the IMSI, because of TMSI corruption or database failure, the SGSN should request the MS to identify itself by means of IMSI on the radio path. Furthermore, when the user roams and the new serving network cannot contact the previous (the old serving network) or cannot retrieve the user identity, then, the new serving network should also request the MS to identify itself by means of IMSI on the radio path. This fact may lead an active attacker to pretend to be a new serving network, to which the user has to reveal his/her permanent identity. In addition, in both cases the IMSI that represents the permanent user identity is conveyed in cleartext over the radio interface violating user identity confidentiality.

Subscriber Authentication

The authentication mechanism used in GPRS also exhibits some weak points regarding security. More specifically, the authentication procedure is one way and thus, it does not assure that a mobile user is connected to an authentic serving network. This fact enables active attacks using a false BS identity. An adversary, who has the required equipment, may masquerade as a legitimate network element mediating in the communication between the MS and the authentic BS. This is also facilitated by the absence of a data integrity mechanism on the radio access network of GPRS, which defeats certain network impersonation attacks. The results of this mediation may be the alternation or the interception of signaling information and communication data exchanged.

Another weakness of the GPRS authentication procedure is related to the implementation of the A3 and A8 algorithms, which are often realized in practise using COMP128. COMP128 is a keyed hash function, which uses two 16-byte (128 bits) inputs and produces a hash output of 12 bytes (96 bits). While the actual specification of COMP128 was never made public, the algorithm has been reverse engineered and cryptanalyzed (Barkan, Biham, & Neller, 2003). Thus, knowing the secret key, Ki, it is feasible for a third party to clone a GSM/GPRS SIM-card, since its specifications are widely available (ETSI TS 100 922, 1999).

The last weakness of the GPRS authentication procedure is related to the network ability of re-using authentication triplets. Each authentication triplet should be used only in one authentication procedure in order to avoid man-in-the-middle and replay attacks. However, this depends on the mobile network operator (home and serving) and cannot be checked by mobile users. When the VLR of a serving network has used an authentication triplet to authenticate an MS, it shall delete the triplet or mark it as used. Thus, each time that the VLR needs to use an authentication triplet, it shall use an unmarked one, in preference to a marked. If there is no unmarked triplet, then the VLR shall request fresh triplets from the home HLR. If fresh triplets cannot be obtained, because of a system failure, the VLR may reuse a marked triplet. Thus, if a single triplet is compromised, a false BS can impersonate a genuine GPRS network to the MS. Moreover, as the false BS has the encryption key, Kc, it will not be necessary for the false BS to suppress encryption on the air interface. As long as the genuine SGSN is using the compromised authentication triplet, an attacker could also impersonate the MS and obtain session calls that are paid by the legitimate subscriber.

Data and Signalling Protection

An important weakness of the GPRS security architecture is related to the fact that the encryption of signalling and user data over the highly exposed radio interface is not mandatory. Some GPRS operators, in certain countries, never switch on encryption in their networks, since the legal framework in these countries do not permit that. Hence, in these cases signaling and data traffic are conveyed in cleartext over the radio path. This situation is becoming even more risky from the fact that the involved end users (humans) are not informed whether their sessions are encrypted or not.

As encryption over the radio interface is optional, the network indicates to the MS whether and which type(s) of encryption it supports in the *authentication request* message, during the GPRS authentication procedure. If encryption is activated, the MS start ciphering after sending the *authentication response* message and the SGSN starts ciphering/deciphering when it receives a valid *authentication response* message from the MS. However, since these two messages are not protected by confidentiality and integrity mechanisms (data integrity is not provided in the GPRS radio interface except for traditional non-cryptographic link layer checksums), an adversary may mediate in the exchange of authentication messages. The results of this mediation might be either the modification of the network and the MS capabilities regarding encryption, or the suppression of encryption over the radio interface.

GPRS Backbone

Based on the analysis of the GPRS security architecture (see the *GPRS security architecture* section) it can be perceived that the GPRS security does not aim at the GPRS backbone and the wire-line connections, but merely at the radio access network and the wireless path. Thus, user data and signaling information conveyed over the GPRS backbone may experience security threats, which degrade the level of security supported by GPRS. In the following, the security weaknesses of the GPRS security architecture that are related to the GPRS backbone network for both signaling and data plane are presented and analyzed.

Signaling Plane

As mentioned previously, the SS7 technology used for signaling exchange in GPRS does not support security protection. Until recently, this was not perceived to be a problem since SS7 networks belonged to a small number of large institutions (telecom operator). However, the rapid deployment of mobile systems and the liberalization of the telecommunication market have dramatically increased the number of operators (for both fixed and mobile networks) that are interconnected through the SS7 technology. This fact provokes a significant threat to the GPRS network security, since it increases the probability of an adversary to get access to the network or a legitimate operator to act maliciously.

The lack of security measures in the SS7 technology used in GPRS results also in the unprotected exchange of signaling messages between a VLR and a VLR/HLR, or a VLR and other fixed network nodes. Although these messages may include critical information for the mobile subscribers and the networks operation like ciphering keys, authentication data (e.g., authentication triplets), user subscription data (e.g., IMSI), user billing data, network billing data, and so forth, they are conveyed in a cleartext within the serving network as well as between the home network and the serving network. For example, the VLR of a serving network may use the IMSI to request authentication data for a single user from its home network, and the latter forwards them to the requesting VLR without any security measure. Thus, the exchanges of signaling messages, which are based on SS7, may disclose sensitive data of mobile subscribers and networks, since they are conveyed over insecure network connections without security precautions.

Data Plane

Similarly to the signaling plane, the data plane of the GPRS backbone presents significant security weaknesses, since the introduction of IP technology in the GPRS core shifts towards open and easily accessible network architectures. In addition, the data encryption mechanism employed in GPRS does not extend far enough towards the core network, also resulting in a cleartext transmission of user data in it. Thus, a malicious user, which gains access to the network, may either obtain access to sensitive data traffic or provide unauthorized/incorrect information to mobile users and network components. As presented previously, the security protection of users' data in the fixed segment of the GPRS network mainly relies on two independent and complementary technologies, which are not undertaken by GPRS but from the network operators. These technologies include: (1) firewalls that enforce security policies to a GPRS core network that belongs to an operator; and (2) pre-configured VPNs that protect specific network connections.

However, firewalls were originally conceived to address security issues for fixed networks and thus are not seamlessly applicable in mobile networks. They attempt to protect the cleartext transmitted data in the GPRS backbone from external attacks, but they are inadequate against attacks that originate from malicious mobile subscribers as well as from network operator personnel or any other third party that gets access to the GPRS core network. Another vital issue regarding the deployment of firewalls in GPRS has to do with the consequences of mobility. The mobility of a user may imply roaming between networks and operators, which possibly results in the changing of the user address. This fact in conjunction with the static configuration of firewalls may potentially lead to discontinuity of service connectivity for the mobile user. Moreover, in some cases the security value of firewalls is considered limited as they allow direct connection to ports without distinguishing services.

Similarly to firewalls, the VPN technology fails to provide the necessary flexibility required by typical mobile users. Currently, VPNs for GPRS subscribers are established in a static manner between the border gateway of a GPRS network and a remote security gateway of a corporate private network. This fact allows the realization of VPNs only between a security gateway of a large organization and a mobile operator, when a considerable amount of traffic requires protection. Thus, this scheme can provide VPN services neither to individual mobile users that may require on demand VPN establishment, nor to enterprise users that may roam internationally. In addition, static VPNs have to be reconfigured every time the VPN topology or VPN parameters change.

CURRENT RESEARCH ON GPRS SECURITY

The analyzed security weaknesses of the GPRS security architecture increase the risks associated with the usage of GPRS networks influencing their deployment, which realizes the mobile Internet. In order to defeat some of these risks, a set of security improvements to the existing GPRS security architecture may be incorporated. Additionally, some complementary security measures, which have been originally designed for fixed network and aim at enhancing the level of security that GPRS supports, may be applied (Xenakis, 2006). In the following, the specific security improvements and the application of the complementary security measures are briefly presented and analyzed.

SIM Card

The majority of the security weaknesses that are related to a MS and the SIM card of a mobile user have to do with the vulnerabilities of COMP128. To address these, the old version of COMP128 (currently named as COMP128-1) is replaced by two newer versions COMP128-2 and COMP128-3, which defeat the known weaknesses. There is an even newer version COMP128-4, which is based on the 3GPP algorithm MILENAGE that uses advanced encryption standard (AES). In addition, it is mentioned to the GPRS operators that the COMP128 algorithm is only an example algorithm and that every operator should use its own algorithm in order to support an acceptable level of security (Xenakis, 2006).

User Data

User data conveyed over the GPRS backbone and the public Internet most likely remain unprotected (except for the cases that the operator supports pre-established VPNs over the public Internet) and thus are exposed to various threats. The level of protection that GPRS provides to the data exchanged can be improved by employing two security technologies: (1) the application of end-user security, and (2) the establishment of mobile IPsec-based VPN, dynamically. End-user security is applied by using application layer solutions such as the secure sockets layer (SSL) protocol (Gupta & Gupta, 2001). SSL is the default Internet security protocol that provides point-to-point security by establishing a secure channel on top of TCP. It supports server authentication using certificates, data confidentiality, and message integrity. On the other hand, IPsec protects traffic on a per connection basis and thus is independent from the applications that run above it. An IPsec-based VPN is used for the authentication and the authorization of user access to corporate resources, the establishment of secure tunnels between the communicating parties, and the encapsulation and protection of the data transmitted by the network. On-demand VPNs that are tailored to specific security needs are especially useful for GPRS users, which require any-to-any connectivity in an ad hoc fashion. Regarding the deployment of mobile VPNs over the GPRS infrastructure, three alternative security schemes have been proposed: (1) the end-to-end (Xenakis, Gazis, Merakos, 2002), (2) the network-wide (Xenakis, Merakos: IEEE Network, 2002), and (3) the border-based (Xenakis, Merakos: IEEE PIMRC, 2002). These schemes mainly differ in the position where the security functionality is placed within the GPRS network architecture (MS, SGSN, and GGSN), and whether data in transit are ever in cleartext or available to be tapped by outsiders.

Signaling Plane of the GPRS Backbone

The lack of security measures in the signaling plane of the GPRS backbone gives the opportunity to an adversary to retrieve critical information such as the permanent identities of mobile users (IMSI), temporary identities (TMSI, TLLI), location information, authentication triplets (RAND, SRES, Kc), charging and billing data, and so forth. The possession of this information enables an attacker to identify a mobile user, to track his/her location, to decipher the user data transferred over the radio interface, to over bill him/her, and so forth. To address this inability of GPRS, it has been proposed the incorporation of the network domain security (NDS) features (Xenakis, 2006; Xenakis & Merakos, 2004) into the GPRS security architecture. NDS features, which have been designed for the latter version of UMTS, ensure that signaling exchanges in the backbone network as well as in the whole wire-line network are protected. For signaling transmission in GPRS the SS7 and IP protocol architectures are employed, which incorporate the mobile application part (MAP) (3GPP TS 09.02, 2004) and the GTP protocol (3GPP TS 09.60, 2002), respectively. In NDS both architectures are designed to be protected by standard procedures based on existing cryptographic techniques. Specifically, the IP-based signaling communications will be protected at the network level by means of the well-known IPsec suite (Kent & Atkinson, 1998). On the other hand, the realization of protection for the SS7-based communications will be accomplished at the application layer by employing specific security protocols (Xenakis & Merakos, 2004). However, until now only the MAP protocol from the SS7 architecture is designed to be protected by a new security protocol named MAPsec (3GPP TS 33.200 2002).

CONCLUSION

This chapter has presented the security architecture employed in 2.5G mobile systems focusing on GPRS. This architecture comprises a set of measures that protect the mobile users, the radio access network, the fixed part of the network, and the related data of GPRS. Most of these measures have been originally designed for GSM, but they have been modified to adapt to the packet-oriented traffic nature and the GPRS network components. The operational differences between the application of these measures in GSM and GPRS have been outlined and commented. In addition, the security measures that can be applied by GPRS operators to protect the GPRS backbone network and inter-network communications, which are based on IP, have been explored. Although GPRS has been designed with security in mind, it presents some essential security weaknesses, which may lead to the realization of security attacks that threaten network operations and data transfer through it. These weaknesses are related to: (1) the compromise of the confidentiality of subscriber's identity, since it may be conveyed unprotected over the radio interface; (2) the inability of the authentication mechanism to perform network authentication; (3) the possibility of using COMP128 algorithm (which has been cryptoanalyzed) for A3 and A8 implementations; (4) the ability of reusing authentication triplets; (5) the possibility of suppressing encryption over the radio access network or modifying encryption parameters; and (5) the lack of effective security measures that are able to protect signaling and user data transferred over the GPRS backbone network. To defeat some of these risks, a set of security improvements to the existing GPRS security architecture may be incorporated. Additionally, some complementary security measures, which have been originally designed for fixed network and aim at enhancing the level of security that GPRS supports, may be applied.

ACKNOWLEDGMENT

Work supported by the project CASCADAS (IST-027807) funded by the FET Program of the European Commission.

REFERENCES

3rd Generation Partnership Project (3GPP) TS 03.6 (V7.9.0). (2002). *GPRS service description, Stage 2.* Sophia Antipolis Cedex, France: Author. Retreived from ftp://ftp.3gpp.org/specs/2006-12/R1998/03_series

3rd Generation Partnership Project (3GPP) TS 09.60 (V7.10.0). (2002*). GPRS tunneling protocol (GTP) across the Gn and Gp interface.* Sophia Antipolis Cedex, France: Author.Retrived from ftp://ftp.3gpp.org/specs/2006-12/R1998/09_series

3rd Generation Partnership Project (3GPP) TS 03.03 (v7.8.0). (2003). *Numbering, addressing and identification.* Sophia Antipolis Cedex, France: Author. Retrieved from ftp://ftp.3gpp.org/specs/2006-12/R1998/03_series

3rd Generation Partnership Project (3GPP) TS 01.61 (v7.0.0). (2001). *GPRS ciphering algorithm requirements.* Sophia Antipolis Cedex, France: Author. Retrieved from ftp://ftp.3gpp.org/specs/2006-12/R1999/01_series

3rd Generation Partnership Project (3GPP) TS 09.02 (v7.15.0). (2004). *Mobile application part (MAP) specification.* Sophia Antipolis Cedex, France: Author. Retrieved from ftp://ftp.3gpp.org/specs/2006-12/R1998/09_series

3rd Generation Partnership Project (3GPP) TS 33.200 (v4.3.0), (2002*). 3G security; network domain security; MAP application layer security.* Sophia Antipolis Cedex, France: Author. Retrieved from ftp://ftp.3gpp.org/specs/2006-12/Rel-4/33_series

Barkan, E., Biham, E., & Neller, N. (2003). Instant ciphertext-only cryptanalysis of GSM encrypted communication. In *Proceedings of Advances in Cryptology (CRYPTO 2003)* (LNCS 2729, 600-616).

ETSI TS 100 922 (v7.1.1). (1999). Subscriber identity modules (SIM) functional characteristics.Retrieved from http://pda.etsi.org/pda/queryform.asp

Gleeson, B., Lin, A., Heinanen, J., Armitage, G., & Malis, A. (2000). *A framework for IP based virtual private networks* (RFC 2764). Retrieved from http://www.faqs.org/rfcs/rfc2764.html

GSM 03.20. (1999). Security related network functions. Retrieved from ftp://ftp.3gpp.org/specs/2006-12/R1999/03_series

Gupta, V., & Gupta, S. (2001). Securing the wireless Internet. *IEEE Communications Magazine, 39*(12), 68-74.

Kent, S., & Atkinson, R. (1998). *Security architecture for the Internet protocol* (RFC 2401). Retrieved from http://www.javvin.com/protocol/rfc2401.pdf

Mitchell, C. (2001). *The security of the GSM air interface protocol.* Retrieved August, 2001, from http://www.ma.rhul.ac.uk/techreports/

Pagliusi, P. (2002). A contemporary foreword on GSM security. In *Proceedings of the Infrastructure Security International Conference (InfraSec)* (LNCS 2437, pp. 129-144). Springer-Verlag.

Srisuresh, P., & Holdrege, M. (1999). *IP network address translator (NAT) terminology and considerations* (RFC 2663). Retrieved from http://www.faqs.org/rfcs/rfc2663.html

Xenakis, C. (2006). Malicious actions against the GPRS technology. *Journal in Computer Virology, 2*(2), 121-133.

Xenakis, C., Gazis, E., & Merakos, L. (2002). Secure VPN deployment in GPRS mobile network. In Proceedings of European Wireless, Florence, Italy (pp. 293-300).

Xenakis, C., & Merakos, L. (2002). On demand network-wide VPN deployment in GPRS. *IEEE Network, 16*(6), 28-37.

Xenakis, C., & Merakos, L. (2002). Dynamic network-based secure VPN deployment in GPRS. In *Proceedings of IEEE PIMRC,* Lisboa, Portugal, (pp. 1260-1266).

Xenakis, C., & Merakos, L. (2004). Security in third generation mobile networks. *Computer Communications, 27*(7), 638-650.

KEY TERMS

General Packet Radio Service (GPRS): GPRS is a mobile data service available to users of GSM.

Global System for Mobile Communications (GSM): GSM is the most popular standard for mobile phones in the world.

GPRS Tunneling Protocol (GTP): GTP is an IP-based protocol that carries signaling and user data with the GPRS core network.

International Mobile Subscriber Identity (IMSI): IMSI is a unique number associated with all GSM network mobile phone users.

Second Generation (2G): 2G is a short for second-generation wireless telephone technology.

Second and a Half Generation (2.5G): 2.5G is used to describe 2G systems that have implemented a packet-switched domain in addition to the circuit-switched domain.

Signaling System 7 (SS7): SS7 is a set of telephony signaling protocols which are used to set up the vast majority of the world's public switched telephone network telephone calls.

Subscriber Identity Module (SIM): SIM is a removable smart card for mobile phones that stores network specific information used to authenticate and identify subscribers on the network.

Temporary Mobile Subscriber Identity (TMSI): TMSI is a randomly allocated number that is given to the mobile the moment it is switched on and serves as a temporary identity between the mobile and the network.

Chapter XXIII
End-to-End Security Comparisons Between IEEE 802.16e and 3G Technologies

Sasan Adibi
University of Waterloo, Canada

Gordon B. Agnew
University of Waterloo, Canada

ABSTRACT

Security measures of mobile infrastructures have always been important from the early days of the creation of cellular networks. Nowadays, however, the traditional security schemes require a more fundamental approach to cover the entire path from the mobile user to the server. This fundamental approach is so-called end-to-end (E2E) security coverage. The main focus of this chapter is to discuss such architectures for IEEE 802.16e (Mobile-WiMAX) and major third generation (3G) cellular networks. The E2E implementations usually contain a complete set of algorithms, protocol enhancements (mutual identification, authentications, and authorization), including the very large-scale integration (VLSI) implementations. This chapter discusses various proposals at the protocol level.

INTRODUCTION

Mobile-WiMAX (802.16e) is a fourth generation (4G) candidate for mobility and is expected to address many of the current issues we face in 3G technologies. E2E security scheme is one of the major issues, which is currently addressed in variety of forms using IP security (IPsec), secure socket layer (SSL)/transport layer security (TLS), OpenPGP, and S/MIME (Gallop, 2005). The E2E architectures of major 3G technologies including global system for mobile communications (GSM), general packet radio service (GRPS), and code division multiple access (CDMA) and 802.16e will be discussed in this chapter.

The management of the sections is as follows: the next section will discuss details about the ultimate security features attributed to 3G technologies. The GSM section will discuss the security weakness in GSM's initial draft and the E2E solution to overcome its weakness. The fourth and fifth sections talk about GPRS and CDMA respectively. The Mobile-WiMAX section opens the discussion on 802.16e, the candidate for the 4G wireless systems, which contains the security weakness of 802.16e's initial draft and the E2E solution. A thorough comparison and references will be given in the last two sections.

OBJECTIVES OF SECURITY FEATURES FOR 3G/MOBILE-WIMAX

Before discussing security weaknesses of individual 3G technologies, we briefly discuss the objective of 3G security features. These features are (Campbell, Mckunas, Myagmar, Gupta, & Briley, 2002):

- **Mutual authentication:** Authentication is a method to verify that the claimed identity of an entity is genuine. Authentication is a fundamental security service and other necessary services often depend on proper authentication. Many protocols offer a one-way authentication. That is, only the client has to authenticate itself to the server and the server is not required to authenticate itself to the client. A one-way authentication is prone to an attack, so-called; *impersonation*, in which an illegitimate entity could pose as a legitimate one and start a new communication with another legitimate entity or take control an already started conversation. A two-way authentication scheme (mutual authentication) resolves impersonation attack. An E2E security scheme uses a balanced mutual authentication technique. A balanced technique requires equal effort by both entities for authenticate themselves to other entities. This decreases the chance of attacker's success
- **Data integrity:** This guarantees that the data received has not been altered by an unauthorized entity. One method of doing this is through the application of a hash function to the data stream
- **Security between networks:** Networks are interconnected using secure wired links, mainly using IPSec tunneling mechanism.
- **Secure international mobile subscriber identity (IMSI) usage:** The first-time user is assigned an initial IMSI number by the home network.
- **Stronger security scope:** Security is based within the radio network controller (RNC) rather than the base station (BS). An RNC is responsible for controlling and managing the multiple BSs including the utilization of radio network services.
- **User- and mobile-station authentication schemes:** Both user and mobile station share a secret common key, which is called the PIN. This is used for authentication.
- **Secure services:** These services protect the infrastructure against usage and access misuses.
- **Security in applications:** This is critical for mobile-based application security.
- **Fraud detection:** Mechanisms to detect and combat fraud in roaming situations.
- **Flexibility:** As technologies evolve, security features are extended and enhanced as required by new services and threats.
- **Service availability and configurability:** Users are to be notified whether security is on and the available level of security.
- **Multiple cipher and integrity algorithms:** The mobile user and the network negotiate and agree on the best available cipher and integrity algorithms (e.g., KASUMI).
- **Lawful interception:** Mechanisms should be provided to authorize agencies with certain necessary information about subscribers.
- **GSM compatibility:** GSM subscribers should be able to roam in 3G networks and cope with the extended security needed via GSM security context.

Figure 1. GSM system overview (Adapted from Pagliusi, 2002)

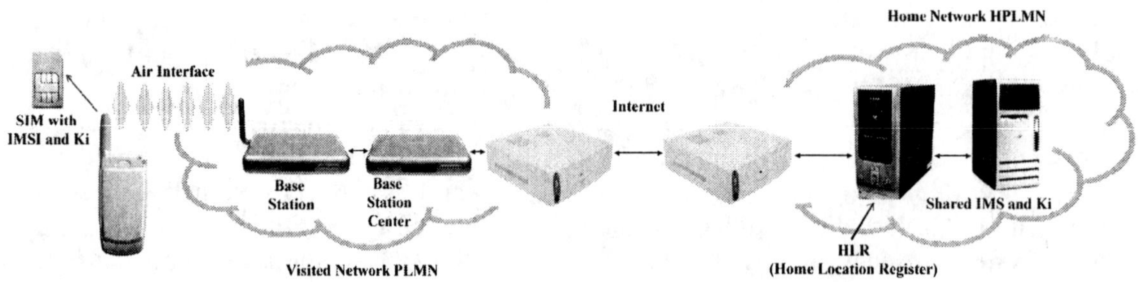

Figure 2: GSM authentication, cipher key generation, and encryption (Adapted from Pagliusi, 2002)

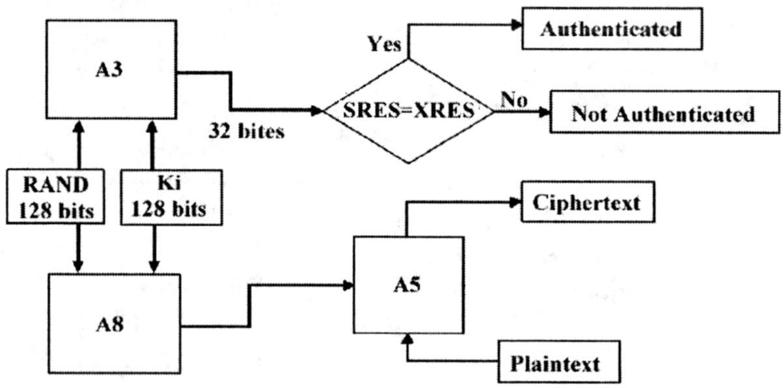

Figure 3. Authentication and encryption in GSM system (Adapted from Pagliusi, 2002)

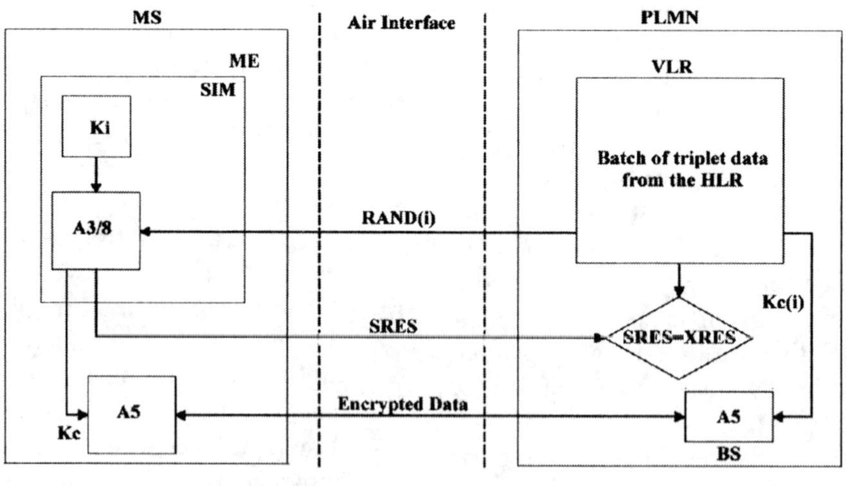

GSM

In this section, the weaknesses associated to GSM security systems (Pagliusi, 2002) are discussed and the E2E security proposals are considered.

GSM Security Features

Figure 1 shows the GSM system overview. The principles behind GSM security scheme are:

- Subscriber identity confidentiality scheme,
- Subscriber identity authentication scheme (Figure 2),
- Stream ciphering of user traffic and user-related control data schemes; and
- Using subscriber identity module (SIM) as a security module scheme.

Figure 2 shows the GSM authentication scheme in which three algorithms (A3, A5, and A8) are used for authentication, key generation, and encryption. The detailed authentication and encryption schemes for GSM are shown in Figure 3, where A3 (authentication algorithm), A8 (stream cipher), and A5 (key agreement algorithm) are performed in the mobile station and the key is verified in the public land mobile network (PLMN).

GSM Security Attacks

The security attacks associated to GSM architecture are (Pagliusi, 2002):

- **SIM/Mobile Equipment (ME) interface:** The SIM/ME interface is unprotected and can be tapped using an unauthorized device.
- **Attacks on the algorithms A3/A8/(A5/1):** Both A3 and A8 heavily reply on the COMP128 authentication algorithm, which have been cryptanalyzed allowing the recovery of shared master key leading to device cloning. A5/1 has also been attacked by Biryukov and Shamir (Pagliusi, 2002).
- **One-way authentication:** A3 is a one way operator-dependent stream-cipher function. Therefore its functionality suffers from being unbalanced
- **Unprotected signaling:** Though nearly all communications between the MS and the BS are encrypted, however in the fixed networks and between GSM central networks, all the communications and signaling are not protected as they are in plaintext most of the time
- **Attack on SIM card:** Interruption could occur on the operation of the smart card's microprocessor by exposing it to an electronic camera flashbulb. These types of attacks are called *optical fault induction*. Another type of attack, which is performed on the execution of COMP128 table lookups is called *partitioning attacks.*
- **False BS:** GSM provides a unilateral authentication (one-way). Because of the unbalanced nature, this allows attacks (such as man-in-the-middle (MITM) attack) where a malicious third party masquerades as a BS to one or more mobile stations.

E2E Scheme for GSM

The security concerns for GSM could be addressed in an E2E fashion. There are two major concerns in the current GSM structure that prevent the E2E communication, one is the fact that authentication is one way (A3/A8) and the fact that data is exposed and unprotected in certain areas. To prevent these flaws and pave the path to go E2E, a strong user authentication along with complete path encryption are proposed (Aydemir & Selcuk, 2005; Mynttinen, 2000).

Strong User Authentication

A strong authentication protocol is achieved through user-based rather than device-based. The GSM authentication algorithm contains three fundamental entities in a session (Pagliusi, 2002):

- The mobile subscriber (MS)
- The visiting location register (VLR)
- The home location register (HLR)

The initial draft of GSM states for the authentication scheme to use a cryptographic authentication key embedded in the SIM card of the device. Through the GSM user authentication protocol (GUAP) approach (Aydemir & Selcuk, 2005), the user can authenticate himself/herself through a password instead of the embedded hard-coded key, which breaks the dependency of the SIM card during authentication. The GUAP is based on three entities and in many cases the third entity is a trusted server whose public key is known by

Figure 4. The GUAP scheme (Adapted from Aydemir & Selcuk, 2005)

1. $MS \rightarrow VLR: IMSI$
2. $VLR \rightarrow MS: RAND$
3. $MS \rightarrow VLR:$ $\{n_1, n_2, n_3, \{RAND\}\Pi\}K_{HLR}, r_a$
4. $VLR \rightarrow HLR:$ $\{n_1, n_2, n_3, \{RAND\}\Pi\}K_{HLR}, \{RAND\}\ K_{VLR}$
5. $HLR \rightarrow VLR:$ $\{k\}\ K_{VLR}, \{n_1, n_2 \oplus k\}\Pi$
6. $VLR \rightarrow MS:$ $\{n_1, n_2 \oplus k\}\Pi, \{r_a\}_k, r_b$
7. $MS \rightarrow VLR:$ $\{r_b\}_k$

all parties. GSM doesn't include synchronized clocks, therefore authentication timestamps are not allowed. This can be remedied through the usage of random nonces. According to Figure 4, through VLR, MS is being authenticated to HLR through the usage of the Π password. The HLR public key, K_{HLR}, is known to all parties, and K_{VLR} is the symmetric encryption key shared among the VLR and the HLR. The GUAP protocol is being depicted in Figure 4.

In regards to GSM authentication, the GUAP's main goal is to break SIM card's dependency for added user flexibility. The GUAP's design includes considerations of the MS's computational restrictions. It also includes provisioning of the VLR authentication to both MS and HLR.

E2E Security of Mobile Data in GSM

In this approach, the E2E security scheme of mobile data in GSM is considered. It focuses on wireless application protocol (WAP) security, which can be broken. The data path protection in WAP is especially important for voicc over IP (VoIP) applications. For this purpose, *WAP Transport Layer E2E Security* is proposed. The E2E security for WAP transport layer is a specification provided by WapForum for supporting WAP E2E security by allowing the WAP clients to establish a straight wireless transport layer security (WTLS) connection with the WAP-based gateway. This gateway no longer encrypts and decrypts the traffic meant for the content-provider's 3rd party. Thus a malicious node is not able to cause problems for the data's confidentiality and integrity. A mechanism, such as; TLS, can be used to provide the required protection. Therefore it is fair to say that the required confidentiality and integrity can be guaranteed. However, non-repudiation property cannot be achieved using this solution. The remedy to this problem, a digital signature can be used in the transaction data, which is able to support integrity and non-repudiation functions. All WAP clients need to have access to digital keys for this to work.

GPRS

GPRS is a data-network-based architecture, which is designed in such a way to integrate well with existing GSM offering MSs "always connected" packet-switched data services. This includes connections to corporate networks and to the Internet. Figure 5 shows a MS logically attached to a serving GPRS support node (SGSN) ("GPRS Security Threats and Solutions," 2002).The SGSN's main functionality is to provide data services to the MS. Through the GPRS tunneling protocol (GTP), the SGSN can logically be connected to the gateway GPRS support node (GGSN). The GTP provides logical connection among the roaming partners of SGSN and GGSN.

GPRS was introduced as a packet service, which provides E2E IP connectivity with similar security options as in GSM. GPRS uses the same A3/A8 algorithms, which is used in GSM but the randomization function is slightly different. The three GPRS encryption algorithms are GEA1, GEA2, and GEA3, which is A5/3.

Figure 5. GPRS architecture (Adapted from "GPRS Security Threats and Solutions," 2002)

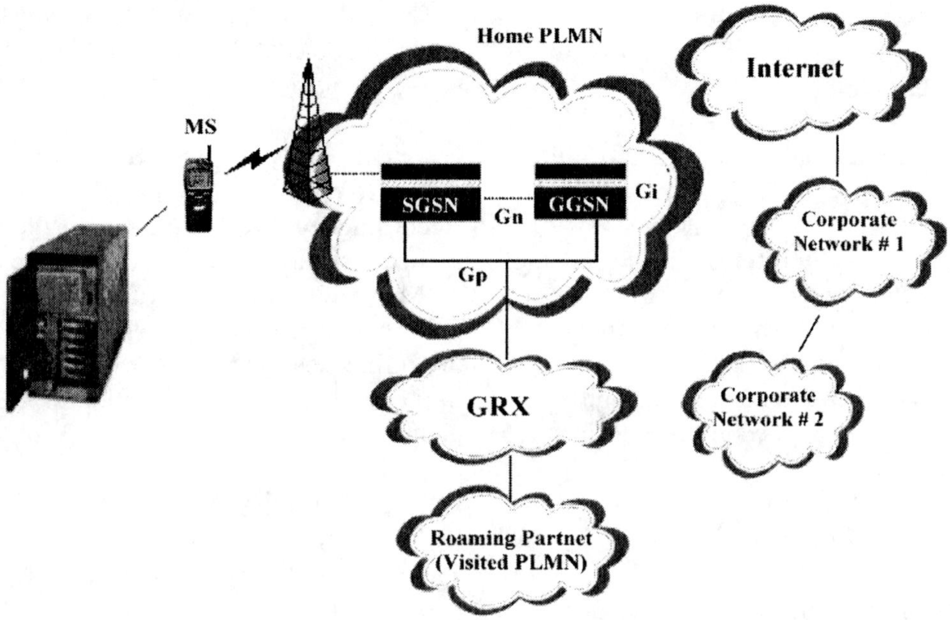

GPRS Classifications of Security Services

Security services provided by GPRS are protections against attacks and providing the following assurances:

- **Integrity:** Integrity is an assurance that data is not altered in an unauthorized manner.
- **Confidentiality:** Confidentiality is protecting data from disclosure to third parties.
- **Authentication:** Authentication provides assurance that all communication parties are really the ones who they claim to be.
- **Authorization:** Authorization is a service, which ensures that only legitimate entities are allowed to take part in any communications.
- **Availability:** Availability means that communication parties and data services are available and usable by any other parties in wireless range.

Data Services Offered on the Gp and Gi Interfaces

Before one can discuss the details about security, it is necessary to discuss the entities related in the data path. There are two main interfaces used in GPRS; *Gp* and *Gi*. G*p* interface is a logical connection among PLMNs. The protocols that deal directly with *Gp* are:

- **GTP:** The logical connection among the roaming partners of SGSN and GGSN.
- **Boarder gateway protocol (BGP):** BGP provides routing for between interfaces.
- **Dynamic name system (DNS):** DNS is a service that translates Internet domain names and computer hostnames to IP addresses.

The GTP provides logical connection among the roaming partners of SGSN and GGSN. If this connection is within the same PLMN, this is called the *Gn* interface. If the connection is between two different PLMNs, then it is known as the *Gp* interface. The G*p* and the G*i* interfaces are the initial and fundamental points of interconnection between the operator's main network and untrusted external networks.

Threats in GPRS

The following threats are associated to GPRS interfaces:

- **Threats on the *Gp* interface:** The specific threats include attacks on the *availability, authentication, and authorization; integrity; and confidentiality.*
- **Threats on the *Gi* interface:** The same types of threats for *Gp* applied to this interface as well.

GPRS Security Features

The SIM contains the individual subscriber authentication key (ISAK) K*i*, the ciphering key generating algorithm (A8), the authentication algorithm (A3), as well as a PIN. The GEA3 algorithm is implemented in the ME. Figure 6 (Kitsos, Sklavos, & Koufopavlou, 2004), shows the block diagram of the GPRS security in the MS. The key K*i* is 128 bits long.

E2E Security in GPRS

According to Figure 7 (Chang, 2002), the E2E security is shown as a sequential process.

Many of the E2E security options adopted by GSM could be used for GPRS due to the extend of similarities, however the major difference is in the hardware implementation of GEA*i* implementations.

GPRS Security Features.

The main architectural units of the system are based on the RIJNDAEL (Soyjaudah, Hosany, & Jamaloodeen, 2004) and KASUMI (Kitsos, Galanis, & Koufopavlou, 2004) block ciphers. These block cipher architectures meet all the GPRS security needs for an E2E security scheme.

Figure 6. GPRS security block diagram (Adapted from Kitsos, Sklavos, & Koufopavlou, 2004)

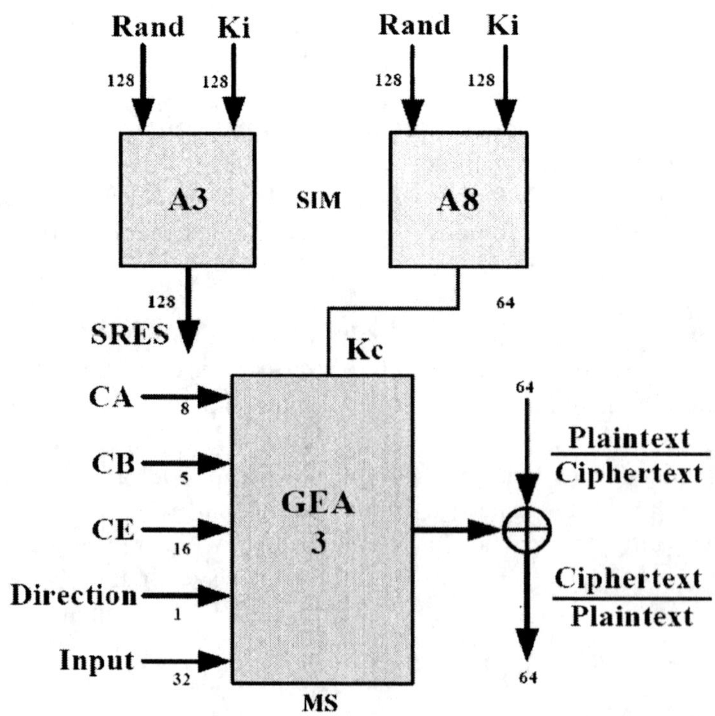

Figure 7: The End-to-End Sequence (Adapted from Chang, 2002)

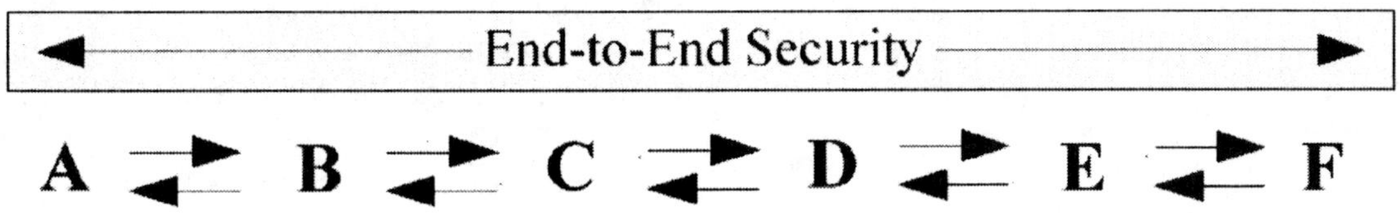

A. Security of the mobile device
B. Security of the radio path
C. Security of the digital cellular network
D. Security of the GPRS network
E. Security of the public Network
F. Security of the corporate network

CDMA

CDMA ("CDMA End-to-end Security Positioning Paper," 2005) air interface is inherently secure and is superior to many other cellular technologies (i.e., analog and time division multiple access [TDMA] systems). Its security strength comes from the fact that CDMA is a spread spectrum technology that uses Walsh codes.

CDMA uses specific spreading sequences and pseudo-random codes for the forward link (i.e., the downlink from BS to MS) and on the reverse link (i.e., the uplink from MS to BS). CDMA also benefits from a unique and soft handoff capability, which allows an MS to be connected to as many as six radios in a network, each with a unique Walsh code. This means that if anyone attempts to eavesdrop on a subscriber's call, they would have to have several devices connected simultaneously in an attempt to synchronize with all signals, which makes it difficult. Therefore it is difficult for a third party to have a stable link for interception of a CDMA voice channel, even with a complete knowledge of a Walsh code. Synchronization is a critical part, since without this synchronization the listener would only hear noise.

The 1x evolution-data optimized (1xEV-DO) version of CDMA was developed by Qualcomm in 1999. This version offered a bandwidth of greater than 2-Mbit/s for downlink stationary communication. The term 1xEV-DO also refers to the fact of being a direct evolution of the 1x (1xRTT) air interface standard, with its channels carrying only data traffic.

Compared to the GPRS and GSM networks, the 1xEV-DO feature of CDMA2000 networks is significantly faster and provides access terminals with air interface speeds of up to 2.4576 Mb/s and 3.1 Mb/s.

1xEV-DO has inherent security that protects the identity of users and makes interception very difficult. In addition, the Media Access Control Identification (MACID), which is assigned to users, is encrypted. Every user packet is assigned a variable time slot and the data rate for all users is controlled by the access terminal based on the radio conditions. Packets are divided into sub-packets using hybrid automatic repeat request (HARQ) and early termination mechanisms. These mechanisms make it virtually impossible to identify the user or correlate user packets. 1xEV-DO standard specification includes the possibility of supporting a security protocol layer suitable for future security implementation.

E2E security. CDMA ("CDMA End-to-end Security Positioning Paper," 2005) supports a number of security features such as OS hardening (file-system security); user access and operation audit; centralized authentication; user profile and group management; privilege-based user groups;

customer authentication; and secure remote access with IP Security Protocol (IPSec). Specific steps in the E2E security follow.

Subscriber Authentication

The key feature control mechanism is to protect the infrastructure and to prevent unauthorized access to network resources. The CDMA's access authentication is performed via an 18-bit authentication signature, which is verified via the user information of the network's database, the HLR and authentication center. CDMA is equipped with over the air service provisioning (OTASP), which is the ability of the signal's carrier to add new types of services to a customer's handset and provision it via wireless network instead of requiring the customer to bring the phone to a carrier's location for reprogramming. The 1xEV-DO uses the same 512-bit algorithm in OTASP for exchanging keys among the mobile device and the access node-authentication authorization accounting (AN-AAA) server. Users are authenticated through the usage of the challenge handshake authentication protocol (CHAP) via the packet data serving node-authentication authorization accounting (PDSN-AAA) server. CHAP is a strong Internet authentication protocol that is leveraged in the wireless network to verify identity.

Packet Core

In CDMA architecture, the packet data network supports:

- **Subscriber stateful firewall (SSF):** SSF protects both subscriber or operator's infrastructure traffic
- **Ingress anti-spoofing (IAS):** IAS prevents any malicious entity from launching attacks based on forged source IP addresses.
- **Filters and virus protection services:** Protects servers from viral infections.

Figure 8. Nortel's threat protection system (TPS) (Adapted from "CDMA end-to-end security; Positioning paper," 2005)

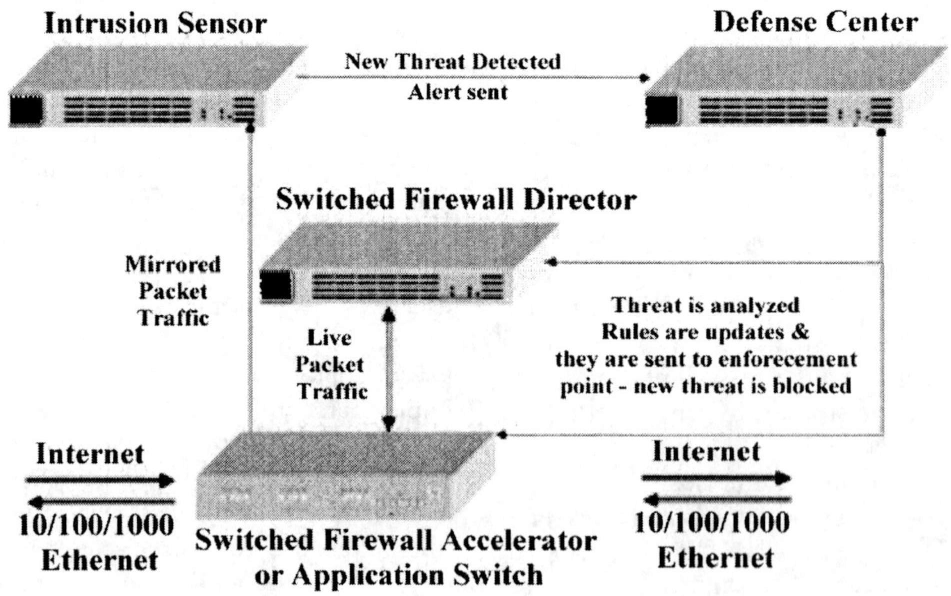

- **Layered protection:** TCP/IP layer to application layer deep packet filtering and inspection.
- **On-board lawful intercept:** For meeting regulatory security requirements
- **Traffic steering:** Services, such as; content filtering and server protections against malicious software.
- **Deep packet inspection:** Including application layers down to TCP/IP layer.

Transport Security (Protecting Traffic in Transit)

This is for protecting the MS or in a layered-security defense mechanism. This offers coverage on virtual private network (VPN) protocols, IPSec, and secure socket layer (SSL). IPSec VPN provides encryption, authentication protection at the IP layer, which protects all IP application/data traffic. SSL VPN provides secure communications for the Web application. Nortel Networks claims to be the first vendor to offer support for both VPN technologies in a single and unique platform. SSL VPNs provide security above the TCP/IP layer, thus ensuring compatibility with existing network address translation (NAT) services and other firewall configurations and proxy settings.

Perimeter Security (PS)

According to Figure 8, Nortel Networks claims to have introduced the threat protection system (TPS) using Snort-based™ intrusion detection system, which is complex but easy to use. This is based on the Perimeter security systems, which have been around for quite some time and are generally referred to as a firewall.

A rule-based, deep-packet inspection method is the heart of the TPS Intrusion Sensor, which examines the protocol and data fields on the incoming packets and checks for possible attack patterns. The sensors are capable of detecting anomalies such as: IP stack fingerprinting, port scan, denial of service (DoS) attack and address resolution protocol (ARP) spoofing. The incoming back packets are analyzed to pin-point threats. This is done at the Defense Center and rules updates are automatically transmitted real-time to firewalls for blocking new detected threat. The Defense Center provides threat analysis along with policy management and control. It also includes: traps and trace capabilities, reports, and event database. It also supports a centralized management for hierarchical sensors grouping.

End Point Compliance

End points, either MS, BS, or the management consoles could be potential sources of threats. These end-points may be infected with viruses or worms, which in case they are allowed to connect to the wireless network, may become the source of an attack. Unprotected end points can carry DoS or distributed DoS attacks. Nortel VPN Tunnel Guard is said to provide a security solution capable of enforcing on both managed (IPSec/SSL) and unmanaged SSL endpoints.

MOBILE-WIMAX (IEEE 802.16E)

Mobile-WiMAX was introduced in 2004 and it is designed to cover a 10-mile range for a data bandwidth of up to 15 MBps. Mobile-WiMAX's security mechanisms claim to have covered the holes in Wi-Fi. These security measures include ("Part 16: Air Interface for Fixed and Mobile Broadband Wireless Access Systems," 2004):

- **Data over cable service interface specification for baseline privacy +interface (DOCSIS BPI+)** for the interoperability equipment certification and device data privacy
- **Privacy key management for extensible authentication protocol (PKM-EAP)**, which is one of the requirements for an E2E authentication using TLS.
- **Counter mode with cipher block chaining message authentication code protocol (CCMP):** Used for encryption using Advanced Encryption Standard (AES) and 3 Data Encryption Standard (DES).

Through PKM-EAP, a mechanism is provided by which both BS and MSS (Mobile Subscriber Station) can mutually be authenticated and can establish a shared secret, which is called the "AAA-key". For completing the EAP-based authentication integration into the 802.16e's protocol, the following terms have to be defined ("Secure Association Establishment for PKM-EAP" 2004), : (1) PKM authorization key (AK) establishment and installation, (2) MSS static Security Association provisioning,and, (3) Ciphersuite signalling.

E2E Security Enhancements

The heart of the authentication protocol for Mobile-WiMAX is around the PKMvl. The only issue about PKMvl is that it offers a one-way authentication. The PKMvl protocol integrates AK and traffic encryption Key (TEK) exchange processes. Two nodes perform authorization process with one another in a one-way fashion, that is, one node only sends the authorization signal to the other node. If authorized and authenticated, the ACK (acknowledgement) signal is sent back and they are allowed to communicate. Together with the ACK, the authorizing node provides the node with the information of security associations (SAs) in which the node is authorized to access. This suffers from variety of attacks, which are described shortly.

To correct this issue, PKMv2 is deployed. This has been shown in Figure 9 (Johnston, 2003). The issues with a one-way authentication are (Puthenkulam, Mandin, 2003) :

- It is prone to false BS attacks (base impersonation).
- It is also prone to MITM attacks.
- It only works when providers control all the devices.

Other security enhancements required for an E2E scheme are:

- **Bi-directional certification exchange:** In order to achieve an E2E security scheme, a

Figure 9. Private key management version 2 (Adapted from Johnston, 2003)

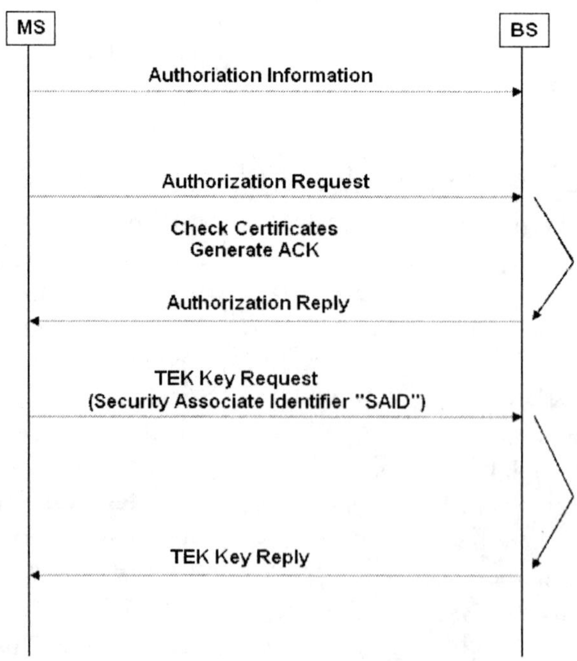

bi-directional certificate exchange for mutual authorization is required. This is achieved by Rivest-Shamir-Adleman (RSA)-based mutual authorization based on PKMv2.

- **Support of key hierarchy:** To support key hierarchy, Temporary Key Integrity Protocol (TKIP) is used through utilization of the master key (MK) and the pairwise master key (PMK) schemes.
- **PKM and EAP messages protections:** EAP has been known to cure security vulnerabilities, such as in the lack of user identity protection and MITM attack. This requires enabling encryption and utilizing PKM EAP messages for user authentication. To fix the previous problems, PKM messages should be bi-directional and EAP messages should use a four-way handshaking scheme
- **Weakness in the X.509 certificates:** X.509 certificate has the following issues:

    - Is restricted to certain business model and flexibility is a major issue.
    - Does not support user-based identity authentication, due to the fact that devices and services are greatly coupled, and
    - Trusting a certificate authority (CA) could become a source of a new attack.
- **Poor IV construction:** Initialization vectors (IVs) often use similar and repetitive structures. Through traffic pattern analysis, IVs can easily be known and broken. to remedy this, more complex IV structures with high key-bits (at least 128-bites) is the remedy to this problem.
- **802.16 key exchange issues:** A 2-key 3DES based key wrap is currently the standard of the initial draft for TEK exchange, which is not as strong (82bits) as the TEK keys (128 bits) it carries. There should be a mechanism to ensure that TEKs do not repeat for frequent exchange of TEKs. This could suffer from replay attacks, since there is no liveliness in the key exchange protocol and it also suffers from MITM attacks. Adding EAP-TLS authentication framework and the AES-CCM cipher suit will solve the problem.
- **Security improvements suggestions:** As a summary, the following security improvements are suggested for Mobile-WiMAX (as well as 802.16d):

    - PKM requires mutual authentication (PKMv2), necessary protections against reply and a stronger than DES-CBC cipher suite.
    - Deployment models should include additional security features for performance related issues (fast roaming, etc).
    - The current security models and solutions are not able to fully utilize the core network AAA infra structures due to the very low PKI support.
    - A single X.509 credential has limitations. To overcome this, it's recommended to use a flexible protocol, such as EAP, which supports multiple user credentials.
    - A scalable security solution is required to be deployed into the existing architecture and infrastructure for 802.16e requirements.

E2E Security Architecture

Figure 10 conceptually displays a client-server-based (i.e., VoIP) E2E AAA on 802.16 networks offering portability and fully mobile operations. The architecture is built around the three-party protocol (PKM v2), as defined in 802.16e (Agis et al., 2004).

Figure 10 shows that the over-the-air security association (authentication and encryption) is established through the PKM-EAP protocol. This is a complete client/server architecture, where EAP carries the AAA backend connectivity using Radius or Diameter. EAP offers a strong support for key-driven cipher mechanisms (i.e., EAP-MSCHAPv2 and EAP-AKA). It is also recommended to use an E2E tunneling protocol such as protected EAP

Figure 10. 802.16 E2E security framework

	SUPPLICANT	AP (ACCESS POINT) AUTHENTICATOR	AUTHENTICATION SERVER
1.	Sends EAP-start message →	Receives the message	
2.	←	Replies with an EAP-request ID	
3.	Protected EAP (PEAP) / Tunneled TLS (TTLS) →		Verifies the client's identity and bound
4.		Receives Accept/Reject message	EAP over RADIUS/Diameter ←
5.	If accepted, start PKMv2/ EAP with AES →	AP forwards the messages to the Authentication Server	
7.	If rejected, the end		

(PEAP) or tunneled TLS (TTLS) for the purpose of mutual authentication and a 128-bit or better TLS encryption method to further fortify the E2E security strength(particularly where weaker EAP methods may be deployed).

CONCLUSION

In this chapter, E2E security schemes were discussed for 3G technologies; GSM, GPRS, and CDMA and for Mobile-WiMAX (802.16e). The weaknesses of the initial drafts were pointed out and different enhancements were suggested. In most cases, mutual authentication as well as strong algorithms for authentication and authorization solved the E2E problems.

Comparing the performance issue, GSM systems have been around for a while and the designed architecture had not taken much security into account due to the security issues not being severe at the time. However as the technology matured, the newer technologies integrated more security aspects in their initial drafts, such as in CDMA. Now Mobile-WiMAX is expected to overcome all security issues and develop a full-scale E2E architecture with high performance E2E availability and security options. The encryption/authentication/authorization strengths of EAP/TLS/CCMP/PKM are unbeatable.

REFERENCES

Agis, E., Mitchel, H., Ovadia, S., Aissi, S., Bakshi, S., Iyer, P., et al. (2004, August 20). Extending WiMAX technology to mobility. Global, InteroperableBroadband Wireless Networks, *Intel Journal, 8*(03). ISSN 1535-864X

Aydemir, O., & Selcuk, A. A. (2005). *GUAP: A strong user authentication protocol for GSM.* Bilkent University, Turkey.

Campbell, R., Mckunas, D., Myagmar, S., Gupta, V., & Briley, B. (2002, June 28). *Analysis of third generation mobile security.* Annual Motorola Project Review.

Chang, D. (2002, January). *Security along the path through GPRS towards 3G mobile telephone network data services.* Bethesda, MD: SANS Institute.

Code division multiple access (CDMA) end-to-end security positioning paper. (2005). Nortel Networks.

Gallop, J. (2005, April 14). The state of the art. *WP2.2 D2.2.1 Volume 2.* Retrieved from http://www.akogrimo.org/modules.php?name=UpDownload&req=getit&lid=16

GPRS Security Threats and Solutions. (2002). A White Paper By NetScreen Technologies Inc. Retrieved from http://www.firewall-reviews.com/documents/ACFKM4dU8.pdf

Johnston, D. (2003, September 3). *IEEE 802.16 security enhancements.* Retrieved from http://www.ieee802.org/16/tgd/contrib/C80216d-03_60.pdf_

Kitsos, P., Galanis, M. D., & Koufopavlou, O. (2004, May 23-26). High-speed hardware implementations of the KASUMI block cipher. In *Proceedings of the 2004 International Symposium on Circuits and Systems, ISCAS '04. Volume 2.*

Kitsos, P., Sklavos, N., & Koufopavlou, O. (2004, May 12-14). An end-to-end hardware approach security for the GPRS. In *IEEE Mediterranean Electrotechnical Conference.*

Mandin, J. *Secure Association Establishment for PKM-EAP*, IEEE 802.16 Broadband Wireless Access Working Group Project, 2004-03-17 Retrieved from http://www.ieee802.org/16/tge/contrib/C80216e-04_46r1.pdf

Mynttinen, J. (2000, November 27). *End-to-end security of mobile data in GSM.* Helsinki University of Technology, Finland.

Pagliusi, P. S. (2002). *A contemporary foreword on GSM security.* Retrieved from http://jazi.staff.ugm.ac.id/IC3-Royal%20Holloway/GSM_Security_v4.pdf

Part 16: Air interface for fixed and mobile broadband wireless access systems. (2004). Draft IEEE Standard for Local and metropolitan area networks. Retrieved from http://ieeexplore.ieee.org/iel5/10676/33683/01603394.pdf

Puthenkulam, J., & Mandin J. *802.16e Security Adhoc Proposal*, IEEE C802.16e-03/70, IEEE 802.16 Presentation Submission Template (Rev. 8.3). Retrieved Nov 13, 2003 from http://www.ieee802.org/16/tge/contrib/C80216e-03_70.pdf

Soyjaudah, K. M. S., Hosany, M. A., & Jamaloodeen, A. (2004, October 24-26). Design and implementation of Rijndael algorithm for GSM encryption. In *Mobile Future, 2004 and the Symposium on Trends in Communications. SympoTIC '04. Joint IST Workshop.*

KEY TERMS

Authentication, Authorization, and Accounting (AAA): AAA is an access control scheme, overseeing the auditing framework and policy enforcement for commercial access and computing systems.

Code Division Multiple Access (CDMA): CDMA is also a 2.5G technology offering codes for multiplexing various cell calls. Therefore it does not divide the channel into time slots (time domain multiple access [TDMA]) or frequency bands (frequency division multiple access [FDMA]). Instead, CDMA encodes data with codes associated with every channel; therefore they do not have any overlaps in time or frequency bands. CDMA is a major improvement in cellular technologies.

Customer-Premises Equipment (CPE): End communication device that local subscribers communicate to. Through CPE, the information transmitted to and from all local subscribers are transmitted back to the centre.

End-to-End (E2E): E2E security covers the system's security functionality and performance from one end to the other and back.

General Packet Radio Service (GPRS): GPRS is an extension to GSM technology, which offers higher data rates compared to GSM. GPRS is considered a 2.5G technology.

Global System for Mobile Communications (GSM): GSM is the most popular standard and one of the oldest technologies still used for cellular networks throughout the world. GSM is considered a 2G cellular technology with digital integration.

Initialization Vector (IV): IV is a block of bit streams that is attached to every security data to produce a unique and independent stream for encryption.

Mobile Subsciber Station (MSS) = Mobile Station (MS): These are end-user devices.

Pairwise Master Key (PMK): PMK is used in peer-to-peer communication schemes for sharing a master key that would last the entire session. This is mainly used for data encryption and integrity.

Privacy Key Management (PKM): PKM is a private key scheme used with EAP and TLS for providing E2E security schemes for wireless technologies.

Third and Fourth Generation (3G/4G): 3G/4G cellular networks are used in the context of mobile standards. The services associated with 3G are capable of transferring both voice and non-voice data simultaneously. Though not official yet, the 4G, however, will be fully IP-based converging wired and wireless access technologies. It is expected to reach bandwidth within a few hundred mega bit per second offering E2E QoS.[1]

Transport Layer Security (TLS): TLS is used mostly in client/server applications, which require endpoint authentication and communications privacy, particularly over the Internet. This is mostly done using cryptographic measures.

Virtual Private Network (VPN): VPN is a communications tunnel uses a pre-existing (and often unsecure, such as the Internet) network to connect a remote user to a corporate network. The information is tunneled, encapsulated, and encrypted when passes through the unsecure network. Once the information reaches the destination, it is decapsulated and decrypted.

Worldwide Interoperability for Microwave Access (WiMAX): WiMAX, which has been defined by the WiMAX Forum, formed in 2001. WiMAX is also known as IEEE 802.16 standard, officially titled; WirelessMAN and is an alternative to DSL (802.16d) and cellular access (802.16e).

ENDNOTE

1 Kim, Y. K., & Prasad, R. *4G roadmap and emerging communication technologies.* Artech House.

Chapter XXIV
Generic Application Security in Current and Future Networks

Silke Holtmanns
Nokia Research Center, Finland

Pekka Laitinen
Nokia Research Center, Finland

ABSTRACT

This chapter outlines how cellular authentication can be utilized for generic application security. It describes the basic concept of the generic bootstrapping architecture (GBA) that was defined by the 3rd generation partnership project (3GPP) for current networks and outlines the latest developments for future networks. The chapter will provide an overview of the latest technology trends in the area of generic application security.

INTRODUCTION

Applications in wireless networks require a very reliable method for user authentication and communication security. We will outline the reason for the security needs for mobile application compared to Internet application security. It starts with the application specific security approach used today by many mobile operators and describes the motivation that lead into the development of the generic bootstrapping architecture (GBA) of 3rd generation partnership project (3GPP).

The main function of GBA and also its dialects and variations are explained. GBA has been adopted by various standardization bodies and used by many applications. GBA was first embraced by mobile applications, like the 3GPP Mobile Broadcast Multicast Service, open mobile alliance (OMA) presence service, OMA broadcast smart card service protection profile, GBA Profile, and so forth.

The ongoing convergence of fixed and mobile network resulted in the adaptation of GBA-based application security for fixed and cable networks. We close with a snapshot of the ongoing work for application security in beyond third generation (B3G) networks.

APPLICATION SECURITY FOUNDATIONS IN MOBILE NETWORKS

Special Requirements for Mobile Application Security

Applications in wireless networks require a very reliable method for user authentication and communication security due to their special environment. There are the following security reasons for this:

- The communication to the service provider goes over the air and can be eavesdropped or modified, if not properly secured.
- The service may be offered over any kind of IP-based channel, then the protection of the service by the underlying bearer protocol can not be taken for granted. The authentication and the service usage may be performed over different channels.
- Username/password authentication is not very secure or user friendly on a mobile device with numeric keypad; hence the temptation to choose too short or easy-to-write passwords is even greater than in the fixed network environment.

Operators or third parties that provide server-based applications in mobile networks have to face additional challenges that reflect on the choice of a potential security solution for application security:

- **Roaming agreements:** The home operator of a user maybe liable to roaming partners or application providers, if an unauthorized user uses a service. This potential liability could even be exploited by malicious application providers. This is no empty threat, as the malicious usage of premium short message service (SMS) usage is an existing problem.
- **Development costs:** Development of mobile applications is much more expensive than that for Internet usage, hence if an application is seriously compromised, then the resulting loss is much higher.
- **Updating problem:** In the fixed network environment security patches and updates are a daily occurrence, this is not that common in the mobile environment. Even if some mobile software platforms offer the possibility to be updated via a PC or over-the-air (OTA) mechanisms.
- **Protection of investments:** Mobile operators make big investments in their network infrastructure; hence reusage of deployed network nodes needs to be taken into account. Especially, when new services are rolled out, these services should work in a harmonic way with the existing nodes.
- **Existing smart card base:** Operators hand out smart cards to their subscribers, these cards have very different capabilities (e.g., subscriber identity module (SIM) cards, universal SIM [USIM], etc.), and operators are unlikely to replace already handed out smart cards. It is more likely, that the user replaces the device. A smart card is replaced, when a user changes operators.
- **Service usage costs:** The cost of browsing is bound to the type of access the user is using. When mobile access is used, this may imply some significant costs for the user. Also, some mobile service fees already include the access cost, that is, the user does not have to pay twice for the content and the delivery. Hence, user authentication comes in mobile applications "earlier" than in the Internet case.
- **Reliability and availability:** If an Internet application does not work, because the authentication database crashed, then in many cases this is not that severe an issue and the user can still use other services (except for those 100% online shops like amazon.com). But if the operator's subscriber database can not be reached, users can not make phone calls, roam, send SMS messages, and, in the end, the operator has no opportunity to offer

any kind of service and obtain revenue. The availability and reliability requirements for mobile network nodes are very high.

- **Scalability:** Scalability of mobile networks security solutions is a critical factor. Solutions are standardized on a global level, for example, for small local operators, as well as for large international operators. Hence, solutions have to work also for millions of people at the same time and it must be possible to extend them gradually depending on the growing subscriber basis. Usage scenarios and scalability requirements, where a whole full soccer arena at once requests one service server and still the service should work and start on time, are not unusual.
- **Convergence:** For operators that run both fixed and mobile networks the issue of convergence gains importance, since it allows a more flexible re-use of the network backend servers and functions.

When the first mobile applications started after voice and SMS the requirements for a more generic application security were not clear. This resulted in a fast to roll-out, but with less generic approach as will be explained next.

Historic Approaches to Application Security

Application nodes in mobile networks in the past tend to have a monolithic security solution that is highly customized to the individual application. This has to do with the fact, that operators like to buy their equipment from various vendors, and re-usage and extensions of existing infrastructure requires standardized interfaces. This standardization takes quite some time and that backward compatibility and integration with the existing nodes is a big challenge. Another argument for having customized security solution was that there were not that many new applications were not expected to come with a fast pace. For application security the return of investment was also an important consideration. The system had first to attract some subscribers and be accepted, before an expensive security solution of higher quality is considered, that is, why spend a lot of money, if it will not be used. This subsection describes how application security is often managed today and how it was managed in the past and what are the problems related to it:

- **Voice:** The terminal authenticates to the network utilizing a shared secret stored in the smart card and the operator's subscriber database. For application security needs, the *authentication vectors* (AVs) are distributed to the corresponding nodes.
- **Early IP multimedia subsystem security (IMS):** The 3GPP early IMS security solution of Release 6 (3GPP, TR33.978, Release 6) uses IP address binding, that is, the IP address assigned by the gateway GPRS support node (GGSN) is used for subsequent user authentication to the IMS service.
- **IMS:** The IMS security is bound to the credentials of the IMS SIM (ISIM) application on the universal integrated circuit card (UICC) smart card and these credentials are used by the mobile terminal for authentication to the IMS network. This is outlined by 3GPP (TS 33.203,Release 6). The user authentication is delegated from the operator's subscriber database towards the IMS network (i.e., the serving call-session-control-function [S-CSCF]).

The first mobile application was voice and few people envisioned the further usage mobile networks would get and were surprised by the popularity of SMS. 3GPP IMS with its wide range of service possibilities has security wise two flavors: (1) IP address binding, which comes quite inexpensive to mobile operators, and (2) the full IMS security, which requires that the subscriber is equipped with a new smart card that contains an ISIM application on it. The early IMS security solution has its cost-wise advantages and allows a roll-out and provisioning of the service also to subscribers with "old" smart cards, but the usage of monolithic and application specific security solutions cause some problems. Additionally, the direct usage of AVs in applications causes some general problems that require a more generic approach. The specific and general issues are listed as follows:

- **Fraud potential:** With the convergence of networks IP address spoofing may become a threat for application specific solutions that utilize IP address binding.
- **Scalability:** The IP address binding solution is not very well scalable for large networks and large amounts of simultaneous service-users. It is difficult to scale for many application scenarios in diverse future networks, where users may roam between network types.
- **Performance and dependability:** The backbone subscriber database, that is, the home subscriber server (HSS) or the home location register-authentication center (HLR-AuC) should not be contacted too often to obtain fresh user credentials, since these databases are very large and if this database is going down due to overload requests from application servers, then the operator can not even offer simple voice calls.
- **Synchronization problems:** If a new application utilizes the smart card based authentication, then they consume AVs from the HSS/HLR-AuC. This can lead to sequence number synchronization problems, which becomes more severe when the number of services increases.
- **User experience:** Different user authentication methods often result in bad user experience and difficulties for the help-desk if the authentication fails for unknown reasons.
- **Combined networks:** If an operator has broadband fixed network access and mobile networks, then the broadband subscriber may not have the ISIM or the IP address assigned from the GGSN.
- **Future proof:** New applications need new security solutions. If the network is structured in separate monolithic pillars, then the re-usage of the existing infrastructure for security is very difficult without modifying the already deployed infrastructure. Since in telecommunication availability is a critical factor, the modification of well-running systems is a very sensitive issue.

There were some early non-standardized approaches to re-use the existing authentication infrastructure of an operator for general application specific security (Gerstenberger, Lahaije, & Schuba, 2004). The aforementioned issues lead in 3GPP and 3GPP2 to the standardization of the GBA.

Generic Bootstrapping Architecture (GBA)

We will now introduce the GBA nodes, the provided functionality, and the interworking. The main goal of GBA is to provide a common shared secret to an application server and a mobile terminal based on the cellular authentication, the details of this protocol are described in the 3GPP technical specification (3GPP TS 33.220, Release 6) and we will focus on this aspect.

The application. The basic setting is that we have an application server that offers services to the user. The application service does not need necessarily to be part of the operator's network. There is a general trend to outsource the application services to external parties and just take selected services into the "operator portal." This service should now be secured using the mobile smart card credentials and hence the existing trust relationship between the user and the operator. For this purpose, the service provider needs to have an agreement with the home operator of the user. This application server has two subcomponents, one that takes care of fetching credentials from the user's home network and is called network application function (NAF) in GBA and then the actual application itself that uses the credentials to secure the communication with the mobile terminal, which we call application service.

The terminal. A 3GPP terminal is called user equipment (UE). In strict 3GPP terms it has two components, the smart card (UICC) and the mobile equipment (ME). We will refine this model slightly, by also distinguishing between the device platform and the application in the terminal. The

Figure 1. Generic bootstrapping architecture

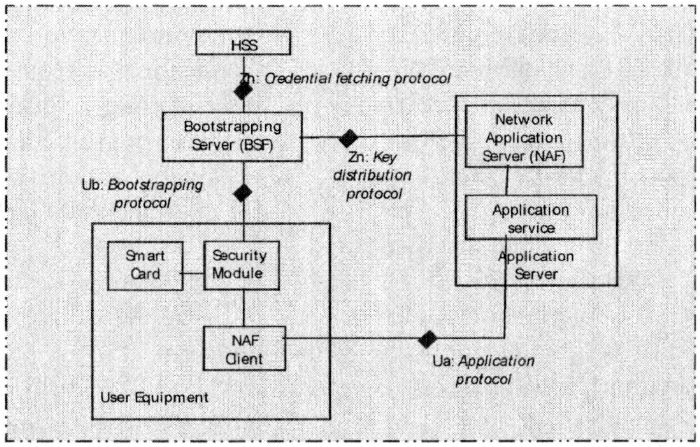

UE can authenticate with the network using cellular second generation (2G) or 3G-based authentication protocols. The intention is to reuse the authentication mechanism for the application communication security. Hence, we have a *security module* (see Figure 1) that communicates with the smart card and the so-called bootstrapping server function (BSF). Then there is the actual client application (NAF client) that communicates with the application server (NAF server) in the network, and uses the application specific keys.

The smart card. 3GPP Release 6 and Release 7 GBA assume the existence of a UICC. The UICC contains an ISIM and/or USIM application. If the operator wishes that the application is really closely bound with the smart card, then he/she can utilize the so-called GBA aware smart card (GBA_U), where the application specific key generation and part of the storage is performed in the UICC. GBA can be used also with SIM cards. This 2G GBA was introduced in Release 7 in the 3GPP technical report (3GPP TR 33.920, Release 7) due to the large market need to allow operators to utilize the existing smart card infrastructure without being forced to hand out immediately new smart cards to the user to use GBA-based services.

The network. The heart of the network is the operators subscriber database the HSS, respectively the HLR with accompanied AuC. This huge database is used to store the subscriber data including the counterpart of the credentials (i.e., master key) stored in the smart card that is handed out to the user and resides in the mobile terminal. This database provides the basic key material (i.e., authentication vector) to the BSF that is under mobile network operator control. This server can be seen as a credential server. Once the user is properly authenticated the BSF generates the application specific keys which are handed out to the application server, that is, the NAF.

The GBA system entities need to interact with each other to provision the application in the terminal and the application server with a shared secret that can then be utilized for various security purposes:

- **Bootstrapping interface (Ub):** The mobile terminal contacts the BSF and authenticates via authentication and key agreement (AKA) and triggers the key generation in the BSF. This interface is called Ub interface and defined in the 3GPP technical specification (3GPP TS 24.109, Release 6).
- **Credential fetching interface (Zh):** The application specific credentials are based on the mobile credentials stored in the subscriber database HSS of the operator. Therefore the BSF needs to obtain the AV to be able to establish an authentication session between the mobile terminal and the BSF and derive

further application specific keys. Also, some operator policies in the form of GBA user security settings (GUSS) can be stored in the HSS and passed to the BSF over this interface. The GUSS can contain application specific USSs and additionally BSF-specific guidance information, like user-specific key lifetimes, and UICC type of the user. The credential interface is defined as Zh interface and specified in the 3GPP technical specification (3GPP TS 29.109, Release 6). This interface is operator internal and specified as being Diameter based (Calhoun, Loughney, Guttman, Zorn, & Arkko, 2003) by the Internet Engineering Task Force (IETF), but since many operators have highly customized HLR/HSS it can be expected that operator-specific adjustments will be made (but likely not standardized).

- **Key distribution interface (Zn):** The application server has a library or a "plug-in" that requests the application-specific credentials, credential-related data, and USS from the credential server (BSF). This key distribution interface in 3GPP Release 6 Diameter based and called Zn interface. In Release 7, an alternative method was specified to support Web services (WS)-based protocol as this makes it easier for application developers to communicate with the credential server. Both implementations of the Zn interface are defined in the 3GPP technical specification (3GPP TS 29.109, Release 7).
- **Application interface (Ua):** The application-specific interface is called Ua interface and specified in (3GPP TS 24.109, 2006). The details of the actual protocol used in the Ua interface depend on the actual use case, for example, browsing, streaming, and so forth. The derived application-specific credentials will be used to secure the communication of this interface, how this is done, is application specific and defined in the application-specific specifications, for example, 3GPP multimedia broadcast/multicast service (MBMS) technical specification (3GPP TS 33.246, Release 6).

The actual application-specific key generation consists of the following basic steps:

1. The user wishes to use a service. The application server wishes to utilize GBA to secure the communication to the terminal. Hence, the terminal is requested to use GBA. This information (i.e., whether GBA needs to be used) can be pre-configured to the NAF client, or the application server may indicate over Ua interface that GBA should be used.
2. The NAF client triggers the security module in the terminal to bootstrap with the BSF utilizing AKA over the Ub bootstrapping interface.
3. The BSF then utilizes the Zh interface to fetch the needed data for the creation of the master session key. The BSF derives the master session key. Based on this master session key NAF specific application keys are derived when a specific NAF requests it over Zn interface later on. (Depending on the GBA type used, one or two application specific keys are derived.)
4. The resulting master session key and transaction ID are stored in BSF server. The security module in terminal also derives the master session key by contacting the smart card. The master session key and the transaction ID are stored in the security module. Note, that here are small differences between the different GBA types. Based on this master session key, NAF-specific application keys are derived. The application-specific key is handed out to the NAF client application in the terminal as response to the initial trigger made in step 2. The application-specific key is used to secure the communication with the application server.
5. The NAF client in the terminal sends transaction identifier to NAF server in the application server over Ua application interface. This transaction ID is needed, so that the NAF can contact the BSF and fetch the correct keys.
6. The NAF server in the application server contacts the BSF to obtain the application-specific session keys from BSF using transaction identifier over Zn key distribution interface.
7. The NAF server in the application server and the client in the terminal now share

Figure 2. HTTP based service request using GBA_ME (and GBA-unaware USIM)

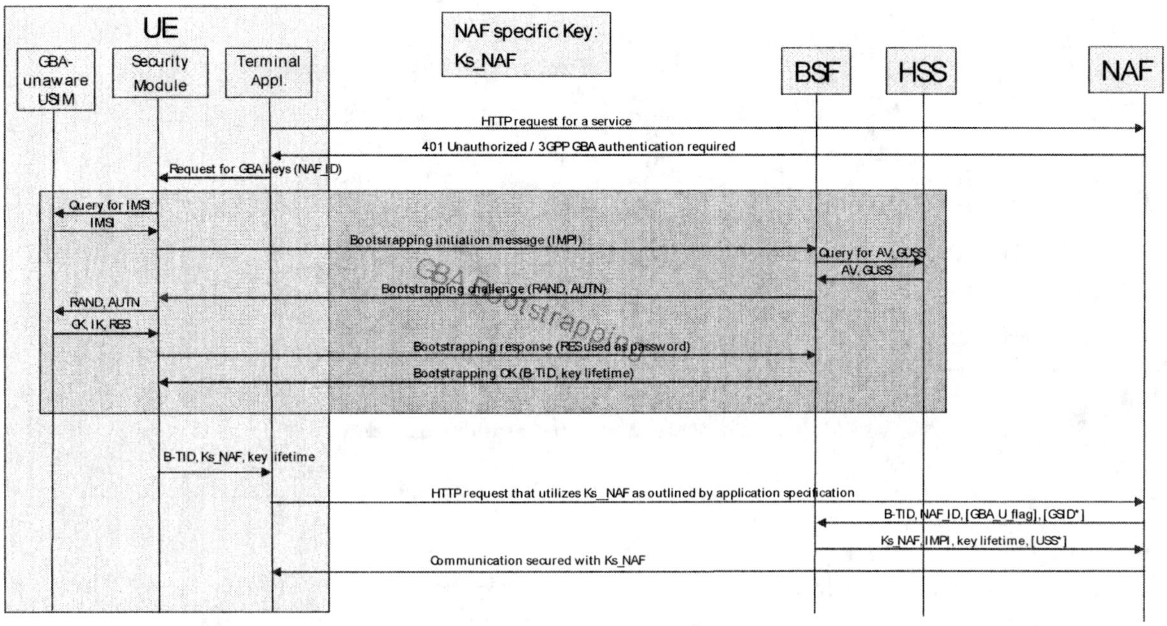

application-specific key(s) that they can use for authentication or other ways to secure the communication between terminal and application server.

In 3GPP there are basically three different ways to generate application-specific credentials in GBA. The previous basic steps are the same for all three types and should be seen as the basic GBA schemes. The variations were caused by different security requirements, business models, and market needs. The three GBA types are:

- **GBA_ME:** This is the terminal-based GBA. It requires a 3GPP UICC smart card, but the smart card does not need to be specially configured to support GBA_ME. The UICC can contain either an ISIM or a USIM application that can be used by GBA. The master session key and the application-specific keys are derived in the terminal. GBA_ME is also sometimes referred to as "normal GBA". Figure 2 outlines the message flow for GBA_ME. GBA_ME is defined in the 3GPP technical specification (3GPP TS 33.220, Release 6).
- **GBA_U:** GBA_U is a GBA type that requires a special smart card that supports GBA and is "GBA aware." The motivation for this was the cryptographic key generation is bound then very closely to the smart card and the issuing operator. The master key and the application-specific keys (Ks_ext_NAF and Ks_int_NAF) are derived in the GBA aware USIM application in an UICC. Only the Ks_ext_NAF application-specific session key is handed out to the terminal. A second application-specific session key Ks_int_NAF is not handed out and is stored and used only in the UICC. Figure 3 outlines the message flow for GBA_U. GBA_U is defined in (3GPP TS 33.220, Release 6). It should be noted, that the BSF modifies the AV received from the HSS. This gives an indication to the GBA-aware USIM, and the USIM returns just the RES and not the secret CK and IK keys to the terminal. The details on how those keys are used are defined by the application-specific documents, like MBMS (3GPP TS 33.246, 2006), HTTPS (3GPP TS 33.222, 2006) or

Figure 3. HTTP-based service request using GBA_U (GBA-aware USIM)

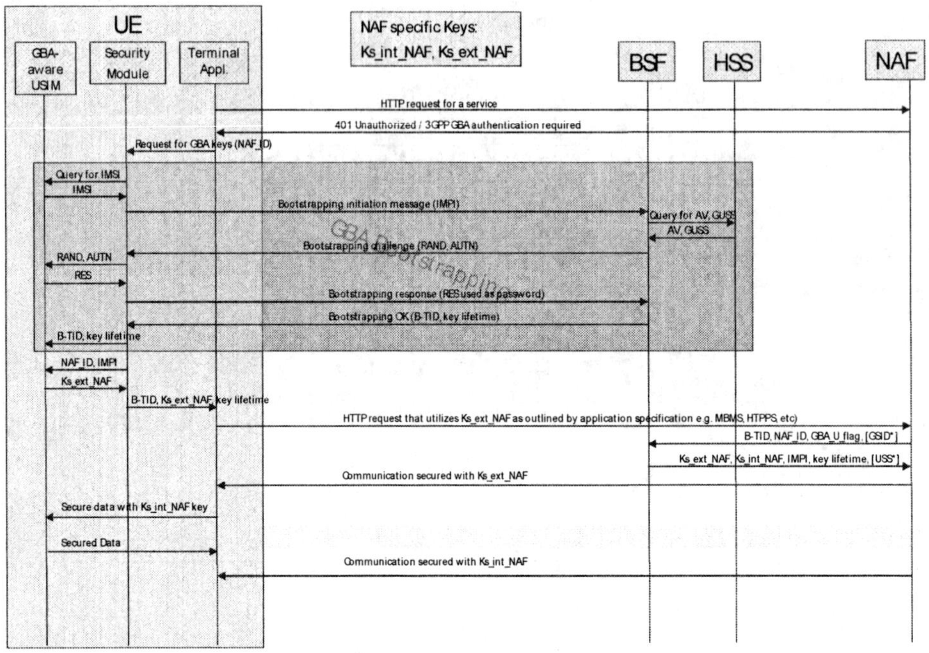

the OMA broadcast (BCAST) smart card profile.

- **2G GBA:** The 2G GBA or legacy GBA is a recent 3GPP GBA feature and defined in the technical report (3GPP TR 33.920, 2006) as an early implementation feature for Release 7. It outlines the usage of the SIM card for GBA. It should be noted, that it does not describe the usage of a legacy network nodes with GBA. The large deployment range of SIM cards created the need for a GBA credential generation solution that is based on legacy SIM cards and does not require immediate handing out of new UICC smart cards to the used. To obtain a similar security level than GBA_ME, the BSF node in the network is authenticated via a transport layer security (TLS) tunnel. The key derivation differs slightly, but the key usage is similar to GBA_ME. Figure 4 outlines the message flow for 2G GBA.

The notation for the Figures 2, 3, and 4 is that the * denotes an optional element.

In all, these three bootstrapping types have in common the basic steps outlined previously, and only the key generation and storage varies slightly. For the application server the usage of GBA_ME and 2G GBA is transparent. The convergence of fixed and mobile networks is, at the time of this writing, raising new GBA variants that will be discussed later in this chapter under Future Trends.

The specification family related to GBA has grown substantially due to new application requirements, further use cases, and new security enablers that were added. This will be outlined in the next section. The GBA can also be utilized to provision a user with a subscriber certificate and also trusted root certificate provisioning for public key infrastructure (PKI) systems. These are outlined in the 3GPP technical specification (3GGP TS 33.221, Release 6).

The term GBA refers typically to the core of GBA, where a master key is established between the mobile terminal (UE), and the network (BSF). Generic authentication architecture (GAA) on the other hand refers typically to the actual usage of the service specific keys that have been derived

Figure 4. HTTP based service request using 2G GBA

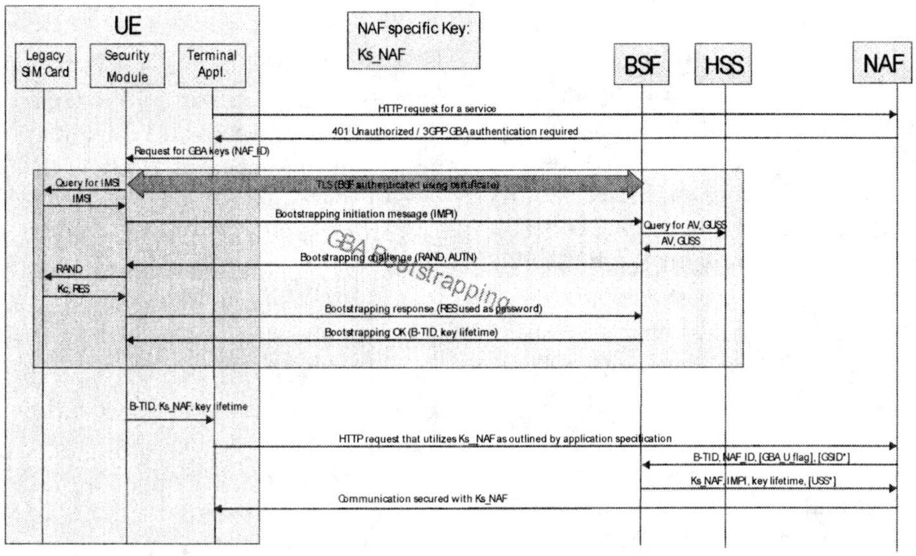

Figure 5. Generic authentication/bootstrapping architecture

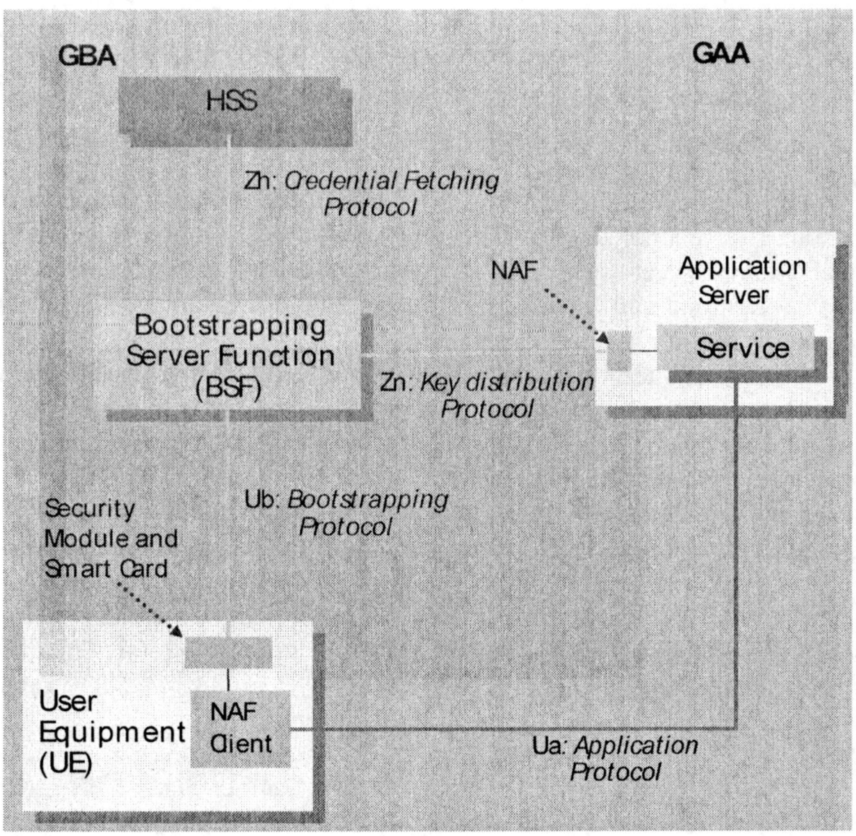

from the master key. Thus, GBA refers to the core functionality and GAA to the actual usage of GBA in use cases, as depicted in Figure 5, but often it is not necessary to differentiate strictly.

GAA and GBA are not only evolving in 3GPP, but also in the American counterpart standardization organization the 3GPP2 (http://www.3gpp2.org/). 3GPP2 utilizes the removable user identity module (R-UIM) as a security baseline for their dialect of GBA. 3GPP2 GBA supports the 3GPP2 legacy algorithms Cellular Authentication and Voice Encryption (CAVE) algorithm, which is used in the American CDMA1x, standard and challenge handshake authentication protocol (CHAP), which is used in American code division multiple access (CDMA) 1xEvDo (evolution data only), but also AKA for the user authentication. For further details on 3GPP2 GBA, please consult the relevant specification (3GPP2 TS S.S0109-0, 2006).

APPLICATION SECURITY BASED ON THE GENERIC BOOTSTRAPPING ARCHITECTURE

In the beginning, GBA was developed to securely provide the user with subscriber certificates where the initial registration of the user to public key interface (PKI) system is authenticated using cellular authentication. The function to provide an application-specific shared secret based on the mobile credentials to a terminal and a network node evolved to a generic enabler for many use cases and service. In this context, terminal refers to 3GPP or 3GPP2 mobile phone.

GBA is not only used for a large range of applications that reside in a mobile network, but also for fixed broadband access security and their access devices (e.g. PC or laptop). The work on GBA for the next 3GPP Release 8 and the integration of GBA into future networks B3G will be discussed in the next section. In this section we outline the different existing applications that use GBA as a security enabler. We will not go into the details of each application, but focus on the usage of GBA.

Mobile Networks Applications Using GBA

GBA was initiated by 3GPP; hence the first applications that utilize GBA were also from 3GPP in their Release 6 and 7. The first service to mandate the usage of GBA is the 3GPP mobile broadcast/multicast service (MBMS) (3GPP TS 33.246, Release 6). The broadcast scenario poses some very special requirements on a key derivation and management system, that is, a content provider specific key that can be linked to a mobile user identity stored on the smart card, protection of the content protection keys (key confidentiality during transport), and the baseline security key should not be transported over the air. This resulted in the fact that MBMS has a quite sophisticated four layer key hierarchy, where the user-specific keys are established using GBA.

Another use case is general authenticated Web browsing. A user browses to a Web page that needs authentication. This is a quite common occurrence in the Internet and there a user typically then has to provide a username/password combination. Inserting a password on a mobile key pad is not very user friendly and would likely result in non-secure passwords, that is, without special characters, very short, no upper/lower case combinations. Many security solutions ignore the usability aspect and try to force the user, which usually results in more or less expensive password recovery systems. In the mobile environment, with a small key pad, inserting long, secure passwords with special characters is not user friendly. The integration of an automatic scheme that provides automatically an application-specific username/password pair to the browser request is therefore desirable for the mobile environment. From a user perspective, the authentication would either be seamless (i.e., the user does not even notice that this is ongoing) or it would be very similar to the user experience, where the password is stored by the browser. The technical side of the procedure runs as follows:

1. User contacts a service that requires HTTP digest authentication.

2. The service triggers the terminal to generate an application-specific shared secret. This is then established using GBA without further user interaction.
3. The transaction ID are put into the username field and the shared application-specific secret is put into the password field.
4. The data is validated and the user can access the service.

The details of this procedure can be found in the 3GPP specifications (3GPP TS 33.220, Release 6) and (3GPP TS 33.222, Release 6).

Web sites that request confidential data are often secured using TLS 1.0 or Secure Socket Layer (SSL) 3.0, which can be considered equivalent. GBA was integrated into the usage of TLS between a mobile terminal and an application server in 3GPP Release 6 (3GPP TS 33.222, Release 6). At the end of 2005 the Internet Engineering Task Force specified the usage of Pre-Shared Key TLS in the IETF (RFC 4279) (Eronen & Tschofenig, 2005). 3GPP integrated the PSK TLS, since pre-shared key computations are very suitable for low capability devices like mobile phones (3GPP TS 33.222, Release 6). It should also be noted, that PSK TLS can also be used with IETF Datagram TLS (Rescorla & Modadugu, 2006).

A user may access a service directly or through an authentication proxy (AP), that takes care of the authentication-related tasks on behalf of the actual application server. If an operator offers many services, then he/she may wish to deploy such an authentication proxy to centralize the user authentication task in one node. An AP is an HTTP reverse proxy which takes the role of the GBA NAF node (the application server) for the terminal. The AP handles the TLS security relation with the terminal and is the TLS end point. GBA is used to ensure for the application server that the service request is coming from an authorized user. The AP has the Zn interface towards the BSF and the Ua interface towards the terminal. When a HTTPS request is sent from the terminal to the application server that resides behind an AP, then the AP terminates the TLS tunnel and performs the terminal authentication. The AP proxies the HTTP requests received from UE to one or many application servers, depending on the request. The AP may add an assertion of identity of the subscriber for use by the application server, when the AP forwards the request from the terminal to the application server.

Operators can also utilize GBA for device management. For this use case, a device management server takes the role of a NAF and establishes a HTTPS tunnel to the UICC as outlined in the 3GPP technical report (3GPP TR 33.918, Release 7). Through this secure tunnel the device management information is then sent.

The European Telecommunications Standards Institute (ETSI) has a Smart Card Platform Group that has defined some use cases, like mobile banking and digital rights management (DRM), which require the existence of a secure channel between the terminal and the UICC smart card. They asked 3GPP to define the key management for this functionality. This was done based on GBA in the 3GPP technical specification (3GPP TS 33.110, Release 7) and is expected to be part of 3GPP Release 7. It remains to be seen which of the use cases defined by the smart card group will be implemented.

Network Agnostic Usage of GBA

GBA is also used outside of the classical mobile environment of 3GPP. The OMA defines bearer agnostic functionalities and services. Since authentication is in most cases bound to the bearer some specifications integrate the authentication of the underlying bearer and provide additional functionality for the case that another access type is used. GBA is used by the following OMA applications:

- **OMA broadcast smart card profile (BCAST)** (2007) defines the usage of a smart card profile for content protection using a four-layer key hierarchy based on GBA (similar to MBMS key hierarchy.
- In **OMA presence and availability working Group (PAG)** (2006) the content server relies on external authentication and authorization done for the presence sources that may reside

on the mobile terminal, and watcher nodes. For this authentication and authorization GBA as defined in 3GPP technical specification (3GPP TS 33.222, Release 6) can be used for that purpose, acting as an AP.

- **OMA secure user plane location (SUPL)** (2006) defines the usage of how the terminal can acquire the location of itself from the network, and this messaging between the terminal and the network can be optionally protected by GBA.
- **OMA XML document management (XDM) and OMA aggregation proxy** were specified by the OMA presence and availability working group (PAG) (2006). These specifications define mechanisms how terminals can manage XML documents in the network servers. The authentication can be optionally by based on GBA, and the authenticating node in the network can be either the XDM server itself, or it can be centralized using aggregation proxy, where all traffic to XDM servers is routed through the proxy.
- **OMA common security functions (CSF)** (OMA Security Working Group, 2005) defines a generic GBA Profile (GBAProfile) that acts as an enabler and that other OMA applications and enablers can use when they are defining the usage of GBA in them.

Another important standardization body, where GBA fits in is the Liberty Alliance Project. The Liberty Alliance Project enables identity federation (alias single sign-on) and Web service security. It is a non-mobile centric consortium that uses the provided user authentication, but does not specify the actual means of authentication and its context. This is left to the standardization bodies, which define the actual authentication method. 3GPP integrated their GBA to be used seamlessly with the Liberty Alliance Project Identity Federation Framework and the Web Service Framework. The details of this interworking are specified in 3GPP technical report (3GPP TR 33.980, Release 6).

These are only some examples of the possible usage of GBA outside of 3GPP, many non-standardized use cases are also enablers. GBA could be utilized for enterprise access or other use cases where a shared secret between a terminal and a network server is needed.

Fixed—Mobile Convergence and GBA

The term converging networks has become a key phrase in latest network evolution work. The trend to merge mobile and fixed network backend systems is caused by several factors:

- **Fewer and larger operators:** There is a general consolidation trend in the industry, which results in large, often international, operators. These operators often have a fixed network and a mobile network. For them it is important that they can use one backend to serve both access types.
- **New players:** The boundaries between technologies are vanishing, as voice over IP shows us. These new players appear and want to utilize the existing technology, but on the other hand want to preserve the investments into infrastructure. Especially, for fixed networks the investments are substantial. Multi-network devices are no longer future, but commercially available. This results in extensions to the existing "pure" mobile specific standards to integrate the new requirements and network types.
- **Seamless services:** The general mobility trend creates high user expectations, when something works with a fixed network, then it is also expected that is works seamlessly in a mobile environment. This can only be provided with a unified backend service system.

The fixed mobile convergence is focused around the IP multimedia subsystem (IMS), but GBA as a general security enabler for applications moved quickly into the scene. The most prominent drivers of mobile and fixed convergence outside of 3GPP are TISPAN and CableLabs.

The telecoms & Internet converged services & protocols for advanced networks (TISPAN) is a standardization body of the ETSI (n.d.). TISPAN focuses on fixed networks and migration from switched circuit networks to packet-based networks with an architecture that can serve in both. TISPAN shares the same IMS specified by 3GPP. The idea is to keep the unity of IMS preserved by specifying the IMS core in 3GPP and TISPAN specific add-ons in TISPAN or 3GPP, depending on the working groups' agreements.

The TISPAN Release 1 which was finalized in 2006 is based upon the 3GPP IMS Release 6 and selected aspects of Release 7 architecture (which finishes early 2007). TISPAN standardizes functionalities or brings them into 3GPP for present and future converged networks, including the next generation networks (NGN). TISPAN utilizes GBA in their security architecture (ETSI TS 187.003, 2006) for the protection of the XML configuration access protocol (XCAP) traffic between a terminal and an application server using 3GPP TS 33.222 (2006) over their Ut interface. Optionally, an AP can be integrated into this interface.

CableLabs (http://www.cablelabs.com/) is a consortium of cable network providers with focus on Northern America. They are interested in utilizing the 3GPP specifications in their systems. In November 2006 a work item description that outlines the intended convergence security work was approved in 3GPP security group document number (3GPP Work Item Description S3-060764, 2006) with the title "IMS Enhancements for Security Requirements in Support of Cable Deployments." This work item proposes, among other extensions, also CableLabs specific extension to the 3GPP GBA "core" specification TS33.220 (Release 6). The outlined extensions include the possibility to bootstrap a shared secret from a username/password. In other words, the user would authenticate in a first step with his/her username/password. This pair would then serve as a baseline for further application-specific key generation. The user would not need to remember a new username/password pair for every service, but would potentially just have one CableLab username/password pair that could also be centrally managed. The details of this are work in progress and the final key derivation methods and further details still have to be fully defined in the near future (i.e., during year 2008).

FUTURE TRENDS AND GENERIC AUTHENTICATION IN BEYOND 3G NETWORKS

The wider deployment of adoption of GBA by the OMA created the need for a solution that an application server can trigger the key generation and send the needed key generation data (not the keys) to the mobile terminal. If this functionality is available it would allow that the continuous service provisioning and a broader usage of GBA-based functionalities for broadcast use cases. 3GPP started working on a Technical Specification GBA Credential Push Specification (3GPP TS 33.223, Release 8) that allows an application server (NAF) to trigger key generation procedure at the network side and push the required data to the terminal. Upon receiving this data the terminal would generate the keys without having a channel back to the network. The specification work for has started in 3GPP, and is targeted to be part of Release 8..

GBA has also been utilized to establish a shared secret between two entities. The main element in this use case is so-called NAF key center, and it can be used establish shared secret between either between the UICC and the ME (3GPP TS 33.110 Release 7) or between two devices where one of the devices is holding the UICC, and the other is not (3GPP TS 33.259, Release 7). This shared secret can then be used to secure the link between those two entities. The secure communication can then be used for any kind of application, for example, streaming.

The work on future networks after 3G has started in 3GPP. The work on the definition of a security framework for the core network system architecture evolution (SAE) and for the radio access network long term evolution (LTE) is ongoing. Sometimes, this work is also referred to as fourth generation (4G) or B3G networks. The work there assumes that there will be larger range of networks and that mobility between them is a key requirement. GBA was chosen as an enabler for mobile IP security in the setting that the related home agent (3GPP HA) resides in the 3G network. The SAE/LTE work is just evolving, so further usage of GBA and additions and modifications can be expected with the progress of the work. On a high level, the basic trust relationship between the Mobile IP communication partners defines the needed security associations independently of the actual protocol version used.

There is the trust relationship between the terminal and the 3GPP authentication, authorization and accounting (AAA) server that resides in the user's home network and is in charge of the user authentication (e.g., using AKA) and authorization. This trust relationship is founded on the user's subscription to his/her home network and secured via a shared secret that can be assumed to be long-lived. The mobile IP authentication is independent of the access authentication, which is analogous to the case, where a user uses a service and requires authentication there. Hence, GBA could be could be used for mobile IP key provisioning.

The second trust relationship is between the 3GPP Mobile IP (MIP) HA and the user's terminal, so that the HA can act on behalf of the terminal for the tasks related to mobility. The relationship between these two entities is established dynamically (in the sense that there is no pre-provisioned shared secret) so the integrity of the MIP signaling can be ensured and depends on the actual mobile IP version used, that is, Mobile IP4 or Mobile IP6 (or DS-MIPv6). 3GPP has at the point of writing only made the decision for Mobile IP4. The decisions if MIPv6 or DS-MIPv6 will be used are not yet taken in 3GPP (status December 2006).

The third trust relationship is between the 3GPP MIP HA and the 3GPP AAA server. The trust between those nodes is high, since they are part of the same network for non-roaming case. For non-roaming cases there exist interoperator security protocols, like network domain security (NDS)/IP securityor IPsec. This trust relationship does not require GBA, since there is no user involvement.

CONCLUSION

The GBA allows secure provisioning of a shared secret to a mobile terminal and an application server based on cellular authentication. This shared secret can then be used for many purposes, like username/password authentication, certificate enrollment, DRM, and so forth. GBA was originally designed by the 3GPP, but has recently been taken up for long term evolution networks, fixed broadband access, and cable networks.

ACKNOWLEDGMENT

Part of this work has been performed in the framework of the IST project System Engineering for Security and Dependability SERENITY and the Service Platform for Innovative Communication Environment (SPICE) project. The authors would like to acknowledge the contributions and review of their colleagues from Nokia Corporation.

REFERENCES

3rd Generation Partnership Project 2 (3GPP2) TS S.S0109-0. (2006). *Generic bootstrapping architecture (GBA) framework, version 1.0.* Retrieved from http://www.3gpp2.org/Public_html/specs/S.S0109-0_v1.0_060331.pdf

3rd Generation Partnership Project (3GPP) Work Item Description S3-060764. (2006, November). *IMS enhancements for security requirements in support of cable deployments.* Retrieved from http://www.3gpp.org/ftp/tsg_sa/WG3_Security/TSGS3_45_Ashburn/Docs/

3rd Generation Partnership Project (3GPP) TS 24.109. (Release 6). *Bootstrapping interface (Ub) and network application function interface (Ua); Protocol details.* Retrieved from http://www.3gpp.org/ftp/Specs/html-info/24109.htm

3rd Generation Partnership Project (3GPP) TS 29.109. (Release 6). *Generic authentication architecture (GAA); Zh and Zn interfaces based on the Diameter protocol; Stage 3.* Retrieved from http://www.3gpp.org/ftp/Specs/html-info/29109.htm

3rd Generation Partnership Project (3GPP) TS 33.110. (Release 8). *Key establishment between*

UICC and a terminal. Retrieved from http://www.3gpp.org/ftp/Specs/html-info/33110.htm

3rd Generation Partnership Project (3GPP) TS 33.203. (Release 7). *3G security; Access security for IP-based services.* Retrieved from http://www.3gpp.org/ftp/Specs/html-info/33203.htm

3rd Generation Partnership Project (3GPP) TS 33.220. (Release 6). *Generic authentication architecture (GAA); Generic bootstrapping architecture.* Retrieved from http://www.3gpp.org/ftp/Specs/html-info/33220.htm

3rd Generation Partnership Project (3GPP) TS 33.221. (Release 6). *Generic authentication architecture (GAA); Support for subscriber certificates.* Retrieved from http://www.3gpp.org/ftp/Specs/html-info/33221.htm

3rd Generation Partnership Project (3GPP) TS 33.222. (Release 6). *Generic authentication architecture (GAA); Access to network application functions using hypertext transfer protocol over transport layer security (HTTPS).* Retrieved from http://www.3gpp.org/ftp/Specs/html-info/33222.htm

3rd Generation Partnership Project (3GPP) TS 33.223. (Release 8). *Generic authentication architecture (GAA); Generic bootstrapping architecture (GBA) push function.* Retrieved from http://www.3gpp.org/ftp/Specs/html-info/33223.htm

3rd Generation Partnership Project (3GPP) TS 33.246. (Release 6). *3G security, security of multimedia broadcast/multicast service (MBMS).* Retrieved from http://www.3gpp.org/ftp/Specs/html-info/33246.htm

3rd Generation Partnership Project (3GPP) TR 33.918. (Release 7). *Generic authentication architecture (GAA); Early implementation of hypertext transfer protocol over transport layer security (HTTPS) connection between a universal integrated circuit card (UICC) and a network application function (NAF).* Retrieved from http://www.3gpp.org/ftp/Specs/html-info/33918.htm

3rd Generation Partnership Project (3GPP) TR 33.920. (Release 7). *SIM card based generic bootstrapping architecture (GBA); Early implementation feature.* Retrieved from http://www.3gpp.org/ftp/Specs/html-info/33920.htm

3rd Generation Partnership Project (3GPP) TR 33.978. (Release 6). *Security aspects of early IP multimedia subsystems (IMS), version 6.5.0.* Retrieved from http://www.3gpp.org/ftp/Specs/html-info/33978.htm

Calhoun, P., Loughney, J., Guttman, E., Zorn, G., & Arkko, J. (2003). *Diameter base protocol* (RFC 3588). Retrieved from http://www.ietf.org/rfc/rfc3588.txt

Eronen, P., & Tschofenig, H. (Eds). (2005). *Pre-shared key ciphersuites for transport layer security (TLS)* (RFC 4279). Retrieved from http://www.ietf.org/rfc/rfc4279.txt

European Telecommunications Standards Institute (ETSI). *Telecoms & Internet converged services & protocols for advanced networks (TISPAN).* Retrieved from http://www.etsi.org/tispan

European Telecommunications Standards Institute (ETSI) TS 187 003. (2006). *Telecoms & Internet converged services & protocols for advanced networks (TISPAN). NGN security—Security architecture, version 1.1.1.* Retrieved from http://www.etsi.org/tispan

Gerstenberger, V., Lahaije, P., & Schuba, M. (2004). Internet ID—Flexible re-use of mobile phone authentication security for service access. In *Proceedings of the 9th (NordSec)*, Helsinki, Finland (pp. 58-64).

Open Mobile Alliance (OMA) BCAST Working Group. (2006). *Broadcast service and content protection for mobile broadcast services, version 1.0.* Retrieved from http://www.openmobilealliance.org/

Open Mobile Alliance (OMA) Location Working Group. (2006). *Secure user plane location architecture (SUPL), version 3.0.* Retrieved from http://www.openmobilealliance.org/

Open Mobile Alliance (OMA) Presence and Availability Working Group (PAG). (2006). *Presence SIMPLE architecture, version 2.0*. Retrieved from http://www.openmobilealliance.org/

Open Mobile Alliance (OMA) Presence and Availability Working Group (PAG). (2006). *XML document management architecture (XDM), version 1.0*. Retrieved from http://www.openmobilealliance.org/

Open Mobile Alliance (OMA) Security Working Group. (2005). *OMA GBA profile, version 1.0*. Retrieved from http://www.openmobilealliance.org/

Rescorla, E., & Modadugu, N. (2006). *Datagram transport layer security* (RFC 4347). Retrieved from http://www.ietf.org/rfc/rfc4347.txt

KEY TERMS

Application Security: Application security encompasses a large range of measures taken to prevent incidents with respect to the security policy of an application or the underlying framework. Application security is realized through design and deployment of the application.

Authentication And Key Agreement (AKA): AKA is a mechanism where a mobile device and mobile network operator authenticate and distribute shared key(s) to be used between them. This process is based on a long-term shared secret that is in the mobile terminal (namely in UICC, e.g., SIM card), and mobile network operators databases (e.g., Home Location Register [HLR]). GBA is based on this process.

Authentication: Authentication is the attempt to verify the digital identity of the sender of an authentication request.

Cellular Authentication: Cellular authentication is the authentication process that is used when a mobile phone is attached to a network (e.g., GSM or UMTS network). This authentication is based on a smart card that is inserted in the mobile phone.

Generic Authentication Architecture (GAA): GAA is an architecture that is built on top of GBA that utilizes the shared secret to gain access to service.

Generic Bootstrapping Architecture (GBA): GBA is an architecture where cellular authentication is used to bootstrap a shared secret between a mobile phone and a network node.

Mobile Application: Mobile application is an application that resides on a server and can be accessed or consumed by a mobile device. The application may require a dedicated software element in the mobile terminal (e.g., for mobile TV).

Second Generation Generic Bootstrapping Architecture (2G GBA): 2G GBA describes the usage of the GBA with legacy SIM smart cards. It does not contain the integration of legacy network nodes.

Universal Integraged Circuit Card (UICC): UICC is the smart card (e.g., SIM card) used in mobile terminals in GSM and UMTS networks.

Chapter XXV
Authentication, Authorization, and Accounting (AAA) Framework in Network Mobility (NEMO) Environments

Sangheon Pack
Korea University, South Korea

Sungmin Baek
Seoul National University, South Korea

Taekyoung Kwon
Seoul National University, South Korea

Yanghee Choi
Seoul National University, South Korea

ABSTRACT

Network mobility (NEMO) enables seamless and ubiquitous Internet access while on-board vehicles. Even though the Internet Engineering Task Force (IETF) has standardized the NEMO basic support protocol as a network layer mobility solution, little studies have been conducted in the area of authentication, authorization, and accounting (AAA) framework that is a key technology for successful deployment. In this article, we first review the existing AAA protocols and analyze their suitability in NEMO environments. After that, we propose a localized AAA framework to retain the mobility transparency as the NEMO basic support protocol and to reduce the signaling cost incurred in the AAA procedures. The proposed AAA framework supports mutual authentication and prevents various threats such as replay attack, man-in-the-middle attack, and key exposure. Performance analysis on the AAA signaling cost is carried out. Numerical results demonstrate that the proposed AAA framework is efficient under different NEMO environments.

INTRODUCTION

With the advances of wireless access technologies (e.g., third generation [3G], IEEE 802.11/16/20) and mobile communication services, the demand for Internet access in mobile vehicles such as trains, buses, and ships is constantly increasing (Ott & Kutscher, 2004). In these vehicles, there are multiple devices constituting a vehicular area network (VAN) or personal area network (PAN) that may access to Internet. This kind of services is referred to network mobility (NEMO) services. Recently, many studies have been conducted for network mobility (Information Society Technologies [IST], 2003; Keio University, 2002). Regarding mobility management, the Internet Engineering Task Force (IETF) has established a working group called NEMO (IETF, 2006) and the NEMO working group has proposed an extended Mobile IPv6 protocol (Johnson, Perkins, & Arkko, 2003), that is, the NEMO basic support protocol (Devarapalli, Wakikawa, Petrescu, & Thubert, 2005). Throughout this chapter, we consider the NEMO basic support protocol as a mobility management framework.

According to the terminologies in Ernst and Lach, 2005), a mobile network (MONET) is defined as a network whose point of attachment to the Internet varies as it moves about. A MONET consists of mobile routers (MRs) and mobile network nodes (MNNs). Each MONET has a home network to which its home address belongs. When the MONET is in the home network, the MONET is identified by its home address (HoA). On the other hand, the MONET configures a care-of-address (CoA) on the egress link when the MONET is away from the home network. At the same time, on the ingress link, the MNNs of the MONET configure CoAs, which are derived from the subnet prefix (i.e., mobile network prefix [MNP]). The MNP remains assigned to the MONET while it is away from the home network. The assigned MNP is registered with the home agent (HA) according to the NEMO basic support protocol.

The main objective of the NEMO basic support protocol is to preserve established communications between the MONET and correspondent nodes (CNs) during movements. Packets sent by CNs are first addressed to the home network of the MONET. Then, the HA intercepts the packets and tunnels them to the MR's registered address, that is, the CoA on the egress link. To deliver packets towards the MR's CoA, the NEMO basic support protocol makes a bi-directional tunnel between the HA and the MR. This tunneling mechanism is similar to the solution proposed for host mobility support, that is, Mobile IPv6 without route optimization.

To make network mobility services feasible in public wireless Internet, well-defined authentication, authorization, and accounting (AAA) protocols should be accompanied. However, to the best of our knowledge, little work has been conducted for AAA protocols in network mobility services. Even though a number of AAA protocols have been proposed for host mobility, all of them are based on per-node AAA operations and therefore they cannot be directly applied to the MONET containing two different types MNNs: local fixed nodes (LFNs) and visiting mobile nodes (VMNs). An LFN belongs to the subnet to the MR and is unable to change its point of attachment, while a VMN is temporarily attached to the MR's subnet by obtaining its CoA from the MNP. The VMN's home network may have different administrative policy (e.g., billing) from the current attached MONET. Therefore, a new AAA procedure for VMNs is required.

In this chapter, we propose a localized AAA protocol that provides efficient AAA procedures for both LFNs and VMNs in NEMO environments. The proposed AAA protocol is consistent with the NEMO basic support protocol. In other words, individual AAA operations for LFNs within a MONET are not performed; instead, the MR is authenticated on behalf of the LFNs. On the other hand, each VMN attached to the MONET performs its AAA operation in an individual manner. The proposed AAA protocol has the following advantages: (1) the proposed AAA protocol localizes the AAA procedure using a local AAA key when the MR hands off within the same foreign network. Therefore, the AAA signaling traffic (also, the AAA latency) can be significantly reduced. We analyze the AAA signaling traffic via an analytical model in the

Signaling Cost Analysis section; (2) the proposed AAA protocol allows mutual authentication and prevents various security attacks such as replay attack and man-in-the-middle attack. The security analysis is given in the fourth section; (3) from the point of view of Internet service providers (ISPs), how to charge a VMN for its network usage is a critical issue. The proposed AAA protocol supports a flexible billing mechanism in which the VMN is informed of a billing agreement between the MR's home network and the new foreign network. Accordingly, the proposed AAA protocol is a suitable solution when the MONET hands off between different networks with different billing or service policies.

The remainder of this chapter is organized as follows. In the next section, an existing AAA protocol for Mobile IPv6 is introduced as a reference protocol. The third section proposes a localized AAA protocol and the fourth section analyzes the security of the proposed AAA protocol. In the fifth section, an analytical model for the AAA signaling cost is developed and numerical results are presented, respectively. The sixth section concludes this chapter.

BACKGROUND

In this section, the AAA protocol in Mobile IPv6 is described as a reference model. Although several AAA protocols have been proposed in the literature, we adopt the Diameter extension for Mobile IPv6 protocol (Le, Patil, Perkins, & Faccin, 2004) because it is the only valid IETF Internet draft as of this writing. The Diameter extension for Mobile IPv6 allows a mobile node (MN) to access a network of a service provider after the AAA procedures based on the Diameter protocol (Calhoun, Loughney, Guttman, Zorn, & Arkko, 2003) is completed.

This protocol considers a network architecture for AAA services, as shown in Figure 1. The AAAv is an AAA server in the visited (foreign) network, while the AAAh is an AAA server in the home network of the MN. Hereafter, we assume that the AAA client is located at each access router (AR). The AAA client performs three tasks: (1) allowing the MN to be authenticated, (2) generating accounting data for the MN's network usage, and (3) authorizing the MN to use network resources. By Le et al. (2004) an MN is identified by its network access identifier (NAI) (Aboda & Beadles,

Figure 1. Mobile IPv6 AAA architecture

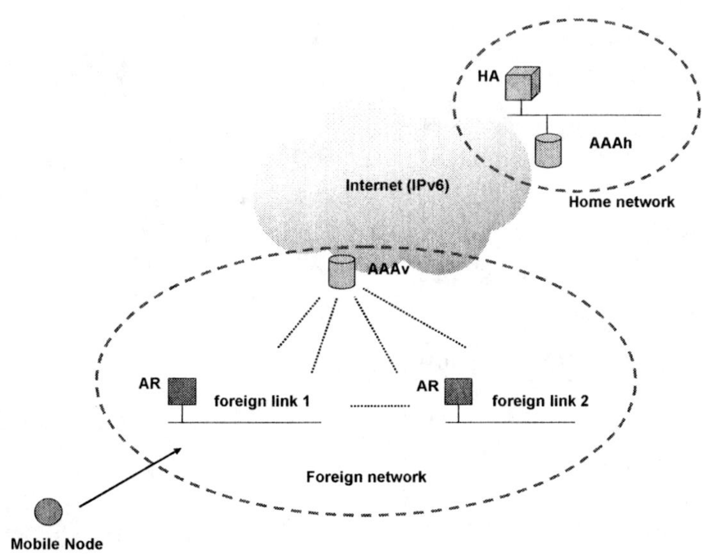

1999), which is globally unique. An MN and its AAAh have a long-term key, and communication between the AAAv and AAAh is secure.

The message flow in the Diameter extension for Mobile IPv6 is illustrated in Figure 2. When entering a new network or at power up, an MN listens to an AR's router advertisement (RA) message which has a local challenge and a visited network identifier. Then, the MN sends an authentication request (AReq) message to the AAA client (i.e., AR) based on the security key shared with its AAAh. When the AAA client receives the AReq message, it creates an AA-Registration-Request Command (ARR) message and sends it to the AAAv. Then, the AAAv relays it to the AAAh of the MN. When receiving the ARR message from the AAAv, the AAAh authenticates the MN by means of the NAI and sends a Home-Agent-MIPv6-Request Command (HOR) message to the MN's HA. Upon receipt of the HOR message, the HA creates a key to establish a security association (SA) with the MN, and replies with a Home-Agent-MIPv6-Answer Command (HOA) message to the AAAh. Then, the AAAh constructs the AA-Registration-Answer Command (ARA) message that has an authentication result and sends it to the AAAv. When receiving the ARA message from the AAAh, the AAAv stores the authentication result locally and then forwards the message to the AAA client. The AAA client converts the ARA message into the authentication reply (ARep) message, in order to inform the MN of the authentication result from the AAAh and deliver the established key (for the SA) to the MN.

Figure 2. Message flow in the AAA protocol for Mobile IPv6

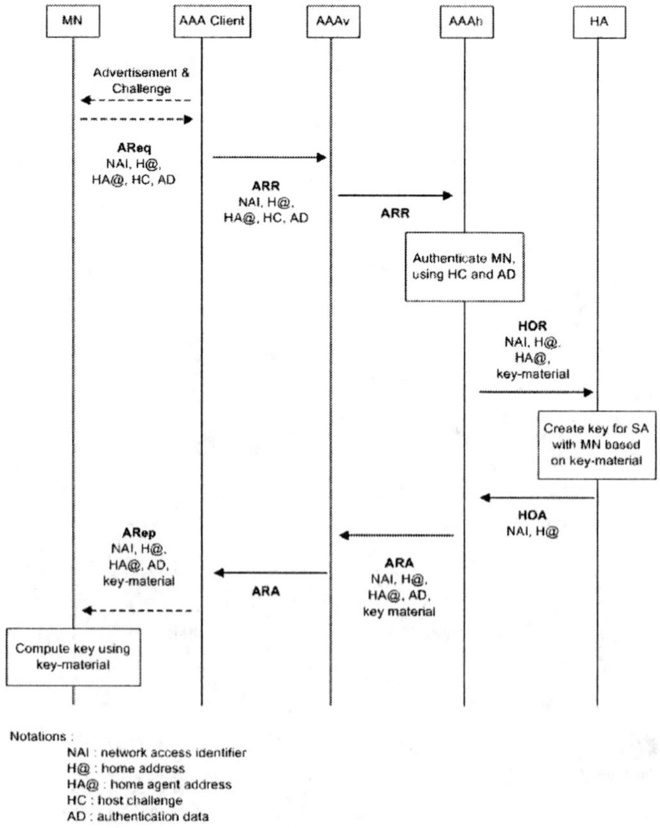

LOCALIZED AAA FRAMEWORK IN NEMO ENVIRONMENTS

System Architecture

In this section, the AAA architecture in NEMO environments is introduced with basic assumptions and concepts (e.g., SA and challenge/response authentication). Figure 3 illustrates the reference AAA architecture in NEMO environments based on the Diameter protocol.

The AAA architecture consists of multiple autonomous wireless networks, each of which is called a domain. Each domain has an AAAH server and/or an AAAL server in order to authenticate any node in a Diameter-compliant manner. The AAAH server of the MR has the profile of the MR and it shares a long-term key with the MR. Likewise, the AAAH server of the VMN shares a long-term key with the VMN. The AAAL server is in charge of an AAA procedure for a visiting MONET (i.e., VMNs and MRs). The trust relationship between the MR's AAAH server and the AAAL server in the visited network is maintained through the Diameter protocol.

When the MONET changes its point of attachment, the MR needs to be authenticated and authorized before it accesses a new domain in the same foreign network (i.e., intra-domain handoff) or a new foreign network (i.e., inter-domain handoff). To accomplish this, the MR and AR authenticate each other through a mutual authentication procedure that involves both the AAAH server of the MR and the AAAL server of the AR. An attendant (which is the same as an AAA client) is an entity that triggers authentication procedures to the AAA system. In Mobile IPv6 networks, ARs normally act as the attendants for an MN. In the proposed AAA protocol, the AR serves as an attendant for the MR's authentication, whereas the MR serves as an attendant for VMN's authentication. In the latter case, the MR broadcasts attendant advertisement messages and receives authentication request messages from VMNs within a MONET. In other words, an attendant (an AR or MR) requests the AAAL server to authenticate the MONET (the MR or VMN). When the AAAL server receives the authentication request, it verifies the identity of the MONET by cooperating with an AAAH server. In terms of SAs, we assume that the MR's AAAH server and the VMN's AAAH server have

Figure 3. AAA architecture in NEMO environments

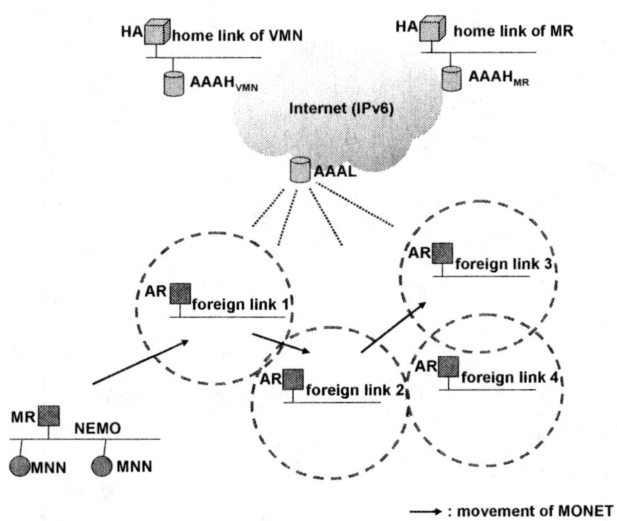

a pre-established SA. In addition, it is assumed that the MR and LFNs have already authenticated each other by a mechanism, which is beyond scope of this chapter.

Notations used in this chapter are summarized in Table 1. A local challenge (LC) is a random number for authentication procedures. An MR or VMN encrypts the LC using a pre-defined SA with its AAAH server. The encrypted value is called a credential (CR), which is used to authenticate an MR that creates it. MRs and VMNs are identified by their NAIs and a replay protection indicator (RPI), which is used to protect from a replay attack. Either a timestamp or a random number can be used as an RPI. The size of the K_{AAA} field is 128 bytes by assuming a public key cryptography algorithm. We adopt a symmetric key cryptography for dynamic keys K_{LOCAL} and K_{HOME}, and their sizes are 32 bytes. Note that a dynamic key is used to establish a dynamic SA while a long-term key is to establish a long-term SA. Other notations will be elaborated later.

In the proposed AAA protocol, we define two Internet Control Message Protocol (ICMP) messages (Conta & Deering, 1998), Attendant Solicit and Attendant Advertisement messages, which are similar to Router Solicit and Router Advertisement messages, respectively. In these messages, we introduce a new Attendant advertisement option and it is used for the authentication of VMNs for an intra-domain handoff. In addition, several Diameter messages, for examples, AA-Mobile-Router-Request and AA-Mobile-Router-Answer, are defined. Their functions will be described later.

Table 1. Notations for the localized AAA protocol

Field	Meaning	Typical Length (bytes)
LC	local challenge	8
MC	mobile challenge	8
NAI	identity of MR or VMN	20
RPI	replay protection indicator	4
H@	home address	16
HA@	home agent address	16
Co@	care of address of MR or VMN	16
K_{AAA}	pre-shared SA between an MR and an AAAH server	128 (public key)
K_{AH}	pre-shared SA between an AAAH server and an HA	128 (public key)
K_{AL}	pre-shared SA between an AAAH server and an AAAL server	128 (public key)
CR	credential	8
CR_L	local credential	8
K_{LOCAL}	dynamic SA between an MR and an AAAL server	32 (symmetric key)
CR_M	mobile credential	8
K_{HOME}	dynamic SA between an MR and its AAAH server	32 (symmetric key)
SP_{LOCAL}	security parameters for constructing K_{LOCAL}	12
SP_{HOME}	security parameters for constructing K_{HOME}	12

Mobile Router Authentication

Inter-Domain AAA Procedure

When a MONET enters a new foreign network domain, an inter-domain AAA procedure is triggered. Since the MR does not have any SA with the AAAL server in the foreign network domain, it should be authenticated with its AAAH server located in its home network domain. The message flows for the inter-domain AAA procedure are illustrated in Figure 4 and the detailed descriptions are as follows:

- **Step 1:** The MR sends an Attendant Solicit message to the attendant, that is, AR.
- **Step 2**: As a response to the Attendant Solicit message, the AR sends an Attendant Advertisement message including an LC. Even without the Attendant Solicit message, the AR broadcasts Attendant Advertisement messages periodically.
- **Step 3:** The MR encrypts the received LC value using its long-term SA with the AAAH server and makes a CR, which enables the MR's AAAH server to authenticate the MR. Then, the MR sends an AReq message that contains the LC and CR to the AR (i.e., attendant). The AReq message also contains the MR's NAI and RPI, which are used for the AAAL server to identify the MR's home domain and to protect from replay attack.
- **Step 4:** When the AR receives the AReq message, it converts it into an AA-Mobile-Router-Request (AMR) message. After then, the AR sends the AMR message to the AAAL server in the foreign domain.
- **Step 5:** The AAAL server detects that it cannot authenticate the MR locally by checking the NAI field and hence forwards the AMR message to the MR's AAAH server. When the AAAH server receives the AMR message, it encrypts the LC using the pre-established SA and compares the result with the CR value. If these two values are identical, the MR is successfully authenticated. Then, the AAAH server generates two dynamic keys: one is a K_{LOCAL} (to be explained later) for intra-domain AAA procedures in the foreign domain and the other is a K_{HOME} for a secure bi-directional tunnel between the MR and the MR's HA. To allow the MR to generate K_{LOCAL} and K_{HOME}, the AAAH server also generates SP_{HOME} and SP_{LOCAL}, and sends them to the MR. These security parameters are encrypted using the long-term key between the MR and AAAH server to avoid the possibility of exposure to other network entities.
- **Step 6:** The AAAH server informs the HA of the MR's NAI and SP_{HOME} using the AA-Home-Agent-Request (AHR) message.
- **Step 7:** The HA constructs K_{HOME} by using SP_{HOME} and replies with an AA-Home-Agent-Answer (AHA) message as confirmation.
- **Step 8:** The AA-Mobile-Router-Answer (AMA) message is used for the AAAH server to notify the AAAL server of the authentication result. When the AAAL server receives the AMA message with authentication approval, the AAAL server decrypts the message using the long-term key (K_{AL}) with the AAAH server, records the MR's NAI, and constructs K_{LOCAL}.
- **Step 9:** The AAAL server re-encrypts the received AMA message from the AAAH server after excluding $E_{KAL}(SP_{LOCAL})$ and sends it to the AR.
- **Step 10:** When receiving the AMA message, the AR learns that the MR is successfully authenticated and grants the MR's network access. Therefore, the AR informs the MR of the result by the ARep message containing SP_{HOME}, SP_{LOCAL}, home agent address, and so forth. On receipt of the ARep message with authentication approval, the MR can access the foreign network. At the same time, the MR generates K_{HOME} and K_{LOCAL} using SP_{HOME} and SP_{LOCAL}, respectively.

Figure 4. MR's AAA procedure for inter-domain handoff

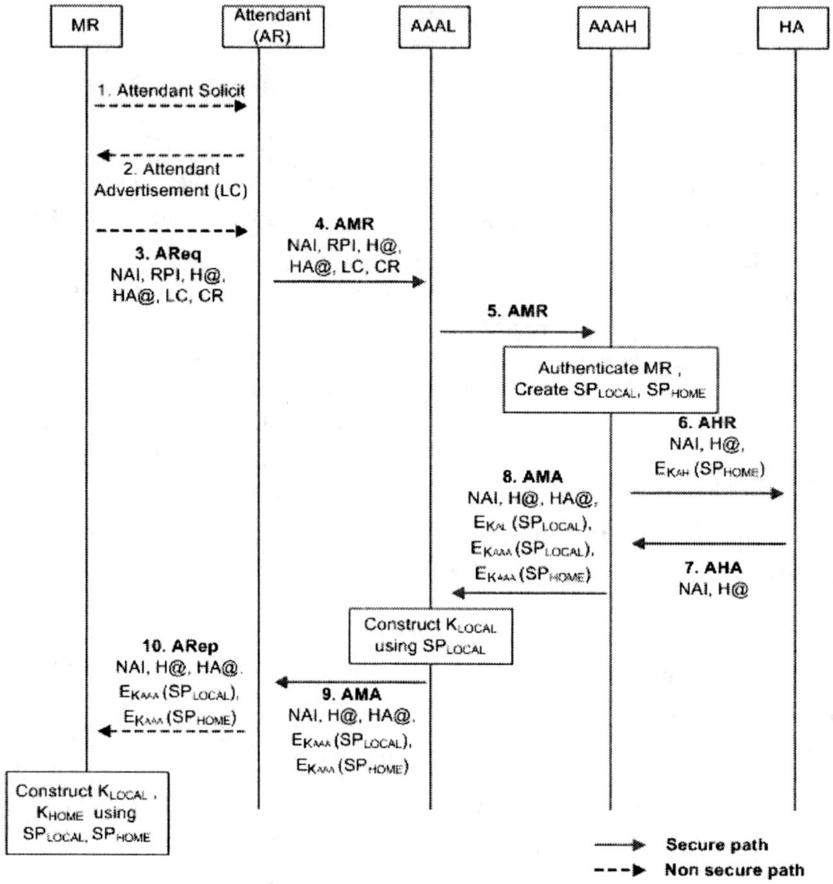

Intra-Domain AAA Procedure

To support real-time multimedia applications in NEMO environments, it is important to reduce the latency for AAA operations. Therefore, when a MONET changes its point of attachment within the same foreign domain, our protocol enables the MR to be authenticated through a localized AAA procedure with the AAAL server in the foreign network without any interaction with its AAAH server. That is, the AAAL server of the foreign network can authenticate the MR using K_{LOCAL}, which was introduced for the inter-domain AAA procedure in the previous section.

Figure 5 illustrates the intra-domain AAA procedure. As a response to the Attendant Advertisement message, the MR sends the AReq message containing CR_L, which is different from CR used in the inter-domain AAA procedure. At this time, the AReq message contains MC for mutual authentication. The CR_L is an authentication code generated using K_{LOCAL}. Then, the attendant constructs an AA-Mobile-Router-Local-Request (AMLR) Diameter message and sends it to the AAAL server. When the AAAL server receives the AMLR message, the AAAL server authenticates the MR by using K_{LOCAL}, which has been already stored at the AAAL server during the inter-domain AAA procedures. Moreover, the AAAL server constructs CR_M by encrypting the MC value and informs the AR of the result via the AA-Mobile-Router-Local-Answer (AMLA) message. Then, the AR transmits the result (i.e., the ARep message) to the MR. The MR receiving the ARep message verifies the CR_M value to authenticate the foreign network, that is, mutual authentication.

Figure 5. MR's AAA procedure for infra-domain handoff

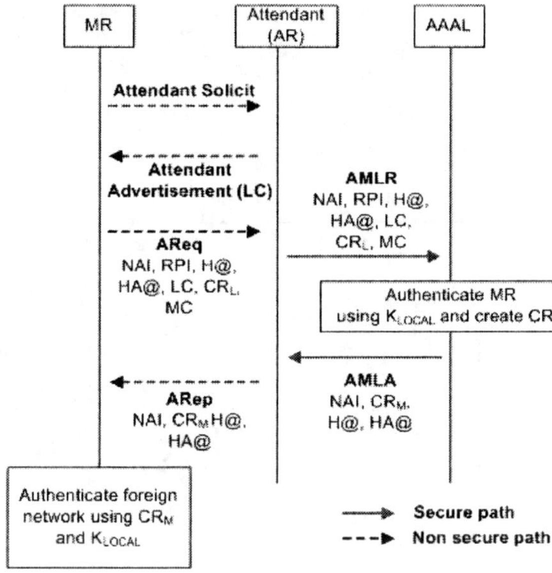

Visiting Mobile Node (VMN) Authentication

A VMN is a visiting MN that accesses the Internet through an MR in a MONET. According to the NEMO basic support protocol, the VMN does not need to know whether its attached router is the AR or the MR. Therefore, the AAA protocol for VMNs should be consistent with this requirement. The VMN in a MONET uses the home network prefix of the MR as its IPv6 network prefix. Accordingly, the VMN will deem it to be in the MR's home network. For VMN authentication, the MR serves as an attendant for VMNs and the MR's AAAH server serves as an AAAL server.

Figure 6 illustrates message flows for the AAA procedure when a VMN is attached to a MONET. As mentioned previously, the MR acts as an attendant. Hence, the MR broadcasts Attendant Advertisement messages periodically or responds to an Attendant Solicit message from the VMN with an Attendant Advertisement message. The VMN creates a CR using a pre-shared SA with its AAAH server ($AAAH_{VMN}$) and sends an AReq message to the MR. Then, the MR converts the AReq message into a Diameter message, AMR, and then sends it to the MR's AAAH server ($AAAH_{MR}$) through a secured bi-directional tunnel. When the $AAAH_{MR}$ receives the AMR message, it sends the AMR message to the $AAAH_{VMN}$ that has a shared SA and requests the AAA procedure for the VMN. Then, the $AAAH_{VMN}$ authenticates the VMN. During these steps, K_{HOME}, K_{LOCAL}, SP_{HOME}, and SP_{LOCAL} are created, which is similar to the inter-domain AAA procedure of the MR. After completion of AAA procedures, the VMN registers its CoA (configured using the MNP) with its HA.

After the initial authentication and binding update procedures, VMNs within a MONET do not need to know whether the MONET changes its point of attachment or not. Thus, VMNs do not have to register their locations to their HAs even though the MONET hands off. This mobility transparency is the key advantage of the NEMO basic support protocol. However, if the mobility transparency is strictly provided, the AAAL server in the foreign network cannot detect the existence of VMNs. In other words, the mobility transparency is beneficial to reduce the binding update traffic, however, it makes the accounting/billing of VMNs' network usages hard. To address this problem, in our protocol, the AAAL server in the

Figure 6. VMN's AAA procedure

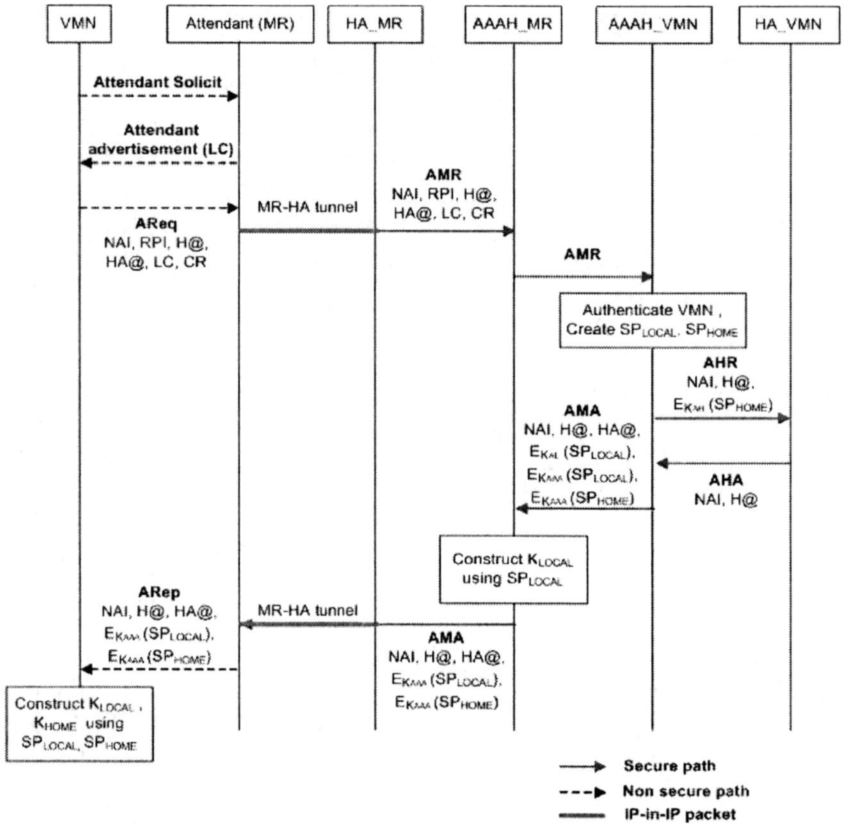

foreign domain accounts the total network usage of the MONET (not individual VMNs) and then this collective accounting/billing information is delivered to the MR's AAAH server. At the same time, the MR's AAAH server maintains the accounting/billing information for the MR as well as individual VMNs.[1] Consequently, the MR's AAAH server can differentiate the accounting/billing information for MRs and VMNs. In addition, we assume that the MR's AAAH server and the VMN's AAAH server have a trust relationship and a shared SA. Therefore, the accounting/billing information collected at the MR's AAAH server is securely transferred to the VMN's AAAH server for suitable billing.

In addition, the mobility transparency causes another problem, that is, how to authorize VMNs when the MONET moves to a foreign domain with a different billing policy. To solve this problem, an MR sends an Attendant Advertisement message with a set R bit when the foreign domain has a different policy and thus a new AAA procedure is required. Hence, from the Attendant Advertisement message, the VMN determines whether it should perform a new AAA procedure or not. We assume that each network domain can have different policies, so that the VMN performs a new AAA procedure for each inter-domain handoff.

SECURITY ANALYSIS

In this section, we analyze the proposed AAA protocol in terms of mutual authentication and security attacks (e.g., key exposure, replay attack, and man-in-the-middle attack).

Mutual Authentication

Mutual authentication is a security feature in which a client (i.e., the MR and VMN) must prove its identity to a service (i.e., network), and the service must prove its identity to the client. To provide mutual authentication in the NEMO AAA protocol, two requirements should be satisfied: (1) the MR or VMN authenticates the foreign network; and (2) the foreign network authenticates the MR or VMN.

Specifically, mutual authentication in the proposed AAA protocol is achieved as follows. First, for the inter-domain authentication, mutual authentication is provided by establishing a session key, K_{LOCAL}. In other words, the objective of inter-domain authentication protocol is that the MR and the AAAL server believe that they share K_{LOCAL} with each other. The MR creates CR as

$$CR = E_{K_{AAA}}(LC) \quad (1)$$

where $E_K(\cdot)$ is an encryption function using a key of K. The AAAH server can verify the MR's identity by comparing with CR sent by the MR with the CR constructed by the AAAH server itself. If two values are identical, the MR is successfully authenticated. Otherwise, the authentication fails. In our protocol, a malicious MR cannot create the correct CR because it does not have K_{AAA}. After verifying the identity of the MR, the AAAH server transmits $E_{K_{AAA}}(SP_{LOCAL})$ and $E_{K_{AL}}(SP_{LOCAL})$ to the AAAL server through a secure path. When the AAAL server receives, it constructs K_{LOCAL} using $E_{K_{AL}}(SP_{LOCAL})$ and forwards $E_{K_{AAA}}(SP_{LOCAL})$ to the MR. At last, the MR constructs K_{LOCAL} using $E_{K_{AAA}}(SP_{LOCAL})$. After this procedure, the MR and the AAAL server share K_{LOCAL}.

For the intra-domain authentication, the AAAL server in the foreign network verifies the identity of the MR by comparing $E_{K_{LOCAL}}(LC)$ constructed by the AAAL server with CR_L sent by the MR. On the other hand, to authenticate the foreign network, the MR uses an MC and CR_M. The AAAL server in the foreign network sends CR_M that is created by

$$CR_M = E_{K_{LOCAL}}(MC) \quad (2)$$

Then, the MR can authenticate the AAAL server in the foreign network by verifying that $E_{K_{LOCAL}}(MC)$ is equal to CR_M. Consequently, a malicious network cannot offer fake services to an MR because it cannot compute CR_M, and mutual authentication is achieved.

Key Exposure

K_{AAA} is a pre-shared key between an MR and the AAAH server, and K_{LOCAL} and K_{HOME} are created using security parameters, SP_{LOCAL} and SP_{HOME}, respectively. Thus, it is desirable not to leak these keys to the other network entities.

With respect to K_{LOCAL}, the AAAH server encrypts SP_{LOCAL} using K_{AL} and sends it the AAAL server. At the same time, the AAAH server send the encrypted SP_{LOCAL} using K_{AAA} to the MR. Therefore, the encrypted SP_{LOCAL} can be decrypted only by the AAAL server and the MR because they have K_{AL} or K_{AAA}. In other words, if K_{AL} and K_{AAA} are not exposed, any other entities except the AAAL server and the MR cannot know SP_{LOCAL} and thus cannot construct K_{LOCAL}. Similarly, SP_{HOME} is encrypted using K_{AH} and K_{AAA}, and delivered to the HA and MR, respectively. Therefore, K_{HOME} derived from SP_{HOME} is not revealed to other entities except the HA and MR.

Replay Attack

Replay attack involves the passive capture of data and its subsequent retransmission to produce an unauthorized effect. A malicious node keeps an AReq message and then it can retransmit an old AReq message to trick the AAAL server for false authentication. In our protocol, LC is created randomly and hence it always changes and therefore the malicious node cannot replay the old AReq message. Even though the same LC is selected by the attendant, RPI (i.e., timestamp) can prevent the replaying attack.

Man-in-the-Middle Attack

A man-in-the-middle attack represents that an attacker is able to read, insert, and modify messages

Figure 7. Man-in–the-middle attack by a malicious MR

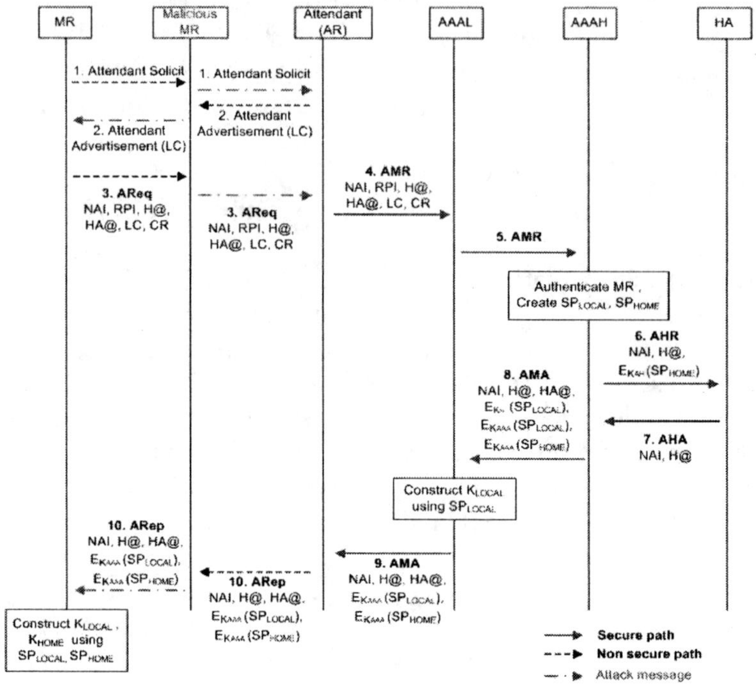

between two parties without either party knowing that the link between them has been compromised. In NEMO environments, we can imagine an attack that a malicious MR relays authentication messages and it intends to use network resource illegally. Figure 7 illustrates the man-in–the-middle attack by a malicious MR for the inter-domain authentication. The malicious MR acts as an AR and relays authentication messages between the victim MR and the AR. After the authentication procedures, the malicious MR still can relay all of the traffic between the victim MR and AR. However, the malicious MR cannot use any network resource because it has no knowledge of K_{LOCAL} and K_{HOME}. Namely, if a fresh session key is established, the malicious MR cannot further compromise the authentication procedure between the MR and the AAAL server.

SIGNALING COST ANALYSIS

Reducing the AAA traffic is an important requirement in NEMO environments where a MONET moves with a high velocity and AAA procedures are frequently performed (e.g., train or car). Therefore, through the analytical model, we quantify the AAA cost (C_{AAA}), which is defined as the volume of AAA-related messages delivered over the network and the unit of C_{AAA} is bytes * hops (Lo, Lee, Chen, & Liu, 2004).

Let i and j be the numbers of intra-domain handoffs and inter-domain handoffs for each session, respectively. It is assumed that the subnet residence time of the MONET follows a general distribution with mean $1/\mu_S$, which probability density function (PDF) is $f_S(t)$ and its Laplace transform is $f^*_S(s)$. In addition, the domain residence time of the MONET follows a general distribution with mean $1/\mu_D$, whose PDF is $f_D(t)$ and its Laplace transform is $f^*_D(s)$. When the inter-session arrival time is assumed to be an exponential distribution with rate λ_I, the PDFs of i and j are respectively given by (Lin, 1997)

$$\alpha(i) = \begin{cases} 1 - \frac{1}{\rho_S}[1 - f_S^*(\lambda_I)] & i = 0 \\ \frac{1}{\rho_S}[1 - f_S^*(\lambda_I)]^2 [f_S^*(\lambda_I)]^{i-1} & i > 0 \end{cases}$$

and

$$\beta(j) = \begin{cases} 1 - \frac{1}{\rho_D}[1 - f_D^*(\lambda_I)] & j = 0 \\ \frac{1}{\rho_D}[1 - f_D^*(\lambda_I)]^2 [f_D^*(\lambda_I)]^{j-1} & j > 0 \end{cases}$$

where $\rho_S = \lambda_I / \mu_S$ and $\rho_D = \lambda_I / \mu_D$.

Since an inter-domain handoff implies that an intra-domain handoff also occurs, i-j represents the number of pure intra-domain handoffs. Therefore, the AAA cost of the MR authentication in the proposed AAA protocol when there are i intra-domain handoffs and j inter-domain handoffs can be computed as

$$C_{AAA}^{MR}(i,j) = (i-j)\cdot C_{intra}^{MR} + j \cdot C_{inter}^{MR} \quad (3)$$

where C_{intra}^{MR} and C_{inter}^{MR} are the costs for intra-domain AAA and inter-domain AAA operations. On the other hand, the AAA cost of the MR authentication without the localized AAA procedure is given by

$$C_{AAA}^{MR}(i,j) = i \cdot C_{non-local}^{MR} \quad (4)$$

where $C_{non-local}^{MR}$ is the cost for an AAA operation without the localized AAA procedure. Then, the average AAA cost of the MR can be expressed as

$$C_{AAA}^{MR} = \sum_i \sum_j C_{AAA}^{MR}(i,j)\cdot\alpha(i)\cdot\beta(j) \quad (5)$$

For the VMN's AAA cost, we consider the AAA cost incurred during the VMN is attached to the MONET. We assume that the VMN's attachment time is drawn from an exponential distribution with mean $1/\eta_A$. Let k be the number of inter-domain handoffs during the attachment time. Then, the PDF of k is given by

Table 2. Parameters for numerical results

Wireless weight	Number of ARs in a domain	λ_1	ε_A	D_1	D_2	D_3
10	49	1	1	2, 5, 10	2, 5	2, 5

Table 3. Message length (bytes)

Attendant Solicit	Attendant advertisement	AReq	ARep	AMR
52	84	116	120	172
AHR	AHA	AMA	AMLR	AMLA
144	136	166	180	152

$$\gamma(k)=\begin{cases}1-\dfrac{1}{\rho_A}[1-f_D^*(\eta_A)] & k=0\\ \dfrac{1}{\rho_A}[1-f_D^*(\eta_A)]^2[f_D^*(\eta_A)]^{k-1} & k>0\end{cases}$$

where $\rho_A = \eta_A / \mu_D$. Hence, the AAA cost of the VMN when there are k inter-domain handoffs during the attachment time can be computed as

$$C_{AAA}^{VMN}(k) = k \cdot C_{AAA}^{VMN}, \qquad (6)$$

where C_{AAA}^{VMN} is the cost for each VMN's AAA operation. Consequently, the average AAA cost of the VMN is expressed as

$$C_{AAA}^{VMN} = \sum_k C_{AAA}^{VMN}(k)\cdot\gamma(k) \qquad (7)$$

In this section, we evaluate the effects of mobility and the distance between a foreign network and a home network on the AAA cost (i.e., C_{AAA}^{MR} and C_{AAA}^{VMN}). The parameters and the size of each AAA message are shown in Tables 2 and 3, respectively, which are based on Calhoun et al. (2003) and Narten, Nordmark, and Simpson (1998). D_1, D_2, and D_3 represent the distances between the AAAL server and the $AAAH_{MR}$ server, between the MR and its HA, and between the $AAAH_{MR}$ server and the $AAAH_{VMN}$ server, respectively. Then, C_{intra}^{MR}, C_{inter}^{MR}, and $C_{non-local}^{MR}$ can be calculated by multiplying the corresponding message length and the distance. For the transmission over a wireless link, a weight value is used and it is set to 10 (Xie & Akyildiz, 2002). The number of subnets in a domain is 49. λ_I and η_A are normalized to 1.0 whereas μ_D equals $\mu_S / \sqrt{N}$ by the fluid flow model (Zhang, Castellanos, & Campbell, 2002), where N is the number of subnets in a domain.

As shown in Figure 8, the proposed AAA protocol has a smaller AAA cost than the non-localized AAA protocol. Also, it can be seen that C_{AAA}^{MR} increases as μ_S increases (i.e., as the subnet residence time of the MONET decreases). This is because the number of inter- or intra-handoffs is reduced when the mobility (i.e., μ_S) is low. Figure 8 also indicates the AAA cost variation for different

Figure 8. The AAA cost of an MR

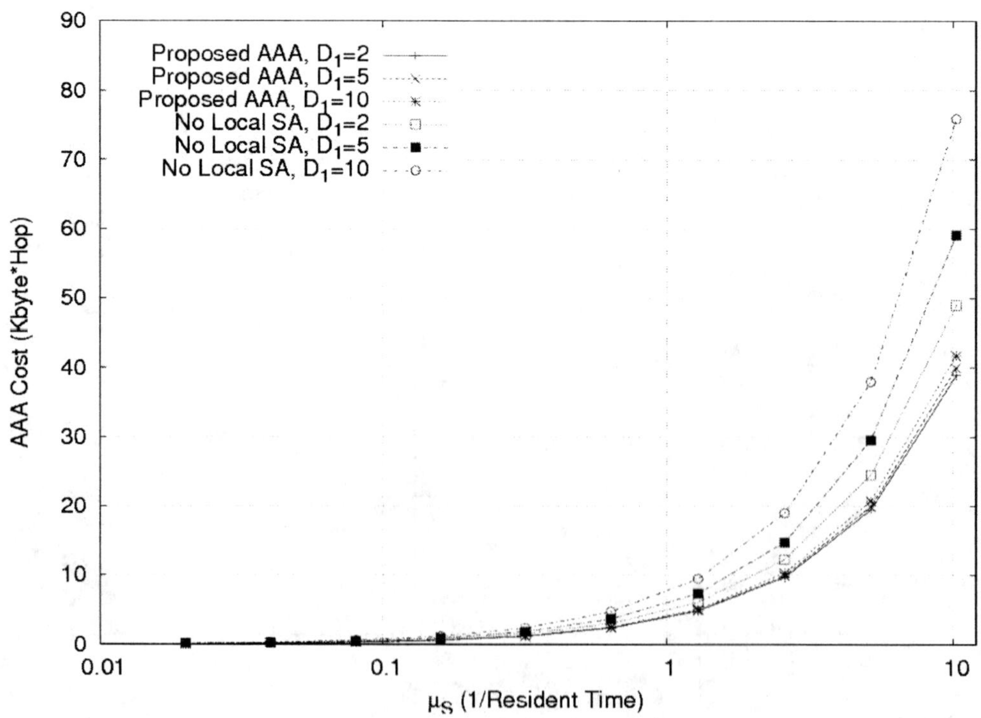

Figure 9. The AAA cost of a VMN

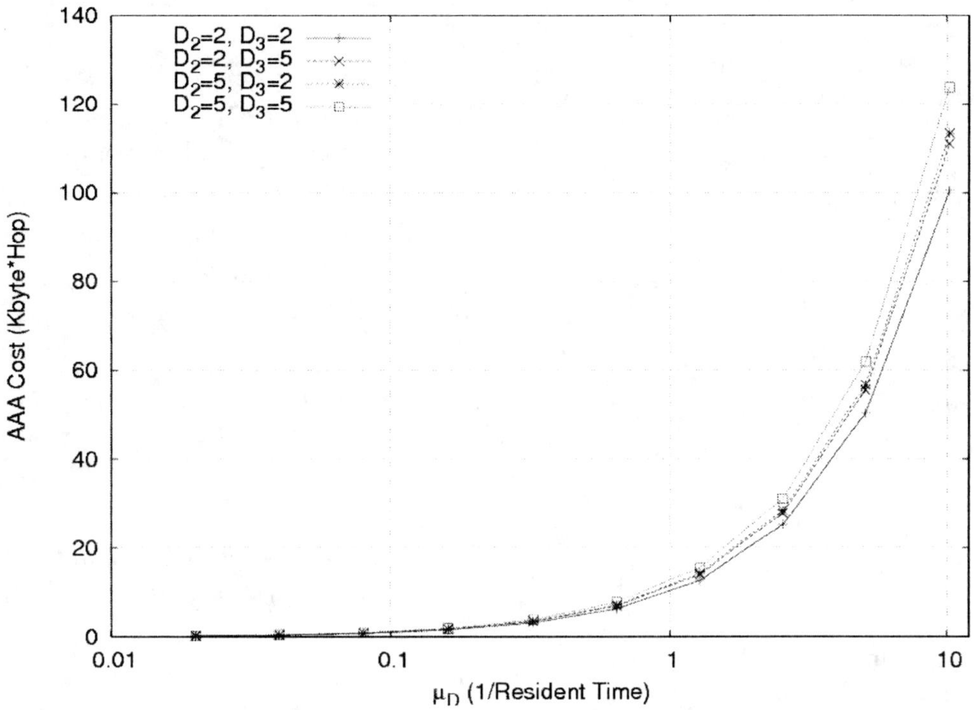

D_1 (i.e., D_1=2, 5, 10). Since C_{intra}^{MR} and C_{inter}^{MR} are proportional to D_1, C_{AAA}^{MR} increases with the increase of D_1 significantly. Especially, the effect of D_1 is more clear in the non-localized AAA protocol because an AAA procedure is always performed at the AAAH server in the non-localized AAA protocol. Therefore, it can be concluded that our protocol is more effective regardless of the distance between the home network and the foreign network. Note that this AAA cost considers only one MR. Hence, as the network mobility services are proliferated, the reduction of the AAA cost by the proposed AAA protocol will be more significant.

Figure 9 shows the AAA cost of the VMN, which exhibits a similar trend to Figure 8. It can be seen that the AAA cost when (D_2, D_3) is (5,2) is higher than the AAA cost of (2,5). This is due to IP-in-IP packet tunneling overhead between the MR and its HA. Namely, as the distance between the MR and its HA D_2 increases, more tunneling overheads incur and then the AAA cost also increases. As similar to Figure 9, the AAA cost of the VMN in our protocol is not highly affected by the distance values. Therefore, it is concluded that out protocol is less sensitive the distance between the home network and the foreign network.

CONCLUSION

In this chapter, we have proposed a localized AAA protocol in NEMO environments. The proposed AAA protocol is consistent with the NEMO basic support protocol where the mobility transparency is supported. The proposed AAA protocol introduces a shared key between the MR and the AAA server in the foreign network, so that the AAA procedure for the MR in intra-domain handoffs can be localized. In addition, we proposed a flexible billing mechanism for VMNs moving across different domains. We analyzed the security concerns in the proposed AAA protocol in terms of mutual authentication, key exposure, replay attack, and man-in-the-middle attack. Performance evaluation results reveal that the localized AAA procedure can reduce the AAA traffic significantly and the localized AAA procedure is less sensitive to the distance between the home network and the foreign network. Consequently, it is expected that the proposed AAA protocol can be widely employed in NEMO environments.

REFERENCES

Aboda, B., & Beadles, M. (1999). *The network access identifier* (RFC 2486). Retrieved from http://tools.ietf.org/html/rfc2486

Calhoun, P., Loughney, J., Guttman, E., Zorn, G., & Arkko, J. (2003). *Diameter base protocol* (RFC 3588). Retrieved from http://www.rfc-editor.org/rfc/rfc3588.txt

Conta, A., & Deering, S. (1998). *Internet control message protocol (ICMPv6) for the Internet protocol version 6 (IPv6)* (RFC 2463). Retrieved from http://www.faqs.org/rfcs/rfc2463.html

Devarapalli, V., Wakikawa, R., Petrescu, A., & Thubert, P. (2005). *Network mobility (NEMO) basic support protocol* (RFC 3963). Retrieved from http://www.ietf.org/rfc/rfc3963.txt

Ernst, T., & Lach, H. (2005). *Network mobility support terminology* (RFC 4885). Retrieved from http: www.ietf.org/rfc/rfc4885.txt

Information Society Technologies (IST). (2003). *Dynamic radio for IP-services in vehicular environments.* Retrieved from http://www.ist-overdrive.org

Internet Engineering Task Force (IETF). (2006). *Network mobility working group.* Retrieved from http://www.ietf.org/html.charters/nemo-charter.html

Johnson, D., Perkins, C., & Arkko, J. (2003). *Mobility support in IPv6* (RFC 3775). Retrieved from http://www.ietf.org/rfc/rfc3775.txt

Keio University. (2002). *InternetCar project.* Retrieved from http://www.sfc.wide.ad.jp/InternetCAR/

Le, F., Patil, B., Perkins, C., & Faccin, S. (2004). *Diameter mobile IPv6 application.* Internet Draft. Retrieved from http://tools.ietf.org/html/draft-ietf-aaa-diameter-mobileip-14

Lin, Y. (1997). Reducing location update cost in a PCS network. *IEEE/ACM Transactions on Networking, 5*(2), 25-33.

Lo, S., Lee, G., Chen, W., & Liu, J. (2004). Architecture for mobility and QoS support in all-IP wireless networks. *IEEE Journal on Selected Area on Communications, 22*(4), 691-705.

Narten, T., Nordmark, E., & Simpson, W. (1998). *Neighbor discovery for IP version 6 (IPv6)* (RFC 2461). Retrieved from http://www.ietf.org/rfc/rfc2461.txt

Ott, J., & Kutscher, D. (2004, March). *Drive-thru Internet: IEEE 802.11b for automobile users.* Paper presented at the IEEE International Conference of the IEEE Communication Society, Hong Kong, China.

Xie, J., & Akyildiz, I. (2002). A distributed dynamic regional location management scheme for mobile IP. *IEEE Transactions on Mobile Computing, 1*(3), 163-175.

Zhang, X., Castellanos, J., & Campbell, A. (2002). P-MIP: Paging extensions for mobile IP. *ACM Mobile Networks and Applications, 7*(2), 127-141.

KEY TERMS

Accounting: Accounting is the action of tracking the consumption of network resources by users.

Authentication: Authentication is the action of confirming that a user who is requesting services is a valid user of the network services requested.

Authorization: Authorization is the action of granting the specific types of service to a user depending on the authentication.

Internet Engineering Task Force (IETF): IETF is an organization to develop, promote, and standardize Internet-related protocols.

Man-in-the-middle Attack: Man-in-the-middle attack is an attack in which an attacker is able to read, insert, and modify messages between two communication parties.

Network Mobility: Network mobility is the mobility of an entire network that changes its point of attachment to the Internet as a single unit.

Replay Attack: Replay attack is an attack in which a valid data transmission is maliciously or fraudulently repeated or delayed.

END NOTE

1 In the NEMO basic support protocol, all packets destined to MNNs are tunneled at the MR's HA, so that the MR's HA can keep track of network usages of individual LFNs and VMNs. Therefore, the MR's HA can report this information to the AAAH server.

About the Contributors

Yan Zhang received the PhD degree in School of Electrical & Electronics Engineering, Nanyang Technological University, Singapore. From August 2004 to May 2006, he worked with the National Institute of Information and Communications Technology (NICT), Singapore. Since August 2006, he has worked with Simula Research Laboratory, Norway (http://www.simula.no/). He is on the editorial board of the *International Journal of Network Security*. He is currently serving as the Book Series Editor for the book series, *Wireless Networks and Mobile Communications* (Auerbach Publications, CRC Press, Taylor, and Francis Group). He is serving as co-editor for several books: *Resource, Mobility and Security Management in Wireless Networks and Mobile Communications; Wireless Mesh Networking: Architectures, Protocols and Standards; Millimeter-Wave Technology in Wireless PAN, LAN and MAN; Distributed Antenna Systems: Open Architecture for Future Wireless Communications; Security in Wireless Mesh Networks; Mobile WiMAX: Toward Broadband Wireless Metropolitan Area Networks; Wireless Quality-of-Service: Techniques, Standards and Applications; Broadband Mobile Multimedia: Techniques and Applications; Internet of Things: From RFID to the Next-Generation Pervasive Networked Systems; Unlicensed Mobile Access Technology: Protocols, Architectures, Security, Standards and Applications; Cooperative Wireless Communications; WiMAX Network Planning and Optimization; RFID Security: Techniques, Protocols and System-On-Chip Design; Autonomic Computing and Networking; Security in RFID and Sensor Networks;* and *Handbook of Research on Wireless Security*. He serves as industrial co-chair for MobiHoc 2008, program co-chair for UIC-08, general co-chair for CoNET 2007, general co-chair for WAMSNet 2007, workshop co-chair FGCN 2007, program vice co-chair for IEEE ISM 2007, publicity co-chair for UIC-07, publication chair for IEEE ISWCS 2007, program co-chair for IEEE PCAC'07, special track co-chair for "Mobility and Resource Management in Wireless/Mobile Networks" in ITNG 2007, special session co-organizer for "Wireless Mesh Networks" in PDCS 2006, and he is a member of Technical Program Committee for numerous international conference, including CCNC, AINA, GLOBECOM, ISWCS, ICC, and so forth. He received the Best Paper Award and Outstanding Service Award as Symposium Chair in the IEEE 21st International Conference on Advanced Information Networking and Applications (AINA-07). His research interests include resource, mobility, energy, and security management in wireless networks and mobile computing. He is a member of IEEE and IEEE ComSoc.

Jun Zheng received the BS and MS degrees in electrical engineering from Chongqing University, China, in 1993, 1996, respectively, the MSE degree in biomedical engineering from Wright State University, Dayton, Ohio, in 2001, and the PhD degree in computer engineering from University of Nevada, Las Vegas in 2005. Currently he is an assistant professor in the Department of Computer Science at Queens

College of The City University of New York. He is also a member of the faculty of the doctoral program in computer science at the Graduate School and University Center of The City University of New York. He is the co-editor for two books: *Security in Wireless Mesh Networks* and *Handbook of Research on Wireless Security*. He served as general co-chair for WAMSNet-07, track co-chair for ITNG 2007, and session co-organizer for PDCS 2006. He also served as TPC member for several international conferences. His research interests are mobility and resource management in wireless and mobile networks, media access control, performance evaluation, network security, computer architectures, fault-tolerant computing, and image processing. He is member of IEEE.

Miao Ma received the BEng. and MEng. degrees in electrical engineering from Harbin Institute of Technology, China, respectively, and the PhD degree in electrical and electronic engineering from Nanyang Technological University (NTU), Singapore. From August 2002 to December 2006, she worked at the Institute for Infocomm Research (I2R), Singapore. Since January 2007, she has been working at the Hong Kong University of Science and Technology (HKUST). She is a member of IEEE. Her research interests include media access control, cognitive radio, security, wireless communications, and networking.

* * * * *

Sasan Adibi received his BSc degree from Amirkabir University, Tehran, Iran, in 1996, first MSc degree from Brunel University, London in 1999, and second MSc degree from University of Windsor, Canada, in 2005. He is currently studying towards his PhD degree at the University of Waterloo and working as a contractor for Siemens Canada as a Verification System's Engineer. His areas of research include security (ad hoc networks, 3G, WiMAX, and Wi-Fi), quality of service (QoS) for MPLS- and wireless-based systems, and optimizations in ad hoc networks. His work experiences include extensive verification in routing (MPLS, OSPF, BGP), quality of service features (DiffServ, 802.11e, WMM, etc.), security and authentication (WPA v2, OKC, 802.1X, 802.11i, and RADIUS), Wi-Fi (802.11a/b/g), and VLAN (802.1D, 802.1Q, and 802.1p).

Gordon B. Agnew received his BASc and PhD in electrical engineering from the University of Waterloo in 1978 and 1982, respectively. He joined the Department of Electrical and Computer Engineering at the University of Waterloo in 1982. In 1984, he was a visiting professor at the Swiss Federal Institute of Technology in Zurich where he started his work on cryptography. Dr. Agnew's areas of expertise include cryptography, data security, protocols and protocol analysis, electronic commerce systems, high-speed networks, wireless systems, and computer architecture. He has taught many university courses and industry sponsored short courses in these areas as well as having authored many articles. In 1985, he joined the Data Encryption Group at the University of Waterloo. The work of this group led to significant advances in the area of public key cryptographic systems including the development of a practical implementation of elliptic curve-based cryptosystems. Dr. Agnew is a member of the Institute for Electrical and Electronics Engineers, a foundation fellow of the Institute for Combinatorics and its Applications, and a registered professional engineer in the Province of Ontario. Dr. Agnew has provided consulting services to the banking, communications, and government sectors. He is also a co-founder of CERTICOM Corp., a world leader in public key cryptosystem technologies.

Sheikh Iqbal Ahamed is an assistant professor in the Department of Mathematics, Statistics, and Computer Science at Marquette University and director of the Ubicomp Research Lab. His research focuses on pervasive security, trust, and privacy for pervasive computing. He received the PhD in computer science from Arizona State University and the BS from the Bangladesh University of Engineering and Technology.

Christer Andersson is a doctoral student at Karlstad University, Sweden, and his main research topic is designing and evaluating technologies for anonymous communication in mobile networks. He has proposed anonymity technologies for both infrastructured and infrastructureless mobile networks. He is furthermore interested in measuring the degree of anonymity and performance in mobile networks, as well as finding an appropriate trade-off between anonymity and performance. He holds a Licentiate in engineering degree from Karlstad University (2005), and a Master Degree in computer science (2002) from Linköping University, Sweden.

AbdelBaset M.H. Awawdeh received the BEng degree in industrial automation engineering from Palestine Polytechnic University in 1999 and the MSE and PhD degrees in electronics engineering from Alcala University in 2004. From 1999 to 2000 he joined the Department of Electrical and Computer Engineering at Palestine Polytechnic University, Palestine. Since 2004, he has held a researcher position in the Department of Electronics at University of Alcala, Spain. His technical interests include multiagent system interaction and design, vehicles on-board electronics, and vehicles fault diagnosis system.

Mohamad Badra is employed by the CNRS (National Center for Scientific Research, France) researching wireless networks security. Badra performs his research activities at the LIMOS Laboratory - UMR6158, University Blaise Pascal. He was a postdoctoral fellow at the Computer Science and Networks Department, ENST-Paris. His research interests include key exchange, wireless security, public-key infrastructures, smart cards, and security algorithms. Badra received a PhD in networks and computer sciences from ENST-Paris. He is a member of the IEEE and of ESRGroups.

Sungmin Baek received his BS degree in School of Information and Communication Engineering, Sung Kyun Kwan University in 2004 and an MS degree from Department of Computer Engineering and School of Computer Science and Engineering, Seoul National University in 2006. Currently, he is a research engineer in information and technology laboratory, LG Electronics Institute of Technology. His research interests include multimedia transmission over wireless network and wireless personal area networks.

Javier A. Barria received the BSc degree in electronic engineering from the University of Chile, Santiago, in 1980, and the PhD and MBA degrees from Imperial College London in 1992 and 1998, respectively. From 1981 to 1993, he was a system engineer and project manager (network operations) with the Chilean Telecommunications Company. Currently, he is a reader in the Intelligent Systems and Networks Group, Department of Electrical and Electronic Engineering, Imperial College London. His research interests include communication networks monitoring strategies using signal and image processing techniques, distributed resource allocation in dynamic topology networks, and fair and efficient resource allocation in IP environments. He has been a joint holder of several FP5 and FP6 European Union project contracts all concerned with aspects of communication systems design and management. Dr. Barria was a British Telecom Research Fellow (2001 – 2002) and a Tan Chin Tuan Research Fellow, NTU Singapore (2003 - 2004). He is a fellow member of IEE, member of IEEE, and a chartered engineer.

Paolo Bellavista graduated from University of Bologna, Italy, where he received PhD degree in computer science engineering in 2001. He is now an associate professor of computer engineering at the University of Bologna. His research activities span from mobile agent-based middleware solutions and pervasive wireless computing to location/context-aware services and adaptive multimedia. He is member of IEEE and Italian Association for Computing (AICA). He is an associate technical editor of the *IEEE Communication Magazine*.

Soong Boon-Hee received his BEng (honors I) degree in electrical and electronic engineering from University of Auckland, New Zealand and a PhD degree from the University of Newcastle, Australia in 1984 and 1990, respectively. He is currently an associate professor with the School of Electrical and Electronic Engineering, Nanyang Technological University. From October 1999 to April 2000, he was a visiting research fellow at the Department of Electrical and Electronic Engineering, Imperial College, UK under the Commonwealth Fellowship Award. He was awarded the Tan Chin Tuan Fellowship to visit the Centre for Advanced Computing and Communications, Duke University in June 2004. He also served as a consultant for mobile IP in a recent technical field trial of next-generation wireless LAN initiated by IDA (InfoComm Development Authority, Singapore). His area of research interests includes mobile ad hoc and sensor networks, mobility issues, mobile IP, optimization of wireless networks, routing algorithms, optimization and planning of mobile communication networks, queuing theory system theory, quality of service issues in high-speed networks, and signal processing. He has served as a reviewer for a number of IEEE top journals and international conferences. He has served on Technical Program Committee for IEEE Globecom 2004, 2005, 2006, and 2007, IEEE WCNC 2005, and IEEE ISWCS 2004, 2005, 2006, and 2007. He is currently organizing co-chair of IEEE Vehicular Technology Conference, Spring 2008 to be held in Singapore. He is currently on the technical committee ISO 204/WG16 that tracks developments in the intelligent transport sector. He is listed in Marquis Who's Who in Science and Engineering 2006-2007. He has published more than one hundred international journals and conferences. He is a senior member of IEEE and a member of ACM.

Noureddine Boudriga is a professor of telecommunication and director of the Communication Networks and Security (CN&S) Research Laboratory at the University of November 7th at Carthage, Tunisia. His research interests have covered several topics including communication networks, network engineering, optical networks, wireless networks, Internet work, and network security. He received his PhD in mathematics from the University of Paris XI and his PhD in computer science from the University of Tunis. Professor Boudriga is the recipient of the Presidential Award for Science and Research in Communication Technologies (Tunisia, 2004).

John Buford is a research scientist with Avaya Labs. Previously he was a lead scientist at the Panasonic Princeton Laboratory, VP of Software Development at Kada Systems, director of Internet Technologies at Verizon, and chief architect-OSS at GTE, Laboratories. Earlier he was tenured associate professor of computer science at the University of Massachusetts Lowell, where he also directed the Distributed Multimedia Systems Laboratory. He has authored or co-authored ninety refereed publications and the book *Multimedia Systems*. He is an IEEE senior member and is co-chair of the IRTF Scalable Adaptive Multicast Research Group. He holds the PhD from Graz University of Technology, Austria, and MS and BS degrees from MIT.

Mihaela Cardei is an assistant professor in the Department of Computer Science and Engineering at Florida Atlantic University, and director of the NSF-funded Wireless and Sensor Network Laboratory. Dr. Cardei received her PhD and MS in computer science from the University of Minnesota, Twin Cities, in 2003 and 1999, respectively. Her research interests include wireless networking, wireless sensor networks, network protocol and algorithm design, and resource management in computer networks. Dr. Cardei is a recipient of the 2007 Researcher of the Year Award at Florida Atlantic University. She is a member of IEEE and ACM.

Luca Caviglione (M.D. 2002, Ph.D. 2006) participated in several research projects funded by the EU, by ESA, and by Siemens COM AG. He is author and co-author of many academic publications about TCP/IP networking, P2P systems, QoS architectures, and wireless networks. He is a member of the Italian IPv6 Task Force and he participates in several TPCs and performance talks about IPv6 and P2P. He is with the Institute of Intelligent Systems for Automation (ISSIA) – Genoa Branch of the National Research Council of Italy.

Symeon Chatzinotas has a BSc in electrical and computer engineering from Aristotle University of Thessaloniki and a MSc in microwave engineering and wireless subsystem design from University of Surrey. Since 2006 he has been working on his PhD at the Centre for Communication Systems Research, University of Surrey. His current research interests include mobile networking, wireless security, and network information theory.

Hsiao-Hwa Chen is currently a full professor in Institute of Communications Engineering, National Sun Yat-Sen University, Taiwan. He received BSc and MSc degrees with the highest honor from Zhejiang University, China, and a PhD degree from the University of Oulu, Finland, in 1982, 1985 and 1990, respectively, all in electrical engineering. He worked with Academy of Finland as a Research Associate from 1991 to 1993 and the National University of Singapore as a lecturer and then a senior lecturer from 1992 to 1997. He joined Department of Electrical Engineering, National Chung Hsing University, Taiwan, as an associate professor in 1997 and was promoted to a full professor in 2000. In 2001 he joined National Sun Yat-Sen University, Tai-Wan, as a founding director of the Institute of Communications Engineering of the University. Under his strong leadership the institute was ranked second in the country in terms of SCI journal publications and National Science Council funding per faculty member in 2004. In particular, National Sun Yat-Sen University was ranked first in the world in terms of the number of SCI journal publications in wireless LANs research papers during 2004 to mid-2005, according to a research report (www.onr.navy.mil/scitech/special/354/technowatch/textmine.asp) released by The Office of Navel Research, USA. He was a visiting professor to the Department of Electrical Engineering, University of Kaiserslautern, Germany, in 1999, the Institute of Applied Physics, Tsukuba University, Japan, in 2000, Institute of Experimental Mathematics, University of Essen, Germany in 2002 (under DFG Fellowship), the Chinese University of Hong Kong in 2004, and the City University of Hong Kong in 2007.

Thomas Chen is an associate professor at Southern Methodist University. Prior to joining SMU, he worked on ATM research at GTE Laboratories (now Verizon). He has been the editor-in-chief of *IEEE Communications Magazine* since 2006. He also serves as senior technical editor for *IEEE Network*, and was the founding editor of *IEEE Communications Surveys*. He co-authored *ATM Switching Systems* (Artech House 1995). He received the IEEE Communications Society's Fred W. Ellersick Best Paper Award in 1996.

Yifan Chen received the BEng and PhD degrees in electrical and electronic engineering from Nanyang Technological University (NTU), Singapore, in 2002 and 2006, respectively. He is presently with the Biomedical Engineering Research Centre, NTU, as a research fellow. His current research interests involve ultra-wideband (UWB) radar system for biomedical applications including microwave imaging of human tissues and noncontact vital-signs monitoring, statistical modeling of mobile radio channels, UWB signal processing for wireless communications and geolocation systems, multiple-antenna system performance analysis, and wireless networks.

Zhijia Chen is currently a PhD student in Department of Computer Science and Technology, Tsinghua University. He is a visiting graduate student at School of Engineering of Stanford in Spring 2007. His research area is in P2P networking and media streaming. He has published four academic papers in area of P2P streaming, protocol modeling, and so forth. He is the International First Prize winner in American Mathematical Contest in Modeling (MCM 2004 Meritorious Winners). He is also the network session chair in 1st Beijing-Hong Kong Doctoral Forum on Network and Media 2006.

Yanghee Choi received BS in electronics engineering from Seoul National University, MS in electrical engineering from Korea Advanced Institute of Science, and Doctor of Engineering in computer science from Ecole Nationale Superieure des Telecommunications (ENST) in Paris, in 1975, 1977, and 1984, respectively. He was with the Electronics and Telecommunications Research Institute (ETRI) during 1977-1991. He is now leading the Multimedia and Mobile Communications Laboratory in Seoul National University. He is also director of Computer Network Research Center in Institute of Computer Technology (ICT). He is vice-president of Korea Information Science Society. His research interest lies in the field of multimedia systems and high-speed networking.

Mohammad M. R. Chowdhury is working toward the PhD degree in the University Graduate Center at Kjeller (UniK)/University of Oslo, Norway in the area of user mobility and service continuity. He received his MSc from Helsinki University of Technology in radio communications. His current areas of interest are identity and identity based service interactions, seamless user experience in heterogeneous wireless networks, and development of innovative service concepts for mobile operators.

Tomasz Ciszkowski received MSc degree in electronics and computer engineering from Faculty of Electronics and Information Technology of Warsaw University of Technology (WUT), Poland, in 2004. Currently, he is working toward a PhD degree in telecommunications at WUT on reputation service in anonymous ad hoc networks. Since 2004 he has been working for Polish Telecom in multimedia services division. His research activities are reflected in European research projects on next generation Internet (EuroNGI) and end-to-end QoS support over heterogeneous networks (EuQoS).

Amitabha Das obtained his BTech (honors) degree in electronics and electrical communication engineering from the Indian Institute of Technology, Kharagpur in 1985, and his PhD in computer engineering from the University of California, Santa Barbara, in 1991. Currently he is an associate professor in the School of Computer Engineering in Nanyang Technological University, Singapore. His research interests include wireless and mobile networks, network security, and intrusion detection. He is a senior member of IEEE.

Robert H. Deng received his Bachelor from National University of Defense Technology, China, and his MSc and PhD from the Illinois Institute of Technology. He has been with the Singapore Management University since 2004, and is currently a professor, associate dean for Faculty & Research, and director of SIS Research Center, School of Information Systems. Prior to this, he was principal scientist and manager of Infocomm Security Department, Institute for Infocomm Research, Singapore. He has 26 patents and more than 200 technical publications in international conferences and journals in the areas of computer networks, network security, and information security. He served as general chair, program committee chair, and member of numerous international conferences, including PC co-chair of the 2007 ACM Symposium on Information, Computer and Communications Security. He received the University Outstanding Researcher Award from the National University of Singapore in 1999 and the Lee Kuan Yew Fellow for Research Excellence from the Singapore Management University in 2006.

Mieso Denko is an associate professor of computing and information science at the University of Guelph, Ontario, Canada. He received his BSc degree in statistics and mathematics from Addis Ababa University. He received his MSc degree form the University of Wales, UK, and his PhD degree from the University of Natal, South Africa, both in computer science. His current research interests include wireless ad hoc networks, wireless mesh networks, wireless sensor networks, pervasive computing, and networking. He has published numerous research papers in international journals, conferences, workshops, and contributed to book chapters. Currently he is co-editing three books in the above areas. Dr. Denko has been actively involved in professional services as organizer or co-organizer of international conferences, symposiums, and workshops, as well as TPC member for a number of conferences and workshops. Among these, most recently he was the general co-chair of the IEEE PCAC-07, general vice-chair of ISPA-07 and program vice-chair of IEEE AINA-07. Currently he is a program vice-chair of the IEEE AINA-08, and co-organizer and program co-chair of the IEEE MHWMN-07 and IST-AWSN-07. Dr. Denko is a senior member of the ACM, a member of the IEEE, ACM SIGMOBILE, IEEE Communications Society, and IEEE Computer Society. Currently, he is an associate professor of computing and information science at the University of Guelph, Ontario, Canada.

Yacine Djemaiel holds a Master Degree in telecommunications and he is currently preparing his PhD thesis in telecommunications in the Engineering School of Communications (SUP'COM, Tunisia). He is conducting research activities in the area of intrusion detection and tolerance and digital investigation of security incidents. Since September 2006, Mr. Djemaiel has been a teacher assistant in telecommunications.

Felipe Espinosa got the BSc and MSc degrees in telecommunications from Polytechnics University of Madrid (Spain) in 1984 and 1991, respectively. He received the PhD degree in telecommunications from University of Alcala (Spain) in 1998. He was a lecturer from 1985 to 2000 and has been an associate professor since 2000, always in the Electronics Department at the University of Alcalá (Spain). His main research interests include electronic control and communication applied to cooperative guidance of robots and vehicles, as well as intelligent transportation systems.

Simone Fischer-Hübner has been a full professor at the Computer Science Department of Karlstad University since June 2000, where she is the head of the PriSec (Privacy & Security) research group. She received Doctoral (1992) and Habilitation (1999) degrees in computer science from Hamburg University. Her research interests include technical and social aspects of IT-security, privacy, and privacy-enhancing technologies. She was a research assistant/assistant professor at Hamburg University (1988-2000) and a guest professor at the Copenhagen Business School (1994-1995) and at Stockholm University/Royal Institute of Technologies (1998-1999).

J. Antonio Garcia-Macias holds a PhD from the Institut National Polytechnique de Grenoble (INPG), France. He is currently a researcher at CICESE Research Center, working in the Computer Science Department. His current research interests are wireless (ad hoc and sensors) networks, ubiquituous computing, next-generation Internet services and protocols, and distributed collaborative systems.

Kaj J. Grahn, Dr. Tech. from Helsinki University of Technology, is presently a senior lecturer in telecommunications at the Department of Business Administration, Media, and Technology at Arcada Polytechnic, Helsinki, Finland. His current research interests include mobile and wireless networking and network security.

Stefanos Gritzalis holds a BSc in physics, an MSc in electronic automation, and a PhD in informatics, all from the University of Athens, Greece. Currently he is an associate professor, the head of the Department of Information and Communication Systems Engineering, University of the Aegean, Greece, and the director of the Laboratory of Information and Communication Systems Security (Info-Sec-Lab). His published scientific work includes several books and more than 150 journal and international conference papers. The focus of these publications is on information and communication systems security. He was a member (secretary general, treasurer) of the Board of the Greek Computer Society.

Yong Guan is an assistant professor in the Department of Electrical and Computer Engineering at Iowa State University. He received his BS (1990) and MS (1996) in computer science from Peking University, China, and his PhD (2002) in computer science from Texas A&M University. His research interests are computer and network forensics, wireless and sensor network security, and privacy-enhancing technologies for the Internet. He received the Best Student Paper Award from the IEEE National Aerospace and Electronics Conference in 1998, won 2nd place in the graduate category of the International ACM Student Research Contest in 2002, and was named the Litton Assistant Professor by Iowa State University in 2007.

Mohamed Hamdi received his Engineering Diploma, Master Diploma, and PhD in telecommunications from the Engineering School of Communications (Sup'Com, Tunisia) in 2000, 2002, and 2005, respectively. From 2001 to 2005 he worked for the National Digital Certification Agency (NDCA, Tunisia) where he was head of the Risk Analysis Team. Dr. Hamdi was in charge in building the security strategy for the Tunisian Root Certification Authority and in continuously assessing the security of the

NDCAs networked infrastructure. He has also served on various national technical committees for securing e-government services. Currently, Dr. Hamdi is serving as a contract lecturer for the Engineering School of Communications at Tunis. He is also a member of the Communication Networks and Security Lab (Coordinator of the Formal Aspects of Network Security Research Team), where he is conducting research activities in the areas of risk management, algebraic modeling, relational specifications, intrusion detection, network forensics, and wireless sensor networks

Munirul M. Haque is currently a PhD student at Purdue University. He received the MS degree in computer science at Marquette University where he researched pervasive computing, security, and privacy in the Ubicomp Research Lab. He completed the BS in computer science and engineering from Bangladesh University of Engineering and Technology.

Jahan Hassan is a research fellow at the School of Information Technologies, University of Sydney. She received her PhD in 2004 from University of New South Wales, Sydney, and Bachelor degree in 1995 from Monash University, Melbourne, both in computer science. She is published widely in peer-reviewed conferences and journals. She was a member of the Technical Program Committee of IEEE LCN 2006, IEEE ICC 2007, IEEE ISWPC 2007, IADIS AC 2006, and IADIS WAC 2007. She served as a reviewer for many conferences and journals. Her research interests include mobile and wireless networking architectures and wireless network security. Her current project focuses on the fast authentication techniques for multiprovider access networks.

Artur Hecker received a diploma in computer science (Dipl.inform.) from the University of Karlsruhe (TH), Germany in 2001. In 2005, he received a PhD degree in computer science and networking from the ENST, France. After his thesis, he worked as CTO of Wavestorm SAS, which he co-founded in 2003. Since 2006, Dr. Hecker holds a position as associate professor at the INFRES department at the ENST. His present research interests are wireless access security, security assurance of complex systems, network and service management, and autonomous networking. Dr. Hecker is actively involved in several IST FP6 and EUREKA CELTIC research activities.

Silke Holtmanns received her PhD in mathematics from the University of Paderborn (Germany), Department of Computer Science and Mathematics. She has been a senior researcher at Nokia Research Center since 2004. Before that, she was working in Ericsson Research Lab Aachen (Germany) as a master research engineer and at the University of Paderborn as a scientific assistant. She has more then 30 publications and co-authored several books on mobile security. She is also rapporteur of six 3GPP security specifications and reports and involved in various standardization activities.

Ismail Khalil Ibrahim is a senior researcher and lecturer at the Institute of Telecooperation- Johannes Kepler University Linz, Austria, where he teaches, consults, and conducts research in mobile multimedia applications and services, agent technologies, and information integration. He received his MSc and PhD in computer engineering and information systems from Gadja Mada University, Indonesia. Dr. Ibrahim previously served as a research fellow at Intelligent Systems Group in the Netherlands and as project manager at the Software Competence Center Hagenberg, Austria. He is the editor-in-chief of *Advances in Next Generation Mobile Multimedia* book series and *Journal of Mobile and Multimedia Communications*, and co-editor in chief of the *International Journal of Web Information Systems* (*JWIS*) and the *Journal of Mobile Multimedia* (*JMM*). His research interests also include business, social, and policy implications associated with the emerging Web technologies.

Biju Issac is a lecturer in the School of IT and Multimedia in Swinburne University of Technology (Sarawak Campus), Malaysia. He is also the head of Network Security Research Group in the Information Security Research Lab at Swinburne University Sarawak. He is an electronics and communication engineer with a post graduate degree in computer applications. Currently he is doing part-time PhD in networking and mobile communications in UNIMAS, Malaysia. His research interests are in wireless and network security, wireless mobility, and IPv6 networks.

Tao Jiang is a research scientist at the Department of Electronic and Computer Engineering, University of Michigan, Dearborn. He received BS and MS degrees in applied geophysics from China University of Geosciences, Wuhan in 1997 and 2000, respectively, and a PhD degree in information and communication engineering from Huazhong University of Science and Technology, Wuhan, P. R. China in April 2004. From August 2004 to August 2005, he worked at Brunel University, London, as an academic visiting scholar, and then moved to University of Puerto Rico in 2006. His current research interests include the areas of wireless communications and corresponding signal processing, especially for OFDM and MIMO systems, cooperative networks, cognitive radio, and ultra wideband communications.

John Felix Charles Joseph is currently pursuing PhD in computer science from Nanyang Technological University, Singapore. His research interests include security in wireless and ad hoc networks, computational intelligence, multicast routing security, and multimedia. He received his Bachelor in engineering, computer science from Madras University, India in 2002 and MS from Anna University, India in 2005. His current work involves design of an intrusion detection algorithm for mobile wireless ad hoc network environment.

Admela Jukan is a professor in electrical and computer engineering at the Technical University Carolo Wilhelmina in Braunschweig, Germany. Prior to coming to TU Braunschweig, she was with University of Illinois at Urbana Champaign (UIUC), Georgia Tech (GaTech), University of Quebec (EMT-INRS), and Vienna University of Technology (TU Wien). From 2002-2004, she served as program director in computer and networks system research at the National Science Foundation (NSF) in Arlington, VA. While at NSF, she was responsible for funding and coordinating US-wide university research and education activities in the area of network technologies and systems. She received the MSc degree in information technologies and computer science from the Polytechnic of Milan, Italy, and the PhD degree (cum laude) in electrical and computer engineering from the Vienna University of Technology (TU Wien), Austria. Dr. Jukan is the author of numerous papers in the field of networking, and she has authored and edited several books. She serves as a member of the Quality Assurance Committee for the EU Network of Excellence, ePhoton/One. Dr. Jukan has chaired and co-chaired several international conferences, including IFIP ONDM, IEEE ICC, and IEEE GLOBECOM. She serves on the editorial board of the IEEE Communications Surveys and Tutorials. She is a senior member of the IEEE.

György Kálmán is a graduate student at UniK, University Graduate Center in Kjeller, Norway. His research area covers personal and device authentication, security, and privacy in wireless systems. He got his MSc degree in the area of communication networks from the Budapest University of Technology and Economics. He was research fellow at Telenor R&I at the Media Platforms group.

Georgios Kambourakis received the diploma in applied informatics from the Athens University of Economics and Business and the PhD in information and communication systems engineering from the Department of Information and Communications Systems Engineering of the University of Aegean. He also holds a MEd from Hellenic Open University. Dr. Kambourakis is a lecturer in the Department of Information and Communication Systems Engineering of the University of the Aegean, Greece. His research interests are in the fields of mobile and ad hoc networks security, VoIP security, security protocols, and PKI, and he has more than 35 publications in the above areas.

Jonny Karlsson has a BSc in information technology from Arcada Polytechnic, Helsinki Finland. Since May 2002 he has been working at Arcada Polytechnic as a course assistant and course teacher in programming and network security related courses and as a research assistant. His current research interests include wireless and mobile network security.

Paris Kitsos received the BSc degree in physics in 1999 and a PhD in 2004 from the Department of Electrical and Computer Engineering, both at the University of Patras. Currently is research fellow with the Digital Systems & Media Computing Laboratory, School of Science & Technology, Hellenic Open University (HOU). His research interests include VLSI design, hardware implementations of cryptographic algorithms, and security protocols for wireless communication systems. Dr. Kitsos has published more than 60 scientific articles and technical reports, as well as is reviewing manuscripts for books, international journals, and conferences/workshops in the areas of his research. He has participated in international journals and conferences organization, as program/technical committee member and guest editor.

Giorgos Kostopoulos received his diploma in electrical and computer engineering from the Electrical & Computer Engineering Department, University of Patras, Greece in 2003. Since then he has been working as a researcher engineer in the Department of Electrical and Computer Engineering of the University of Patras. His research interests include security in wireless networks, new generation networks architectures, security management in new generation networks, and communication networks. Giorgos Kostopoulos has published more than 15 technical papers and book chapters in these areas. He has also participated as senior engineer in European Research Projects.

Zbigniew Kotulski received his MSc in applied mathematics from Warsaw University of Technology and PhD and DSc degrees from Institute of Fundamental Technological Research of the Polish Academy of Sciences. He is currently professor at IFTR PAS and professor and head of Security Research Group at Department of Electronics and Information Technology of Warsaw University of Technology, Poland. He is the author and co-author of three books and more than 150 research papers on applied mathematics, cryptology, and information security.

Odysseas Koufopavlou received the Diploma of Electrical Engineering in 1983 and the PhD degree in electrical engineering in 1990, both from University of Patras, Greece. From 1990 to 1994 he was at the IBM Thomas J. Watson Research Center, Yorktown Heights, NY. He is currently a professor with the Department of Electrical and Computer Engineering, University of Patras. His research interests include computer networks, high performance communication subsystems architecture and implementation, VLSI low power design, and VLSI crypto systems. Dr. Koufopavlou has published more than 150 technical papers and received patents and inventions in these areas. He has participated as coordinator or partner in many Greek and European R&D programs. He served as general chairman for the IEEE ICECS'1999.

Geng-Sheng (G.S.) Kuo worked with R&D laboratories of the communications industry in the United States, such as AT&T Bell Laboratories. In August 2000, he joined National Chengchi University, Taipei, Taiwan as a professor. Since 2001, he has been invited as chair professor of Beijing University of Posts and Telecommunications (BUPT) in Beijing, China. His current research interests include mobile communications, wireless communications, and IP-networks. From 2001 to 2002, he was editor-in-chief of *IEEE Communications Magazine*, whose impact factor in 2002 was 3.165. Currently, he is area editor for *Networks Architecture of IEEE Transactions on Communications*, editor and ComSoc representative to *IEEE Internet Computing*, editor of *European Transactions on Telecommunications*, and so forth.

Taekyoung Kwon is an assistant professor in Multimedia & Mobile Communications Lab., School of Computer Science and Engineering, Seoul National University He received his PhD, MS, and BS degrees in computer engineering from Seoul National University in 2000, 1995, and 1993, respectively. He was a visiting student at IBM T. J. Watson Research Center in 1998 and a visiting scholar at the University of North Texas in 1999. His recent research areas include radio resource management, wireless technology convergence, mobility management, and sensor network.

Pekka Laitinen received his MSc degree in information sciences in Helsinki University of Technology, Department of Engineering Physics and Mathematics. He is principal engineer in Nokia Research Center where he has been working since 1996. His research interests include identity management and applied security.

Björn Landfeldt received a BSc equivalent from the Royal Institute of Technology in Sweden. He received his PhD from The University of New South in 2000. Afterwards he joined Ericsson Research in Stockholm as a Senior Researcher. In 2001, Dr. Landfeldt took up a position as a CISCO senior lecturer in Internet Technologies at the University of Sydney. He has published more than 50 publications in international books, journals, and conferences. Dr. Landfeldt is serving on the editorial boards of international journals and as a program committee member of many international conferences. His research interests include wireless systems, systems modeling, mobility management, and QoS.

Peter Langendoerfer received his doctoral degree in 2001. Since 2000 he has been with the IHP in Frankfurt (Oder) where he is leading the mobile middleware group. He has published more than 55 refereed technical articles, filed seven patents in the security/privacy area, and worked as guest editor for the *Journal of Super Computing* (Kluwer), *Computer Communications* (Elsevier), *Wireless Communications and Mobile Computing* (Wiley), and *ACM Transactions on Internet Technology.* He is/was also a TPC member/chair of many conferences. His research interests include mobile communication (especially privacy and security issues), protocol engineering, and automated protocol implementation.

Shahram Latifi, an IEEE fellow, received the Master of Science degree in electrical engineering from Fanni, Teheran University, Iran in 1980. He received the Master of Science and the PhD degrees both in electrical and computer engineering from Louisiana State University, Baton Rouge in 1986 and

1989, respectively. He is currently a professor of electrical engineering at the University of Nevada, Las Vegas and director of the Center for Information and Communication Technologies (CICT). Dr. Latifi has designed and taught graduate courses on security, image processing, computer networks, fault tolerant computing, and data compression in the past 16 years. He has given seminars on the aforementioned topics all over the world. He has authored over 120 technical articles in the areas of image processing, document analysis, computer networks, fault tolerant computing, parallel processing, and data compression. His research has been funded by NSF, NASA, DOE, Boeing, Lockheed, and Cray Inc. Dr. Latifi is an associate editor of the IEEE Transactions on Computers and co-founder and general chair of the IEEE International Conference on Information Technology. He is also a registered professional engineer in the State of Nevada.

Bu-Sung Lee received his BSc (honors) and PhD from the Electrical and Electronics Department, Loughborough University of Technology, UK in 1982 and 1987, respectively. He is currently associate chair (research) with the School of Computer Engineering, Nanyang Technological University. He is also the founding president of Singapore Research and Education Networks (SingAREN). He has been an active member of several national standards organization such as the National Grid Pilot Project. His research interests are in network management, broadband, distributed, ad hoc and mobile networks, network optimization, as well as grid computing.

Supeng Leng is an associate professor in the School of Communication and Information Engineering, University of Electronic Science and Technology of China (UESTC). He received his BEng degree from UESTC in 1996, and PhD degree from Nanyang Technological University (NTU), Singapore in 2005. He has experience as a R&D engineer in the field of computer communications, and as a research fellow in the Network Technology Research Center, NTU. His research focuses on ad hoc/sensor networks, wireless mesh networks, and broadband wireless networks.

Mo Li received the BE from Beijing University of Posts and Telecommunications. Then, he worked for Lucent Technologies and Computer Associates (CA), where he has been involved in the design of system architectures for DWDM/SDH/IP Backbone O&M systems. He is currently working toward the PhD at the Faculty of Engineering, University of Technology, Sydney. His research interests include handover management and trust-assisted networking.

Xinghua Li obtained his ME and Ph D degrees in computer architecture and computer application from Xidian University (Xi'an) in 2004 and 2006, respectively. Currently, Xinghua Li is the lecturer of the School of Computer of Xidian University. His research interests include information and network security.

Zheng-Ping Li received the BE degree at Department of Electronics and Information, Lanzhou Railway University, Lanzhou, China, in 2000. He is currently working toward the PhD degree in the School of Telecommunication Engineering, Beijing University of Posts and Telecommunications, Beijing, China. His research interests include medium access control, routing, and intrusion detection in wireless mesh networks.

Shiguo Lian, member of IEEE, SPIE, and EURASIP, got his PhD degree in multimedia security from Nanjing University of Science and Technology in July 2005. He was a research assistant at City University of Hong Kong from March to June in 2004, studying on multimedia encryption. He has being with France Telecom R&D Beijing since July 2005, focusing on multimedia content protection, including digital rights management (DRM), image or video encryption, watermarking and authentication, and so forth.

Chuang Lin is a professor and the former head of the Department of Computer Science and Technology, Tsinghua University, Beijing, China. He received his PhD degree in computer science from Tsinghua University in 1994. Professor Lin is a senior member of the IEEE, the Chinese Delegate in TC6 of IFIP, and has served as associate editor for several journals. His current research interests include computer networks, performance evaluation, logic reasoning, and Petri net theory and its applications. He has co-authored more than 200 papers in research journals and IEEE conference proceedings in these areas and has published three books.

Bin Lu is an assistant professor in the Department of Computer Science at West Chester University of Pennsylvania. Dr. Lu received her BS (1996) and MS (1998) degrees in computer science from Harbin Institute of Technology, China, and her PhD (2005) in computer science from Texas A&M University. Her research interests include network security, quality of service, and wireless networks.

Jianfeng Ma received his BS degree in mathematics from Shaaxi Normal University (Xi'an) in 1985, and obtained his ME and PhD degrees in computer software and communications engineering from Xidian University (Xi'an) in 1988 and 1995, respectively. Professor Ma is a member of the executive council of the Chinese Cryptology Society. Currently, Professor Ma is the director of the Ministry of Education Key Laboratory of Computer Networks and Information Security, and he is the dean of the School of Computer of Xidian University. His research interests include information security, coding theory, and cryptography.

Ismat K. Maarouf obtained his BS degree in computer engineering in 2005 and an MS degree in computer networks in 2007 from King Fahd University of Petroleum and Minerals (KFUPM), Dhahran, Saudi Arabia. He is currently working as a research assistant in the Computer Engineering Department in KFUPM. His main research interests include mobile ad hoc and wireless sensor networks, computer networks security, reputation systems, and WLAN-Cellular networks integration.

Michael Maaser received his Master's degree in computer science from Brandenburg University of Technology Cottbus in 2004. After his thesis about negotiation of privacy he started as a research scientist at IHP. His research focuses on privacy preserving techniques mainly, but not limited to, the field of location based services. Throughout the recent 2 years he has seven publications thereof five in the area of privacy and two filed patents.

Ashraf S. Hasan received the BSc degree in electrical and computer engineering from Kuwait University in 1990, and the MEng in engineering physics (computer systems) from McMaster University, Hamilton, Canada in 1992. He received his PhD in systems and computer engineering from Carleton University, Ottawa, Canada in 1997. During 1997-2002, he was with Nrtel Networks Research and Development where he focused on development and evaluation of radio resource management algorithms for broadband and 3G networks. Since 2002, he has been with the Computer Engineering Department at King Fahd University of Petroleum and Minerals, Dhahran, KSA as an assistant professor. His research interests include radio resource management for 3rd and 4th G networks, wireless local area networks, and integration of heterogeneous networks.

Amel Meddeb Makhlouf received the engineering eegree (in 2001) and the Master degree in communications (in 2003) from the Engineering School of Communications (SUP'COM, Tunisia). She is member of the Communication Networks and Security (CN&S) Research Laboratory (University of November 7th, Carthage, Tunisia). Since September 2004, she has joined the Engineering School of Communications (SUP'COM, TUNISIA) as a teacher assistant in telecommunications.

Leonardo A. Martucci is a doctoral student at Karlstad University, Sweden, where he works with research on privacy enhancing technologies for wireless environments. He is involved in education, research, deployment, and industrial projects in the field of wireless network security and privacy since 2001. Mr. Martucci's research is focused especially in privacy problems in dynamic and distributed environments, such as mobile ad hoc networks. He holds a Licentiate in engineering from Karlstad University (2006), a Masters in electrical engineering (2002), and an electrical engineer degree (2000) from University of São Paulo, Brazil.

Geyong Min is a senior lecturer in the Department of Computing at University of Bradford, United Kingdom. He received the PhD degree in computing science from University of Glasgow, UK, in 2003. His research interests include performance modeling and simulation, network traffic engineering, mobile computing and wireless networks, multimedia systems, and information security. Dr. Min has published over 100 research papers in the well-established journals and conferences. Dr. Min serves on the editorial board of the *International Journal of Wireless and Mobile Computing* and *Journal of Simulation Modeling Practice and Theory*, and serves as the guest editor for 10 international journals.

Lawan A. Mohammed is currently an assistant professor in Computer Science and Engineering Technology Department at King Fahd University of Petroleum and Minerals (HBCC Campus), Saudi Arabia. His main research interests are in the design of authentication protocols for both wired and wireless networks, wireless mobility, group oriented cryptography, smartcard security, and mathematical programming.

Rebecca Montanari graduated from University of Bologna, Italy, where she received PhD degree in computer science engineering in 2001. She is now an associate professor of computer engineering at the University of Bologna. Her research primarily focuses on policy-based networking and systems/service management, mobile agent systems, security management mechanisms, and tools in both traditional and mobile systems. She is member of IEEE and AICA.

Luminita Moraru is currently a PhD candidate in the TCS-sensor lab of the Computer Science Department of the University of Geneva. She received a BS degree in electrical engineering and computer science from the Polytechnic University of Bucharest, in 2004, and a MS degree in computer science (embedded systems) from the University of Science and Technology of Lille, in 2005. Her research interests are in sensor networks, mobile ad hoc networks, security, and reputation based trust. Her current research focuses on security and QoS of routing protocols for sensor networks.

A. R. Naseer is an assistant professor in Department of Computer Engineering at King Fahd University of Petroleum and Minerals, Dhahran, KSA. He received the PhD degree in computer science and engineering from Indian Institute of Technology (IIT), Delhi, India, in 1996. He is a recipient of "Best Student Paper Award" at the IEEE/ACM 7th International Conference on VLSI Design, 1994 for his doctoral research paper in the area of FPGA based synthesis. His current research interests include wireless sensor networks security, reputation systems, computer networks, design automation of digital systems, FPGA based synthesis, computer architecture, parallel computing, and multicore processor architectures. He has published several refereed journal and conference papers on related topics.

Huansheng Ning received BS degree from Anhui University, China, in 1996, and a PhD degree from Beihang University, China, in 2001. From 2002 to 2003, he was the CTO of Aerospace Golden Card Company. Since 2004, he has been an associate professor in Beihang University. His current research interests include RFID, EM computing, ITS, and so forth.

Josef Noll holds a professor stipend from the University of Oslo in the area of mobile services. Working areas include mobile authentication, wireless broadband access, personalized services, and the evolution to 4G systems. He is also senior advisor in Movation, Norway's leading innovation company for mobile services. Previously he was senior advisor/group leader at Telenor R&I, project leader of "Operators' Vision on Systems Beyond 3G" and other international projects, use-case leader in the EU "Adaptive Services Grid (ASG)" project, and has initiated a.o. the EU's 6th FP ePerSpace and several ITEA and Eurescom projects.

Christoforos Ntantogian received his BSc degree in computer science and telecommunications from the Department of Informatics and Telecommunications, University of Athens, Greece. In 2006 he finished his postgraduate studies in computer systems technology in the same department and currently he is a PhD student. Since 2004 he has been working for the Communication Networks Laboratory of the University of Athens and he is a member of the Security Group.

Sangheon Pack received the BS (2000) and PhD (2005) degrees from Seoul National University, both in computer engineering. Since March 2007, he has been an assistant professor in the School of Electrical Engineering, Korea University, Korea. From July 2006 to February 2007, he was a postdoctoral fellow at Seoul National University. From 2005 to 2006, he was a postdoctoral fellow in the Broadband Communications Research (BBCR) Group at University of Waterloo, Canada. His research interests include mobility management, multimedia transmission, and QoS provision issues in next-generation wireless/mobile networks. He is a member of the ACM and the IEEE.

Luis E. Palafox received his BS in computer engineering from the University of Baja California in 1997. He also received his MS degree in digital systems from the National Polytechnic Institute of Mexico in 2002. In 2004, he enrolled in the PhD program in computer science program at the CICESE Research Center in Ensenada. He is a faculty member of the School of Chemical Science and Engineering at the University of Baja California since 1999. His areas of interest are computer networking, embedded systems, wireless sensor networks, and digital signal processing.

Cyrus Peikari, MD, is a practicing physician and author of several leading technical security books, including *Security Warrior* from O'Reilly and *Maximum Wireless Security* from SAMS. In his work with Airscanner Corporation he pioneered some of the first antivirus solutions for handheld computing devices. His main area of research is in reverse engineering of "airborne viruses." Dr. Peikari has been a popular speaker and keynote at several major security conferences.

Steffen Peter received his diploma in computer science from the Brandenburg University of Technology at Cottbus (BTU) in 2006. In 2006 he joined the IHP in Frankfurt (Oder), where he was also involved in developing a hardware TCP accelerator as a student. In his diploma thesis he was developing hardware cryptography accelerators. He is a member of the mobile middleware group, working on the research of solutions for security issues in wireless sensor networks. He has filed three patents and has authored two technical papers. His research interests include security and privacy in mobile environments focusing on efficient hardware implementation.

Krzysztof Piotrowski received his Master in computer science from the University of Zielona Gora (Poland) in 2004. Since 2004, he has been with the IHP in Frankfurt (Oder) where he is a member of the mobile middleware group. He published 15 refereed technical articles in the area of security and privacy. His research interests include mobile/wireless communication (focus on privacy and security issues), especially on resource-constrained devices (wireless sensor networks).

Olivier Powell is a senior researcher at the Computer Science Department of the University of Geneva in Switzerland. He was previously a Swiss National Research Foundation fellow at the Research and Academic Computer Technology Institute and the University of Patras in Greece. Previously, he was a post-doctoral research associate at the TCS-sensor lab of the University of Geneva. He received a PhD in computer science in the field of complexity theory from the University of Geneva and a MSc degree in mathematics from the same university. His current research interest is algorithmic aspects of wireless sensor networks.

Göran Pulkkis, Dr. Tech. from Helsinki University of Technology, is presently senior lecturer in computer science and engineering at the Department of Business Administration, Media, and Technology at Arcada Polytechnic, Helsinki, Finland. His current research interests are network security, applied cryptographic, and quantum informatics

Slim Rekhis holds a PhD and a Master degree in telecommunications from the Engineering School of Communications (SUP'COM, Tunisia). He is conducting research activities in the area of digital investigation of security incidents, formal modelling, intrusion detection and tolerance, and wireless security. Since September 2005, Dr. Rekhis has been an assistant professor in telecommunications.

Angelos Rouskas received the Diploma in Electrical Engineering from the National Technical University of Athens (NTUA), the MSc in communications and signal processing from Imperial College, London, and the PhD in electrical and computer engineering from NTUA. He is an assistant professor in the Department of Information and Communication Systems Engineering of the University of Aegean, Greece, and director of the Computer and Communication Systems Laboratory. Dr. Rouskas has been involved in several European and Greek funded research projects and has published extensively in the field of mobile and wireless communication networks.

Miguel A. Ruiz was born in Valdepeñas (Ciudad Real), Spain. He received the Technical Telecommunication Engineering and Telecommunication Engineering degrees from the Polytechnic School at the University of Alcala (Madrid), Spain, in 1999 and 2003, respectively. He is currently working toward the PhD degree in telecommunications at University Alcala. Since 2000, he has been working in the Electromagnetic Compatibility Laboratory as technical manager at the High Technology and Homologation Center (CATECHOM), research support center of the University Alcala. Furthermore, he is an assistant lecturer at the Electronic Department of the same university. His main research interest is EMC effect on electrical and electronic automotive systems.

Kumbesan Sandrasegaran holds a PhD in electrical engineering from McGill University (Canada) (1994), a Masters of Science degree in telecommunication engineering and information Systems from Essex University (UK) (1988), and a Bachelor of Science (honors) degree in electrical engineering (first class) (UZ) (1985). Dr Sandrasegaran is a professional engineer (Pr.Eng) (ECSA) and has more than 20 years experience working either as a practitioner, researcher, consultant, and educator in telecommunication networks. During this time, he has focused on the planning, optimization, forecasting, security, and network management of telecommunication networks. At present, he is program head of ICT Engineering at the Faculty of Engineering, University of Technology Sydney (UTS).

David Sanguino was born in Talavera de la Reina (Toledo), Spain. He received the technical telecommunication engineering degree from the Polytechnic School at the University of Alcala (Madrid), Spain, in 2004. He is currently working toward the telecommunication engineering degree at University Alcala (UAH). Since 2005, he has been working in the Electromagnetic Compatibility Laboratory as Technician at the High Technology and Homologation Center (CATECHOM), research support center of the University of Alcala.

Boot-Chong Seet received his PhD in 2005 from the School of Computer Engineering, Nanyang Technological University (NTU), where he is currently serving as an instructional faculty. Prior to joining NTU, he was with the Singapore-MIT Alliance (SMA), National University of Singapore, where he worked as a research fellow for a pilot project on adaptive location-aware computing. His current research interests include ad hoc, mesh, and sensor networks, mobile peer-to-peer computing, vehicular communications, and emerging broadband wireless technologies. He has over 20 refereed publications and one patent pending. He is a member of IEEE and ACM SIGMOBILE.

Jean-Marc Seigneur is a senior researcher and lecturer at the University of Geneva. He received his MSc and PhD in computer science from Trinity College Dublin. His more than 30 international scientific publications cover ubiquitous computing security, trust, reputation, and privacy. He is an international expert reviewer for French ANR security research projects and the European Commission. He worked in Hewlett-Packard in France and China. He leads the http://www.trustcomp.org online community on computational trust management with now more than 190 academic and industrial members. He has provided technical consulting and presentations to many companies, among them, Philips, Ericsson, SAP, and Amazon.

Moushumi Sharmin is currently a PhD student at University of Illinois. She received the MS degree in computer science at Marquette University where she researched pervasive computing, security, and privacy in the Ubicomp Research Lab. She completed the BS in computer science and engineering from Bangladesh University of Engineering and Technology.

Nicolas Sklavos received the PhD degree in electrical and computer engineering, and the diploma in electrical and computer engineering, in 2004 and 2000, respectively, both from the Electrical & Computer Engineering Department, University of Patras, Greece. His research interests include cryptography, wireless communications security, computer networks, and VLSI design. He holds an award for his PhD thesis on "VLSI Designs of Wireless Communications Security Systems" from IFIP VLSI SOC 2003. He was the general co-chair of MobiMedia'07. He has participated to international journals and conferences organization as program committee member and guest editor. Dr. N. Sklavos is a member of the ACM, IEEE, IEE, the Technical Chamber of Greece, and the Greek Electrical Engineering Society. He has authored or co-authored up to 90 scientific articles, books chapters, tutorials, and reports in the areas of his research.

Nilothpal Talukder is a graduate student in computer science at Marquette University where he researches pervasive computing, security, and privacy in the Ubicomp Research Lab. He completed the BS in computer science and engineering from Bangladesh University of Engineering and Technology.

Daniela Tibaldi graduated from University of Bologna, Italy, where she received her PhD degree in computer science engineering in 2006. Her research activity is focused on middleware solutions for supporting the secure service provisioning in mobile and heterogeneous environments. Since 2002 she works at the DSAW – Direction and Development of Web Activities of the University of Bologna with both technical and quality management responsibilities. One of the DSAW main tasks is to build the University Web sites, services, and the corresponding technological, informative, and organizational infrastructure to fully support University educational, academic, and administrative activities.

Tom Tofigh is a principal and technical member of the AT&T architecture team. He is responsible for architecture studies and vendors technology evaluation. Currently, he supports the AT&T labs advanced services and architecture group. Tom has worked in semiconductor companies as director of product management, director of software development, and has consulted and worked for a number of start-ups and had responsibility for architecture and developments of switches and access products. In addition Tom attended George Washington University and completed his doctoral course work in electrical engineering and computer science graduate school. Furthermore, Tom has a judicial doctoral degree from Northern Virginia Law School with emphasis in intellectual properties. Currently, Tom is the founder and chair of the WiMAX Forum's Application Architecture Working Group.

Alessandra Toninelli graduated from University of Bologna, Italy, where she is currently a PhD student in computer science engineering. Her research interests focus on semantic-based middleware supports for service provisioning, context-aware services, security solutions for pervasive environments, policy-based service management, and mobile agent systems. She is a member of IEEE and ACM.

Denis Trĉek is principal investigator at Jozef Stefan Institute and has been involved in the field of computer networks, security, and privacy for almost 20 years. He has taken part in various European projects, as well as domestic projects in government, banking, and insurance sectors. His bibliography includes over one hundred titles, including works published by renowned publishers like Springer and Wiley. D. Trcek has served (and still serves) as a member of various international boards, from editorial to professional ones. He is inventor of a patented family of light-weight cryptographic protocols. His interests include e-business, security, trust management, privacy, and human factor modelling.

Yu Wang received the PhD degree in computer science from Illinois Institute of Technology in 2004, and the BEng degree and the MEng degree in computer science from Tsinghua University, China, in 1998 and 2000. He has been an assistant professor of computer science at the University of North Carolina at Charlotte since 2004. His current research interests include wireless networks, ad hoc and sensor networks, mobile computing, and algorithm design. He has published more than 50 papers in peer-reviewed journals and conferences. Dr. Wang is a recipient of Ralph E. Powe Junior Faculty Enhancement Awards from Oak Ridge Associated Universities.

Yawen Wei is a PhD candidate in the Department of Electrical and Computer Engineering at Iowa State University. She obtained her BEng (2004) in electronic engineering from Tsinghua University, China. Since then she has been doing research on localization security issues and location-based services in wireless sensor networks.

Bing Wu is an assistant professor in the Department of Mathematics and Computer Science at Fayetteville State University. Dr. Wu received his PhD and MS in the Department of Computer Science and Engineering at Florida Atlantic University. His research interests include wireless ad hoc and sensor networks, mobile computing, and network security. He has worked as research assistant for four years at Motorola. He has published more than ten papers including refereed journal, book chapter, and conference proceedings. He is a member of IEEE.

Jie Wu is a distinguished research professor at the Department of Computer Science and Engineering, Florida Atlantic University and a program director at US National Science Foundation. He has published over 350 papers in various journals and conference proceedings. His research interests are in the areas of wireless networks and mobile computing, routing protocols, fault-tolerant computing, and interconnection networks. Dr. Wu was on the editorial board of *IEEE Transactions on Parallel and Distributed Systems* and was a co-guest-editor of *IEEE Computer and Journal of Parallel and Distributed Computing.* He served as the program co-chair for MASS 2004, program vice-chair for ICDCS 2001, and program vice-chair for ICPP 2000. He was also general co-chair for MASS 2006 and is general chair for IPDPS 2008. He is the author of the text *Distributed System Design* published by the CRC press. He was also the recipient of the 1996-97, 2001-02, and 2006-07 Researcher of the Year Award at Florida Atlantic University. Dr. Wu has served as an IEEE Computer Society Distinguished Visitor and is the chairman of IEEE Technical Committee on Distributed Processing (TCDP). He is a member of ACM and a senior member of IEEE.

Christos Xenakis received his BSc degree in computer science in 1993 and his MSc degree in telecommunication and computer networks in 1996, both from the Department of Informatics and Telecommunications, University of Athens, Greece. In 2004 he received his PhD from the University of Athens (Department of Informatics and Telecommunications). Since 1996 he has been a member of the Communication Networks Laboratory of the University of Athens and, currently, he is the head of the Security Group. In addition, he is a lecturer (faculty of the Department of Technology Education and Digital Systems) in the University of Piraeus, Greece.

Lu Yan is a research fellow at University College London and a Visiting Fellow at University of Cambridge. Previously, he was with Department of Information Technologies in Åbo Akademi University, Distributed Systems Design Laboratory in Turku Centre for Computer Science (TUCS), Institute of Microelectronics (IME) in Peking University. He holds visiting professor positions in both École Supérieure d'Ingénieurs généralistes (ESIGELEC) and École Supérieure de Commerce de Rouen (ESC).

Laurence T. Yang is a professor in computer science at St Francis Xavier University, Canada. His research includes high performance computing and networking, embedded systems, ubiquitous/pervasive computing, and intelligence. He has published around 250 papers in refereed journals, conference proceedings, and book chapters in these areas. He has been involved in more than 100 conferences and workshops as a program/general conference chair and more than 200 conference and workshops as a program committee member. He served as the vice-chair of IEEE Technical Committee of Supercomputing Applications (TCSA) until 2004. Currently he is on the executive committee of IEEE Technical Committee of Scalable Computing (TCSC), of IEEE Technical Committee of Self-Organization and Cybernetics for Informatics, of IFIP Working Group 10.2 on Embedded Systems, and of IEEE Technical Committee of Granular Computing. He is also the co-chair of IEEE Task force on Intelligent Ubiquitous Computing. In addition, he is the editors-in-chief of nine international journals and a few book series. He is serving as an editor for around 20 international journals. He has been acting as an author/co-author or an editor/co-editor of 30 books from Kluwer, Springer, Nova Science, American Scientific Publishers, and John Wiley & Sons. He has received three Best Paper Awards, as well as: the IEEE 20th International Conference on Advanced Information Networking and Applications (AINA-06); one IEEE Best Paper Award, 2007; one IEEE Outstanding Paper Award, 2007; Distinguished Achievement Award, 2005; Distinguished Contribution Award, 2004; Outstanding Achievement Award, 2002; Canada Foundation for Innovation Award, 2003; and University Research/Publication/Teaching Award 00-02/02-04/04-06.

Hao Yin, is currently an associate professor with the Department of Computer Science and Technology, Tsinghua University, Beijing, China He received Ph.D. degrees in electrical engineering from Huazhong University of Science and Technology, China in 2002. His research interests span broad aspects of network architecture, P2P technology, wireless network, video coding, multimedia communication over wireless network, and network security. He has published over 50 papers in refereed journal and conferences. He is on editorial boards of *Advances in Multimedia* and *AD HOC NETWORKS Journal*, and has been involved in organizing over 12 conferences.

Rong Yu was born in Guangdong, China, in 1979. He received his BE degree in communications engineering from Beijing University of Post and Telecommunications (BUPT), Beijing, China, in 2002. After that, he joined the Electronic Engineering Department of Tsinghua University, Beijing, China, where he received his PhD degree at July 2007. His research interests include protocol design and performance analysis of wireless sensor networks and board-band wireless multimedia networks.

Zhen Yu is a PhD candidate in the Department of Electrical and Computer Engineering at Iowa State University. He obtained his BEng (1995) and MEng (2001) in electrical engineering from Shanghai Jiao Tong University, China. He also received his MS in electrical engineering from Iowa State University in 2003. Since then he has been researching security issues in wireless networks and distributed systems.

Said Zaghloul is currently a PhD candidate at the Technical University Carolo-Wilhelmina in Braunschweig, Germany. Prior to his PhD studies, he was with Sprint-Nextel as a telecommunication design engineer mainly focusing on wireless IP infrastructures. During his employment at Sprint-Nextel, he submitted two patents in the area of telecommunication protocols and received excellence awards. In 2002, he received the first IEE award for his BSc graduation project in UMTS capacity planning. In 2003, he was granted a Fulbright Scholarship to pursue his MSc studies at the University of Kansas. In 2005, Mr. Zaghloul received his MSc degree with honors. His research interests include wireless protocols, IP technologies, and wireless communications.

Guo-Mei Zhu received the BE degree in communication engineering from ChongQing University of Posts and Telecommunications, Chongqing, China, in 2002. She is currently pursuing her PhD degree at the School of Telecommunication Engineering, Beijing University of Posts and Telecommunications, Beijing, China under the supervision of Professor Geng-Sheng Kuo. Her current research interests include distributed intrusion detection for wireless networks, cross-layer communication protocol design for wireless networks, next generation wireless networks, and wireless mesh networks.

Albert Y. Zomaya is currently the head of school and the CISCO systems chair professor of internetworking in the School of Information Technologies, The University of Sydney. He is the author/co-author of more than 300 publications and serves as an associate editor for several leading journals. Professor Zomaya is the recipient of the Meritorious Service Award (in 2000) and the Golden Core Recognition (in 2006), both from the IEEE Computer Society. He is a chartered engineer (CEng), a fellow of the American Association for the Advancement of Science, the IEEE, and the Institution of Electrical Engineers (U.K.), and a distinguished engineer of the ACM.

Aneta Zwierko holds MSc and PhD in telecommunications from Warsaw University of Technology, Poland. Her doctoral thesis "Cryptographic Protocols for Mobile Agent Systems with Applications" concerned application of cryptographic protocols in mobile environment for providing integrity, anonymity, and more complex services such as secure e-voting. Her current interest include zero-knowledge proofs and its application, identification, and authentication protocols, anonymity and privacy, security issues of the agent systems, E/M-voting protocols, electronic payments, and AI and its application in security.

Index

Symbols

A

B

C

D

E

F

G

K

L

M

N

O

P

Q

R

S

T

U

V

W

X

Y

Z